Fisheries Processing

Fisheries Processing

Biotechnological applications

Edited by

A.M. Martin

Department of Biochemistry
Memorial University of Newfoundland
St John's
Canada

CHAPMAN & HALL
London · Glasgow · Weinheim · New York · Tokyo · Melbourne · Madras

Published by Chapman & Hall, 2–6 Boundary Row, London SE1 8HN, UK

Chapman & Hall, 2–6 Boundary Row, London SE1 8HN, UK

Blackie Academic & Professional, Wester Cleddens Road, Bishopbriggs, Glasgow G64 2NZ, UK

Chapman & Hall GmbH, Pappelalle 3, 69469 Weinheim, Germany

Chapman & Hall USA, One Penn Plaza, 41st Floor, New York NY 10119, USA

Chapman & Hall Japan, ITP Japan, Kyowa Building, 3F, 2-2-1 Hirakawa-cho, Chiyoda-ku, Tokyo 102, Japan

Chapman & Hall Australia, Thomas Nelson Australia, 102 Dodds Street, South Melbourne, Victoria 3205, Australia

Chapman & Hall India, R. Seshadri, 32 Second Main Road, CIT East, Madras 600 035, India

---

First edition 1994
Reprinted: 2009

Typeset in 10/12 Palatino by Acorn Bookwork, Salisbury, Wiltshire
Printed in India by Saurabh Printers Pvt. Ltd.

ISBN 0 412 58460 3

A catalogue record for this book is available from the British Library

Library of Congress Catalog Card Number: 94-70930

Contents

Contributors

P.M. Aasjord (deceased), NorBio A/S, Bergen High Technology Center, Thormøhlengst. 55, N-5008 Bergen, Norway.

S. Arason, Iceland Fisheries Laboratory, Skulagata 4, PO Box 1405, 121 Reykjavik, Iceland.

E. Bárzana, Departamento de Alimentos y Biotecnología, Facultad de Química, Universidad Nacional Autónoma de México, México DF 04510, México.

R. Dillon, Department of Biochemistry, Memorial University of Newfoundland, St John's, Newfoundland, Canada A1B 3X9.

D.Y.C. Fung, Department of Animal Science and Industry, Kansas State University, Manhattan, KS 66506–1600, USA.

N. Gagne, Food Science and Agricultural Chemistry Department, McGill University, MacDonald Campus, Ste Anne de Bellevue, Quebec, Canada H9X 3V9.

M. García-Garibay, Departamento de Biotecnología, Universidad Autónoma Metropolitana, Iztapalapa, México DF 09340, México.

S. Ghazala, Department of Biochemistry, Memorial University of Newfoundland, St John's, Newfoundland, Canada A1B 3X9.

Ma.P. González, Instituto de Investigacións Mariñas (CSIC), r/Eduardo Cabello 6, Vigo, 36208 Galicia, Spain.

N.F. Haard, Institute of Marine Resources, Department of Food Science and Technology, University of California, Davis, CA 95616, USA.

A. Illanes, Escuela de Ingeniería Química, Universidad Católica de Valparaíso, Casilla 4059, Valparaíso, Chile.

J.M. Lema, Department of Chemical Engineering, University of Santiago de Compostela, 15706 Santiago de Compostela, Spain.

R.E. Levin, Department of Food Science, University of Massachusetts, Amherst, MA 01003, USA.

A.M. Martin, Department of Biochemistry, Memorial University of Newfoundland, St John's, Newfoundland, Canada A1B 3X9.

R. Méndez, Department of Chemical Engineering, University of Santiago de Compostela, 15706 Santiago de Compostela, Spain.

M.A. Murado, Instituto de Investigacións Mariñas (CSIC), r/Eduardo Cabello 6, Vigo, 36208 Galicia, Spain.

T.R. Patel, Department of Biochemistry, Memorial University of Newfoundland, St John's, Newfoundland, Canada A1B 3X9.

L. Pastrana, Instituto de Investigacións Mariñas (CSIC), r/Eduardo Cabello 6, Vigo, 36208 Galicia, Spain.

P. Saisithi, Dean of the College of Biotechnology, Rangsit University, Muang-Ake, Paholyotin Road, Patumatani 12000, Thailand.

B.K. Simpson, Food Science and Agricultural Chemistry Department, McGill University, MacDonald Campus, Ste Anne de Bellevue, Quebec, Canada H9X 3V9.

M.V. Simpson, Food Science and Agricultural Chemistry Department, McGill University, MacDonald Campus, Ste Anne de Bellevue, Quebec, Canada H9X 3V9.

E. Slinde, MATFORSK, Norwegian Food Research Institute, Osloveien 1, N-1430 Ås, Norway.

A. Tangerås, Department of Biochemistry and Molecular Biology, University of Bergen, Årstadveien 19, N-5009 Bergen, Norway.

M.C. Veiga, Department of Industrial and Fundamental Chemistry, University of La Coruña, Campus de Zapateira, 15071 La Coruña, Spain.

V. Venugopal, Bhabha Atomic Research Centre, Food Technology and Enzyme Engineering Division, Trombay, Bombay 400 085, India.

D.R. Ward, Department of Food Science, College of Agriculture and Life Sciences, North Carolina State University, Raleigh, NC 27695, USA.

M.E. Zúñiga, Escuela de Ingeniería Química, Universidad Católica de Valparaíso, Casilla 4059, Valparaíso, Chile

Preface

The fish processing industry is still far from the levels of scientific and technological development that characterize other food processing operations. It has also been slow in finding uses for by-products and processing wastes, compared with the meat and poultry industries. The utilization of fisheries by-products or wastes constitutes an area in which the application of modern techniques could potentially improve profitability. At present, increased attention is being focused on the application of new biotechnological methods to operations related to the seafood industry, with the objective of increasing its general efficiency.

Because fish processing operations are commonly carried out in the vicinity of the sea, most of the resulting fish wastes have been disposed of by returning them to it. Pollution control measures and a better understanding of the valuable composition of the products extracted from the sea are expected to encourage their recovery and the development of new products from them. In the past, fisheries wastes and species not used for food have been generally utilized through technological processes with a low level of sophistication, such as those for the production of animal feed and fertilizer.

Limited economic success has accompanied the application of physical and chemical processes for the recovery of non-utilized fisheries biomass and for the production of quality products from them. However, there is evidence that the inclusion of biological operations in fisheries processing could be essential for the total economy of the process. The current rate of development of fisheries biotechnology should contribute to closing the present technological gap which exists between the seafood industry and other biochemical and food industries.

This book concentrates on the application of bioprocessing methods for the enhancement of the seafood industry, and for the recovery and processing of the marine biomass not utilized in traditional fisheries operations. In addition, it also covers the study of other modern applications, such as novel analytical techniques for quality control, and new sources of feed components for aquaculture processes, among others.

It is expected that this book will be useful to both academic and industrial specialists.

Antonio M. Martin
St John's, Newfoundland, Canada

PART ONE

Processing Quality Control

CHAPTER 1

Microbiological quality of fishery products

D.R. Ward

1.1 INTRODUCTION

Twenty-years ago, a seafood processor commented that the most significant technological advance in the seafood industry over the past 50 years had been the replacement of the wood-handled knives with stainless steel-handled knives. Obviously, many would take issue with this statement. Nonetheless, it does portray an industry in which change has occurred slowly. Currently, it is doubtful that one would describe the changes occurring in the industry as slow. In fact, the seafood industry is in the process of such rapid change that it has been characterized as an 'industry under siege'. Regulatory inspections, safety, quality, resource availability, expansion of aquaculture, labeling, packaging, labor, waste utilization and disposal, and conflicts with recreational and environmental interests are just a few of the issues necessitating change.

Many of the issues mentioned have become 'issues' as a consequence of increased consumer interest in the consumption of seafoods. Over the past decade, the consumption of seafood in the US has risen approximately 23%. One of the repercussions of this dramatic increase has been increased scrutiny from consumer groups, regulatory agencies and the media. Microbiological issues, as they relate to quality and safety of seafoods, have been an extremely visible aspect of consumer, regulatory and media focus. This has resulted in the seafood processing industry becoming much more aware of the microbiological issues surrounding their industry. It is important to appreciate the fact that microbiological issues associated with foods are very sensitive subjects and are often misunderstood by a large portion of society. In the minds of many people microorganisms are 'germs', and, since germs make

Fisheries Processing: Biotechnological applications.
Edited by A.M. Martin.
Published in 1994 by Chapman & Hall, London. ISBN 0 412 58460 3

people sick, germs are to be avoided. While this is an over-simplification it represents the framework in which many people view microorganisms on food. In their minds, the food is contaminated. Consequently, when the public reads that a food product contains millions of bacteria, reaction can be expected to be swift and critical. It is in this context that the seafood industry must consider the 'problems' associated with their products. To suggest that this issue is not important is a mistake and does not serve the best interest of the seafood industry or consumers. On the other hand, to exaggerate the issue out of proportion to its true dimensions also serves no one, other than those seeking attention.

1.1.1 Process flow

The route by which seafood reaches the consumer may be as direct as the consumer catching their own, or highly complex with a variety of exchanges. The scheme below is an illustrative model:

Harvest → Processor → Distributor → Retailer → Consumer

The model is simple and, in fact, looks like a model that could be used for most any food product. Nonetheless, the methods used to handle and process seafoods are unique, particularly at the harvester and processor levels. Consequently, comparisons of seafoods and seafood processing procedures to other muscle protein foods is often inappropriate. Frequently, the only valid comparison is that at some point all were 'living'.

1.1.2 Microflora and spoilage

Finfish are generally regarded as being much more perishable than other muscle protein foods. Although actual spoilage mechanisms are a complex series of inter-related processes, the highly perishable nature of seafood is often attributed to the large concentration of non-protein nitrogenous compounds present in fish muscle. These compounds, which include free amino acids and volatile nitrogen bases such as ammonia, trimethylamine, creatine, taurine, the betaines, uric acid, anserine and carnosine, are utilized actively by bacteria during spoilage (Jay, 1986). Another factor which contributes to the perishability of finfish is the composition of the native microflora, which appears to be a function of the water temperature from which the fish are harvested. The bacterial flora of cold water (0–15°C) fish are dominated by psychrotrophic Gram-negative genera (Shewan, 1977), which are not effectively inhibited by refrigeration temperatures. Conversely, the bacterial flora found on fish from tropical water (20–30°C) are often dominated by Gram-positive genera. Thus, when handled properly, tropical fish are generally less prone to rapid spoilage and exhibit a longer refrigerated shelf-life than cold water species (Disney, 1976; Poulter, Curran and Disney, 1981; Sumner *et al.*, 1984; Gram *et al.*, 1989). According to Lima dos Santos (1981), this generalization should be reconsidered, suggesting that there are too many variables for direct comparisons; and that there are instances of cold water fish such as halibut and grenadiers having shelf-life, on ice, for up to 3 weeks.

In this era of sophisticated research at the genetic and sub-molecular level, it is interesting to note that the mechanisms associated with food spoilage remain fertile research areas. This seems to be especially true in seafoods. However, given the vast diversity of animal genera consumed as seafood, the product compositional differences associated with this diversity, as well as the range of environmental extremes (fresh to salt water; polar to tropical water temperatures) from which seafoods are harvested, perhaps diversity in spoilage mechanisms should not be unexpected.

It is generally accepted that the internal flesh of live, healthy fish is sterile. The natural bacterial flora reside mainly in the outer slime layer of the skin, on the gills and in the intestines of feeding fish. Bacterial numbers range from 10^2 to 10^6 colony forming units (c.f.u.)/cm^2 on the skin, on the gills from 10^3 to 10^5/g and in the intestine from very few in non-feeding fish to 10^7 or greater in feeding fish (Liston, Stansby and Olcott, 1976). This initial microflora is directly related to the environment, while the total microbial load is subject to seasonal variation (Liston, 1956; Shewan, 1961). Shewan (1977) indicated that warm water fish seem to have a more mesophilic, Gram-positive microflora (micrococci, bacilli, coryneforms), while cold water fish harbor a predominately Gram-negative psychrophilic population (*Moraxella*, *Acinetobacter*, *Pseudomonas*, *Flavobacterium* and *Vibrio*). Regardless of the differences in the initial microflora, the spoilage patterns of finfish during iced storage are usually quite similar and are caused by *Pseudomonas* spp. and *Alteromonas putrefaciens* (Barile *et al.*, 1985b). It is the factors which affect the growth and invasion of these spoilage bacteria during harvesting and processing that this chapter will discuss.

1.2 HARVESTING

Seafoods are harvested with a variety of methods and gear: nets (gill, trawl, trap, purse, trammel, etc.), hooks (longline), pots, dredges, harpoons, etc. The initial quality and microbial load of fresh finfish is affected by the method of harvesting. It is at this point that quality maintenance must begin. Abusive handling at harvest will be detrimental to subsequent quality and shelf-life at the retail level.

While little quantitative data are available to compare the microbial load of freshly landed fish by different harvesting methods, Shewan (1949) demonstrated that trawled fish usually carry microbial loads which are 10–100 times greater than line-caught fish. This increase is attributed to dragging along the ocean bottom, which stirs up the mud and contaminates the fish, and to compaction of the fish, which causes gut contents to be expressed. When trawling for fish, it is generally accepted that longer tows will result in lower quality (Costakes, Connors and Paquette, 1982). During periods of heavy fishing the cod end of the net becomes very full and the resulting catch, which may have been dead for hours, is bruised and crushed from compression. Furthermore, larger catches take longer to stow properly. The fish are therefore subjected to the physical abuse of sliding on the boat deck and exposure to ambient temperature and sunlight. In this case fishing for quantity adversely effects quality.

The amount of stress that the fish endures during capture, just prior to death, has also been shown to affect postharvest quality. Fish which are normally very active, such as tuna and mackerel, may become excited and die in a frenzied state when harvested by purse-seining. Inherent physiological features of tuna make them unique from other fish species. They are among the fastest swimming fish with burst speeds for Yellowfin (*Thunnus albacores*) reported as 21 body lengths per second (Walters and Fierstine, 1964). Tuna have very high metabolic rates and some species have the ability to adjust their body temperature. When tuna are captured in a highly stressed state, the build up of lactic acid in the muscle combined with elevated muscle temperature results in a serious flesh defect known as burnt flesh (Goodrick, 1987). The flesh is no longer bright red and the flavor is acidic with a metallic aftertaste. In this state the tuna is still acceptable for canning but unacceptable for the highly lucrative Japanese sashimi market. For this reason, longlining is the desired harvesting method for minimizing stress and maintaining postharvest quality of tuna.

The effect of stress on postharvest quality has been documented with other fish species as well. Salmon, harvested by gill netting, die after an exhausting struggle. As a result rigor mortis sets in quickly and earlier signs of deterioration occur during icing (Dassow, 1976). Harvesting by hook and line, where the fish is quickly brought aboard the vessel and killed, minimizes stress and its associated quality deterioration. This concept of 'clean kill' is well-known in the slaughter of poultry and livestock.

A study on the harvesting of Atlantic cod indicated that the extent of struggling significantly affected protein and caloric content of the flesh by reducing moisture (Botta, Kennedy and Squires, 1987). A concurrent study by the same researchers indicated that the method of catching Atlantic cod was more significant than the time of season in its effect on sensory quality (Botta, Bonnell and Squires, 1987). Studies by Herborg and Villadsen (1975) indicated that the quality and shelf-life of rainbow trout were adversely affected by physical stress. They also stated that the bacterial infection level in the fish muscle increased with stress.

1.3 ONBOARD HANDLING

The significance of proper onboard handling cannot be overemphasized. Since it is not always possible to be selective in the harvesting methods utilized for many commercial fish species, it is critical that the fish be handled in a quality conscious manner as soon as they are landed on the vessel. Handling of fish with gaff hooks, picks or forks should be avoided, or at least limited to the head region. Breaks in the skin and unsightly holes in the flesh quickly introduce spoilage bacteria and accelerate quality deterioration. Likewise, stepping or standing on the fish should be avoided as this may result in bruising.

Unlike other muscle protein food industries, where animals in good physiological condition are brought live to the processing facility, finfish are usually harvested in remote locations and must be stored for several hours or days aboard the vessel prior to processing. Consequently, the quality of the finfish arriving at the processing plant can vary enormously. Often the phrase 'top of the catch' is used to refer to fish stored on top in the storage bins. Ostensibly these fish represent those caught at the end of the trip and thus should be of higher quality than those caught during the beginning of the trip.

In 1973, the Food and Agricultural Organization (FAO) published its *Code of Practice for Fresh Fish* (FAO 1973), the primary focus being the handling of fish at sea. Section 3.1.1 states:

> 'Fish are extremely perishable food, and should be handled at all times with great care and in such a way as to inhibit the growth of microorganisms. Fish quality deteriorates rapidly and potential keeping time is shortened if they are not handled and stored properly. Much of the fish landed for human consumption is subjected to fairly rough handling treatment. Fish should not be exposed to direct sunlight or to the drying effect of wind, but should be carefully cleaned and cooled down to the temperature of melting ice, 0°C, as quickly as possible. Any careless treatment or delay in reducing the temperature of fish will have a marked effect on their potential keeping time.'

Note that one of the suggestions in the quote is that the fish should be carefully 'cleaned'. Several states require that certain ground fish be bled, gutted and gilled on board the fishing vessel. Shewan (1961) points out that the main advantage to gutting is to prevent autolytic spoilage, rather than bacterial decomposition. Indeed, the FAO (1973) states that 'bad gutting might be worse than no gutting' inasmuch as it can facilitate entry of bacteria into the flesh. Stansby and Lemon (1941) reported that gutting of fresh mackerel could increase rather than diminish bacterial numbers. In another study on board a New England trawler, Samuels *et al.* (1984) found that gutted cod and haddock actually had lower psychrotrophic counts than did cod immediately after catch, although the differences were not significant. Furthermore, they found that the practice of washing the gutted fish in circulating seawater prior to ice storage had no effect on bacterial numbers, although this practice did remove mud and other visible extraneous matter.

Work on spiny dogfish demonstrated that microbial numbers on fillets were a function of the level of handling and processing prior to removal of the fillet (Ravesi *et al.*, 1985). In other words, dogfish stored whole had lower counts than gutted fish, which in turn had lower counts than those headed and gutted. However, the study recommended gutting if the fishing trip is longer than 2 days, since it results in increased shelf-life of the fillets, increases in bacterial counts notwithstanding. On the other hand, if the trips are short (less than 2 days) the fishermen's time would be best spent on adequately icing, and thus rapidly chilling, the whole dogfish. The researchers reported that heading and gutting was not helpful, since removing the head creates another cut surface for potential bacterial contamination and does not benefit shelf-life.

A study by Scott *et al.* (1986) compared microbiological and sensory assessment of whole and headed and gutted orange roughy during iced storage. Prior to storage the whole fish and the headed and gutted fish were both washed with seawater. Microbiological results indicated that there were no significant differences between the bacterial counts of the two groups. The shelf-life of whole orange roughy stored in ice, as determined by sensory analysis, was between 11 and 13 days with only a slight increase resulting from heading and gutting. The authors concluded that the slight increase in the shelf-life of the headed and gutted fish was due to reduced autolysis rather than to reduced microbial activity. This autolytic spoilage by digestive enzymes is more prevalent in fish which are actively feeding at the time of catch (Dassow, 1976).

A number of sources recommend that fish be bled, as well as gutted, prior to storage aboard the fishing vessel (Costakes, Connors and Paquette, 1982; Valdimarsson, Matthiasson and Stefansson, 1984; Strom and Lien, 1984). The Canadian grading standards for Atlantic groundfish specifies that the color of the flesh should be characteristic of bled fish (DFO, 1983). There is some question, however, whether the bleeding of fish at harvest actually benefits quality in all situations. A cooperative project by the Maine Groundfish Association and the Maine

Department of Marine Resources concluded that the bleeding of live-caught cod, before gilling and gutting, appeared to have no significant effect on quality or shelf-life when compared with identically harvested, but unbled cod samples from the same tow (Moser, 1986). In this study samples were subjected to color measurement with a color difference meter, sensory panel evaluations, torrymeter readings, surface bacterial load tests, and determination of trimethylene and hypoxanthine (spoilage indicators) concentrations. In all cases no significant differences were found between the bled and unbled fillets. The author pointed out that Maine boats tend to fish deeper waters and/or make longer tows than boats in the New Bedford fishery, where prior information on bleeding was generated. The Maine boats bring up a combination of live, stunned and dead fish, whereas in the New Bedford fishery more fish are live-caught.

A study by Botta, Squires and Johnson (1986) revealed that the bleeding of northern cod, which were caught by Canadian Otter trawlers, was beneficial to quality only if conducted within 1–2 h of the fish being brought aboard. These observations were based on either a one- or two-step bleeding procedure. The authors stressed that the benefits of bleeding may vary with location of catch and time of year.

Mayer *et al.* (1986) reported that onboard processing (bleeding, gutting and gilling) did not extend the shelf-life of bluefish, inasmuch as fillets from these fish were very similar in sensory scores to the fillets cut from fish boxed and stored whole. However, the authors pointed out that since the whole bluefish were only on ice for 1–2 days before processing, there may not have been sufficient time lapse to see any quality benefit from onboard processing. This agrees with the comments of Ravesi *et al.* (1985).

It is obvious from the conflicting literature that the universal application of bleeding and gutting procedures may not be applied in all fish harvesting operations. Each fishery will have to make their own decision based on the species, method of harvest and other pertinent factors. Whatever onboard processing is conducted it must not subject the fish to lengthy delays in icing. The use of ergonomic design principles has been suggested for improving onboard fish handling and processing procedures (Rodman, 1987).

The FAO (1973) report recommended that fish should be cooled down to the temperature of melting ice, 0°C, as quickly as possible. The rationale for this is apparent. Barile *et al.* (1985a) found that the shelf-life of Faughn's mackerel, on ice, was reduced by 1 day for each hour delay in icing or exposure to ambient temperatures of 28–30°C. Moreover, at the point of rejection by a trained taste panel, the fish samples had the following aerobic plate counts at 20°C: fish iced immediately, 10^9 organisms/g; fish iced after a 3 h delay, 10^7; fish iced after a 6 h delay, 10^6; and fish iced after 9 and 12 h delays, 10^5. There was also a distinct change in the spoilage microflora. The principal spoilage organisms after 0–6 h of delay in icing were *Pseudomonas* spp. and *Alteromonas putrefaciens*, while prolonged delay (9–12 h) resulted in a final spoilage flora of *Bacillus* spp., *Aeromonas hydrophila* and *Pseudomonas* spp. These fish were purse-seined off the west coast of the Philippines.

Work done on trench sardines (Chinivasagam and Vidanapathirana, 1986) indicated little difference between the quality of the fish iced immediately after landing and after a 5 h delay. Other studies (DeSilva, 1978; Jayaweera, 1980) have also reported no detrimental effects from delayed icing, for specific fish, for up to 12 h. Chinivasagam and Vidanapathirana offered two possibilities as to why delayed icing was not detrimental. One is the low incidence of *Pseudomonas* organisms on tropical fish and the second is the observation of Poulter *et al.* (1981) that when warm water fish are kept at high temperatures, the onset of rigor is slow and of longer duration, thus delaying bacterial growth. However, the work in which Barile *et al.* (1985a) found that shelf-life was reduced by 1 day for every hour delay in icing was done on a tropical species. Once again, we see evidence of seemingly contradictory research findings. The concern regarding 'broad generalizations' (Lima dos Santos, 1981) expressed earlier is also worth noting here. Broad generalizations are useful and are probably unavoidable; however, diversity renders specific applications risky.

Most large, commercial fishing vessels in the mid-Atlantic and New England regions store their catch on ice in the ship's hold. An exception to this would be smaller vessels which leave the docks early in the morning and return later in the day. These vessels, commonly referred to as day-boats, store fish in the holds but often without ice. Samuels *et al.* (1984) reported significant increases in the psychrotrophic counts on the surface of non-iced bluefish, stored in the hold of these vessels, from the time of catch to the arrival at the dock. However, there were no significant increases on fish caught at the same time but held in iced coolers.

Delays in the icing of fresh fish is of particular concern with fish of the Scombridae family (tuna, mackerel, bonita), due to the possibility of histamine poisoning. Other fish which have been implicated, however, include mahi-mahi and bluefish. Histamine poisoning is a chemical intoxication resulting from the ingestion of foods which contain unusually high levels of histamine. Histamine is formed by the decarboxylation of the amino acid histidine by the microbial enzyme histidine decarboxylase. Histidine is found at high levels in the above fish. Since the production of histamine is a function of microbial enzymatic activity, the simplest method of prevention is rapid chilling of harvested fish and maintenance of low temperature through consumption. Chen *et al.* (1987) reported that *Morganella morganii, Klebsiella pneumoniae* and *Hafnia alvei* are the only histamine-producing bacteria that have actually been isolated from fish causing scombroid or histamine poisoning. There are, however, numerous fish isolates that have been reported to display histidine decarboxylase activity. These include *Pseudomonas putrefaciens, Aeromonas hydrophila, Proteus vulgaris, Clostridium perfringens, Vibrio alginolyticus* and *Enterobacter aerogenes* (Middlebrooks *et al.*, 1988).

Apart from the traditional method of stowing fish in bulk on ice in the hold of the vessel, fisherman have also used short-shelving and boxing at sea. When fish are stowed in bulk they are often stacked 5–6 feet deep with ice. The fish near the bottom are under a tremendous amount of weight. This pressure causes poorer quality, decreased shelf-life and lower yields. The benefit of short-shelving is the reduction of this pressure on the fish. Short-shelving consists of layering fish and ice between shelves that help support the weight. Shelving requires the use of removable shelves that can be easily cleaned, such as corrugated aluminum. The shelves should be placed 18–22 inches apart. When economically feasible, boxing at sea can offer an excellent method of maintaining quality. Plastic boxes of various sizes and design are available and presently in use. Boxing of fish at sea is done, at least to some extent, in almost every fishing nation in the world. While boxing is still a relatively new concept in North America, it has been a common practice in countries such as Norway and Iceland. In addition to reduced compression, boxing provides less abusive handling and quicker unloading at dockside. According to a study by the New England Fisheries Development Foundation (Costakes, Connors and Paquette, 1982), boxing at sea provided a 2–6 day increase in shelf-life and a 7–15% increase in landed weight over traditionally bulk stowed fish. In general the longer the trip, the more dramatic the difference in quality.

While the use of ice is beneficial in the preservation of fish quality, it is not without its disadvantages. Holston and Slavin (1965) suggested that some of the undesirable attributes of ice are a tendency to injure and bruise the flesh, and a leaching of flavor components and nutritionally desirable minerals, as well as water-soluble proteins. Also, icing of fish can be very labor intensive and expensive, in view of the large catches that are typical of some fisheries (Lee and Kolbe, 1982; Reppond *et al.*, 1985;). Consequently, systems using some form of mechanically refrigerated seawater (RSW) have begun to replace ice in certain fisheries, especially in the Pacific Northwest (Lee and Kolbe, 1982).

Concerns have been expressed regarding the effects of these systems not only on the microbiological quality of products, but also on physical, chemical and sensory changes. RSW and CO_2-modified refrigerated seawater (MRSW) have been shown to delay spoilage in shrimp and fish, compared with storage on ice (Bullard and Collins, 1978; Reppond *et al.*, 1979; Reppond and Collins, 1983). Roach *et al.* (1967) stressed the importance of proper sanitation, especially on RSW installations:

> 'While with iced fish only a part of a load may spoil, with fish held in tanks of RSW even under proper conditions there is a danger that all, or nearly all, the fish will be rejected if spoilage occurs.'

The implications of this statement are underscored in work performed by Lee and Kolbe (1982). In their study, aerobic plate counts (at 25°C) of RSW from a commercial vessel were found to be 7.7×10^4 organisms/ml on day 1, rising to 1.5×10^6/ml on day 8. The authors reported that the recirculating seawater seemed to be quickly contaminated by inadequately cleaned surfaces of the fish hold and by debris remaining in the pipes and on the wooden bin boards. The authors' contention seemed to be further supported by the finding that ceiling scrapings of the fish hold yielded a microbial count of 4.5×10^8 organisms/g, even after the hold had been cleaned and sanitized.

Bronstein *et al.* (1985) examined the storage of dressed salmon in refrigerated freshwater, diluted seawater, seawater and ice. Although differences between storage media were not consistently statistically significant, bacterial growth in fish held on ice appeared to be more rapid than on fish held in chilled-water systems. The authors suggest that the differences may be due to the more aerobic conditions present during iced storage, which allow the normal aerobic spoilage flora to grow more rapidly.

Several studies have reported the MRSW system, which has CO_2 injected to control bacterial growth by reducing the pH of the brine (Nelson and Barnett, 1973), to be superior to both ice and RSW in slowing product deterioration (Barnett *et al.* 1971; Longard and Reiger, 1974; Lemon and Reiger, 1977; Bullard and Collins, 1978). Tomlinson *et al.* (1974), however, determined that while the addition of CO_2 to refrigerated seawater did help to restrict bacterial growth it was detrimental to certain sensory attributes of the stored fish. In their study eviscerated salmon and lingcod were stored in RSW and MRSW. The authors stated that the addition of CO_2 promoted salt uptake, increased the susceptibility of the fish to rancidity during frozen storage and, in the case of salmon, promoted the flesh color to fade. Reppond and Collins (1983) found that the quality of Pacific cod was acceptable to 6 days in ice and 9 days in MRSW, but they also stated that the absorption of salt during MRSW storage may present a problem.

1.4 FISHING VESSEL SANITATION AND HACCP

Within the food industry, sanitation of food-contact surfaces and food handling equipment are generally given high priority. On board fishing vessels, however, sanitation as practiced by most food industries is not always practical. Because of the naturally occurring high bacterial loads on fish, counts on board the vessels are quite high. Samuels *et al.* (1984) reported the $\log_{10}$ psychrotrophic counts per square inch of fish-contact surfaces to be: hand gaffs, 8.2 organisms; decks, 7.3; fish hold, 6.5; and shovels/pushes, 7.7. The researchers went on to report that while the sanitary environment on board commercial fishing vessels could be improved, it was not shown to be detrimental to a well-iced catch. This suggestion is in agreement with the work performed by Huss *et al.* (1974). While strict sanitary control may not benefit the quality and shelf-life of well-iced whole fish, it is still important to avoid gross contamination of the stowed fish with bilge oil or dirty ice from previous trips. If onboard processing is performed, however, an unsanitary environment will contribute to reduced quality and shelf-life by exposing the previous sterile flesh to high numbers of spoilage bacteria. Fish holds should be constructed or so altered to make them conducive to sanitary practices. Wood, for example, cannot be properly sanitized.

It is important to note that there is current interest in extending the Hazard Analysis Critical Control Point (HACCP) concept to fishing vessels. Obviously, one could think of several factors that would make 'enforcement' of a HACCP-based inspection program very difficult on board fishing vessels. On the other hand, it is not difficult to envision the impact such a program could have on the quality and safety of seafoods. While safety, especially problems associated with scombrotoxic fish, may be enhanced with implementation of a fishing vessel HACCP-based program, it is the opinion of the author that the most pronounced benefit would be in the area of quality.

Although a thorough discussion of HACCP is outside the scope of this chapter, suffice it to suggest that a HACCP-based fishing vessel program would focus primarily on two areas, vessel sanitation and onboard handling practices such as time on deck and adequate icing (chilling).

I think it fair to suggest that an industry HACCP program that extends back to the harvest vessels would yield an additional, all be it intangible, benefit – improved image.

1.5 PROCESSING

Given the large diversity of commercially exploited species and the products which can be derived from them, the term 'processing' can encompass operations which range from the very simple to the complex. Moreover, when it comes to processing, as with most food industries, evidence indicates that the sanitary conditions of the seafood processing plant do correlate well with the microbial quality of the finished product (Phillips and Peeler, 1972; Wentz *et al.*, 1985).

Work by Samuels *et al.* (1984) in finfish processing plants indicated that in some instances processing lines seem to reach a contamination 'saturation point'. Fish from the boats quickly contaminated the processing equipment (sorting tables, weigh scales, scalers, totes, knives, cutting boards) with significant levels of psychrotrophic spoilage bacteria (1.0^5 c.f.u./inch^2). The authors theorized that establishing an effective in-plant sanitation program could possibly lead to improved product quality and shelf-life. Effective sanitation would, however, be contingent on decontaminating or reducing microbial loads on fish before they enter the processing lines. To accomplish this they used a high pressure wash on the whole fish to remove surface slime and its associated microflora. This treatment reduced psychrotrophic counts by 99%. Moreover, taste panel evaluations judged flounder fillets processed under the recommended procedures to have a shelf-life of 11–12 days, as opposed to 7–8 days for fillets processed in the traditional manner.

One area of concern, with respect to microbial contamination of finfish during processing, is the use of wash tanks. While wash tanks are effective in removing blood and physical debris, they can be a significant source of microbial contamination. A study by Mayer *et al.* (1986) revealed that when similar lots of fish were processed at nearby processing plants, the microbial load differed greatly depending on the use of wash tanks or high pressure washing. Aerobic plate counts (APCs; 20°C) of dressed Atlantic mackerel, which were passed through a wash tank, were more than 4.0×10^5 c.f.u./inch^2 higher than high pressure washed fish. Similar results were also obtained with dressed sea bass and dressed porgy. Additional studies indicated that finfish fillet quality can be best maintained by high pressure washing whole fish, prior to filleting, rather than high pressure washing the fillets themselves. The physical appearance of the fillets, especially those from soft fleshed fish, is easily abused by the high pressure spray.

The effectiveness of high pressure washing with surfactants on reducing surface microflora was also tested by Mayer *et al.* (1986). High pressure washing whole croaker, with a solution of 0.1% cetylpyridinium chloride, provided a 1.08 log c.f.u/inch reduction over high pressure washing with tap water alone. By scaling the fish, prior to high pressure washing with 0.1% cetylpyridinium chloride, the APC was reduced an additional 0.88 log (average APC 2.43 log c.f.u/inch^2).

To the author's knowledge, the use of high pressure wash technology is not being actively explored by the processing industry. With the current level of interest in producing high quality seafoods with extended shelf-life characteristics, this technology appears to hold promise. The apparent lack of activity is unfortunate.

Kosak and Toledo (1981) also investigated the effect of microbial decontamination on the storage stability of finfish. Chlorine dip was used for microbiological decontamination of the fish prior to packaging.

The authors reported a doubling of shelf-life by using the combined effects of a 3.5 min dip in a 1000 μg/ml free chlorine solution followed by packaging under vacuum or in polyethylene bags. However, Samuels *et al.* (1984) found that dipping in a 200 p.p.m. hypochlorite solution for 4 min followed by overwrapping in an oxygen-permeable film was only marginally advantageous over dipping in water.

Mayer *et al.* (1986) reported that a spray application of 2000 p.p.m. of free chlorine solution reduced the APC of gray sea trout from 4.18 to 2.98 log c.f.u./g. The total reduction, even at this high concentration, was only 1.20 logs. It appears that chlorine application is only marginally effective at reducing the spoilage microflora on fish surfaces.

The effect of ozone on the iced storage life of fresh gutted Atlantic cod was investigated by Ravesi *et al.* (1987). The ozone was incorporated in either the ice, rinse water or chilled seawater. The authors reported that the shelf-life, as assessed by sensory, chemical and microbiological tests, was not sufficiently extended by any of the treatments.

Because of the unique nature of seafood and seafood processing operations, it can be a difficult task to maintain extremely rigid standards with respect to microbial numbers on processing plant equipment. One of the plants reported by Samuels *et al.* (1984) to be operating under good sanitary conditions had psychrotrophic concentrations on equipment surfaces ranging from 1.3×10^1 to 2.5×10^7 organisms/inch^2. Because of gross contamination resulting from contact with these heavily contaminated surfaces, there has been research regarding the benefits of delayed processing. Shaw *et al.* (1984), working with cod, found that maximum overall shelf-life could be attained by delayed filleting, thereby reducing the time stored as fillets. The data indicated a 2 log difference in total APC (21°C) from fillets cut from 7 day post mortem fish. The $\log_{10}$ counts were 7.3 and 5.3 c.f.u./g, respectively. (These fillets were from cod which had been eviscerated at sea.)

Townley and Lanier (1981) looked at the possibility of delayed evisceration. Working with Atlantic croaker and grey trout, they reported no advantages to delayed processing. In fact, fish eviscerated immediately upon landing maintained class 1 quality 7–10 days longer than uneviscerated fish or iced fish eviscerated 3 days after harvest. Microbial populations were reported to remain lower in eviscerated fish, for both species, throughout a 2 week storage period (0–1°C with top icing).

Mayer *et al.* (1986) investigated the effect of delayed processing of boxed at sea bluefish. Bluefish which were processed immediately had a 10 day shelf-life (tray-packed fillets, 33°F), while bluefish which were stored whole on ice for a period of 4 and 7 days before processing lasted approximately 11 and 13 days from day of catch. The authors stated that the increase in shelf-life would be dependent upon whether the fish were actively feeding prior to harvest. Actively feeding fish have accelerated quality deterioration when held uneviscerated.

A discussion on seafood processing can not be concluded without some mention of HACCP. Within the continuum from 'harvest to consumption', the processing sector is the most likely area for the initial implementation of HACCP. The US Food and Drug Administration has recently published its proposed rules for the mandatory use of HACCP in the processing of fish and fishery products. This regulation has the potential for initiating significant changes in the fundamental manner in which fish are processed. Although a thorough discussion of HACCP is outside the scope of this chapter, two excellent articles on HACCP and seafood processing have been written by Huss (1992) and Garrett and Hudak-Roos (1991).

1.6 PRESERVATIVES

Given the perishable nature of seafood, a satisfactory method for extending the shelf-life of chilled products that ensures quality and a continuity of supply with a minimum of waste has been the ultimate goal. Research with certain preservatives to aid in this effort has met with varying degrees of success. With the present negative attitude that consumers have against the use of chemical preservatives in food products, it is unlikely that their use will become widespread for fresh finfish in the immediate future.

1.7 SUMMARY

If fishermen, processors and retailers understand the factors which impact seafood quality and make a conscientious effort to control these quality factors, problems associated with microbial growth could be significantly minimized. It is important to remember that every stage of handling from harvest to consumption affects quality. On the fishing vessels fishermen must avoid abusive handling. In the processing plants and at retail, good sanitary conditions and strict temperature conditions must be maintained. It will take a combined effort to provide the quality finfish that the consumer demands.

REFERENCES

Barile, L.E., Milla, A.D., Reilly, A. and Villadsen, A. (1985a). Spoilage patterns of mackerel (*Rastrelliger faudhni* Matsui). 1. Delays in icing. *ASEAN Food J.*, **1**(2), 70.

Barile, L.E., Milla, A.D., Reilly, A. and Villadsen, A. (1985b). Spoilage patterns of mackerel (*Rastrelliger faudhni* Matsui). 2. Mesophilic and psychrophilic spoilage. *ASEAN Food J.*, **1**(3), 00.

Barnett, H.J., Nelson, R.W., Hunter, P.J. *et al.* (1971) Studies on the use of carbon dioxide dissolved in refrigerated brine for the preservation of whole fish. *Fish. Bull.*, **69**, 433.

Botta, J.R., Kennedy, K. and Squires, B.E. (1987) Effect of method of catching and time of season on the composition of Atantic cod (*Gadus morhua*). *J. Food Sci.*, **52**(4), 922.

Botta, J.R., Bonnell, G. and Squires, B.E. (1987) Effect of method of catching and time of season on sensory quality of fresh raw Atantic cod (*Gadus morhua*). *J. Food Sci.*, **52**(4), 928.

Botta, J.R., Squires, B.E. and Johnson, J. (1986) Effect of bleeding/gutting procedures on the sensory quality of fresh raw Atlantic cod (*Gadus morhua*). *Can. Inst. Food. Technol. J.*, **19**(4), 186.

Bronstein, M.N., Price, R.J., Strange, E.M. *et al.* (1985) Storage of dressed chinook salmon, *Oncorhynchus tshawytscha*, in refrigerated freshwater, diluted seawater, seawater, and in ice. *Mar. Fish. Rev.*, **47**(1), 68.

Bullard, F.A. and Collins, J. (1978) Physical and chemical changes of pink shrimp, *Pandalus borealis*, held in carbon dioxide modified refrigerated seawater compared to pink shrimp held in ice. *Fish. Bull.*, **76**, 73.

Chen, C.-M., Marshall, M.R., Koburger, J.A. *et al.* (1988) Determination of minimal temperatures for histamine production by five bacteria, in *Proceedings from the First Joint Conference with Atlantic Fisheries Technological Society*, 1988, Tropical and Subtropical Fisheries Technological Society of the Americas, Florida Sea Grant, University of Gainesville, W.T. Otwell (compiler), pp. 365–376.

Chinivasagam, H.N. and Vidanapathirana, G.S. (1986) Quality changes and bacterial flora associated with trench sardines (*Amblygaster sirm*) under delayed icing conditions. *FAO Fisheries Report 317*, Food and Agricultural Organization, Rome, Italy.

Costakes, J., Connors, E. and Paquette, G. (1982) *Quality at sea – recommendations for on-board quality improvement procedures*, New Bedford Seafood Producers Association and New England Fisheries Development Foundation.

Dassow, J.A. (1976) Handling fresh fish, in *Industrial Fishery Technology*, (ed. M.E. Stansby), Robert E. Krieger, New York, pp. 45–64.

DeSilva, G.T.K. (1978) Storage life of Soora puraw (*Selas leptolepsis*) with delayed icing of 9 and 12 hours. Cited in Chinivasagam and Vidanapathirana (1986).

DFO (1983) *Grading Standard for Fresh and Frozen Atlantic Groundfish Products*. Department of Fisheries and Oceans. Ottawa, Ontario. Minister of Supply and Services. Cat. No. Fs 38–2/1–1983F.

Disney, J.G. (1976) The spoilage of fish in the tropics, in *Proceedings of the First Annual Tropical and Subtropical Fisheries Technology Conference*, (eds B.F. Cobb and A.B. Stockton), Texas A&M University Sea Grant, College Station, TX, Vol. 1, p. 23.

FAO (1973) Code of practice for fresh fish. *FAO Fisheries Circular C318*, Food and Agricultural Organization, Rome, Italy.

Garret, E.S. and Hudak-Roos, M. (1991) US Seafood Inspection and HACCP, in *Microbiology of Marine Food Products*, (eds D.R. Ward and C.R. Hackney), Van Nostrand Reinhold, New York, p. 111.

Goodrick, B. (1987) Postharvest quality of tuna meat, a question of technique. *Food Technol. Australia*, **39**, 343.

Gram, L., Oundo, J.O. and Bon, J. (1989) Storage life of Nile Perch (*Lates niloticus*) in relation to temperature and initial bacterial load. *Trop. Sci.*, **29**, 221.

Herborg, L. and Villadsen, A. (1975). Bacterial infection/invasion of fish flesh. *J. Food Technol.*, **10**, 507.

Holston, J. and Slavin, J.W. (1965) Technological problems in the preservation of fish, iced fish requiring more knowledge from fundamental research, in *Technology of Fish Utilization*, (ed. R. Kreuzer), Fishing News (Books) Ltd, London, p. 41.

Huss, H.H. (1992) Development and use of the HACCP concept in fish processing. *Int. J. Food Microbiol.* **15**, 33.

Huss, H.H., Dalsgaard, D., Hansen, L. *et al.* (1974) The influence of hygiene in catch handling on the storage life of iced cod and plaice. *J. Food Technol.*, **9**, 213.

Jay, J.M. (1986) *Modern Food Microbiology*. Van Nostrand Reinhold, New York, p. 227.

Jayaweera, V. (1980) Storage life of silverbelly (*Leiognathus* spp.) with delayed icing. *Bull. Fish. Res. Sta., Sri Lanka*, **30** (1/2), 53.

Kosak, P.H. and Toledo, R.T. (1981) Effects of microbial decontamination on the storage stability of fresh fish. *J. Food Sci.*, **46**, 1012.

Lee, J.S. and Kolbe, E. (1982) Microbiological profile of Pacific shrimp, *Pandalus jordani*, stowed under refrigerated seawater spray. *Mar. Fish. Rev.*, **44** (3), 12.

Lemon, D.W. and Regier, L.W. (1977) Holding of Atlantic mackerel (*Scomber scombrus*) in refrigerated sea water. *J. Fish. Res. Board Can.*, **34**, 439.

Lima dos Santos, C.A.M. (1981) The storage of tropical fish in ice – a review. *Trop. Sci.*, **23** (2), 97.

Liston, J. (1956) Quantitative variations in the bacterial flora of flatfish. *J. Gen. Microbiol.*, **15**, 305.

Liston, J., Stansby, M.E. and Olcott, H.S. (1976) Bacteriological and chemical basis for deteriorative changes, in *Industrial Fishery Technology*, (ed. M.E. Stansby), Robert E. Krieger, New York, pp. 345–358.

Longard, A.A. and Regier, L.W. (1974) Color and some compositional changes in ocean perch (*Sebastes marinus*) held in refrigerated sea water with and without carbon dioxide. *J. Fish. Res. Board Can.*, **31**, 456.

Mayer, B., Samuels, R., Flick, G. *et al.* (1986) *A Seafood Quality Program for the Mid-Atlantic Region, Part II*, A report submitted to the Mid-Atlantic Fisheries Development Foundation. Virgina Polytechnic Institute and State University, Sea Grant, Blacksburg.

Middlebrooks, B.L., Toom, P.M., Douglas, W.L. *et al.* (1988) Effects of storage time and temperature on the microflora and amine development in Spanish mackerel (*Scomberomorus maculatus*). *J. Food Sci.*, **53** (4), 1024.

Moser, M.D. (1986) *Maine Groundfish Association Vessel Quality Handling Project*, A report submitted to the New England Fisheries Development Foundation.

Nelson, R.W. and Barnett, H.J. (1973) Fish preservation in refrigerated seawater modified with carbon dioxide. *Proc. 13th Int. Cong. Refrig.*, **3**, 57.

Phillips, F.A. and Peeler, J.T. (1972) Bacteriological survey of the blue crab industry. *Appl. Microbiol.*, **24**, 958.

Poulter, R.G., Curran, C.A. and Disney, J.G. (1981) Chilled storage of tropical and temperate water fish – Differences and similarities. Advances in refrigerated treatment of fish, especially underutilized species. *Bull. Int. Inst. Refrig.*, **49**, 111.

Ravesi, E.M., Licciardello, J.J., Tuhkunen, B.E. and Lundstrom, R.C. (1985) The effects of handling or processing treatments on storage characteristics of fresh spiny dogfish, *Squalus acanthias*. *Mar. Fish. Rev.*, **47**(1), 48.

Ravesi, E.M., Licciardello, J.J. and Racicot, L.D. (1987) Ozone treatments of fresh Atlantic cod, *Gadus morhua*. *Mar. Fish. Rev.*, **49**(4), 37.

Reppond, K.D., Collins, J. and Markey, D. (1985) Walleye pollock (*Theragra chalcogramma*): Changes in quality when held in ice, slush-ice, refrigerated seawater, and CO_2-modified refrigerated seawater then stored as block of fillets at −18°C. *J. Food Sci.*, **50**, 985.

Reppond, K.D. and Collins, J. (1983) Pacific cod (*Gadus macrocephalus*) : change in sensory and chemical properties when held in ice and CO_2 modified refrigerated seawater. *J. Food Sci.*, **48**, 1552.

Reppond, K.D., Bullard, F.A. and Collins, J. (1979) Walleye pollock (*Theragra chalcogramma*): physical, chemical and sensory changes when held in ice and in carbon dioxide modified refrigerated seawater. *Fish. Bull.*, **77**, 481.

Roach, S.W., Tarr, H.L.A. and Tomlinson, N. (1967) Chilling and freezing salmon and tuna in refrigerated sea water. *Bulletin 160*, Fisheries Research Board of Canada, Ottawa.

Rodman, W.K. (1988) On board fish handling systems for offshore wetfish trawlers 'work smarter, not harder', in *Proceedings from the first joint conference with Atlantic Fisheries Technological Society*, 1988, Tropical and Subtropical Fisheries Technological Society of the Americas, Florida Sea Grant, University of Gainesville, W.T. Otwell (compiler), pp. 332–342.

Samuels, R.D., DeFeo, A., Flick, G.J. *et al.* (1984) *Demonstration of a Quality Maintenance Program for Fresh Fish Products*. A report submitted to Mid-Atlantic Fisheries Development Foundation. Virginia Polytechnic Institute and State University, Sea Grant, Blacksburg.

Scott, D.N., Fletcher, G.C., Hogg, M.G. and Ryder, J.M. (1986) Comparison of whole and headed and gutted orange roughy stored in ice: sensory, microbiology and chemical assessment. *J. Food Sci.*, **51** (1), 79.

Shaw, S.J., Bligh, E.G. and Woyewoda, A.D. (1984) Effect of delayed filleting on quality of cod fish. *J. Food Sci.*, **49**, 979.

Shewan, J.M. (1949) Some bacteriological aspects of handling, processing and distribution of fish. *J. Royal Sanit. Inst.*, **59**, 394.

Shewan, J.M. (1961) The microbiology of sea-water fish, in *Fish as Food*, (ed. G. Borgstrom), Academic Press, New York, Vol. 1, pp. 487–560.

Shewan, J.M. (1977) The bacteriology of fresh and spoiling fish and the biochemical changes induced by biochemical action, in *Proceedings of the Conference of Handling, Processing, and Marketing of Tropical Fish*, Tropical Products Institute London, p. 51.

Stansby, M.E. and Lemon, J.M. (1941) Studies on the handling of fresh mackerel (*Scomber scombus*). *Research Report 1*, US Fish and Wildlife Service, Washington, DC.

Strom, T. and Lien, K. (1984) Fish handling on board Norwegian fishing vessels, in *Fifty Years of Fisheries Research in Iceland. Proceedings of a Jubilee Seminar*, 1984, (ed. A. Moller), p. 15.

Sumner, J.L., Gorczyca, E., Cohen, D. and Brady, P. (1984) Do fish from tropical waters spoil less rapidly in ice than fish from temperate waters? *Food Technol. Australia*, **36**, 328.

Tomlinson, N., Geiger, S.E., Boyd, J.W. *et al.* (1974) Comparison between refrigerated sea water (with or without added carbon dioxide) and ice as storage media for fish to be subsequently frozen. *IIF–IIR Commissions B2, D3*, Tokyo, p.163.

Townley, R.R. and Lanier, T.C. (1981) Effect of early evisceration on the keeping quality of Atlantic croaker (*Micropogon undulatus*) and grey trout (*Cynoscion regalis*) as determined by subjective and objective methodology. *J. Food Sci.*, **46**, 863.

Valdimarsson, G., Matthiasson, A. and Stefansson, G. (1984) The effect of onboard bleeding and gutting on the quality of fresh, quick frozen, and salted products, in *Fifty Years of Fisheries Research in Iceland. Proceedings of a Jubilee seminar*, 1984, (ed. A. Moller), p.61.

Walters, V. and Fierstine, H.L. (1964) Measurement of swimming speeds of Yellowfin tuna and Whoo. *Nature*, **202**, 208.

Wentz, B.A., Duran, A.P., Swartzentruber, A. *et al.* (1985) Microbiological quality of crabmeat during processing. *J. Food Protect.*, **48**, 44.

CHAPTER 2

Rapid methods and automation for seafood microbiology

D.Y.C. Fung

2.1 INTRODUCTION

Rapid methods and automation in microbiology are dynamic fields of study that address the utilization of microbiological, chemical, biochemical, biophysical, immunological and serological methods for the study of improving isolation, early detection, characterization and enumeration of microorganisms and their products in clinical, food, industrial and environmental samples. Although early developments in this field were directed to clinical microbiology, in the past 5–10 years, food microbiologists have started to adapt rapid and automated methods in their laboratories. Conventional methods of detection, enumeration, identification and characterization of microbes are described in reference books such as *Compendium of Methods for the Microbiological Examination of Foods* (Vanderzant and Splittstoesser, 1992), *Official Methods of Analysis of the Association of Official Analytical Chemists* (AOAC, 1990), *Bacteriological Analytical Manual* (FDA, 1992), *Standard Methods for the Examination of Dairy Products* (APHA, 1985) and *Modern Food Microbiology* (Jay, 1992).

The advantages of conventional methods are their long history of development and usage and recognition by regulatory agencies, nationally and internationally. The disadvantages of some of the conventional methods are time required in performing the tests, time delay in obtaining results, labor intensive, cumbersome and costly in terms of usage of large number of tubes, plates, bottles and incubation space, etc.

The search for rapid methods started almost immediately with the development of applied microbiology at the turn of the century.

Fisheries Processing: Biotechnological applications.
Edited by A.M. Martin.
Published in 1994 by Chapman & Hall, London. ISBN 0 412 58460 3

Interest in the field of rapid methods and automation in microbiology has been growing steadily on an international scale in recent years. The *First International Symposium on Rapid Methods and Automation in Microbiology* was held in Stockholm, Sweden in 1973. Subsequently symposia were held in Cambridge, UK (1976), Washington, DC (1981), Berlin, Germany (1984), Florence, Italy (1987) and Helsinki, Finland (1990). The *Seventh International Congress on Rapid Methods and Automation in Microbiology and Immunology* was held in September, 1993 in London, UK. The proceedings of these symposia have been published in book form (Heden and Illeni, 1975a,b; Johnson and Newsom, 1976; Tilton, 1982; Habermehl, 1985; Balows, Tilton and Tuxano 1989; Vaheri, Tilton and Balows, 1991).

Other important publications on the subject of rapid methods for medical specimens, water, food, industrial and environmental samples are in a series of papers by Fung and colleagues (Cox *et al.*, 1984, 1987a,b; Fung, Goldschmidt and Cox, 1984; Fung, Hart and Chain, 1987; Fung, Cox and Bailey, 1988; Fung *et al.*, 1989; Fung, 1991, 1992) and books such as *Mechanizing Microbiology* (Sharpe and Clark, 1978), *Foodborne Microorganisms and Their Toxins: Developing Methodology* (Pierson and Stern, 1986), *Rapid Methods in Food Microbiology* (Adams and Hope, 1989) and *Instrumental Methods for Quality Assurance in Foods* (Fung and Matthews, 1991).

The purpose of this chapter is to review the basic principles and practical applications of a variety of instruments and procedures directly and indirectly related to improved methods for microbiology in quality assurance and research in food science and technology, and when appropriate address their usefulness in seafood microbiology.

2.2 IMPROVEMENTS IN SAMPLING AND SAMPLE PREPARATION

In order to carry out a microbiological analysis, the scientists must first prepare the food sample properly. The subject of food sampling plans and initial sample handling as well as preparation of food homogenates by conventional methods are described in the aforementioned reference books.

2.2.1 Stomacher

One of the most useful instruments developed for sample preparation is the Stomacher (Tekmar, Cincinnati, OH). This instrument is designed to massage food samples in a sterile bag. The food sample (such as seafood and fish) is first placed in the sterile disposable plastic bag and appropriate sterile diluents are added. The bag with the food is placed in the open chamber. After the chamber is closed, the bag is then

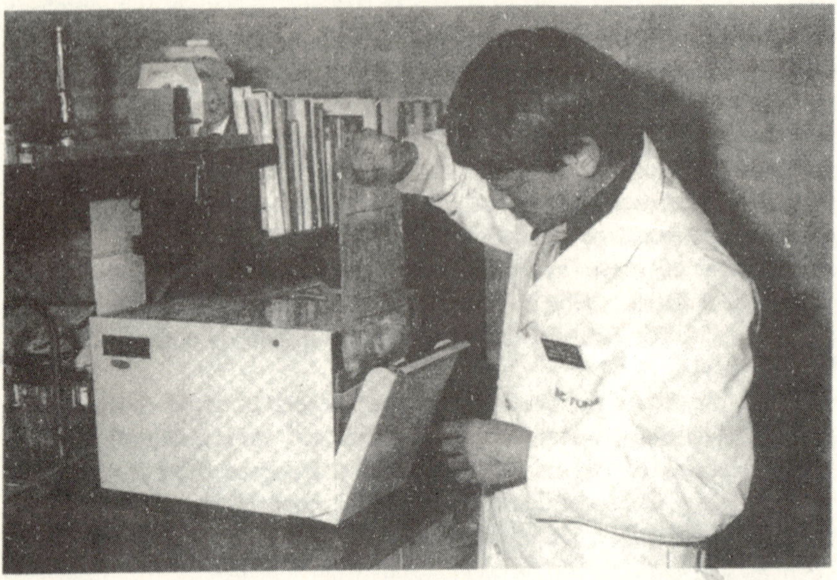

Figure 2.1 Author removing a sample in a stomacher bag from the Stomacher instrument.

massaged by two paddles for a suitable time period, usually from 1–5 min. No contact occurs between the instrument and the sample. During massaging, microorganisms are dislodged into the diluent for further microbiological manipulation. Massaged slurries are then used for microbiological analysis (Figure 2.1). Sharpe and Jackson (1975) and Emswiler, Pierson and Kotula (1977) have shown that satisfactory results can be obtained by this method compared with the conventional blending of foods. The advantages of the Stomacher instrument compared with conventional blending are:

1. Use of disposable sterile bags, thus eliminating the need for large numbers of glass or metal jars to be cleaned and resterilized continuously.
2. No generation of aerosols, such as those created by conventional blending.
3. No heat generation during 'stomaching' as compared with blending, which may generate considerable amounts of heat in prolonged operation.
4. The bag with the homogenized sample can be used as a storage bag for time course studies.
5. Ease of operation.

The disadvantages include possible breakage of the bag by sharp objects in samples such as bones, hard fibers, nut shells, metal pieces, wood chips, straws, etc. This may be an important concern for seafood microbiology as many seafood products have shells, bones, fins and hard objects. The initial cost may also be a concern. However, if soft tissues are used, this is an excellent instrument.

2.2.2 Gravimetric Diluter or Diluflo

Another new instrument that holds promise for sample preparation is the Gravimetric Diluter (Spiral Biotech, Bethesda, MD). Food and meat microbiologists almost always have to dilute a food sample before microbiological testing. One of the most time-consuming procedures of routine microbiological work is to aseptically measure a sample of food (e.g. 25 g of meat) and then aseptically add an exact amount of sterile diluent (e.g. 225 ml) to make a desired dilution (1:10). With the Gravimetric Diluter, the analyst needs only to aseptically place an amount of food (e.g. 5.3 g) into a pretared Stomacher bag or a sterile blending jar, set a desired dilution (1:10 dilution) and set the instrument to deliver the appropriate amount of sterile diluent (e.g. 47.7 ml). Thus, the dilution operation can be automatically done. The dilution factor can be chosen by the analyst to satisfy the need (1:10, 1:50, 1:100, etc.) simply by programming the instrument. Manninen and Fung (1992a) evaluated this instrument and found that depending on the volume tested, the accuracy of delivery for most samples was found to be in the range 90–100%. A new version of this instrument is called Diluflo and has been in use satisfactorily in the author's laboratory since 1992. Both instruments can be used simply as an accurate dispenser of liquid into containers as well.

2.2.3 Other methods

A review by Lee and Fung (1986) documented many useful ways to obtain surface samples from meats and other foods by stainless steel templates, surgical dermatome-like assemblage and coring devices. Fung, Lee and Kastner (1980) developed an adhesive tape method that can effectively 'peel' viable cells from meat surfaces and later deposit them on agar surfaces to obtain viable cell counts of meat surfaces.

Truly automated sample preparation systems belong to the field of robotics. The excellent book *Laboratory Robotics: A Guide to Planning, Programming and Applications* by Hurst and Mortimer (1987) should be consulted on this subject.

2.3 ALTERNATIVE METHODS FOR THE VIABLE CELL COUNT PROCEDURE

The conventional viable cell count or Standard Plate Count method is time consuming both in terms of operation and collection of data.

Several methods have been explored to improve the efficiency of operation of the viable cell count procedure.

2.3.1 The spiral plating method

The spiral plating method is an automated system for obtaining viable cell counts (Spiral Biotech, Bethesda, MD). The instrument can spread a liquid sample on the surface of agar contained in a Petri dish in a spiral shape (the Archimedes spiral) with a concentration gradient starting from the center and decreasing as the spiral progresses outward on the rotating plate. The volume of liquid deposited at any segment of the agar plate is known. After the liquid containing microorganisms is spread, the agar plate is incubated overnight at an appropriate temperature for the colonies to develop. The colonies appearing along the spiral pathway can be counted either manually or electronically.

Figure 2.2 shows the basic spiral system. The stylus in the center is first placed in a liquid sample contained in a small cup to the left and a sample is obtained through suction by a vacuum source shown on the right. The stylus is then placed on the agar surface. At the start of the operation, the platform holding the Petri dish starts to rotate at a constant rate and the stylus moves outward. Simultaneously the liquid sample is forced out at a decreasing volume as the stylus moves outward. After the stylus reaches the edge of the Petri dish, it is

Figure 2.2 The Spiral System for rapid plating of bacteria from foods (used with permission of Spiral Biotech, Bethesda, MD).

automatically lifted. The plate is then incubated. Figure 2.3 shows the results of growth of colonies on agar plates. The colonies along the spiral can be counted manually by placing the plate on top of a template marked with segments. The analyst counts a segment of the template and multiplies the number with a designated factor for the particular area, converting the count into organisms per milliliter of sample. This method is simple, but it requires some training and judgment of the analyst. To make counting even more automated, a laser counter has been developed along with appropriate computer software for operations. The plate is placed on the stage of the counter, the instrument is set, and the laser counter will automatically count the number of colonies and convert the number to organisms per milliliter. The spiral system has been used in the US for a variety of foods with satisfactory results (Schalkowsky, 1986). The laser counter is also designed to count

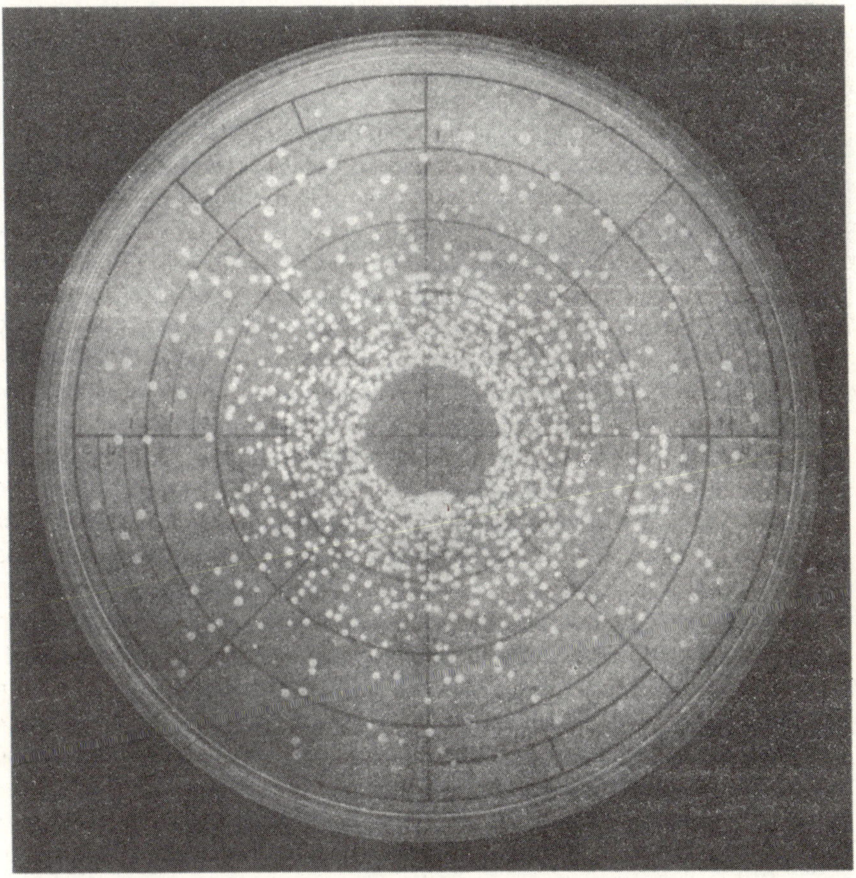

Figure 2.3 Growth of bacterial along the Archemidas Spiral. The grid is for counting colonies (used with permission of Spiral Biotech, Bethesda, MD).

Table 2.1 Comparison of pour plate and spiral plate counted manually and by laser for bacterial cultures ($\log_{10}$ c.f.u./ml)

Test cultures	Pour plate *manual*	Pour plate *laser*	Spiral plate *manual*	Spiral plate *laser*
Escherichia coli	8.86	8.85	8.73	8.85
Salmonella enteritidis	8.76	8.66	8.78	8.92
Pseudomonas aeruginosa	8.00	8.00	8.00	8.00
Staphylococcus aureus	8.04	7.78	8.18	8.18
Lactobacillus plantarum	9.48	9.40	9.60	9.69
Streptococcus spp.	7.73	7.66	8.00	8.08
Bacillus cereus	7.26	7.15	7.15	7.26
Micrococcus luteus	7.40	7.32	7.51	7.57

Data from Manninen, Fung and Hart (1991).

colonies on conventional agar plates, making the instrument very versatile in a food microbiology laboratory.

Manninen, Fung and Hart (1991) evaluated the spiral system and laser colony scanner for enumeration of microorganisms and found excellent correlation between pour plate and spiral plate both using manual and laser counting procedures for bacteria (Table 2.1). Yeast counts were also very comparable but mold counts are not as reliable (Table 2.2), especially for fast growing mold such as *Rhizopus oligosporus*. Manninen and Fung (1992b) also evaluated the spiral plater and laser colony scanner for enumeration of microorganisms in meat and devised a convenient protocol combining the conventional aerobic plate count method with the spiral system to evaluate surface microbial loads of pork loins. In this procedure, a surface (25, 50 or 125 cm^2) is swabbed and then placed in 9 ml of sterile buffer. For each sample both the conventional and the spiral plating method were used (Table 2.2). For conventional plating 1 ml was used. For spiral plating the liquid was plated without further dilution. In this protocol when the number of cells was too numerous to count for the conventional plating (more than 300 per plate), the spiral plated sample will provide the accurate count. Conversely, when the number is very low (1–250 cells per plate), the conventional pour plate sample will provide accurate count while the spiral plated sample will have virtually no colonies. Table 2.3 shows that when samples had very low counts the pour plate provided more accurate data. However, samples with higher counts gave better results using spiral, rather than pour, plates, as shown in Table 2.4. Using four plates (two plates were really needed), a large range of microbial load on meat surfaces can be counted effectively.

Table 2.2 Comparison of pour plate and spiral plate counted manually and by laser for yeast and mold cultures ($\log_{10}$ c.f.u./ml)

	Pour plate		*Spiral plate*	
Test cultures	*manual*	*laser*	*manual*	*laser*
Candida macedoniensis	7.11	6.97	7.30	7.38
Hansenula subpelliculosa	5.00	4.99	5.11	5.15
Hansenula wingei	7.04	7.04	7.15	7.00
Saccharomyces cerevisiae	6.38	6.43	6.28	6.46
Saccharomyces fragilis	7.34	7.36	7.23	7.34
Aspergillus flavus	7.04	7.41	7.23	7.56
Pennicillium camemberti	6.89	6.86	7.04	7.48
Rhizopus oligosporus	6.51	7.54	6.59	7.49

Data from Manninen, Fung and Hart (1991).

Table 2.3 Psychrotropic counts/plate and c.f.u./cm² of pork surfaces (20 samples) after storage using two plating techniques (APC and Spiral)

Swab sample	*Pour plates*			*Spiral plates*		
	colonies/plate		*c.f.u./cm²*	*colonies/plate*		*c.f.u./cm²*
1	7	3	2	1	0	3
2	1	3	1	0	0	0
3	1	3	1	0	0	0
4	2	0	<1	0	0	0
5	209	206	67	10	17	86
6	12	14	4	1	0	3
7	6	5	2	1	0	3
8	1	0	<1	0	0	0
9	8	9	3	1	0	3
10	4	3	1	0	0	0
11	0	0	0	0	0	0
12	63	80	11	4	2	10
13	17	17	3	2	1	5
14	3	11	1	1	0	2
15	1	0	<1	1	0	2
16	4	1	<1	0	0	0
17	5	12	1	1	1	3
18	43	36	6	3	0	5
19	3	1	<1	0	0	0
20	49	38	7	1	2	5

Data from Manninen and Fung (1992b).

Table 2.4 Psychrotropic counts/plate and c.f.u./cm^2 of pork surfaces (20 samples) after storage using two plating techniques (APC and Spiral)

Swab sample	*Pour plates*			*Spiral plates*		
	colonies/plate		*c.f.u./cm^2*	*colonies/plate*		*c.f.u./cm^2*
1	TNTC	TNTC	–	16	22	24
2	TNTC	TNTC	–	17	21	24
3	TNTC	TNTC	–	22	16	24
4	TNTC	TNTC	–	14	21	22
5	TNTC	TNTC	–	41	31	46
6	TNTC	TNTC	–	16	16	20
7	TNTC	TNTC	–	27	25	33
8	TNTC	TNTC	–	33	51	54
9	TNTC	TNTC	–	22	23	29
10	TNTC	TNTC	–	38	25	40
11	TNTC	TNTC	–	56	41	155
12	TNTC	TNTC	–	93	70	261
13	TNTC	TNTC	–	488	488	781
14	TNTC	TNTC	–	63	66	206
15	TNTC	TNTC	–	34	40	118
16	TNTC	TNTC	–	93	98	306
17	TNTC	TNTC	–	66	38	166
18	TNTC	TNTC	–	43	57	160
19	TNTC	TNTC	–	24	18	67
20	TNTC	TNTC	–	12	21	118

Data from Manninen and Fung (1992b).
TNTC, too numerous to count.

A new version of the spiral plater is recently introduced named 'Autoplater'. In the instrument an analyst needs only to present the liquid sample and the instrument completely and automatically processes the sample, including resterilizing the unit for the next sample.

One of the problems of the spiral system is clogging of the dispensing stylus by food particles. Konuma and Kurata (1982) modified a Stomacher bag by placing a filter in the bag so the homogenized liquid poured from the bag will be free of particles. Since then commercial companies have manufactured stomacher bags with filters in place. The liquid, after filtration, when presented to the spiral system will not clog the stylus. The spiral system is listed as an alternate method for the examination of foods in the *Standard Methods for the Examination of Dairy Products* (APHA, 1985).

2.3.2 The Isogrid system

Another alternative method for viable cell count is the Isogrid system (QA Laboratories Ltd, San Diego, CA). This system consists of a square filter with hydrophobic grids printed on the filter to form 1600 squares for each filter. Food samples are weighed, blended and enzyme treated before passage through the membrane filter containing the hydrophobic grids by vacuum. The filter is then placed on agar containing a suitable nutrient for growth of the bacterium, yeast or mold. The hydrophobic grids prevent colonies from growing further than the square grids; thus all colonies have a square shape. This facilitates counting of the colonies both manually and electronically. This method has been successfully used to make viable cell counts for a variety of foods, including milk, meat, black pepper, flour, peanut butter, mushrooms, rice, fish, fish shrimp and fish oyster (Entis, Brodsky and Sharpe, 1982; Entis, 1983, 1984, 1986; Sharpe and Peterkin, 1988). Lin, Fung and Entis (1984) developed a Trypan blue agar for use in conjunction with the Isogrid system for effective enumeration of yeasts and molds from foods. Other applications of the Isogrid system include determination of total coliforms, fecal coliforms, *Escherichia coli*, *Salmonella* spp., etc. (Sharpe, 1991).

2.3.3 The Petrifilm system

Rehydratable nutrients are imbedded into a series of films in the Petrifilm (3M Co., St Paul, MN). The top layer of the protective cover is lifted and 1 ml of liquid sample is introduced to the center of the unit, and the cover is then replaced. A plastic template is placed on the cover to make a round mold. The rehydrated medium with nutrient will support the growth of microorganisms after suitable incubation time and temperature. The colonies can be counted directly in the unit. The unit is about the size and thickness of a plastic credit card; thus, providing great savings of space in storage and incubation. Petrifilm units have been developed for total bacterial count, coliform count, fecal coliform count, and yeast and mold counts. Petrifilm has been successfully used to count organisms from milk and meat (Ginn, Packard and Fox, 1986; Smith, Fox and Busta, 1986). Fung, Hart and Chain (1987) obtained a 0.99 correlation coefficient comparing Petrifilm SM with the conventional viable cell count method for seafood analysis (shrimp, perch, cod, whiting) of mesophiles (Table 2.5). Bishop and Juan (1988) found that the Petrifilm technique was not statistically different from the agar pour plate method for enumeration of psychrotrophs in raw milk samples.

The advantages of the Petrifilm system include its ease of operation, savings of storage and incubation space, its long shelf-life because of the use of dehydrated medium in the film, and independence from heat treatment compared with the conventional agar pour plate method. The Petrifilm can be rehydrated with sterile distilled water and after solidification can be used directly to contact surfaces for surface bacterial count.

Table 2.5 Comparison of the standard plate count method and the Petrifilm method for viable cell counts of shrimp, perch, cod and whiting

Seafood	*Sample*	*Standard plate count (c.f.u./g)*	*Petrifilm SM (c.f.u./g)*
Shrimp	1	2.1×10^4	0.9×10^4
	2	9.3×10^3	6.8×10^3
	3	1.5×10^4	1.2×10^4
	4	4.4×10^4	2.0×10^4
	5	2.0×10^4	0.9×10^4
Perch	1	1.4×10^3	1.2×10^3
	2	5.4×10^2	3.7×10^2
	3	8.0×10^2	4.8×10^2
	4	5.0×10^2	3.2×10^2
	5	1.0×10^3	1.2×10^3
Cod	1	1.2×10^2	1.1×10^2
	2	2.2×10^2	2.4×10^2
	3	1.0×10^4	1.0×10^4
	4	2.1×10^4	2.3×10^4
	5	1.2×10^4	1.2×10^4
Whiting	1	3.0×10^2	3.1×10^2
	2	5.8×10^2	5.6×10^2
	3	1.2×10^3	1.1×10^3
	4	5.4×10^2	5.2×10^2
	5	8.8×10^2	8.4×10^2

Samples were massaged in a Stomacher for 1 min in sterile diluent. Viable cell counts were made according to standard method. Incubation time was 48 h at 32°C. All samples were done in duplicate. Correlation coefficient between the two methods is $r = 0.99$.
Data from Fung, Hart and Chain (1987).

2.3.4 The Redigel system

Redigel system (RCR Scientific, Inc., Goshen, IN) is another convenient viable cell count system. This system consists of sterile nutrients with a pectin gel in a tube. The tube is ready to be used any time and no heat is needed to 'melt' agar. A 1 ml food sample is first pipetted into the tube. After mixing, the entire content is poured into a special Petri dish previously coated with calcium. When liquid comes in contact with the calcium a calcium pectate gel is formed and the complex swells to resemble conventional agar. After an appropriate incubation time and temperature, the colonies can be counted exactly like a conventional standard plate count method. Fung and Chain (1991) compared 17 different foods (pasteurized milk, raw milk, cheddar cheese, chocolate chips, rice, wheat germ, corn meal, whole wheat, flour, peanuts, pecans, ground beef, chicken, ground black pepper, thyme, broccoli, mushrooms and turkey pot pie, 20 samples each) and obtained a correlation of 0.964. Roth (1988) reported a comparative analysis of nine different non-dairy and dairy food products, and found that Redigel provided statistically significant results compared with the conventional method. Roth and Bontrager (1989) also evaluated the coliform count of cream, cheddar cheese, cottage cheese, homogenized milk, raw milk, sour cream and yogurt by the Redigel method and the conventional pour plate method, and again found statistically significant results.

The four aforementioned methods have potential as an alternative to the conventional agar pour plate method. Chain and Fung (1991) made a comprehensive analysis of all four methods against the conventional method on seven different foods (skinless chicken breast, fresh ground beef, fresh ground pork, packaged whole shelled pecans, raw milk, thyme and whole wheat flour, 20 samples each) and showed that the new systems and the conventional method were highly comparable and exhibited a high degree of accuracy and agreement ($r = 0.95+$). It should be noted that these newer methods need some training and experience before satisfactory results can be obtained consistently. They are good methods if operated carefully.

2.3.5 The Direct Epifluorescent Filter Technique (DEFT)

In the DEFT method, the liquified sample is first passed through a filter that retains the microorganisms, the filter is then stained with acridine orange and the slide observed under ultraviolet microscopy. 'Live' cells usually stain orange–red, orange–yellow or orange–brown, whereas 'dead' cells fluoresce green. The slides can be read by the eye or by a semiautomated counting system marketed by Bio-Foss. A 'viable cell count' can be made in less than an hour. With the use of an image analyzer, an operator can count 50 DEFT slides per hour (Pettipher, 1986). This method has been used satisfactorily for counting viable cells in milk and other food samples such as fish (Pettipher and Rodrigues, 1980, 1982; Pettipher *et al.*, 1980; Pettipher, 1989). Recently the Nordic countries have been using this method for quality assurance of ground beef.

Although viable cell counts have been satisfactorily made by the DEFT test for the milk samples, some investigators noted that cells treated by heat, irradiation, etc., can also fluoresce orange even if they are dead (Sjoberg *et al.*, 1990). The exact mechanism for the phenomenon is not known.

An interesting method to estimate viability of cells treated with chemicals was described by Czechowski and Banner (1989), who treated *Yersinia enterocolitica* and *Salmonella typhimurium* cells with hypochlorite for a few minutes, after which the cells were exposed to 0.004% nalidixic acid and incubated for 18 h at 25°C. Because nalidixic acid allows live cells to grow but not to divide, one can estimate the population of live cells versus dead cells by staining these cells with acridine orange and observing them under fluorescence microscopy. Elongated orange cells in this system are those that were not killed by the hypochlorite solution.

These alternative methods for viable cell counts of microorganisms in foods provide an intermediate step toward replacing the cumbersome conventional method. These methods need to be tested in more laboratories with more classes of foods before they can completely replace the time-honored standard plate count method. Although many examples in this section were in food other than seafood, there is no reason to believe that these methods cannot be adapted to seafood microbiology.

2.4 NEW METHODS FOR ESTIMATION OF MICROBIAL POPULATIONS AND BIOMASS

Many methods have been developed in recent years to estimate the total number of microorganisms by parameters other than the viable colony count as described in the previous section. In order that a new method be acceptable, it must have some direct correlation with the total viable cell count. Thus, standard curves correlating parameters such as adenosine triphosphate (ATP) level, detection time of electrical impedance or conductance, generation of heat, radioactive CO_2, etc., with viable cell counts of the same sample series must be made. In general, the larger the number of viable cells in the sample, the shorter the detection time of these systems. A scattergram is then plotted and used for further comparison of unknown samples. The assumption is that as the number of microorganisms increases in the sample, these physical, biophysical, and biochemical events will also increase accordingly. This assumption may or may not be exactly correct, but it serves as a functional way to estimate the number of microorganisms or biomass indirectly.

Theoretically, these methods can detect as low as one viable cell in the sample if the incubation period is long enough (days or weeks). On the practical side, usually the limit is 10^4–10^5 cells/ml. When a sample has 10^6–10^7 organisms/ml detection can be achieved in about 4–6 h.

2.4.1 ATP estimation

All living things utilize ATP. In the presence of a firefly enzyme system (luciferase and luciferin system), oxygen and magnesium ions, ATP will facilitate the reaction to generate light. The amount of light generated by this reaction is proportional to the amount of ATP in the sample; thus, the light units can be used to estimate the biomass of cells in a sample. The light emitted by this process can be monitored by a variety of fluorimeters such as LKB-Wallac (Turku, Finland), Turner Designs (Mountain View, CA), Lumac (Landgraaf, The Netherlands), Luminometer (Dynatech Laboratories, Chantilly, VI), BioTrace (San Diego, CA) and Enliten-Promega (Madison, WI). These procedures can be automated for handling of large numbers of samples. Some of the instruments can detect as little as 10^2–10^3 fg. The amount of ATP in 1 colony forming unit (c.f.u.) has been reported as 0.47 fg with a range of 0.22–1.03 fg. It should be noted that different bacteria have different amounts of ATP per cell and for the same bacterium at different stages of the growth cycle the amount of ATP per cell may differ. Yeast cells have about 100 times more ATP than bacterial cells.

Using this principle, many researchers have tested the efficacy of using ATP to estimate microbial cells in foods and beverages. Littel, Pikelis and Spurgash (1986) indicated that the ATP procedure was able to predict bacterial levels within 0.5 $\log_{10}$ of the actual count for beef and chicken samples. Minimum sensitivity is 5×10^4 c.f.u./g of meat sample. Ward, LaRocco and Hopson (1986) also found positive correlation between the ATP method and the conventional method in evaluating fish samples.

Lumac markets several models of ATP instruments and provides customers with test kits with all necessary reagents, such as a fruit juice kit, hygiene monitoring kit, etc. The reagents are injected into the instrument automatically and readout is reported as relative light units (r.l.u.s). By knowing the number of microorganisms responsible for generating known e.l.u.s, one can estimate the number of microorganisms in the food sample. Figure 2.4 depicts the author operating the Lumac system. In some food systems, such as wine, the occurrence of any living matter is undesirable; thus, monitoring of ATP can be a useful tool for quality assurance in the winery. Recently, much interest has been expressed in using ATP estimation not for total viable numbers but as a sanitation check. Reviews on the principles and application of ATP have been made by LaRocco, Littel and Piersen (1986) and Stannard (1989).

2.4.2 Impedance, capacitance and conductance measurements

As microorganisms grow and metabolize nutrients, large molecules change to smaller molecules in a liquid system and cause a change in electrical conductivity and resistance in the liquid as well as at the interphase of electrodes. By measuring the changes in electrical impedance, capacitance and conductance, scientists can estimate the number of microorganisms in the liquid because the larger the number of microorganisms in the fluid, the faster the change in these parameters which can be measured by sensitive instruments. A detailed analysis on

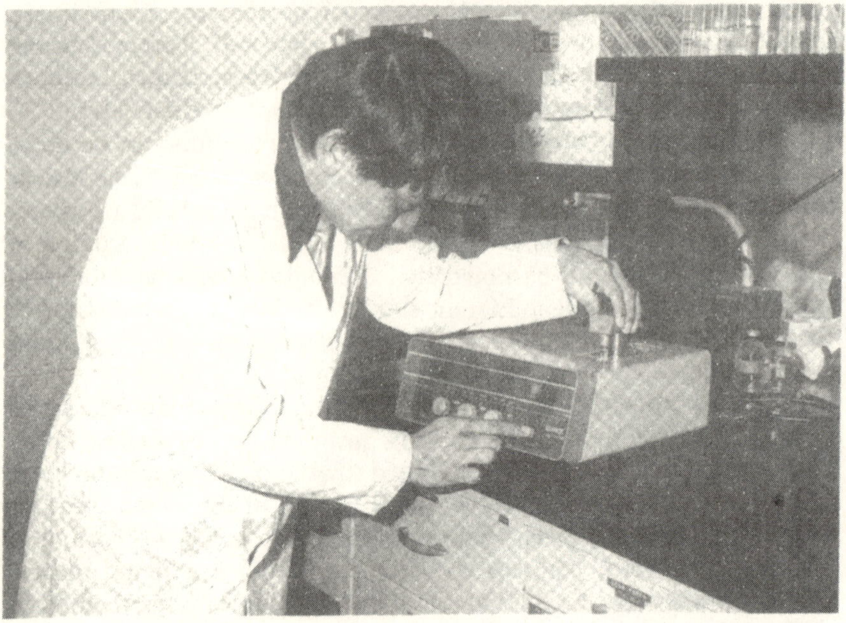

Figure 2.4 Author measuring bioluminescence using the LUMAC ATP instrument.

the subject of impedance, capacitance and conductance in relation to food microbiology has been made by Eden and Eden (1984).

(a) The Bactometer

The Bactometer (bioMerieur Vitek, Inc., Hazelwood, MO) is an instrument designed to measure impedance changes in foods. Samples are placed in the wells of a 16-well module. After the module is completely or partially filled, it is plugged into the incubator unit to start the monitoring sequence. At first, there is a stabilization period for the instrument to adjust to the module, then a base line is established. As the microorganisms metabolize the substrates and reach a critical number (10^5–10^6 cells/ml), change in impedance increases sharply and the monitor screen shows a slope similar to the log phase of a growth curve. The point at which the change in impedance begins is the 'detection time' and this is measured in hours from the start of the experiment. The detection time is inversely proportional to the number of microorganisms in the sample. By knowing the number of microorganisms per milliliter in a series of liquid samples and the detection time of each sample, one can establish a standard curve. From the curve one can decide the cut-off points to monitor certain specifications of the food products. For example, if one finds that meat with 10^6 organisms/g when applied to the instrument will result in a detection time of 5 h, then one can use 5 h as a cut-off point for an indicator that meat samples have fewer or more than 10^6 organisms/g. A similar meat sample with a detection time of 3 h would indicate that the meat has more than 10^6 organisms/g. Conversely, a similar meat with a detection time of 7 h would indicate that the meat has less than 10^6 organisms/g. In newer models of the Bactometer, the screen displays bars with one of three colors instead of an impedance curve. A red bar signals that the food being analyzed is 'out of spec', a green bar signals that the food is 'in spec' and a yellow bar signals a 'intermediate caution spec'. These new developments are designed to be 'user friendly'. Hardware and software are provided for users to monitor their food products conveniently as far as impedance microbiology is concerned. In order to use this instrument effectively users must generate standard curves for their specific products. This process may be quite time consuming.

Impedance methods have been used to estimate bacteria in milk, dairy products, meats and other foods (Eden and Eden, 1984; Waes and Bossuyt, 1984; Zindulis, 1984; Bishop and White, 1985). Of particular interest is the application of this method for determining the shelf-life potential of pasteurized whole milk by Bishop, White and Firstenberg-Eden (1984).

(b) The Malthus system

An instrument somewhat similar in principle and operation is the Malthus system (Malthus Instrument, Crawley, UK). The Malthus system works by measuring the conductance of the fluid as the organisms grow in the system. It also generates a conductance curve similar to the impedance curve of the Bactometer and it also uses detection time in monitoring the density of the microorganisms in the food. Furthermore, the screen also displays red, green, and yellow bars to signify 'fail', 'pass' and 'caution' levels of microorganisms in particular food products. The major difference between the two systems, besides the scientific principle (impedance versus conductance), is the incubation units. In the Bactometer system, the size of the well (about 2 ml in capacity) in the 16-well module is fixed. No modification is possible because the module is designed to fit into the incubator chamber. The unit (with two separate chambers) can be operated in two different temperatures at the same time. The temperature is controlled by heating air. The Malthus system, however, allows analysts to choose three sizes ranging from 2 to 100 ml samples, depending on the sample involved. The temperture unit is controlled by heating water. Another important difference is that the modules of the Bactometer are disposable, whereas the jars, tubes and electrodes of the Malthus system are autoclavable and reusable. In terms of performance, the systems are equivalent in sensitivity and detection time.

The Malthus system has been used for microbial monitoring of brewing liquids (Day, 1983; Kilgour and Day, 1983; Evans, 1985), milk (Visser and de Groote, 1984) and hygiene monitoring (McMurdo and Whyward, 1984).

Gibson and colleagues (Gibson and Ogden, 1980; Ogden, 1986; Gibson and Hobbs, 1987), have done a considerable amount of work using the Malthus system to study seafood microbiology.

Besides estimating viable cells in foods, both the Bactometer and the Malthus systems can detect specific organisms by the use of selective and differential liquid media. New developments of these two systems are constantly being made. For example, the Malthus system has developed a tube system to detect CO_2 production by yeast using indirect conductance measurements. Disposable units have also been introduced in the system.

2.4.3 Reflectance colorimetry

A new instrument called the 'Omnispec bioactivity monitor system' (Wescor, Inc., Logan, UT) is a tri-stimulus reflectance colorimeter that monitors dye pigmentation changes mediated by microbial activity. Dyes can be used that produce color changes as a result of pH changes, changes in the redox potential of the medium or the presence of compounds with free amino groups. Samples are placed in microtiter wells or other types of containers and are scanned by an automated light source with computer interface during the growth stages (0–24 h). The change of color or hue (a*, b*, L) can be monitored like impedance or conductance curves. Manninen and Fung (1992c) evaluated this system in a study of pure cultures of *L. monocytogenes* and food samples, and found high correlation coefficients (*r*) of 0.90–0.99 for pure bacterial cultures and 0.82 for minced beef between the colony counts predicted by the colorimetric technique and the results of the traditional plate count method. They also showed that detection times for bacterial cultures such as *Enterobacter aerogenes, E. coli, Haffnia alvei* and several strains of *L. monocytogenes* were substantially (2–24 h) shorter using the instrument than using the traditional method, and concluded that the colorimetric detection technique employed by the 'Omnispec' system simplifies the analyses, saves labor and materials, and provides a high sampling capacity. More work is needed to confirm the usefulness of this new instrument.

2.4.4 Limulus amoebocyte lysate (LAL) and catalase tests

(a) The LAL method

The LAL method is a sensitive method for determining the endotoxin level of lipopolysaccharides of Gram-negative cell walls. Levin and Bang (1964) first mentioned the use of amoebocyte lysate of horseshoe crab (*Limulus*) to detect the presence of endotoxins. The test was first developed for medical uses. Gradually, this method has been applied to detect Gram-negative bacteria in foods. Jay (1989) made a detailed account of all aspects of the LAL test, including its history, the mechanism of the LAL reaction and the various methods by which the LAL test can be performed. He reported the use of the LAL test in milk, dairy products, sugar and meat samples. One advantage of this test is that it can be completed within an hour and requires very little skill. This test is quite useful in estimating Gram-negative bacterial populations especially the psychrotrophs in foods, such as meat, fish and poultry, that are cold stored under aerobic conditions. However, this test is not suitable for foods that contain mainly Gram-positive organisms, such as fermented foods.

(b) The catalase test

The catalase test is another rapid method for estimation of microbial populations in certain foods. Catalase is a very reactive enzyme. Figure 2.5 shows five types of reactions of catalase in capillary tubes. From left to right, extremely active (4+), very active (3+), active (2+), slightly active (1+), negative (0) and very active (3+). Microorganisms can be divided into catalase positive and catalase negative. Both groups are important in food microbiology; however, under certain food-storage conditions, a certain group predominates. Most perishable foods (commercial as well as domestic) are cold stored under aerobic conditions. The organisms causing spoilage of these foods are psychrotrophs. The predominant psychrotrophic bacteria are *Pseudomonas* spp., which are strongly catalase positive. Other important psychrotrophs such as *Micrococcus*, *Staphylococcus* and a variety of enterics are also catalase positive. Thus, one can make use of the presence of catalase to estimate the bacterial population. Catalase activity can be detected by a simple capillary tube method (Fung, 1985) or by an instrument called Catalasemeter (Wang and Fung, 1986). The capillary tube method is simple to use and can detect organisms when the level of catalase positive population reaches about 10^5 cell/ml. The Catalasemeter, although it showed some promise, is no longer in production and not available commercially at this moment.

Another way to use the catalase test is as an index of cleanliness of meat-processing areas. Preliminary data from the authors laboratory

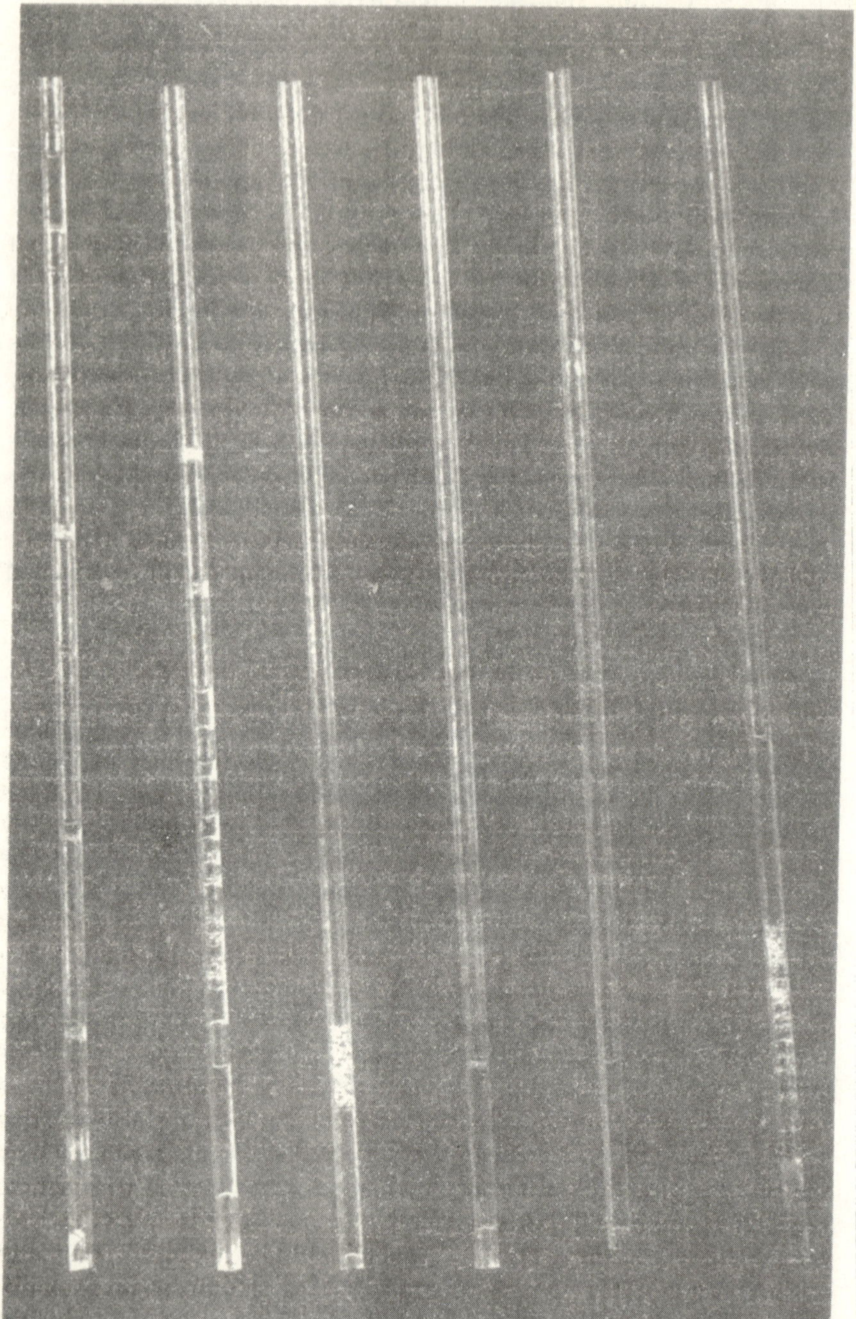

Figure 2.5 Catalase activity of six different bacterial cultures. Capillary tubes contain 3% H_2O_2 and the bottom opening interacted with a bacterial colony. From left to right: 4+, 3+, 2+, 1+, negative reaction and 3+.

indicate that a simple swab-catalase test can determine the degree of cleanliness after clean-up of a meat-processing plant. A moist swab was applied to a 2 × 2 inch area and then the swab was placed in 2 ml of 3% H_2O_2 in a test tube. The amount of bubbles generated in the tube can be scaled as 0, 1+, 2+, 3+, 4+ and 5+ depending on the activity of the sample. A bacterial count of the immediate adjacent area was also taken. The results showed that prior to slaughter, the room had very low catalase activity, which indicated that bacteria as well as blood or other particles that may carry catalase activity were present in low numbers. During the slaughter operation, the catalase activity was high because of meat particles, blood and bacteria in the environment. After proper clean-up, the catalase activity in the area fell to the original level. Bacterial counts did not directly correlate with the catalase activity. This is to be expected because the swab picks up material other than bacteria that can generate bubbles in the presence of H_2O_2. Further work is in process to ascertain the value of this simple test for environmental sanitation monitoring.

Currently the author is testing the use of catalase to estimate bacterial count in ground beef and other food. The use of catalase for seafood quality assurance certainly deserves further examination.

2.5 MINIATURIZED MICROBIOLOGICAL TECHNIQUES

Identification of microorganisms is an important part of quality assurance and control programs in the food and seafood industry. The author has developed many miniaturized methods to reduce the volume of reagents and media (from 5–10 to about 0.2 ml) for microbiological testing in Microtiter plates. The basic components of the miniaturized system are the Microtiter plates for test cultures, a multiple inoculation device and containers to house solid media (large Petri dishes) and liquid media (another series of Microtiter plates). The procedure involves placing liquid cultures (pure cultures) to be studied into sterile wells of a Microtiter plate to form a master plate. Each Microtiter plate can hold up to 96 different cultures, 48 duplicate cultures or various combinations as desired. The cultures are then transferred by a sterile multipoint inoculator (96 needles protruding from a template) to solid or liquid media. Sterilization of the inoculator is by alcohol flaming. Each transfer represents 96 separate inoculations in the conventional method. After incubation at an appropriate temperature, the growth of cultures on solid media or liquid media can be observed and recorded, and the data can be analyzed. Figure 2.6 shows a control plate where all 96 organisms grew on the agar. Figure 2.7 shows a dye-containing agar plate inoculated with 96 bacterial cultures. However, only 46 cultures grew and the other 50 cultures were suppressed by the dye. These

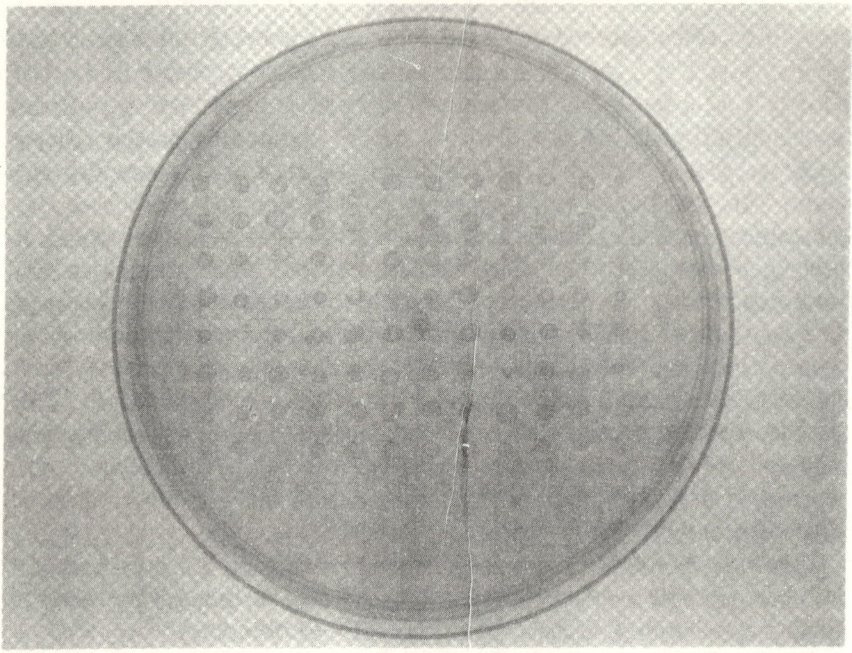

Figure 2.6 Growth pattern of 96 bacterial cultures on a non-selective agar plate after mass inoculation and incubation.

miniaturized procedures save a considerable amount of time in operation, effort in manipulation, materials, labor and space. These methods are ideal for studying large numbers of isolates or for research involving challenging large numbers of microbes against a host of test compounds.

The miniaturized methods have been used to study large numbers of isolates from foods (Fung and Miller, 1970, 1971; Fung and Hartman, 1975; Lee, Fung and Kastner, 1982, 1985) and to develop bacteriological media and procedures (Fung and Miller, 1973; Fung and Petrishko, 1973; Fung and Neimeic, 1977; Chein and Fung, 1991;). Miniaturized methods for studying food yeast were also developed in the author's laboratory (Lin, Fung and Entis, 1984; Lin and Fung, 1985, 1987; Fung and Liang, 1989). Currently, these miniaturized methods are being used to study food mycology (Hart and Fung, 1990; Hart *et al.*, 1991).

Fung and Kraft (1968, 1969) and Fung and LaGrange (1969) utilized miniaturized concepts to develop viable cell procedures and most probable number (MPN) tests. This involves using Microtiter loops (0.025 ml) to dilute liquid samples in Microtiter wells (each containing a known volume of sterile buffer) and then aseptically transferring a small volume

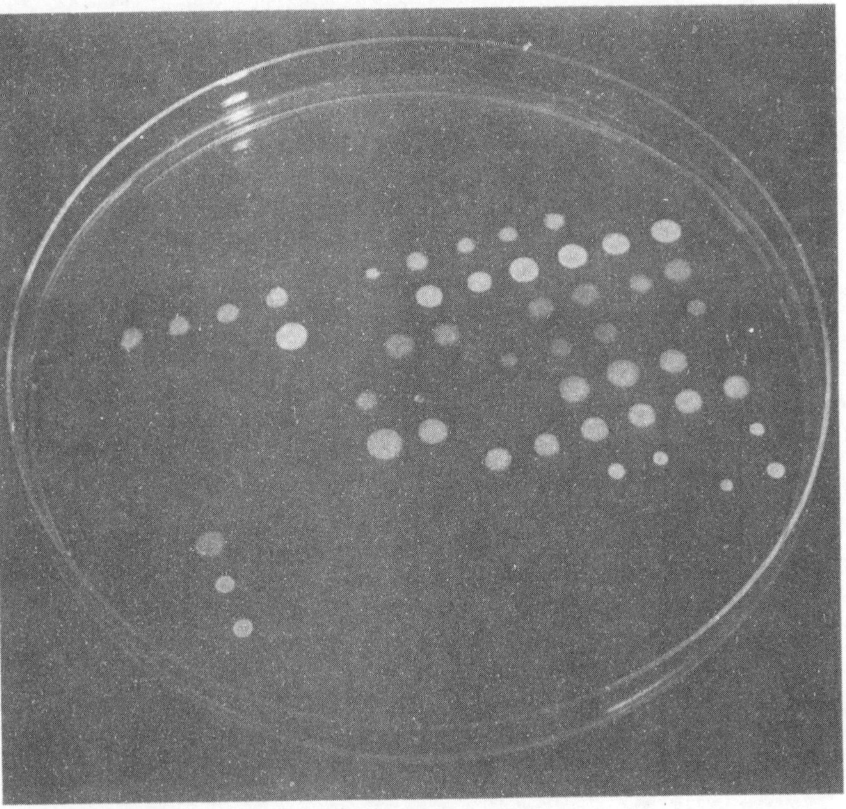

Figure 2.7 Growth pattern of 46 out of 96 cultures inoculated on to agar containing a selective dye.

(0.025 ml) of diluted samples onto agar surfaces to form spots. The agar plates (each holding four or six spots) are then incubated and colonies are counted after bacterial growth. This method was successfully tested against the conventional method in a collaborative study involving four laboratories (Fung *et al.*, 1976). Figure 2.8 shows growth of 1:100 (2) and 1:1000 (3) dilution of a liquid sample on agar. The spots from 1:1000 (3) are countable (46 and 47) and after conversion factors (40 × 1000) provide an average count of 1.5 × 10^5 cells/ml.

Another concept is to dilute a sample of liquid food (e.g. raw milk) in a three-well series in nutrient broth in the Microtiter plate to form a miniaturized three-tube MPN. These tests were developed for efficient operation of the viable cell count as an alternative to the standard plate count method and the standard MPN count procedure.

On the commercial side, many diagnostic kits to identify microorganisms have been developed and marketed since the 1970s. Currently,

API, Enterotube, R/B, Minitek, MicroID and IDS are available. Most of these systems were first developed for the identification of enterics (*E. coli, Salmonella, Shigella, Proteus, Enterobacter* spp., etc.). Later many of the companies expanded the capacity to identify nonfermentor, anaerobes, Gram-positive organisms, and even yeasts and molds. Most of the early comparative analyses centered around evaluation of these kits for clinical specimens. Cox, McHan and Fung (1977) and Fung and Cox (1981) studied these systems from the standpoint of food microbiology

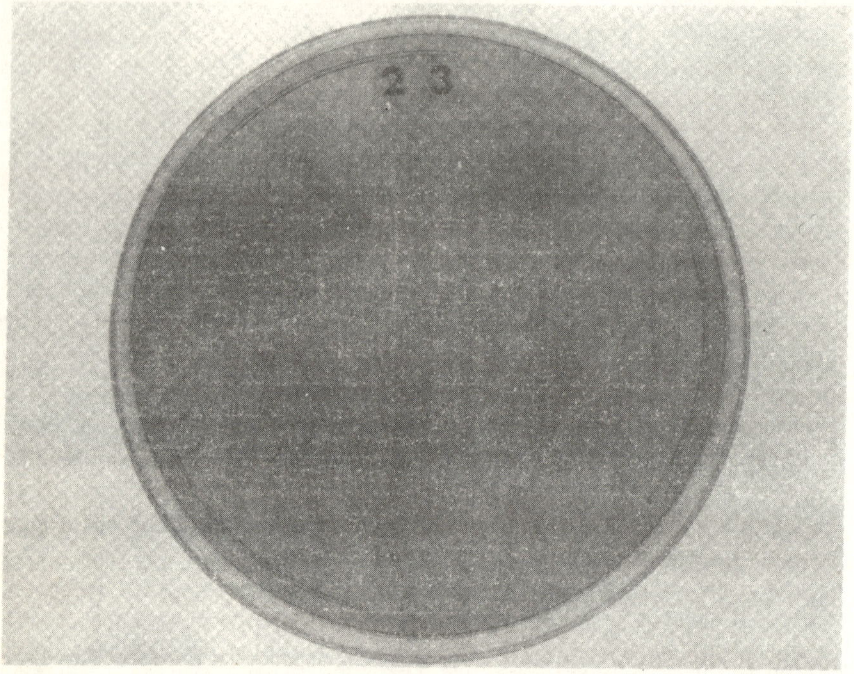

Figure 2.8 Miniaturized viable cell method. Each drop represents 0.025 ml of diluted culture. There are two drops from 10^2 dilution and two drops from 10^3 dilution. The numbers of colonies (46 and 47) developed from 10^3 dilution drops were used to determine the viable cell count of the original sample.

and concluded that these systems generally provide 90–95% accuracy when compared with conventional methods. Comparative analysis of diagnostic kits and selection criteria for miniaturized systems were made by Fung, Goldschmidt and Cox (1984) and Cox *et al.* (1984). They concluded that these miniaturized systems are accurate, efficient, labor saving, space saving and cheaper than the conventional procedure. Their usefulness in clinical and food microbiological laboratories will continue to be important.

2.6 NEW AND NOVEL TECHNIQUES

2.6.1 Vitek system

Many sophisticated instruments have been developed to identify isolates from clinical specimens, such as Sensititre (Radiometer Amer, Westlake, OH) and Biolog (Hayward, CA). One of the most automated systems for the identification of isolates (clinical and foods) is the Vitek system. The system depends on the growth of target organisms in specially designed media housed in tiny chambers in a plastic 'card'. The card is then inserted into the incubation chamber. The instrument periodically scans the wells of the cards and sends information to the computer, which then matches the data base and identifies the unknown cultures in the cards. The system is entirely automated and computerized and provides hard copies for record keeping. Most evaluations of the usefulness of the Vitek system had previously used clinical specimens. Bailey *et al.* (1985) were successful in utilizing the Vitek system to identify Enterobacteriaceae from foods. The system is capable of identifying enterics, yeast, *Bacillus*, selected Gram-positive pathogens and other organisms.

2.6.2 DNA probes

Many miniaturized kits and procedures have been developed to detect the presence or absence of target food pathogens. These new methods have cut the detection time for negative screens from 7 days to 48 h. However, if the test is positive, the conventional method must be used to confirm the presence of target food pathogens. Concepts and applications of miniaturized kits, immunoassays and DNA hybridization for recognition and identification of foodborne bacteria are reviewed by Cox *et al.* (1987b). The DNA probe (Genetrak, Farmingham, MA) is a sensitive method to detect pathogens such as *Campylobacter*, *Salmonella*, *Listeria* and *E. coli*. At first, the system utilized radioactive compounds for assay. The second generation of probes uses enzymatic reactions to detect the presence of pathogens. Fitts (1985) and Flowers (1985) showed that the DNA probe technique is as sensitive or more sensitive than the conventional method for *Salmonella* detection. Another major change in this area is the development of probes to detect target RNA. In a cell there is only one copy of DNA, however there may be 1000 to 10 000 copies of ribosomal RNA. Thus, the new generation of probes are designed to probe target RNA. Currently, kits are available for *Salmonella*, *Listeria*, *Campylobacter*, *Yersinia*, etc. As the need arises more organisms will be added to the list.

2.6.3 Enzyme linked immunosorbent assay (ELISA) tests

The ELISA system commercialized by Organon Teknika (Durham, NC) utilizes two monoclonal antibodies specific for *Salmonella* detection. In a comparative study involving 1289 samples, Eckner *et al.* (1987) found that there was no significant difference between the conventional method and the ELISA method for foods sampled, except cake mix and raw shrimp. Another ELISA system, the Tecra system (International BioProducts, Redmond, WA), was developed in Australia and uses polyclonal antibodies to detect *Salmonella*. These methods have been used to detect *Listeria* and *E. coli* also. Many companies are providing a host of monoclonal and polyclonal antibodies for a variety of diagnostic tests, some including food pathogens (Fung, Bennett and Lehleitner, 1988a).

2.6.4 The VIDAS immunoanalysis system

The VIDAS™ (Vitek ImmunoDiagnostic Assay System) is a multi-parametric immunoanalysis system developed and marketed by bio-Merieux Vitek, Inc. (Hazelwood, MO). The system utilizes the enzyme-linked fluorescent immunoassay (ELFA) method. According to the manufacturer:

> The end result of the test protocol is a fluorescent product and the VIDAS reader utilizes a special optical scanner that measures the degree of fluorescence. From the moment the solid phase receptacles and the reagent strips are placed in the instrument, the VIDAS is fully automated.

This is a revolutionary development because one of the drawbacks of the ELISA test is the many steps necessary for adding reagents and washing test samples. In the VIDAS system, all intermediate steps are automated.

2.6.5 The polymerase chain reaction (PCR)

One of the newest developments in rapid methods is PCR technology (McNamara, 1992). With this technology, theoretically, one can amplify one piece of target DNA to many millions of copies in several hours, thus enabling a scientist to detect target microorganisms rapidly. To be successful with this technique, one must know the sequence of two known regions of the target DNA and use two oligonucleotides to serve as 'primers' for a series of polymerization reactions that will result in millions of copies of the DNA. Typical PCR protocol involves denaturation of DNA at 96°C (30 s to several minutes), primer annealing at 55°C (30 s) and primer extension at 72°C (1.5 min). The reaction components include target DNA, bases, DNA polymerase (Taq), Mg^{2+} or Mn^{2+}, primers, buffer (pH 8.0–8.3) and mineral oil. The PCR products can be detected by gel electrophoresis, Southern blot, dot blot or ELISA tests. Automation and improvements of the entire system are constantly being made, especially by Perkin-Elmer Cetus (Norwalk, CT). This highly sophisticated procedure started in research laboratories for molecular biology and is now slowly finding its way to applied microbiology. Hill *et al.* (1991) reported identification of *Vibrio vulnificus* in artificially contaminated oysters by PCR. Ferreira *et al.* (1992) used PCR for improved assay for identification of Type A *Clostridium botulinum*. In the 1992 annual meeting of the American Society for Microbiology in New Orleans, a seminar entitled *Nuclear Acid Amplification and Other Innovative Detection Systems* was held and the use of PCR in food microbiology was discussed in detail. Certainly, PCR will be used in all areas of applied microbiology, including meat microbiology in the near future.

2.6.6 Motility enrichment

Many microbes, including those important to seafood microbiology, are motile. Motility enrichment is a very useful concept in rapid isolation and identification of food pathogens. Fung and Kraft (1970) described a motility flask system for rapid detection and isolation of *Salmonella* spp. from mixed cultures and poultry products. The system involves a flask with a side arm that contains several agar layers. Lactose broth is placed in the flask, and then a sample (with or without salmonellae) is inoculated into the lactose broth. When salmonellae are present, they will swim through the first level of agar, which contains selenite cysteine and sodium lauryl sulfite that inhibits other organisms but allows salmonellae to pass through. Once salmonellae pass the first layer, they can grow and metabolize compounds in the second and third layers. By looking at the color changes in the second and third agar layers, one can make an assumption that salmonellae were in the sample that was put into the lactose broth. Further serological tests can then be performed to confirm the presence of *Salmonella*.

In more recent years, a commercial system called *Salmonella* 1–2 test (BioControl, Bothell, WA) was developed that utilizes motility as a form of selection. The food sample is first pre-enriched for 24 h in lactose broth, and then 0.1 ml is inoculated into one of the chambers in an L-shaped system. The chamber contains selective enrichment liquid medium. There is also a small hole connecting the liquid chamber with the rest of the system, which has a soft agar through which salmonellae can migrate. An opening on the top of the second chamber allows the analyst to deposit a drop of polyvalent anti-H antibody. If the sample contains salmonellae from the lower side of the L unit, salmonellae will migrate through the hole and up the agar column. Simultaneously, the antibody against flagella of salmonellae will move downward by gravity. When the antibody meets the salmonellae they will form a visible 'immunoband'. The presence of an immunoband in this system is a positive test for *Salmonella* spp. The system is easy to use and has gained popularity because of its simplicity.

2.6.7 Oxyrase enzyme system

'Oxyrase' (Oxyrase Inc., OH), a membrane fraction of *E. coli*, was found in the author's laboratory to stimulate the growth of a large number of important facultative anaerobic food pathogens. In the presence of a hydrogen donor such as lactate, 'Oxyrase' can convert O_2 to H_2O, thereby reducing the oxygen tension of the medium and creating anaerobic conditions that favor the growth of facultative anaerobic organisms. In a medium containing 0.1 units/ml of the enzyme, the growth of *L. monocytogenes*, *E. coli* 0157:H7, *S. typhimurium*, *Streptococcus faecalis* and *Proteus vulgaris* were greatly enhanced; colony counts were greater by 1–2 log units, depending on the initial count and the strain studied, after incubation in the presence of the enzyme for 5–8 h at 35–42°C compared with control without oxyrase.

By combining the oxyrase enzyme and a unique 'U'-shaped tube, Yu and Fung (1991a,b, 1992) developed an effective method to detect *L. monocytogenes* and *Listeria* spp. from laboratory cultures and meat systems. Currently much work is being done in the author's laboratory to stimulate the growth of *Campylobacter* from foods.

There are many other systems that involve modern biochemistry, chemistry and immunology. For example, one can use protein profiles for microbial 'fingerprinting' (AMBIS system, San Diego, CA) or cell composition as a way to identify bacterial cultures (Hewlett-Packard, Palo Alto, CA).

2.7 CONCLUSIONS

This chapter describes a variety of methods that are designed to improve current methods, explore new ideas, and develop new concepts and technologies for the improvement of applied microbiology. Although many of these methods were first developed for clincial microbiology they are being used for food microbiology and seafood microbiology. A new journal named *Journal of Rapid Methods and Automation* was started in 1992 to encourage rapid dissemination of information concerning current developments in rapid methods and automation. This field will certainly grow and many seafood microbiologists will find these new methods very useful in their routine work in the immediate future.

Many methods described here are already being used by applied microbiologists nationally and internationally.

ACKNOWLEDGMENT

Contribution No. 92–676-J, Agricultural Experiment Station, Kansas State University, Manhattan, KA 66506. This material is based on work supported by the Cooperative State Research Service, US Department of Agriculture, under agreement no. 889034187 4511.

REFERENCES

Adams, M.R. and Hope, C.F.A. (1989) *Rapid Methods in Food Microbiology*, Elsevier, Amsterdam.

AOAC (1990) *Official Methods of Analysis of the Association of Official Analytical Chemists*, Vols I and II, Association of Official Analytical Chemists, Arlington, VA.

APHA (1985) *Standard Methods for the Examination of Dairy Products*, American Public Health Association, Washington, DC.

Balows, A., Tilton, R.C. and Tuxano, A. (1989) *Rapid Methods and Automation in Microbiology and Immunology*, Brixia Academic Press, Bregcia, Italy.

Bailey, J.S., Cox, N.A., Thomson, J.E. and Fung, D.Y.C. (1985) Identification of Enterobacteriaceae in foods with the Automicrobic System. *J. Food Protect.*, **48**, 147.

Bishop, J.R. and Juan, J.Y. (1988) Improved methods for quality assessment of raw milk. *J. Food Protect.*, **51**, 955.

Bishop, J.R. and White, C.H. (1985) Estimation of potential shelf-life to cottage cheese utilizing bacterial numbers and metabolites. *J. Food Protect.*, **48**, 663.

Bishop, J.R., White, C.R. and Firstenberg-Eden, R. (1984) A rapid impedimetric method for determining the potential shelf-life of pasteurized whole milk. *J. Food Protect.*, **47**, 471.

Chain, V.S. and Fung, D.Y.C. (1991) Comparison of Redigel, Petrifilm, Spiral Plate System, ISOGRID and standard plate count for the aerobic count on selected foods. *J. Food Protect.*, **54**, 208.

Chein, S.P. and Fung, D.Y.C. (1991) Acriflavin violet red bile agar for the isolation and enumeration of *Klebsiella pneumoniae*. *Food Microbiol.*, **7**, 73.

Cox, N.A., McHan, F. and Fung, D.Y.C. (1977) Commercially available minikits for the identification of *Enterobacteriaceae*: a review. *J. Food Protect.*, **40**, 86.

Cox, N.A., Fung, D.Y.C., Goldschmidt, M.S., Bailey, J.S. and Thomson, J.E. (1984) Selecting a miniaturized system for identification of *Enterobacteriaceae*. *J. Food Protect.*, **47**, 74.

Cox, N.A., Bailey, J.S., Fung, D.Y.C. and Hartman, P.A. (1987a) Rapid methods for the detection and identification and microorganisms in foods, in *Food Protection Technology* (ed. C.W. Felix), Lewis Chelsea, MI, pp. 125–31.

Cox, N.A., Fung, D.Y.C., Bailey, J.S. *et al.* (1987b) Miniaturized kits, immunoassays and DNA hybridization for recognition and identification of foodborne bacteria. *Dairy Food Sanitat.*, **7**, 628.

Czechowski, M.H. and Banner, M.J. (1989) Detection of surface-attached, viable *Salmonella* and *Yersinia* bacteria, in *Proceedings of the Annual Meeting of the Institute of Food Technologists.*, Chicago IL, June 25–29.

Day, A. (1983) Application in brewing. *Lab. Prac.*, **January**, 17.

Eckner, K.F., Flowers, R.S., Robinson, B.J. *et al.* (1987) Comparison of *Salmonella* Bio-EnzaBead immunoassay method and conventional culture procedure for detection of *Salmonella* in foods. *J. Food Protect.*, **50**, 379.

Eden, R. and Eden, G. (1984) *Impedance Microbiology*, Research Studies Press, Letchworth, UK.

Emswiler, B.S., Pierson, C.J. and Kotula, A.W. (1977) Stomaching vs. blending. *Food Technol.*, **31**(10), 40.

Entis, P. (1983) Enumeration of coliforms in nonfat dry milk and canned custard by hydrophobic grid membrane filter method: collaborative study. *J. Ass. Off. Anal. Chem.*, **66**, 897.

Entis, P. (1984) Enumeration of total coliforms and *Escherichia coli* in foods by hydrophobic grid membrane filters collaborative study. *J. Ass. Off. Anal. Chem.*, **67**, 812.

Entis, P. (1986) Membrane filtration systems, in *Foodborne Microorganisms and Their Toxins: Developing Methodology*, (eds M.D. Pierson and N.J. Stern), Marcel Dekker, New York, pp. 91–106.

Entis, P., Brodsky, M.H. and Sharpe, A.N. (1982) Effect prefiltration and enzyme treatment on membrane filtration of foods. *J. Food Protect.*, **45**, 8.

Evans, H.A.V. (1985) A note on the use of conductivity in brewery microbiology control. *Food Microbiol.*, **2**, 19.

FDA (1992). *Bacteriological Analytical Manual*, 6th edn, Association of Official Analytical Chemists. Food and Drug Administration. Arlington, VA.

Ferreira, J.L., Hamdy, M.K., McCay, S.G., and Baumstark, B.R. (1992) *J. Rapid Methods Automation Microbiol.*, **1**, 29.

Fitts, R. (1985) Development of a DNA–DNA hybridization test for the presence of *Salmonella* in foods. *Food Technol.*, **39**(3), 95.

Flowers, R.S. (1985) Comparison of rapid *Salmonella* screening methods and the conventional culture method. *Food Technol.*, **39**(3), 103.

Fung, D.Y.C. (1985) Procedures and methods for one-day analysis of microbial loads in foods, in *Rapid Methods and Automation in Microbiology and Immunology*, (ed. K.-O. Habermehl), Springer-Verlag, Berlin, pp. 656–64.

Fung, D.Y.C. (1991) Rapid methods and automation for food microbiology, in *Instrumental Methods for Quality Assurance in Foods*, (eds D.Y.C. Fung and R.F. Matthews), Marcel Dekker, New York.

Fung, D.Y.C. (1992) Historical development of rapid methods and automation in microbiology. *J. Rapid Methods Automation Microbiol.*, **1**, 1.

Fung, D.Y.C. and Cox, N.A. (1981) Rapid identification systems in the food industry: present and future. *J. Food Protect.*, **44**, 877.

Fung, D.Y.C. and Chain, V.S. (1991) Comparative analysis of Redigel and aerobic plate count methods for viable cell count of selected foods. *Food Microbiol.*, **8**, 299.

Fung, D.Y.C. and Hartman, P.A. (1975) Miniaturized microbiology techniques for rapid characterization of bacteria, in *New Approaches to the Identification of Microorganisms*, (eds C.G. Heden and T. Illeni), Wiley, New York, pp. 347–370.

Fung, D.Y.C. and Kraft, A.A. (1968) Microtiter method for the evaluation of viable cell in bacterial cultures. *Appl. Microbiol.*, **16**, 1036.

Fung, D.Y.C. and Kraft, A.A. (1969) Rapid evaluation of viable cell counts using the Microtiter system and MPN technique. *J. Milk Food Technol.*, **32**, 408.

Fung, D.Y.C. and Kraft, A.A. (1970) A rapid and simple method for the detection and isolation of *Salmonella* from mixed cultures and poultry products. *Poul. Sci.*, **49**, 146.

Fung, D.Y.C. and LaGrange, W. (1969) Microtiter method for bacterial evaluation of milk. *J. Milk Food Technol.*, **32**, 144.

Fung, D.Y.C. and Liang, C. (1989) A new fluorescent agar for the isolation of *Candida albicans*. *Bulletin d'information des Laboratoires des Services Veterinaries (France)*; **29/30**, 1–2.

Fung, D.Y.C. and Matthews, R.F. (eds) (1991). *Instrumental Methods for Quality Assurance in Foods*, Marcel Dekker, New York.

Fung, D.Y.C. and Miller, R.D. (1970) Rapid procedure for the detection of acid and gas production by bacterial cultures. *Appl. Microbiol.*, **20**, 527.

Fung, D.Y.C. and Miller, R.D. (1971) Miniaturized techniques for IMViC tests. *J. Milk Food Technol.*, **35**, 328.

Fung, D.Y.C. and Miller, R.D. (1973) Effects of dyes on bacterial growth. *Appl. Microbiol.*, **25**, 793.

Fung, D.Y.C. and Neimcic, M. (1977) Acriflavin violet red bile medium for the isolation of *Klebsiella*. *Lab. Health Sci.*, **14**, 273.

Fung, D.Y.C. and Petrishko, D.I. (1973) Capillary tube catalase test *Appl. Microbiol.*, **26**, 631.

Fung, D.Y.C., Donahue, R., Jensen, J.P. *et al.* (1976). A collaborative study of the microtiter count method and standard plate count method on viable cell count of raw milk. *J. Milk Food Technol.*, **39**, 24.

Fung, D.Y.C., Lee, C.Y. and Kastner, C.L. (1980) Adhesive tape method for estimation of microbial load on meat surface. *J. Food Protect.*, **43**, 295.

Fung, D.Y.C., Goldschmidt, M.C. and Cox, N.A. (1984) Evaluation of bacterial diagnostic kits and systems at an instructional workshop. *J. Food Protect.*, **47**, 68.

Fung, D.Y.C., Hart, R.A. and Chain, V. (1987) Rapid methods and automated procedures for microbiological evaluation of seafood, in *Seafood Quality Determination*, (eds D.E. Kramer and J. Liston), Elsevier, Amsterdam, pp. 247–253.

Fung, D.Y.C., Bennett, R. and Lehleitner, G.C. (1988) Rapid diagnosis in bacteriology: Contribution of polyclonal and monoclonal antibodies, in *Biologie Prospective, le Colloque de Pont-e-Mousson*, (eds M.M. Gaeteau, J. Henry and G. Siest), John Libbey, London, pp. 21–6.

Fung, D.Y.C., Cox, N.A. and Bailey, J.S. (1988) Rapid methods and automation in microbiology. *Dairy Food Sanit.*, **8**, 292.

Fung, D.Y.C., Cox, N.A., Goldschmidt, M.C. and Bailey, J.S. (1989) Rapid methods and automation: A survey of professional microbiologists. *J. Food Protect.*, **52**, 65.

Gibson, D.M. and Hobbs, G. (1987) Some recent developments in microbiological methods in seafood quality, in *Seafood Quality Determination*, (eds D.F. Kramer and J. Liston), Elsevier, Amsterdam, pp. 283–298.

Gibson, D.M. and Ogden, I.D. (1980) Assessing bacterial quality of fish by conductance measurement. *J. Appl. Bacteriol.*, **49**, XII.

Ginn, R.E., Packard, V.S. and Fox, T.L. (1986) Enumeration of total bacteria and coliforms in milk by dry rehydratable film method: collaborative study. *J. Ass. Off. Anal. Chem.*, **69**, 27.

Habermehl, K.O. (ed.). (1985) *Rapid Methods and Automation in Microbiology and Immunology*, Wiley, New York.

Hart, R.A. and Fung, D.Y.C. (1990) Evaluation of dye media selective for *Aspergillus* and/or *Penicillium*, in *Proceedings of the Annual Meeting of the American Society of Microbiologists*, p. 279.

Hart, R. A., Mo, O., Borius, F. and Fung, D.Y.C. (1991) Comparative analysis of trypan blue agar and congo red agar for the enumeration of yeast and mold using HGMF system. *J. Food Safety.*, **11**, 227.

Heden, C.G. and Illeni, T. (1975a) *Automation in Microbiology and Immunology*, Wiley, New York.

Heden, C.G. and Illeni, T. (1975b) *New Approaches to the Identification of Microorganisms*, Wiley, New York.

Hill, W.E. Keasler, S.P., Trucksness, M.W., *et al.* (1991) Polymerase chain reaction identification of *Vibrio vulmificus* in artificially contaminated oysters. *Appl. Environ. Microbiol.* **57**, 707–70.

Hurst, W.J. and Mortimer, J.W. (1987) *Laboratory Robotics: A Guide to Planning, Programming and Applications*, VCH Publishers, New York.

Jay, J.M. (1989) The Limulus amoebocyte lysate (LAL) test, in *Rapid Methods in Food Microbiology*, (ed. M.R. Adams. and C.F.A. Hope), Elsevier, Amsterdam, pp. 101.

Jay, J.M. (1992) *Modern Food Microbiology*, 4th edn, Van Nostrand Reinhold, New York.

Johnson, H.H. and Newsom, S.W.B. (1976) *Rapid Methods and Automation in Microbiology*, Learned Information, Oxford.

Kilgour, W.J. and Day, A. (1983) The application of new techniques for the rapid determination of microbial contamination in brewing, in *Proceeding of the 19th Congress of the European Brewing Convention*, London.

Konuma, H.A. and Kurata, H. (1982) Improved Stomacher 400 bag applicable to the spiral plate system for counting bacteria. *Appl. Environ. Microbiol.*, **44**, 765.

LaRocco, K.A., Littel, K.J. and Pierson, M.D. (1986) The bioluminescent ATP assay for determining the microbial quality of foods, in *Foodborne Microorganisms and Their Toxins: Developing Methodology*, (ed. M.D. Pierson and N.J. Stern), Marcel Dekker, New York, pp. 145–74.

Lee, J.Y. and Fung, D.Y.C. (1986) Surface sampling techniques for bacteriology. *J. Environ. Health*, **48**, 200.

Lee, C.Y., Fung, D.Y.C. and Kastner, C.L. (1982) Computer-assisted identification of bacteria on hot-boned and conventionally processed beef. *J. Food Sci.*, **47**, 363.

Lee, C.Y., Fung, D.Y.C. and Kastner, C.L. (1985) Computer-assisted identification of Microflora on hot-boned and conventionally processed beef: effect of moderate and slow initial chilling rate. *J. Food Sci.*, **50**, 553.

Levin, J. and Bang, F.B. (1964) The role of endotoxin in the extracellular coagulation of *Limulus* blood. *Bull. Johns Hopkins Hosp.*, **115**, 265.

Lin, C.C.S. and Fung, D.Y.C. (1985) Effect of dyes on growth of food yeast. *J. Food Sci.*, **47**, 770.

Lin, C.C.S. and Fung, D.Y.C. (1987) Critical review of conventional and rapid methods for yeast identification. *CRC Crit. Rev. Microbiol.* **14**, 273.

Lin, C.C.S., Fung, D.Y.C. and Entis, P. (1984) Growth of yeast and mold on trypan blue agar in conjunction with the ISOGRID system. *Can. J. Microbiol.*, **30**, 1405.

Littel, K.J., Pikelis, S. and Spurgash, A. (1986) Bioluminescent ATP assay for rapid estimation of microbial numbers in fresh meat. *J. Food Protect.*, **49**, 18.

Manninen, M.T. and Fung, D.Y.C. (1992a) Use of the Gravimetric diluter in microbiological work. *J. Food Protect.*, **55**, 59.

Manninen, M.T. and Fung, D.Y.C. (1992b) Use of spiral plater and laser colony scanner for enumeration of microorganisms in meat. *J. Rapid Methods Automat. Microbiol.* **1**, 117.

Manninen, M.T. and Fung, D.Y.C. (1992c) Estimation of microbial numbers from pure bacterial cultures and from minced beef samples by reflectance colorimetry with Omnispec 4000. *J. Rapid Methods Automat. Microbiol.*, **1**, 41.

Manninen, M.T., Fung, D.Y.C. and Hart, R.A. (1991) Spiral system and laser colony scanner of enumeration of microorganisms. *J. Food Safety*, **11**, 177.

McMurdo, I.H. and Whyward, S. (1984) Suitability of rapid microbiological methods for the hygiene management of spray drier plant. *J. Soc. Dairy Technol.* **34**, 4.

McNamara, A.M. (1992). Polymerase Chain Reaction: Techniques and Application, in *Handbook for Rapid Methods and Automation in Microbiology Workshop*, (ed. D.Y.C. Fung), Kansas State University, Manhattan, KS.

Ogden, I.D. (1986) Use of conductance methods to predict bacterial counts in fish. *J. Appl. Bacteriol.*, **61**, 236.

Pettipher, G.L. (1986) Review: the direct epifluorescent filter technique. *J. Food Technol.* **21**, 535.

Pettipher, G.L. (1989) The direct epifluorescent filter technique, in *Rapid Microbiological Methods*, (eds M.R. Adams and C.F.A. Hope), Elsevier, New York.

Pettipher, G.L. and Rodrigues, V.M. (1980). Rapid enumeration of bacteria in heat treated milk and milk products using a membrane filtration-epifluorescent microscopy technique. *J. Appl. Bacteriol.*, **50**, 157.

Pettipher, G.L. and Rodrigues, V.M. (1982). Rapid enumeration of microorganisms in foods by the direct epifluorescent filter technique. *Appl. Environ Microbiol.*, **44**, 809.

Pettipher, G.L., Mansell, R., McKinnon, C.H. and Cousins. C.M. (1980) Rapid membrane filter epifluorescent microscopy technique for the direct enumeration of bacteria in raw milk. *J. Appl. Environ. Microbiol.*, **39**, 423.

Pierson, M.D. and Stern, N.J. (1986) *Foodborne Microorganisms and Their Toxins: Developing Methodology*, Marcel Dekker, New York.

Roth, J.N. (1988) Temperature independent pectin gel method for aerobic plate count in dairy and nondairy food products: collaborative studies. *J. Ass. Off. Anal. Chem.*, **71**, 343.

Roth, J.N. and Bontrager, G.L. (1989) Temperature-independent pectin gel method for coliform determination in dairy products: Collaborative study. *J. Ass. Off. Anal. Chem.*, **72**, 298.

Schalkowsky, S. (1986) Plating systems, in *Foodborne Microorganisms and Their Toxins: Developing Methodology*, (eds M.D. Pierson and N.J. Stern), Marcel Dekker, New York, pp. 16–26.

Sharpe, A.N. (1991). Rapid methods: consideration to adoption by regulatory agencies, in Proceedings of the 105th AOAC Annual International Meeting, Phoenix, AZ.

Sharpe, A.W. and Clark, D.S. (eds) (1978). *Mechanizing Microbiology.*, Thomas, Springfield, Il.

Sharpe, A.W. and Jackson, A.K. (1975) Automation requirements in microbiological quality control of foods, in *Automation in Microbiology and Immunology* (eds C.G. Heden and T. Illeni), Wiley, New York, pp. 117–24.

Sharpe, A.W. and Peterkin, P.I. (1988) *Membrane Filter Food Microbiology*, Research Studies Press, Letchworth, UK.

Sjoberg, A.M., Manninen, M., Pinnioja, S. and Harmala, P. (1990) Methods for detection of irradiation of spices. 2. *Lebensm. Unters. Forsch.*, **190**, 99.

Smith, L.B., Fox, T.L. and Busta, F.F. (1986) Comparison of a dry medium culture plate (Petrifilm SM plates) method to the aerobic plate count method for enumeration of mesophilic aerobic colony-forming units in fresh ground beef. *J. Food Protect.*, **48**, 1044.

Stannard, C.J. (1989) ATP estimation, in *Rapid Methods in Food Microbiology*, (eds M.R. Adams and C.F.A. Hope), Elsevier, Amsterdam, pp. 1–18.

Tilton, R.C. (ed.) (1982) *Rapid Methods and Automation in Microbiology*, American Society Microbiology, Washington, DC.

Vaheri, A., Tilton, R.C. and Balows, A. (eds) (1991) *Rapid Methods and Automation in Microbiology and Immunology: Proceedings of the 6th International Symposium*, Springer-Verlang, New York.

Vanderzant, C. and Splittstoesser, D. (eds) (1992) *Compendium of Methods for the Microbiological Examination of Foods*, American Public Health Association, Washington, DC.

Visser, I.J.R. and deGroote, J. (1984) Prospects for the use of conductivity as an aid in the bacteriological monitoring of pasteurized milk. *Antonin van Leewenhoek*, **50**, 202.

Waes, G.M. and Bossuyt, R.G. (1984) Impedance measurements to detect bacteriophage problems in cheddar cheese. *J. Food Protect*. **47**, 349.

Ward, D.R., LaRocco, K.A. and Hopson, D.J. (1986) Adenosine triphosphate bioluminescent assay to enumerate bacterial numbers on fresh fish. *J. Food Protect*. **49**, 647.

Yu, L.S.L. and Fung, D.Y.C. (1991a) Oxyrase enzyme and motility enrichment Fung-Yu tube for rapid detection of *Listeria monocytogenes* and *Listeria* spp. *J. Food Safety*, **11**, 149.

Yu, L.S.L. and Fung, D.Y.C. (1991b) Effect of oxyrase enzyme in *Listeria monocytogenes* and other facultative anaerobes. *J. Food Safety*, **11**, 163.

Yu, L.S.L. and Fung, D.Y.C. (1992) Growth kinetics of *Listeria* in the presence of oxyrase enzyme in a broth model system. *J. Rapid Methods Automation Microbiol.*, **1**, 15.

Zindulis, J. (1984) A medium for the impedimetric detection of yeasts in foods. *Food Microbiol.*, **1**, 159.

CHAPTER 3

Microbiological control for fish smoking operations

R. Dillon, T.R. Patel and A.M. Martin

3.1 INTRODUCTION

In industrialized countries, the smoking of fish is done for the enhancement of flavor and texture (Robinson, 1983). This form of processing sometimes provides little protection against microbiological, enzymatic and chemical deteriorative alterations, and, in some instances, smoked foods spoil as readily as non-smoked foods. It is nonetheless used as a method of food preservation in developing countries (Hsu, Fisher and Daun, 1979).

3.1.1 Historical overview of fish smoking operations

Smoking technology arose long ago, and has been used as a method of long-term preservation (Burgess and Bannerman, 1963). The advantages of exposing food to smoke are several: a more preserved food, enhanced flavor, and the prevention of infestation by insects. Fish smoking can be traced back to as early as 1349 (Cutting, 1962) and traditional smoking can still be found in practice today. Although fish smoking processes can be divided into two distinct types, hot and cold, and these arose in different geographical areas, the distinct boundaries of their use have today vanished (Anonymous, 1965).

Until the early 20th century, fish products were heavily salted (almost to the point of saturation), thoroughly dried and heavily smoked (sometimes for up to 3 weeks). The finished product was totally preserved for long periods (Anonymous, 1965). Refrigeration technology has reduced the need for harsh curing and allowed for milder cures. This has coincided with the modern consumer becoming more health-conscious

Fisheries Processing: Biotechnological applications.
Edited by A.M. Martin.
Published in 1994 by Chapman & Hall, London. ISBN 0 412 58460 3

and desiring milder cured fish. Modern smoked fish products are thus salted, dried and smoked more lightly.

The developments in mechanical and automated processing equipment, including automated briners and controllable mechanical kilns, have been important innovations in quality control in the fish smoking industry. The introduction in 1939 of the mechanical 'Torry Kiln' (Figure 3.1), for example, which represented a radical digression from the traditional method of smoking fish in oversized brick chimneys, overcame problems encountered with traditional kilns, such as inconsistency in product quality, the influence of weather conditions and requirements for intensive labor (Burgess and Bannerman, 1963; Anonymous, 1965). Since then, more modern kilns have become available with computer-programmable control of, for example, smoke velocity, temperature and relative humidity, and with automated smoke generators. These smokers represent a further reduction in the requirement for manual operation and operator expertise.

Typical products of the early 1900s were bloaters, kippers and red herring; these can still be found processed in a similar manner today (Cutting, 1962; Burgess and Bannerman, 1963). The term 'kipper', originally applied to spent Atlantic salmon caught in European rivers and smoked (Whelan, 1982), can now be applied to any species of fish that is smoked. In such cases it precedes the name of the species, as in kippered mackerel.

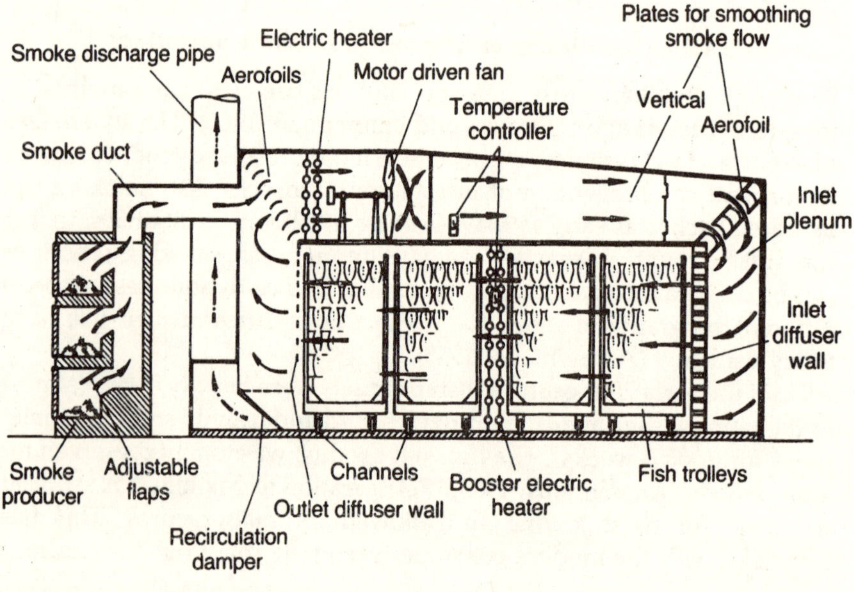

Figure 3.1 Sectional view of Torry kiln and smoke producer (Burgess and Bannerman, 1963). Reproduced with permission from the Torry Research Station, Aberdeen, UK.

Europe is the biggest producer of smoked fish, but there is growing interest in it in North America and the Far East. Salmon is the most prized smoked product in Europe with smoked eel following. Smoked herring, although less expensive, is the most common smoked product. Various other types of smoked fish products include schillerlocken, cod and cod roe, black halibut, haddock and trout; in general, any good-quality seafood can be smoked.

3.2 SMOKING OF FISH: TECHNOLOGICAL ASPECTS

The processing of smoked fish involves five major steps:

1. selection, handling, and preparation of raw material;
2. brining/salting;
3. racking/hanging and drying;
4. smoking;
5. packaging.

3.2.1 Smoking procedure

(a) Selection, handling and preparation

Top-quality fish are needed to produce a top-quality smoked product. All frozen products should be thawed slowly for 12–18 h at chill temperatures (4°C). Great care should be taken when handling to prevent damage to the product. Once the fish to be smoked is prepared in form (fillets, steaks, whole, split, etc.), the fish are salted.

(b) Brining/salting

Salting is the removal of water with its partial replacement by salt and is done in either of two ways; dry salting, in which the product is layered with salt at 5–10% of the weight of the fish to be salted; or liquid brining, in which the product is immersed in a prepared salt solution of a particular degree of salt saturation (°s). The general trend is towards 70–80°s and towards the use of liquid brine as opposed to dry salting (Churchill, Mitchell and Thompson, 1989). The salting process plays an especially important role with cold smoked products. The salt denatures the surface protein and, with drying, an artificial, glossy skin forms on the cut surface, known as the 'pellicle'. This finish helps seal in natural juices and flavors.

For a bactericidal effect, a salt concentration in excess of 9% would be ideal, but the smoked products of today usually contain 2–4% salt, just enough to impart flavor and texture. The length of time in brine is dependent on various factors: increased flesh thickness and fattiness, higher fish quality, and lower temperatures (10–15°C being recommended) slow the uptake of salt. As well, a fish which has been frozen and thawed will take up salt twice as fast as a fresh one.

At the start of the process, the ratio of fish to brine should be 1:1. With use, the brine will lose its strength and will require additional salt or complete replacement. Dyes, sugars, spices, liquors and liqueurs may be added at this stage of the smoking process.

(c) Racking/drying

Racking and drying aids in the formation of the pellicle, reduces drying time during smoking and provides maximum exposure of the fish to the smoke. Racks are more widely used today, but some products still require the use of speats, or spits, to be threaded through the neck, eyes or the gill and mouth of the fish for hanging. This procedure is known as 'speating'. Cold smoking requires a certain amount of drying prior to smoking to help produce the pellicle, but hot smoking can occur immediately after racking, since a pellicle would only be destroyed by the heat.

Some factors affecting drying would be: the velocity of smoke passing over the fish, the amount of moisture in the fish, the temperature and the amount of moisture held in the smoker, and the amount of salt in the fish. The warmer the air the more moisture it can hold. A relative humidity of 60–70% is most satisfactory for cold smoking of fish. If the relative humidity is lower than this, drying occurs rapidly with consequent 'case hardening'; if it is above 70%, then very little drying occurs. Rate of drying can be controlled by allowing more or less fresh air into the kiln (Burgess and Bannerman, 1963).

(d) Smoking

Smoke is produced through the incomplete combustion of any essentially carbonaceous material; for fish smoking this is usually wood in the form of sawdust or woodchips. Wood smoke consists of two overall fractions: visible tarry droplets and invisible vapors. The tarry droplets are considered of no significant value and may be removed by electrostatic precipitation. In addition to imparting characteristic color and flavor, and having bactericidal and antioxidative properties, smoke also contributes potentially toxic substances and causes some loss in nutritive value (Daun, 1979). One study found smoking to lower the pH of the food as well as the amounts of amino nitrogen, lysine and free sulfhydryl groups it contained (Sink, 1979). In general, the application of heat to foods, as in the case of smoking, causes a reduction in the quality of constituent proteins. The amino acid lysine seems the most

Table 3.1 Compounds identified in wood smoke[a]

Compounds	*Number of types*
Acids	48
Alcohols	22
Carbonyls	130
Esters	22
Furans	46
Lactones	16
Phenols	75
Miscellaneous	50
Total	409

[a]Adapted from Maga (1988), with permission from CRC Press, Boca Raton, FL.

negatively affected by smoking, with destruction exceeding 55% (Aminullah Bhuiyan, Ackman and Lall, 1986; Clifford, Tang and Eyo, 1980).

Smoke from wood is a highly complex mixture of chemicals including organic acids, alcohols, ammonia, carbon dioxide, carbon monoxide, carbonyls, esters, furans, hydrocarbons, lactones, nitrogen oxides, particulates, phenols, sulfur dioxides and other miscellaneous compounds such as benzene, indene, naphthalene, styrene, toluene, thymol, etc. (Daun, 1979). Over 400 volatiles have been identified in wood smoke (Maga, 1987) (Table 3.1).

Although many researchers believe that phenols are the primary contributors to the smoke aroma (Maga, 1987), others have found different classes of compounds to also produce typical smoke aroma. The design of more advanced protocols to ensure the purity of each smoke fraction is needed to resolve this question. According to Kernreich and Issenberg (1972), phenols, guaiacol, 4-methyl guaiacol and 2,6-dimethoxy phenol in the smoke vapor phase are all thought to be responsible for odor and flavor. Maga and Chen (1985) have reported that smoke pyrazines, thermally induced potent aroma compounds, also contribute to the overall smoky aroma. Flavor chemists, in general, believe that the overall characteristic flavor of wood smoke is probably due to the contribution of compounds from several classes, each of which contributes to the overall sensory properties of wood smoke.

Phenols also play a role in color formation (Daun, 1979). The color development in fish being smoked is believed to be caused by the interaction of smoke carbonyls with amino components on the flesh surface (Ruiter, 1979).

The acceptability of smoke depends on the type of wood used (Table 3.2). Woods are composed primarily of celluloses (40–60%), hemicelluloses (20–30%) and lignins (20–30%). Both hardwood and softwood can

Table 3.2 Wood source and smoke acceptability[a]

Wood	*Smoke*
Aspen	poor
Alder	poor
Beech	good
Birch	fair
Fir	poor
Lime	poor
Maple	good
Oak	good
Pine	fair

[a]Maga (1988), Tilgner (1958). Adapted from Maga (1988), with permission from CRC Press, Boca Raton, FL.

be used for smoking, although hardwood is much preferred because it imparts a milder flavor, whereas softwoods such as fir and pine impose a more resinous flavor but with a more rapid color development. Hardwoods produce a lower pH in the food being smoked (Clifford, Tang and Eyo, 1980). Special blends of various woods may be formulated to impart specific flavors. The largest amounts of pyrazines are found in hickory woods (Maga and Chen, 1985). Asita and Campbell (1990) concluded that the smoke from different wood sources affects microorganisms differently. For example, smoke produced from burning white mangrove, mahogany and abura inhibited *Escherichia coli, Staphyloccous aureus* and *Saccharomyces cerevisiae*, whereas smoke from other woods did not inhibit all of these organisms. Most sawdust is soaked prior to use to prevent flare-ups during smoking. Liquid smoke, a new innovation, has also been reported to be effective and will be discussed later in this work.

The surface of the product to be smoked is damp and tacky; hence, during initial stages of smoking, a high rate of drying as well as heavy deposition of smoke may occur. As smoking progresses, smoke deposition and drying rate will decrease. The ideal smoking process would have minimized drying and maximized smoke density during initial stages (Churchill, Mitchell and Thompson, 1989). Case hardening, or over-drying of the surface, may occur when there is excessive drying during the initial stages.

(e) Post-processing

All smoked products, whether hot or cold smoked, require slow cooling to room temperature of the plant, immediately followed by chilling to

4°C, and, if required, quick freezing and storing at −18°C (Bannerman, 1980). Smoked products can be vacuum-packaged (removal of oxygen has the added advantage of increasing shelf-life), shrink packaged (the product is placed on a cardboard liner and wrapped in packaging material, which is then heat-sealed and heat-shrunk to conform to the shape of the product), canned or bottled (heat treatment of the product inside a rigid container), packaged in retortable foil pouches (heat-sealed, heat-treated foil pouches of food not needing refrigeration) and packed in wooden boxes (involves shrink packaging and placing the final product in a wooden box) or master cartons (bulk- or individually-packaged products can be placed in a master carton). Cold smoked products darken in color with heat treatment and therefore initially require a light smoking process.

3.2.2 Cold smoking

The procedures for cold smoking and hot smoking differ only in the temperature at which the smoking occurs. In both cases, a top-quality raw material is required, and is brined, racked and dried until the surface is tacky to the touch. Cold smoking is the process of applying a smoky flavor to a food at temperatures below 30°C, producing a raw finished product. Further cooking of these smoked products occurs prior to consumption, with the exception of cold smoked salmon. The formation of a pellicle is essential for cold smoked products. Most cold smoked products are laid on racks and a heavy smoke initially deposited, tapering off toward the end. The drying is minimal and increases near the end of the cycle. The temperature may be raised to 40°C for the last 15 min of the cycle for fatty fish such as herring and mackerel, to bring the natural oils to the surface to be spread evenly over it, adding to the glossy appearance. Products are cooled to room temperature, chilled to 4°C and packaged appropriately.

3.2.3 Hot smoking

Hot smoking uses a temperature of 70–80°C at some time in the smoking process in order to cook the flesh (Bannerman, 1980). Hot smoking consists of the same preparatory procedures as cold smoking, except for a reduction in the brining stage, because the product will lose much more moisture during cooking. Hot smoked products are often hung rather than racked. Products can also be smoked immediately after brining without a drying stage, since the formation of a pellicle is not necessary. There are three stages to the actual smoking process. The fish are first smoked for 0.5–1 h at 30°C, often referred to as the tempering stage. This process allows for drying and toughening of the skin. The second stage is called the heating step, in which the temperature of the kiln is raised and the products continue to be smoked at 50°C for 1 h. Finally, at the cooking stage, the temperature is raised to between 70 and 90°C and the product is cooked to the desired level, usually for 1–2 h. Smoke amounts may be altered at this stage, depending on the product desired. As well, air inlets and outlets should be closed or set at a minimum to increase the temperature. The hot smoked products are then cooled slowly to room temperature and chilled to 4°C prior to packaging.

With the introduction of more modern distribution systems, cold smoked products are becoming more extensively used in technologically advanced countries than the hot smoked.

3.2.4 Application of smoking to specific fish species

*(a) Cod (*Gadus morhua*) and haddock*

Cod are smoked as skinless boned fillet, skin-on bone-in fillet or as headless split fish; haddock usually have the skin left on. Cod fillets are brined at 70°s for 5–10 min (dye is usually added to the brine bath at this stage), hung or racked, smoked accordingly (2–5 h) and post-processed. When hot smoking a lean fish such as cod, oil can be brushed on to the racks to prevent the fish from sticking. The brined fish should be allowed to stand and drip for at least 2 h before smoking. Smoking time varies from 2 to 5 h according to the fish size. Since white fish fillets are dyed, the smoking process is intended to give flavor and texture and not color (Burgess and Bannerman, 1963).

(b) Cod roe

Roes are washed in cold water, layered with dry salt to a height of 60 cm for 6–8 h, and then placed in mesh baskets and thoroughly washed. The baskets of washed roe are dipped into boiling water for 1–2 min to cause the roe to swell and have a plump appearance. Dye may be added to the boiling water to give uniform color. The roe are then laid on mesh trays and smoked at 18–24°C for 6–8 h (Burgess and Bannerman, 1963).

*(c) Eel (*Anguilla anguilla*)*

Smoked eel is a special delicacy in Europe, particularly in Germany and Holland. Eels are always hot smoked and are either alive or live-frozen. Live and live-frozen eels are immersed in a 2% v/v ammonia solution equal to the volume of the fish, or dry salted, which kills the eels in a few minutes and helps deslub (deslime) them. The eels are scrubbed and rinsed in fresh water. Eels are totally gutted, taking care to remove the gall bladder intact to prevent discoloration and flavor problems. Machines are available to cut and gut eels and are in use in Germany. Small eels (shorter than 1 m) are brined for 0.5–1 h in 80°s and larger eels (longer than 1 m) are brined for 1–2 h. Eels are speated through the neck, rinsed in cold water and allowed to drain. The bellies are opened for maximum exposure during smoking by either using small wooden sticks between the belly flaps, immersing the upper half of the eel into almost boiling water for 5–10 s or by placing the eels in a preheated smoker (80°C) for 10 min, then removing until the smoker cools to 30°C. Eels undergo a slightly different hot smoking because of the toughness of the skin. They are smoked for 0.5 h at 30–35°C, then for 1 h at 60°C and finally at 70°C until their internal temperature reaches 58°C (Churchill, Mitchell and Thompson, 1989). The products are then post-processed as previously described.

(d) Finnan haddock

Split haddock are cleaned of all entrails, brined at 80°s for 7–15 min, and the fillets speated onto metal rods and dried for a short period to develop a pellicle. Brined fillets are cold smoked for 3–6 h and removed from the kiln when the product is slightly less colored than desired, as the full color develops over a period of time (Burgess and Bannerman, 1963).

(e) Glasgow pales

Small haddock (less than 0.75 lb) are split, brined in 80°s for 4–5 min, cold smoked for 2–2.5 h and removed when they are just beginning to develop color (Burgess and Bannerman, 1963). Dye is not used.

*(f) Herring (*Clupea harengus*)*

Herring are considered fatty fish, a very important characteristic for smoke processors. A fat content of more than 13% is often desired, but some processors will except herring with fat contents of 10%. Smoked products from herring include bloaters, buckling, kippered herring ('kippers') and kipper fillets. Kippers can be prepared head-on backbone-in, head-off backbone-in, head-off backbone removed, as butterfly fillets or individual fillets, but are usually prepared with head off. Kippers are made by cutting down the backbone and exposing the belly when removing the entrails. Kippers are brined at 70°s for 7–12 min, racked or speated, cold smoked 2–6 h and post-processed. Bloaters are prepared from whole, mature herring, dry salted overnight (salt at approximately 10% of the weight of the fish). They can also be prepared from pickled herring soaked in water for 3–4 h prior to racking.

Herring are threaded on rods through the gills, mouth or eyes and placed in the smoker immediately. In modern smokers, bloaters are ready within 6 h, but in traditional kilns they are smoked for several days with inconsistent fires. In modern smokers the first half of the process is for drying only and therefore smoke is introduced intensely during the second half of the process. Bloaters are most often packaged as bulk. Buckling are prepared as head-off herring with roe or milt remaining. They are brined for 1 h at 70°s, speated through the shoulder region or placed on racks, and hot smoked and post-processed as described.

*(g) Mackerel (*Scomber scombrus*)*

Mackerel are also regarded as a fatty fish and, hence, require special handling to prevent spoilage. Cold smoking of mackerel kippers and fillets is done the same as described for herring kippers and fillets. Hot smoking involves brining mackerel fillets in 70°s for 7–12 min, followed by spreading a spice mixture consisting of coarsely-ground black pepper, whole mustard seed and coarsely-ground red peppers over the surface of the fillets, before proceeding with the smoking.

*(h) Mussels (*Mytilus edulis*)*

Mussels are bivalve mollusc filter-feeders that grow in size to over 5 cm within 3 years. Fresh, live mussels are required at the beginning of the process. The live mussels are either steamed for 2–4 min or placed in boiling water (with 3–4% salt) for 5–7 min, immediately cooled to prevent meat toughening and removed from the shells, leaving the beard (byssus). They are brined in 70°s for 1–2 min, placed on oiled racks as is or on half-shells (also sprayed or dipped in edible oils), cold smoked for 1–2 h and post-processed. They can be vacuum packaged, canned and heat processed or placed in glass jars.

(i) Oysters

Cleaned oysters are steamed for 20–30 min, removed from their shells and brined for 5 min in 50°s. They are quickly immersed in oil, placed on wire mesh and hot smoked at 68°C for 30 min, being turned over halfway through the smoking for uniform penetration. The smoked meats are packed in small jars with top-quality edible oil and sterilized (Burgess and Bannerman, 1963).

(j) Salmon, Arctic char and trout

Blast-freezing salmon suspended by the tails renders top-quality frozen salmon, which is often graded in two-pound intervals (2–4, 4–6 lb, etc.).

Frozen salmon should be thawed slowly in a chill room and handled with great care, although fillets can also be cut from the frozen salmon using a bandsaw and then thawed before trimming. This method is meant to result in much higher product quality by preventing gapping (openings in the muscle) and ragged fillets. Fillets are immersed in a 80°s brine for 1–2 h. Alternate treatments are dry salting for 8–12 h for cold smoking or placing in 80°s brine for 15–20 min for hot smoking. For larger sides of salmon, slits in the lateral line can enhance salt uptake. Dry-salted fillets are rinsed and then racked on wire meshes without any overlapping. Drying may take place at room temperature for 2–3 h or overnight at 4°C for cold smoked products, but for hot smoked products the fish can be immediately smoked after the excess water is drained. Some modern smokers have drying cycles, but excess liquid should be permitted to drip off before smoking. The salmon sides are either hot or cold smoked for a period of time, depending on the type of smoker used. The ideal temperature range is 22–28°C. Products may be packaged as whole sides, pieces, D-slices, or long slices. The same procedures are used for Arctic char and trout.

(k) Smokies

These products were first made in Arbroath, Scotland. Haddock (0.2–0.3 kg) are beheaded, gutted and tied together in pairs by the tails. They are brined for 0.5–0.75 h in 80°s, hung over wooden sticks and hot smoked.

(l) Sprats

Sprats for canning are washed, soaked in 80°s brine for 15 min, threaded through the eyes on thin speats, hot smoked for 1 h, and beheaded before packing into cans.

Continental hot smoked sprats (Kielersprotten) are washed and brined as for canning but are speated through the gill and mouth. They are smoked for 1.25–1.5 h; first for 0.5 h at 18°C and then at 68°C until golden brown and cooked (Burgess and Bannerman, 1963).

*(m) Turbot (*Reinhardtius hippoglossoides*)*

Turbot, also known as Greenland halibut, Greenland turbot, blue halibut and black halibut, may be hot or cold smoked. For cold smoking the turbot is filleted, skinned, brined at 70°s for 10 to 15 min, dried on racks, cold smoked and post-processed. Hot smoking of turbot involves preparing the fish as steaks 5 cm thick, brining at 70°s for 10–15 min, placing on racks and smoking immediately. They are post-processed and usually packed in bulk.

3.2.5 Recent technological innovations

Recently, adaptations of the fish smoking process have included the use of more complex mixtures of salts and other ingredients than just salt and water to impart specific flavors. Such ingredients include peppercorns, wines, and such spice blends as curry, paprika and pepper. Orange peels, lemon peels or juniper berries may be added to the sawdust to modify the flavor of the smoke (Philip, 1981).

Preventing the addition of carcinogens during the smoking process has become a major issue in recent years. Several methods have been investigated. Smoke generation is a two-part process consisting of pyrolysis and oxidation. It has been suggested that by maintaining the temperatures of pyrolysis and oxidation at the optimum 425 and 200°C, respectively, the production of some carcinogens can be kept to a minimum (Hardy and McGill, 1979). Electrostatic filtration is another option, whereby solid and liquid particles are trapped and only the gas phase is used for smoking. This smoke is lower in desirable effective substances and produces a certain adverse affect on the color and flavor of the product, but the content of carcinogens is reduced to 2% of the original level. Alternately, a steam smoke method uses smoke generated by the action of superheated steam, with a temperature of 280–380°C, on a compressed sawdust mixture. The product is then infused with the vapors in a special ventless unit.

A fourth option is to treat the product with liquid smoke (Kyzlink, 1990). Since carcinogens are primarily associated with the crude tar fraction of smoke, the solubilization of smoke ingredients in water allows removal of the crude tar by sedimentation and filtering. Smoke solutions have been reported to carry less than detectable levels of benzo [*a*] pyrene and to have no mutagenic activity when the solutions were tested in the Ames Standard Plate Test (Hollenbeck, 1979). Liquid smoke can be applied by either of four methods: drenching, i.e. showering with the liquid smoke; atomizing, whereby liquid smoke is made into an aerosol using high-pressure air through nozzles; regenerating, i.e. atomized liquid smoke is passed over a heating system, converting the liquid to vapor to be then introduced to the smoker; and, finally, liquid smoke may be added as an ingredient to canned products prior to heat treatment. Liquid smoke performs all the desired functions of smoke flavoring, allows more rigid flavor control and has the added advantages of lowered costs, less environmental damage, and greater availability and variety of application methods. The use of refined liquid smoke flavorings as a substitute for vaporous smoking is expanding in the US and in other countries.

New methods of brining the fish prior to smoking have also been investigated. Automated briners are now available as well as pickle injectors. Pickle injectors use needles to penetrate the fish, while they move continuously on a conveyor belt, and inject the brine deep into the flesh, providing a uniform salt penetration. The automated briners make use of a conveyor belt upon which the fish move through the brine or from the briner to collection lines. Alternately, mechanized conveyor and paddle systems may be used to pass fish through a brine tank at a set speed. Furthermore, an automatic brine strength controller can maintain correct brine strength and a static filter mesh screen ensure cleanliness of the recirculated brine solution (Anonymous, 1987a). One of the latest developments is an automatic conveyorized system whose variable speed drive allows total control of the rate of production.

Packaging materials for these smoked fish products has also become a major area of research. Vacuum packaging has become one of the more important means of marketing smoked fish products, whether whole or sliced (Bradfield, 1984). However, other forms of packaging that enhance the appearance of the product are also available; wooden boxes for individual products or vacuum packaging with foil backings are examples. More recent advances are smoke-permeable synthetic food packaging films in which the product is packaged prior to being smoked, and afterwards requires no further handling or packaging (Hisazumi, Funabashi and Tomioka, 1988). Vacuum packaging, modified atmospheric packaging and controlled atmospheric packaging have increased the shelf-life of such products. Modified atmospheric packaging involves displacing the air with carbon dioxide, nitrogen and oxygen, alone or in combination. Carbon dioxide appears to have inhibitory effects on bacteria. Oxygen is added to prevent discoloration that may be caused by carbon dioxide, and nitrogen is added to prevent pack collapse (Ward and Baj, 1988). Under controlled atmosphere packaging, a constant-composition atmosphere is provided for the packaged product throughout storage (Cann, 1984). For smoked fish the recommended gas mixtures are 60% carbon dioxide and 40% nitrogen dioxide (Cann, 1984).

3.3 SMOKING INDUSTRY

The smoking industry comprises the producers of smoked products on a commercial scale for local and international markets, caterers of restaurants and small private businesses, as well as the individual who smokes fish for his own personal use. Smoking units are available in all sizes and designs (Anonymous, 1983) to accommodate each particular requirement (Figures 3.2 and 3.3.). However, for the most part the general preparation of products for smoking are similar in all cases.

To meet the new demands for small-scale smoking of gourmet foods, a range of excellent mini-kilns was introduced in the early 1980s. Mini-kilns may lack some of the technical refinements of full-scale industrial

Figure 3.2 Example of a modern modular smokehouse. Reproduced with permission from AFOS Ltd, Manor Estate, Anlaby, Hull, UK.

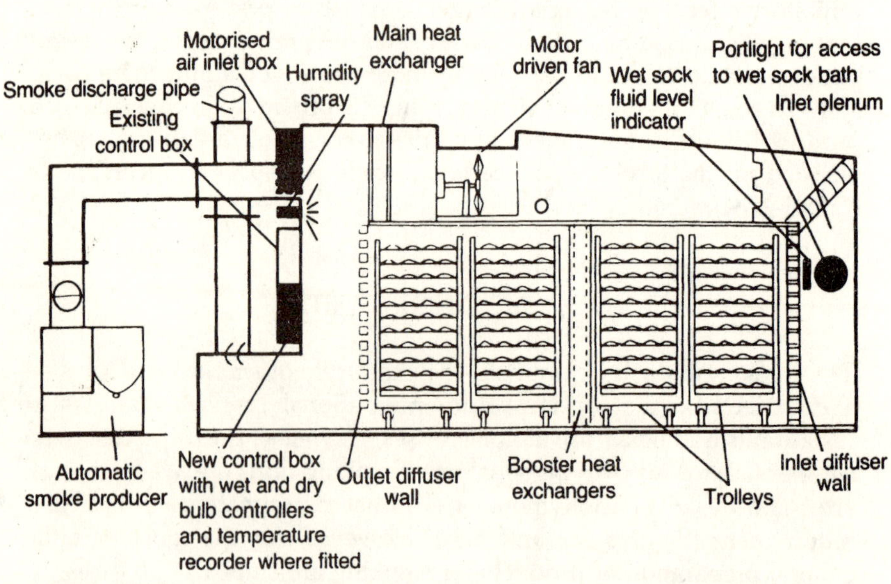

Figure 3.3 Sectional view of a modern modular smokehouse. The darkened area shows where the control and humidity spray can be retrofitted. Reproduced with permission from AFOS Ltd, Manor Estate, Anlaby, Hull, UK.

smoking units, but they are well-proven devices capable of producing excellent smoked fish and are straightforward to operate and control. Philip (1981) has described some of the miniature models. A new smoker, the 'Räuchermiester' (smoke master), developed in Germany, uses a liquid smoke solution and allows retailers to hot-smoke fish in the presence of the customer (Anonymous, 1988) or to cold-smoke fish in 120 min. Small smokers with integrated smoke producers, ranging in size from 12 to 25 kg capacity, are also available, as are large-scale models ranging up to multidoor, high-tonnage factory installations. Commercial smoker manufacturers offer various types of smokers to meet the needs of different processors (Figures 3.2 and 3.3).

3.3.1 International outlook

Smoked fish is one of the oldest of all processed foods and it is currently undergoing considerable growth in popularity. The traditional smoked varieties like haddock, cod and kippers appeal to a new demand for 'old-fashioned' food, particularly as processing is now being done without chemicals or other additives. Smoked eel and trout fillets in vacuum packs have gained popularity in the UK, France and other European countries. Smoking of under-utilized species such as dogfish (Shiau and Chai, 1985) and shark fillets (Davis, 1986) is becoming an area of great interest. Peppered and flavored mackerel, patés, and roulades are now being added to this list (Anonymous, 1987a). Other markets for smoked fish include Italy, Austria, West Germany, Malaysia, Singapore, Australia and the US (Anonymous, 1989).

Once a luxury for the rich, smoked salmon (and trout) are now competitively priced to retail to a much wider market. This change in marketing strategy is a result of a worldwide increase in the availability of fresh-caught and farmed salmon (Anonymous, 1985). As salmon of many kinds from around the world are now available to smokers, custom-designed and custom-packaged salmon has begun to appear in many European markets. The development of pliable hanging packs for the packaging of smoked fish, giving it a shelf-life of 21 days (Anonymous, 1985), means such products can be hung on pegs in self-service cabinets of shops, next to quick-selling delicatessen products. The development of the cryovac packaging system has also facilitated wider disribution of smoked fish products.

3.3.2 Markets for smoked fish products

The world's fish and shellfish catch for human consumption has increased dramatically over the past four decades, increasing the need for preservation capacities. With the complementary technological developments in refrigeration and transportation, and with increasing health consciousness and changing tastes, the dependence on curing as a means of preservation, and the percentage of fish that is cured, has decreased markedly. In 1988, over 80% of fish caught in Canada for human consumption was processed as fresh or frozen (Anonymous, 1990a). The overall pattern that emerges is that newer preservation technologies (e.g. freezing) are displacing the older technologies (e.g. curing). A Department of Fisheries and Oceans (Anonymous, 1990a) report predicts a gradual but steady decline in the curing industry worldwide until at least the year 2000.

Nonetheless, there has been an increase in the number of curing processing operations in Canada from 470 in 1979 to 854 in 1987. Some fish processors in the UK have expanded (Anonymous, 1984, 1985), grown very profitable and expect to grow further in the future as the market for high-quality 'luxury' fish products continues to grow (Anonymous, 1980; Lee, 1983; Tuley, 1985) and as such products become more marketable and available. Some popular smoked fish products in demand there are hot and cold smoked salmon, smoked mackerel, trout, and eel, as well as other speciality products being introduced to the market such as dill-cured Scottish salmon (gravadlax), spiced smoked mackerel, smoked sea trout, crevettes and fish bites (Anonymous, 1987a). Young (1984) examined the demand for fresh, cured and frozen fish in Great Britain using regression analysis.

3.3.3 Economics of smoking products

Smoked salmon and char are primary food-service products, while smoked cod, herring/kippers, haddock and mackerel are primary retail products. A specific case, based on data for the Province of Newfoundland, Canada, is presented as an example in Table 3.3, which shows the

Table 3.3 Percent distribution of smoked fish sold by species for Newfoundland producers, 1986[a]

Species	*Volume sold (%)*	*Value sold (%)*
Caplin	4	3
Char	1	2
Cod	19	26
Cold cod	–[b]	–[b]
Herring/kippers	66	43
Mackerel	5	4
Salmon	5	22

[a]From Anonymous (1987b). Reproduced with permission from the Department of Fisheries, Government of Newfoundland and Labrador, St John's, Newfoundland, Canada.
[b]Less than 0.5%.

percent distribution of smoked fish sold by species and the percentage in their sale value in 1986 (Anonymous, 1987b).

3.3.4 Preservation of smoked products

The smoking process provides fish and other smoked products some protection against spoilage, compared with the fresh state of the raw material. Modern smoked products, however, still need to be stored at temperatures of 0–5°C, where they can keep for from 1 to 4 weeks or frozen and kept at −20°C until needed (Philip, 1981). The main purpose of smoking in this case is to impart flavor and texture. For traditional types of smoked products, however, heavy smoking may circumvent the need for refrigeration. The heat and dry air associated with smoking reduces the water activity of foods, thereby depriving organisms of a prerequisite for growth (Daun, 1979; Asita and Campbell, 1990). Also, the chemical constituents of smoke itself provide bactericidal action, inhibiting some spoilage bacteria (Draudt, 1963; Dawn, 1979). Some components of smoke have been found to prevent spore production and fungal growth, and inhibit viral activities on the surface of the product (Daun, 1979). Smoke penetrates to the interior of food slowly, and so the bactericidal effect is progressively weakened beneath the surface of the flesh (Daun, 1979).

3.4 MICROBIOLOGY OF FISH SMOKING PROCESSES

The food industry utilizes numerous effective measures which inactivate critical bacteria so as to control and limit potential hazards. Such measures involve heat processing (pasteurization, canning, hot smoking), dehydration, freezing/refrigeration, specialized packaging and antimicrobial additives. Extensive quality control programs are implemented to ensure these procedures are effective. However, the new consumer trend towards healthy eating habits places restrictions on the processor in the measures that can be used for preservation. Hence, the control and safety precautions used to maintain a safe food supply are now being altered to suit consumer desires. This has created a need for monitoring of the microbiological aspect of these alterations in food processing.

Bacteria, yeasts (Magnússon and Traustadóttir, 1982) and molds (Nieto and Magno-Orejana, 1984) are microorganisms associated with smoked fish. Many species of bacteria cause food poisoning either by infection or intoxication. An infection results from the ingestion of food contaminated with bacteria which then invade body tissues, resulting in disease. Intoxication is caused by the consumption of a food containing toxins produced by bacteria; it is the toxins which cause the illness.

Organisms of concern that cause infections are *Salmonella* spp., and organisms of concern causing intoxication are *Staphylococcus aureus* and *Clostridium botulinum*.

3.4.1 Microbiological problems

Though it can be very costly, it is critically necessary to maintain high quality standards, monitor microbial levels, and avoid further contamination in processing facilities. Microorganisms tend to increase on a typical product at each stage of the smoking process. Alexander and Austin (1986), studying haddock processing, found an increase in bacterial counts after filleting, a decrease after smoking and an increase again with storage. Some typical bacteria present on the haddock muscle were *Acinetobacter* spp., *Alcaligenes* spp., *Pseudomonas* spp. and *Vibrio* spp., as well as coryneforms. In another study, high aerobic plate counts occurred after brining baltic herring; however, the counts were lower if the brine was chilled (Korkeala and Pakkala, 1988). Food service establishments in the US accounted for 77% of the known sources of foodborne disease outbreaks in 1986, while homes and food processing plants accounted for 20 and 3%, respectively (Duxbury, 1986). Also of microbiological concern are two parasites, *Anisakis* spp. and *Diphyllobothrium* spp., which have been detected in cold smoked salmon that have not been previously frozen, and which can result in intestinal discomfort and other more severe symptoms. Many cases of these parasites have been reported in Japan (Oshima, 1972; Rodrick and Cheng, 1989) and there have been two known outbreaks in the US (Anonymous, 1981). Freezing fish to −23°C or below for 60 h is sufficient to destroy the parasites (Ching, 1984). Cold smoking of Pacific herring and salmon does not destroy these parasites (Hauck, 1977; Gardiner, 1990). The most common and obvious causes of smoked fish spoilage are mold growth, with *Penicillium* spp. and *Aspergillus* spp. encountered most often. These molds grow well at refrigeration temperatures. The most common source of mold contamination in smoked fish is the sawdust (Chichester and Graham, 1973).

Moisture contents greater than 50% in smoked fish are favorable to mold growth. The growth is often very rapid and fish can easily be covered with fungal mycelium once the process is initiated. Some kinds of molds produce aflatoxin (fungal toxin or mycotoxin). The practice of scraping the mold off smoked fish is not acceptable since it leaves the toxin in the product and increases the consumers' risk of food poisoning.

(a) Listeria monocytogenes

Listeria monocytogenes is a facultatively anaerobic, Gram-positive rod (Seeliger and Jones, 1986), which was primarily a veterinary concern, being associated with abortions and encephalitis in sheep and cattle, until the 1980s (Lovett and Twedt, 1988). It was recognized as a food-borne pathogen in 1981 (Schlech *et al.*, 1983) and has a zero-tolerance classification in ready-to-eat foods. Its target populations are immuno-suppressed individuals, elderly, infants and pregnant mothers and their fetuses (Bille and Doyle, 1991). The major symptoms of listeriosis are meningitis, septicemia, premature birth, stillbirth and abortions of infected infants in severe cases (Bille and Doyle, 1991) with an average fatality rate of 30%. Milder cases have flu-like symptoms. It can grow in temperatures of −0.4 to 50°C (Anonymous, 1990b), has a normal pH range of 5–9.5 but has been found at much lower pH levels (Farber *et al.*, 1989; Parish and Higgins, 1989) and can survive in salt concentrations up to 30.5% at 4°C (Lovett and Twedt, 1988). The organism is ubiquitous in nature and has been isolated from vegetation, soil, sewage, water, animal feed, meat, seafoods, dairy products and poultry, as well as from the processing plants of these substances (Lovett and Twedt, 1988).

The *Listeria* bacterium has now become the greatest problem facing fish smokers exporting smoked fish to the US (Hadeler, 1990). Smoking processors in Seattle, Vancouver and Montreal have already been shut down due to product contamination with *L. monocytogenes*. Several documented outbreaks have occurred (Schlech *et al.*, 1983; Ho *et al.*, 1986; Linnan *et al.*, 1988) including one in New Zealand in which shellfish and fish were epidemiologically implicated as the vehicle of contamination (Lennon *et al.*, 1984). *Listeria* has been detected in various types of raw and processed seafoods, including smoked seafoods (Jemmi, 1990a, b; Farber, 1991; Hartminke and Georgsson, 1991; Dillon and Patel, 1992; Dillon, Patel and Ratnam, 1992). Guyer and Jemmi (1991) reported that cold smoking had no significant effect on the organism's viability in experimentally inoculated salmon fillets; the organism increased in cell number upon storage at 4 or 10°C. The same trend was observed in a similar study of the smoking process with deliberately inoculated cod fillets; organisms grew well at 4°C and remained viable at −20°C (Dillon and Patel, 1993). Weagant *et al.* (1988) have detected *Listeria* in frozen seafood products, confirming the organism's ability to survive freezing temperatures. Farber (1991) as well as Miller, Smith and Somkuti (1990) reported that *L. monocytogenes* grows well in inoculated seafoods such as smoked salmon. As well, the latest packaging methods, controlled atmospheric packaging, modified atmospheric packaging and vacuum packaging, among others, have no effect on the organism's ability to proliferate (Wimpfheimer, Altman and Hotchkiss, 1990; Kallander *et al.*, 1991; Grau and Vanderlinde, 1992). Liquid smoke was reported to have inhibitory effects on *L. monocytogenes* in frankfurters, causing a more than 99.9% reduction of *L. monocytogenes* after 72 h at 4°C (Messina *et al.*, 1988). The method of detection (Warburton and Farber, 1990) takes approximately 10 days for confirmation. This is too long for products to be held in waiting, so more rapid methods of detection are currently being investigated.

It is essential that processors observe good manufacturing practices (GMPs) in order to help reduce any contamination. The majority of smoked products are cooked prior to consumption, and this reduces the risk of contracting listeriosis. However, some smoked products are ready-to-eat and, therefore, may pose a consumer health risk. Production flow must be designed to separate *Listeria*-free environments from potentially contaminated environments. Personnel must also have limited movement within the plant and there should be no cross-contaminations between raw and finished product. Packaging and storage also requires strict monitoring since *Listeria* can grow under very harsh environmental conditions.

(b) Clostridium botulinum

Clostridium botulinum is the organism of greatest concern in smoked foods (Eklund *et al.*, 1982) because ingestion of food containing the toxin produced by this organism results in up to 60% fatalities. These organisms grow and produce toxins in canned and vacuum-packed foods because they are strict anaerobes (Cato, George and Finegold, 1986). Botulism, the disease they cause, leads to neuroparalysis in man and animals. Other symptoms include vomiting, constipation, ocular paresis, double vision, pharyngeal paralysis, abdominal distention, extremely sore throat, secretion of a thick fluid and often dysphonia or difficulty speaking (Nickerson and Sinskey, 1972). It is theorized that the toxin prevents the formation in its victims of a compound called sympathin, which is required for the synthesis of acetyl choline, which is necessary for nerve function. Therefore, breathing, blood circulation and other functions cease. Approximately 0.2 μg of toxin will kill a 200 lb man (Nickerson and Sinskey, 1972). Symptoms develop 12–72 h after ingestion of the toxin-containing food, duration of illness may range from 1 to 10 days or more and recovery may take weeks to months (Pierson and Reddy, 1988).

Botulism has been recognized as a foodborne disease for more than 100 years, with marine and freshwater fish being the primary disease-causing foods (Huss, 1981). The primary habitat of *C. botulinum* is the soil. The organism has eight toxigenic types (A, B, C_1, C_2, D, E, F and G) based on serological differentiation of the neurotoxin (Hackney and Dicharry, 1988; Hauschild, 1990; Allen and Baron, 1991). These eight types have been divided into four groups based on proteolytic activity, of which Groups 1 and 2 are most significant with respect to human botulism. Group 1 consists of type A and proteolytic strains of types B and F, while Group 2 contains type E and nonproteolytic strains of types B and F (Hackney and Dicharry, 1988). However, type E is the one most commonly found in freshwater and marine environments and is most associated with seafoods (Eklund and Poysky, 1967; Eklund *et al.*, 1982), although other types have also been noted in these environments (Eklund and Poysky, 1967). Type E is supposedly more heat-resistant than types A, B and F; it is nonproteolytic and so its presence in smoked foods often goes undetected. For nonproteolytic *C. botulinum* types B, E and F, the minimum temperature for growth and toxin production is 3.3°C, but for the proteolytic types A and B, it is 10°C. Therefore, if smoked products were stored at a temperature of less than 3.3°C, toxin production would be impossible. Most refrigeration cabinets at the retail level have a temperature range of 2–10°C allowing the possibility of toxin production if the organism was present in the smoked product. Type E spores are inhibited at pH levels less than 4.6, water-phase salt concentrations of 5–6%, and water activities of less than 0.96.

The contribution of fishery products to botulism intoxication has been documented in Russian literature since 1818 and the history of the disease resulting from fish was well recorded up to 1963 (Dolman, 1964). Between 1899 and 1977, there have been a total of 41 documented botulism outbreaks caused by fishery products (Anonymous, 1979) with the largest outbreak being from smoked chubs processed in the Great Lakes region (Anonymous, 1963). Outbreaks in 1932 and 1934 in New York were caused by canned smoked salmon (Dolman and Kerr, 1947). In 1970 in Germany, smoked trout was responsible for a type E botulism outbreak (Baumgart, 1970). Control of botulism is possible through heat sterilization, refrigeration and freezing. An additional safeguard would be to impose a maximum shelf-life of about one week until methods can be determined to eliminate this problem from the food. The toxin is also quite readily destroyed by most cooking procedures and, therefore, a reminder to cook such products should perhaps be attached to the product. Spores can be eliminated by heating to 80°C for 30 min, 90°C for 10 min or 100°C for 5 min. Faulty handling and processing is almost always the cause of contamination.

Addition of greater than 9% sodium chloride (salt) to smoked fish, or a water-phase salt content of at least 3% will inhibit the growth and toxin production of *C. botulinum* (Kosak and Toledo, 1981). However, this salt level is unacceptable to the consumer and furthermore, brining procedures do not often result in uniform salt penetration. Substitution of potassium chloride for some of the sodium chloride was found feasible for the inhibition of outgrowth and toxin production by *C. botulinum* type E (Pelroy *et al.*, 1982). Sodium nitrite in conjunction with NaCl inhibits toxin production as well (Chichester and Graham, 1973; Pelroy *et al.*, 1982; Cuppett *et al.*, 1987). Sodium nitrite is not permitted in fishery products because of its association with *N*-nitrosamine formation in the finished product, although conflicting data exist (Malins,

Roubal and Robisch, 1970; Sen *et al.*, 1970; Fazio *et al.*, 1971; Gadbois *et al.*, 1975). However, in the US, sodium nitrate plus sodium nitrite at 500 and 200 p.p.m., respectively, are permitted as a general preservative in smoked salmon, sable and shad. As well, sodium nitrite in the range of 100–200 p.p.m. in combination with 3.5% sodium chloride with a process temperature of 57°C for 30 min is permitted in smoked chub for inhibition of *C. botulinum* type E (Chichester and Graham, 1973). It has been reported that the growth of *C. botulinum* and the production of spores can be inhibited during storage of smoked fish by the addition of an appropriate amount of hypophosphorous acid and/or salt (Thompson and Jadlocki, 1981, 1982). Salt water species contain high levels of trimethylamine, which upon degradation to dimethylamine, may react with the nitrite salt to form *N*-nitrosodimethylamine, a carcinogen, but it has been reported in one study that it is unlikely that volatile nitrosamines will form in the stomach, after ingestion of smoked foods (Sen *et al.*, 1985). Eklund *et al.* (1982) found liquid smoke, used with less than 3% salt in the brining step, effective in preventing *C. botulinum* outgrowth and toxin production in hot smoked whitefish, chub and carp.

(c) Salmonella *spp.*

Salmonella grow at temperatures between 5 and 47°C, with a pH range of 4–11. It is not normally found in seafoods, but can contaminate them when the processing water is contaminated with animal or human waste. Salmonellosis is the result of bacteria invading body tissues, growing and causing typical symptoms in the host. Antibiotics are available, but some deaths still do occur. Typical symptoms include abdominal pain, diarrhea, chills, frequent vomiting and prostration. More severe symptoms include septicemia with leucopenia, endocarditis or pericarditis, meningitis, osteomyelitis, etc. The incubation period is 7–72 h. Carriers of salmonellosis exist and excrete the organism in their feces, from which it can be transmitted to foods. The carrier stage for ordinary salmonellosis usually does not persist for longer than 12 weeks (Nickerson and Sinskey, 1972). However, the carrier stage may continue for months or years (Le Minor, 1984). Treatment is by means of chloramphenicol. *Salmonella* is a facultative anaerobe with more than 1200 serovars, all of which are pathogenic to man. Shellfish, especially clams and oysters, have had a history of involvement in transmission of salmonellosis (Old and Gill, 1940). Fresh crabmeat has also been implicated, as have smoked fish products. The only means of preventing such outbreaks is to follow GMPs. To minimize risk to the consumer, frying of the smoked fish prior to consumption is highly recommended in order to kill pathogenic organisms. This is particularly important in the case of commercial smoked fish, and also when the product has been stored for some time before consumption.

(d) Staphylococcus aureus

Staphylococcal food intoxications are regarded as the second most prevalent foodborne illness in the US (Newsome, 1988). The enterotoxin may cause illness within 0.5–8 h of ingestion of the toxin-containing food, resulting in such symptoms as nausea, abdominal cramps, vomiting, chills, sweating and diarrhea, and requires 2–3 days for recovery (Newsome, 1988). Six toxins have been identified as A, B, C, C_2, D and E, with enterotoxin A being the most toxic and the one most often involved in staphylococcal food poisoning outbreaks. The primary habitat of *S. aureus* is the skin, skin glands and mucous glands (Kloos and Lambe, 1991). Coughing and sneezing by individuals with respiratory infections, as well as touching one's hair, can result in staphylococcal contamination when foods are handled. Approximately 50–70% of the population are carriers of the organism. The most common location of the organism on man is the nasal–pharynx region, and by touching this area it can be spread to the hands. *S. aureus* is a facultative anaerobe and therefore can grow in the presence and absence of oxygen (Kloos and Schleifer, 1986). This species can grow in a temperature range of 6.7–46°C and once the toxin is formed in food it is heat stable. The organism produces no pronounced odour or taste. *S. aureus* can grow in up to 20% salt, has a pH range of 4.2–9.3, and can grow in foods which have a water activity as low as 0.86 under aerobic conditions and as low as 0.90 under anaerobic conditions.

Outbreaks have occurred in cases where the toxin was present in the food, having resisted processing conditions, but the bacteria were killed. Smoked and charred herrings in Finland have been reported to be a common vehicle of staphylococcal food poisoning outbreaks (Hirn, 1982). The major pathogen found on smoked dry herring in fish sampled in Ghana was *S. aureus* (Lu, Pace and Plahar, 1991). Proteinaceous foods such as meat and fish best support growth of this species (Newsome, 1988). Foods that have caused staphylococcal poisoning were found to have a staphylococcal count of 10^7–10^8 cells per gram of food. Usually, mishandling of foods causes outbreaks, indicating that at some point the product was introduced to temperatures greater than 6.6°C. Hence, proper refrigeration at all times would alleviate this problem. Personnel with hand infections should also be prevented from coming into direct contact with foods (Nickerson and Sinskey, 1972).

(e) Escherichia coli

This organism is a facultative anaerobe and can be found in the lower part of the intestinal tracts of warm-blooded animals (Ørskov, 1984). This organism is a significant cause of diarrhea in developing countries and other areas having poor sanitation practices. There are four known subgroups of pathogenic *E. coli*: enterotoxigenic, enteroinvasive, hemorrhagic and enteropathogenic, all of which cause various symptoms with extreme diarrhea being the most common (Frank, 1988). Feces and untreated water are the most likely modes for the contamination of foods. Humans are believed to be the only carriers of toxigenic and invasive strains of *E. coli* isolated from foods. Foods of animal origin may be contaminated with the hemorrhagic strain (0157:H7) by slaughter methods or post-process contamination, since such animals as poultry, sheep and swine are the reservoirs of this particular strain of *E. coli* (Frank, 1988). Monitoring critical control points for the presence of coliforms is essential in food processing plants to avoid any health hazards. These organisms are also found in smoked fish (Nieto and Magno-Orejana, 1984).

(f) Campylobacter *spp.*

These are aerobic/microaerophilic bacteria found in the reproductive organs, intestinal tracts and oral cavities of humans and animals; they are widely found in nature (Penner, 1991). *Campylobacter jejuni*, which causes fever, vomiting, diarrhea, abdominal pain and enteritis in man (Penner, 1991), has emerged as an important human intestinal pathogen, and has become as common as *Salmonella* and *Shigella* in more recent years. The organism infects all ages but more frequently infects children, with seasonal incidences higher in the summer and fall. Transmission of infection is mostly oral. Food, milk and water present a problem since the infectionous dose (the number of organisms that result in infection) is unknown (Smibert, 1984).

(g) Yersinia *spp.*

A facultative anaerobe found occurring in a wide spectrum of habitats, live and inanimate. The organism grows at temperatures of 4–42°C and has a pH range of 4.0–10. Most species are found worldwide, with wild animals acting as the reservoir in many cases. Humans and animals are contaminated orally either by direct contact with sick or asymptomatic animals or through food contaminated by the excrement of animals. *Y. intermedia* and *Y. frederiksenii* have been isolated mainly from fresh water, fish and foods. *Y. kristensenii* have been isolated mainly from soil, foods and asymptomatic animals. *Y. ruckeri* has only been encountered in Canada and the US, is a natural component of freshwater eco-systems and is a fish pathogen. *Y. entercolitica* is also transmitted to adults orally (Bercovier and Mollaret, 1984).

3.5 CONCLUSIONS

3.5.1 Future prospects and directions for smoking operations in the fisheries industry

Some future prospects for smoking operations include the use of complete monitoring controls to preclude hazards, automatic briners, and automated kilns equipped for drying cycles and with smoke generators. This type of equipment will reduce the labor and handling as well as provide a more consistent and uniform product. Modern equipment will cut down on processing time, and it is possible that freshly smoked fish will be available to the consumer upon ordering. Furthermore, the removal of carcinogens from smoke will require much more investigation to provide a more healthy product to the consumer. Trends will have to match consumer demands, but the market appears to be thriving. Smoking of under-utilized species promises to have great potential.

3.5.2 Microbiological concerns

Every food commodity has microorganisms associated with it that could potentially cause safety problems. Hence, it is important that scientists, processors, distributers and consumers of foods be aware of the problems that could arise with improper processing and handling conditions. Cold smoked seafoods have a longer shelf-life than hot smoked, but the microflora of the more modern smoked products are still often in high numbers. Hot smoked seafoods provide a greater risk of causing disease because the normal competitive microflora are killed and the opportunity for pathogens to invade and grow is much better. Post-processing contamination is a major factor and can be controlled by practising GMPs and by implementing a quality management program. Continuous water testing should be a major priority as well as removing sick personnel from handling foods. In general, smoked seafoods have the potential to cause disease and infection and, therefore, appropriate control measures are required. However, smoked seafoods can also provide a very nutritious addition to our diet and so promotion of the product as a healthy food would increase markets.

ACKNOWLEDGMENTS

The authors would like to thank Mr P. Bemister, Department of Biochemistry, Memorial University of Newfoundland, for his assistance with the manuscript.

REFERENCES

Alexander, B. and Austin, B. (1986) Bacterial microflora associated with a commercial fish smoker. *FEMS Microbiol. Lett.* **34**, 309–12.

Allen, S.D. and Baron, E.J. (1991) *Clostridium*, in *Manual of Clinical Microbiology*, 5th edn., (eds A. Balows, W.J. Hausler, Jr, K.L. Herrman *et al.*), American Society for Microbiology, Washington, DC, pp. 505–21.

Aminullah Bhuiyan, A.K.M., Ackman, R.G. and Lall, S.P. (1986) Effects of smoking on protein quality of atlantic mackerel (*Scomber scombrus*). *J. Food Process. Preserv.*, **10**, 115–26.

Anonymous (1963) Outbreak of food poisoning associated with commercially smoked fish. *Morbid. Mortal. Weekly Rep.*, **12**, 337–41.

Anonymous (1965) The smoking of fish, in *Fish Handling and Processing*, (eds G.H.O. Burgess, C.L. Cutting, J.A. Lovern and J.J. Waterman), Ministry of Technology, Torry Research Station, Aberdeen, UK, pp. 70–101.

Anonymous (1979) *Annual Summary of Foodborne Diseases 1978*, Centres for Disease Control, US Department of Health and Human Services, Atlanta, GA.

Anonymous (1980) Can smoking preserve the mackerel market? *Food Flavourings, Ingredients, Packaging and Processing*, **1**(11), 38–9.

Anonymous (1981) Diphyllobothriasis associated with salmon – United States. *Morbid. Mortal. Weekly Rep.*, **30**(27), 331–2, 337–8.

Anonymous (1983) Smoking is good for you! *Food Manufact.*, **58**(10), 53, 55.

Anonymous (1984) Smoked fish. *Froz. Chill. Foods*, **37**(8), 14, 16, 18, 20, 21.

Anonymous (1985) Smoked fish. *Froz. Chill. Foods*, **38**(8), 29–30, 33, 35–36, 39–40, 42.

Anonymous (1987a) Smoked fish. *Froz. Chill. Foods*, **41**(8), 12, 14–15, 18, 20, 23.

Anonymous (1987b) *Smoked fish product market study. Industry Support Services Report 35*, Department of Fisheries, Government of Newfoundland and Labrador, St John's, Newfoundland, Canada.

Anonymous (1988) Salmon 'finesmoked'. *Fish Int.*, **4**, 40–1.

Anonymous (1989) *Fish Product Export Market Opportunities Guide 1988–1989*, 2nd edn, Government of Canada External Affairs, Fisheries Division, Ottawa, Canada.

Anonymous (1990a) Market trends in cured fish: implications for Canada's cured fish industry. Department of Fisheries and Oceans Report. *Econ. Commer. Anal. Rep.*, **51**, 1–15.

Anonymous (1990b) *Listeria monocytogenes – An Information Handbook*, Health and Welfare Canada / Agriculture Canada, Ottawa, Canada.

Asita, A.O. and Campbell, I.A. (1990) Anti-microbial activity of smoke from different woods. *Lett. Appl. Microbiol.*, **10**, 93–5.

Bannerman, A. McK. (1980) Hot smoking of fish. *Torry Advisory Note 82*, Torry Research Station, Ministry of Agriculture, Fisheries and Food, UK, pp. 3–12.

Baumgart, J. (1970) Nachweiss von *Clostridium botulinum* Type E bei handelsfertigen forellen. *Fleischwirtsch*, **50**, 1545–7.

Bercovier, H. and Mollaret, H.H. (1984) Genus XIV. *Yersinia* Van Loghem 1944, 15AL, in *Bergey's Manual of Systematic Bacteriology*, (eds N.R. Krieg and J.G. Holt), Williams and Wilkins, Baltimore, MD, Vol. 1, p. 498–506.

Bille, J. and Doyle, M.P. (1991) *Listeria* and *Erysipelothrix*, in *Manual of Clinical Microbiology*, 5th edn, (eds A. Balows, W.J. Hausler, Jr, K.L. Herrmann *et al.*), American Society for Microbiology, Washington, DC, pp. 287–95.

Bradfield, D. (1984) Clearly a better pack for the Holden du Cros luxury product. *Froz. Chill. Foods*, **37**(1-2), 52–3.

Burgess, G.H.O. and Bannerman, A. McK. (1963) *Fish Smoking, A Torry Kiln Operators' Handbook*, Torry Research Station, Ministry of Agriculture, Fisheries and Food, UK.

Cann, D.C. (1984). Packing fish in a modified atmosphere. *Torry Advisory Note 88*, Torry Research Station, Ministry of Agriculture, Fisheries and Food, UK.

Cato, E.P., George, W.L. and Finegold, S.M. (1986) Genus *Clostridium* Prazmowski 1880, 23AL, in *Bergey's Manual of Systematic Bacteriology*, (eds P.H.A. Sneath, N.S. Mair, M.E. Sharpe and J.G. Holt), Williams and Wilkins, Baltimore, MD, Vol. 2, pp. 1141–200.

Chichester, C.O. and Graham, H.D. (1973) *Microbial Safety of Fishing Products*, Academic Press, New York, pp. 105–07, 190–202.

Ching, H.L. (1984) Fish tapeworm infections (Diphyllobothriasis) in Canada, particularly British Columbia. *Can. Med. Assoc. J.*, **130**, 1125–8.

Churchill, G.L., Mitchell, J.-A. and Thompson, M. (1989) *Smoked Fish Production Technology for the Food Processing Industry*, Inshore Fisheries Development Agreement, St John's, Newfoundland, Canada.

Clifford, M.N., Tang, S.L. and Eyo, A.A. (1980) Smoking of foods. *Process Biochem.*, **15**(5), 8, 10, 11, 17, 26.

Cuppett, S.L., Gray, J.I., Pestka, J.J. *et al.* (1987) Effect of salt level and nitrite on toxin production by *Clostridium botulinum* Type E spores in smoked Great Lakes whitefish. *J. Food Protect.*, **50**(3), 212–17.

Cutting, C.L. (1962) Historical aspects of fish, in *Fish as Food*, (ed. G. Borgstrom), Academic Press, New York, Vol. 2, pp. 1–28.

Daun, H. (1979) Interaction of wood smoke components and foods. *Food Technol.*, **33**(5), 66, 68, 70–1, 83.

Davis, P. (1986) Smoked shark and market possibility. *Australian Fisheries*, **45**, 41.

Dillon, R. and Patel, T. (1992) *Listeria* in seafoods: a review. *J. Food Protect.*, **55**(12), 1009–15.

Dillon, R., Patel, T. and Ratnam, S. (1992) Prevalence of *Listeria* in smoked fish. *J. Food Protect.*, **55**(11), 866–70.

Dillon, R. and Patel, T. (1993) Effect of cold smoking and storage temperature on *Listeria monocytogenes* in inoculated cod (*Gadus morhua* fillets. *Food Res. Int.*, **26**, 97–101.

Dolman, C.E. (1964) Botulism is a world health problem, in *Botulism 1964*, (eds K.H. Lewis and K. Cassel, Jr), US Department of Health, Education and Welfare, Public Health Service, pp. 5–32.

Dolman, C.E. and Kerr, D.E. (1947) Botulism in Canada, with a report of type E outbreak at Nanaimo (B.C.). *Can. J. Publ. Health*, **38**, 48.

Draudt, H.N. (1963) The meat smoking process: a review. *Food Technol.*, **17**, 1557–62.

Duxbury, D.D. (1986) Quality has new meaning for processed, refrigerated foods. *Food Process.*, **47**(9), 62–8.

Eklund, M.W. and Poysky, F. (1967) Incidence of *Clostridium botulinum* type E from the Pacific Coast of the United States, in *Botulism 1966* (eds M. Ingraham and T. A. Roberts), Chapman & Hall, London, UK, pp. 49–55.

Eklund, M.W., Pelroy, G.A., Paranjpye, R. *et al.* (1982) Inhibition of *Clostridium botulinum* Types A and E toxin production by liquid smoke and NaCl in hot-process smoke-flavored fish. *J. Food Protect.*, **45**(10) 935–41.

Farber, J.M. (1991) *Listeria monocytogenes* in fish products. *J. Food Protect.*, **54**, 922–24.

Farber, J.M., Sanders, G.W., Dunfield, S. and Prescott, R. (1989) The effect of various acidulents on the growth of *Listeria monocytogenes*. *Lett. Appl. Microbiol.*, **9**, 181–3.

Fazio, T., Damico, J.N., Howard, J.W. *et al.* (1971) Gas chromatographic determination and mass spectrophotometric confirmation of N-nitrosodimethylamine in smoke-processed marine fish. *J. Ass. Off. Anal. Chem.*, **19**, 250–3.

Frank, J.F. (1988) Enteropathogenic *Escherichia coli*. *Food Technol.*, **42**(4), 192–3.

Gadbois, D.F., Ravesi, E.M., Lundstrom, R.C. and Maney, R.S. (1975) N-nitrosodimethylamine in cold-smoked sable fish. *J. Agric. Food Chem.*, **23**, 665–71.

Gardiner, M.A. (1990). Survival of *Anisakis* in cold smoked salmon. *Can. Inst. Food Sci. Technol. J.*, **23**(2-3), 143–4.

Grau, F.H. and Vanderlinde, P.B. (1992) Occurrence, numbers, and growth of *Listeria monocytogenes* on some vacuum-packaged processed meats. *J. Food Protect.*, **55**, 4–7.

Guyer, S. and Jemmi, T. (1991) Behaviour of *Listeria monocytogenes* during fabrication and storage of experimentally contaminated smoked salmon. *Appl. Environ. Microbiol.*, **57**, 1523–7.

Hackney, C.R. and Dicharry, A. (1988) Seafood-borne bacterial pathogens of marine origin. *Food Technol.*, **42**(3), 104–9.

Hadeler, H. (1990) There's a bug in the smokehouse. *Can. Aquacult.*, **July/August**, B1–3, B6, B16.

Hardy, R. and McGill, A.S. (1979) Smoking of foods: methods and some toxicological aspects. *Process. Biochem.*, **14**, 2–5.

Hartemink, R. and Georgsson, F. (1991) Incidence of *Listeria* species in seafood and seafood salads. *Int. J. Food Microbiol.*, **12**, 189–96.

Hauck, A.K. (1977) Occurrence and survival of the larval nematode *Anisakis* sp. in the flesh of fresh, frozen, brined, and smoked Pacific herring, *Clupea harengus pallasi*. *J. Parasitol.*, **63**, 515–19.

Hauschild, A.H.W. (1990) *Clostridium botulinum* toxins. *Inter. J. Food Microbiol.*, **10**, 113–24.

Hirn, J. (1982) Ruokamyrkytysten ja elintarvikeinfektioiden esiintyminen Suomessa. (Foodborne infection and intoxications in Finland). *Duodecim*, **98**, 1705–11.

Hisazumi, J., Funabashi, S. and Tomioka, W. (1988) Smokable synthetic food-packaging film. *US Patent 4 764 406*.

Ho, J.L., Shands, K.N., Friedland, G., *et al.* (1986) An outbreak of type 4b *Listeria monocytogenes* infection involving patients from eight Boston hospitals. *Arch. Intern. Med.*, **146**, 520–4.

Hollenbeck, C.M. (1979) Liquid smoke flavoring – status of development. *Food Technol.*, **33** (5), 88, 90.

Hsu, L.A., Fisher, R.W. and Daun, H. (1979) Meat preservation by smoke curing in developing countries. *Food Technol.*, **33** (5), 86, 88.

Huss, H.H. (1981) *Clostridium botulinum Type E and botulism*, Technological Laboratory, Ministry of Fisheries, Technical University of Denmark, Copenhagen, Denmark.

Jemmi, T. (1990a) Actual knowledge of *Listeria* in meat and fish products. *Mitt. Geb. Lebensmittelunters. Hyg.*, **31**, 144–57.

Jemmi, T. (1990b) Occurrence of *Listeria monocytogenes* in imported smoked and fermented fish. *Arch. Lebensmittelhygiene*, **41**, 107–9.

Kallander, K.D., Hitchins, A.D., Lancette, G.A. *et al.* (1991) Fate of *Listeria monocytogenes* in shredded cabbage stored at 5 and 25°C under a modified atmosphere. *J. Food Protect.*, **54**, 302–4.

Kernreich, M.R. and Issenberg, P. (1972) Determination of phenolic wood smoke components as trimethylsilyl ethers. *J. Agric. Food Chem.*, **20**, 1109.

Kloos, W.E. and Schleifer, K.H. (1986) Genus IV. *Staphylococcus* Rosenbach 1884, 18AL, (Nom. Cons. Opin. 17 Jud. Comm. 1958, 153), in *Bergey's Manual of*

Systematic Bacteriology, (eds P.H.A. Sneath, N.S. Mair, M.E. Sharpe and J.G. Holt), Williams and Wilkins, Baltimore, MD, Vol. 2, pp. 1013–35.

Kloos, W.E. and Lambe, D.W. Jr (1991). *Staphylococcus*, in *Manual of Clinical Microbiology*, 5th edn, (eds A. Balows, W.J. Hausler, Jr, K.L. Herrmann *et al.*), American Society for Microbiology, Washington, DC, pp. 222–37.

Korkeala, H.J. and Pakkala, P.K. (1988) Microbiological changes in smoked and charred baltic herrings during storage. *J. Food Protect.*, **51**(3), 197–200.

Kosak, P.H. and Toledo, R.T. (1981) Brining procedures to produce uniform salt content in fish. *J. Food Sci.*, **46**, 874–6.

Kyzlink, V. (1990) *Principles of Food Preservation*, Elsevier, Amsterdam, pp. 491–9.

Lee, J. (1983) Smoked foods – past, present, and future. *Food Process.*, **52**, 39–41.

Le Minor, L. (1984) Genus III. *Salmonella* Lignières 1900, 389AL, in *Bergey's Manual of Systematic Bacteriology*, (eds N.R. Krieg and J.G. Holt), Williams and Wilkins, Baltimore, MD, Vol. 1, pp. 427–58.

Lennon, D., Lewis, B., Mantell, C. *et al.* (1984) Epidemic perinatal listeriosis. *Pediatr. Infect. Dis.*, **3**, 30–4.

Linnan, M.J., Mascola, L., Lou, X.D. *et al.* (1988). Epidemic listeriosis associated with Mexican-style cheese. *N. Engl. J. Med.*, **319**, 823–8.

Lovett, J. and Twedt, R.M. (1988) *Listeria*. *Food Technol.*, **42**(4), 188–91.

Lu, J.Y., Pace, R.D. and Plahar, W.D. (1991) Storage conditions and microbial quality of smoked dry herring in Ghana. *J. Food Protect.*, **54**, 557–9.

Maga, J.A. (1987) The flavour chemistry of wood smoke. *Food Rev. Int.*, **3**(1/2), 139–83.

Maga, J.A. (1988) *Smoke in Food Processing*, CRC Press, Boca Raton, FL, pp. 61–8, 77.

Maga, J.A. and Chen, Z. (1985) Pyrazine composition of wood smoke as influenced by wood source and smoke generation variables. *Flavour and Fragrance J.*, **1**, 37–42.

Magnússon, H. and Traustadóttir, K. (1982) The microbial flora of vacuum packed smoked herring fillets. *J. Food Technol.*, **17**, 695–702.

Malins, D., Roubal, W.T. and Robisch, P.A. (1970) The possible nitrosation of amines in smoked chub. *J. Agric. Food Chem.*, **18**, 740–1.

Messina, M.C., Ahmad, H.A., Marchello, J.A. *et al.* (1988) The effect of liquid smoke on *Listeria monocytogenes*. *J. Food Protect.*, **51**, 629–31.

Miller, A.L., Smith, J.L. and Somkuti, G.A. (1990) *Foodborne Listeriosis*, Elsevier, Amsterdam, The Netherlands.

Newsome, R.L. (1988) *Staphylococcus aureus*. *Food Technol.*, **42** (4), 194–5.

Nickerson, J.T. and Sinskey, A.J. (1972) *Microbiology of Foods and Food Processing*, American Elsevier, New York, pp. 197–268.

Nieto, M.B. and Magno-Orejana, F. (1984) Smoke-curing of fish. *SAFIS Manual*, **12**, 1–22.

Old, H.N. and Gill, S.L. (1940) Typhoid fever epidemic caused by carrier bootlegging oysters. *Am. J. Publ. Health*, **30**, 623.

Ørskov, F. (1984) Genus 1. *Escherichia* Castellani and Chalmers 1919, 941AL, in *Bergey's Manual of Systematic Bacteriology*, (eds N.R. Krieg and J.G. Holt), Williams and Wilkins, Baltimore, MD, Vol. 1, pp. 420–3.

Oshima, T. (1972) Anisakis and anisakiasis in Japan and adjacent area. *Prog. Med. Parasitol. Jpn.*, **4**, 303–7.

Parish, M.E. and Higgins, D.P. (1989) Survival of *Listeria monocytogenes* in low pH model broth systems. *J. Food Protect.*, **52**, 144–7.

Pelroy, G.A., Eklund, M.W., Paranjpye, R.N. *et al.* (1982) Inhibition of *Clostridium botulinum* Types A and E toxin formation by sodium nitrite and sodium chloride in hot-process (smoked) salmon. *J. Food Protect.*, **45**(9), 833–41.

Penner, J.L. (1991) *Campylobacter, Helicobacter* and related spiral bacteria, in *Manual of Clinical Microbiology*, 5th edn, (eds A. Balows, W.J. Hausler, Jr, K.L. Herrmann *et al.*), American Society for Microbiology, Washington, DC, pp. 402–9.

Philip, M. (1981) Profits up in smoke. *Catering and Hotel Management*, **53**(10), 21–2.

Pierson, M.D. and Reddy, N.R. (1988) *Clostridium botulinum. Food Technol.*, **42**(4), 196–8.

Robinson, R.K. (1983) *The Vanishing Harvest: A Study of Food and its Conservation*, Oxford University Press, New York, pp. 131–4.

Rodrick, G.E. and Cheng, T.C. (1989) Parasites: occurrence and significance in marine mammals. *Food Technol.*, **43**(11), 98–102.

Ruiter, A. (1979) Color of smoked foods. *Food Technol.*, **33**(5), 54, 56, 58–60, 63.

Schlech, W.F., Lavigne, P.M., Bortolussi, R.A. *et al.* (1983) Epidemic listeriosis – evidence for transmission by food. *N. Engl. J. Med.*, **308**, 203–6.

Seeliger, H.P.R. and Jones, D. (1986) Genus *Listeria* Pirie 1940, 383AL, in *Bergey's Manual of Systematic Bacteriology*, (ed. P.H.A. Sneath, N.S. Mair, M.E. Sharpe and J.G. Holt), Williams and Wilkins, Baltimore, MD, Vol. 2, pp. 1235–45.

Sen, N.P., Smith, D.C., Schwinghammer, L. and Howsam, B. (1970) Formation of nitrosamines in nitrite treated fish. *J. Inst. Can. Aliment.*, **3**, 66–9.

Sen, N.P., Tessier, L., Seaman, S.W. and Baddoo, P.A. (1985) Volatile and nonvolatile nitrosamines in fish and the effect of deliberate nitrosation under simulated gastric conditions. *J. Agric. Food Chem.*, **33**(2), 264–8.

Shiau, C.Y. and Chai, T. (1985) Smoked dogfish processing and its refrigerated storage stability. *J. Food Sci.*, **50**, 1348–50.

Sink, J.D. (1979) Effects of smoke processing on muscle food product characteristics. *Food Technol.*, **33**(5), 72, 74, 78, 80, 83.

Smibert, R.M. (1984) Genus *Campylobacter* Sebald and Véron 1963, 907AL, in *Bergey's Manual of Systematic Bacteriology*, (eds N.R. Krieg and J.G. Holt), Williams and Wilkins, Baltimore, MD, Vol. 1, pp. 111–18.

Thompson, J.S. and Jadlocki, J.F., Jr (1981) Method for inhibiting the growth of *Clostridium botulinum* and the formation of enterotoxin in smoked fish products. *US Patent 4 277 507*.

Thompson, J.S. and Jadlocki, J.F., Jr (1982) Smoked fish products which inhibit the growth of *Clostridium botulinum. US Patent 4 344 977*.

Tilgner, D.J. (1958) Analysis and use of smoke from various kinds of wood. *Fleischwirtschaft*, **10**, 751. Quoted in Maga, J.A. (1988) *Smoke in Food Processing*, CRC Press, Boca Raton, FL, p. 77.

Tuley, L. (1985) In the pink. *Food Flavourings, Ingredients, Packaging and Processing*, **7**, 44–45.

Warburton, D.W. and Farber, J.M. (1990) Isolation of *Listeria monocytogenes* from all foods except meats. HPB laboratory procedures: MFLP-59, August 1990 update, in *Compendium of Analytical Methods*, Polyscience Publications, Montreal, Canada, Vol. 3.

Ward, D.R. and Baj, N.J. (1988) Factors affecting microbiological quality of seafoods. *Food Technol.*, **42**(3), 85–9.

Weagant, S.D., Sado, P.N., Colburn, K.C. *et al.* (1988) The incidence of *Listeria* species in frozen seafood products. *J. Food Protect.*, **51**, 655–7.

Whelan, J. (1982) *Smoking Salmon and Trout plus Pickling, Salting, Sausaging and Care*, Aerie Publishing, Vancouver Island, Canada, pp. 26–103.

Wimpfheimer, L., Altman, N.S. and Hotchkiss, J.H. (1990) Growth of *Listeria monocytogenes* Scott A Serotype 4 and competitive spoilage organisms in raw chicken packaged under modified atmospheres and in air. *Int. J. Food Microbiol.*, **11**, 205–14.

Young, T. (1984) A study of demand for fresh, cured and frozen fish in Great Britain. *Occasional Paper Series 2*, Sea Fish Industry Authority, Fishery Economics Research Unit.

CHAPTER 4

New packaging technology for seafood preservation – shelf-life extension and pathogen control

S. Ghazala

4.1 INTRODUCTION

There are significant marketing advantages in offering high-quality fresh, unfrozen food to the consumer. To achieve this for seafood requires chilling facilities and coordination throughout the chain from time of catching to retailing. Potential developments including the combination of chilling with gas packaging will be discussed in this chapter. The consumption of ready meal seafood products employing modified atmosphere packaging (MAP), controlled atmosphere packaging (CAP) and *sous vide* processing technologies has significantly increased in the last decade despite the lack of information on the microbiology of the products. Because of their convenience and high quality, it is likely that the consumption of these types of processed foods will continue to grow.

Chilling preserves the freshness and quality of seafood products compared with traditional processing methods of smoking, canning, drying and freezing (Perkins, 1990). Achieving this requires chilling and refrigerating foods immediately after packaging and processing. MAP, CAP and *sous vide* processing technologies are the tools for producing high quality ready meals. Modified atmosphere packaging refers to a system in which ambient air is replaced either totally or partially by a gas, usually an inorganic gas. The assumption is that the gas will induce

Fisheries Processing: Biotechnological applications.
Edited by A.M. Martin.
Published in 1994 by Chapman & Hall, London. ISBN 0 412 58460 3

a reduction in the activity of antagonistic microbes in the packaged product. However, the atmosphere of the MAP product will be changed during the shelf-life depending on the metabolic activity of the microbes and the permeability of the packaging material. Controlled atmosphere packaging is similar to MAP, but gas composition inside the package is monitored and regulated to maintain a relatively stable gas composition during the storage period. *Sous vide* is a process whereby food is vacuum packaged then cooked, chilled and refrigerated. For MAP, CAP and *sous vide*, moderate temperature thermal processing (MTTP) is used in order to extend the refrigerated shelf-life of high-quality prepackaged seafood products. Because of this MTTP for seafood, concern has grown about the safety and preservation of MAP, CAP and *sous vide* products. Therefore, there is currently growing interest in both the microbiology and the time/temperature combinations for processing these food products. Ooraikul and Stiles (1991) reported the importance of differentiation between MAP and CAP. They also indicated that vacuum packaging is sometimes considered as a simple variation of MAP. Accordingly, the three types of packaging can be considered as different techniques. In this chapter, MAP, CAP and *sous vide* will be regarded as different methods of packaging.

The demand for seafood has steadily increased in the last few decades. while the supply seems to have decreased. The per capita consumption of seafood in the US has gone from 5.35 kg in 1970 to 6.53 kg in 1985, a 22% increase in 15 years. Some claim that the retail value of seafood in 1989 was about $28.8 billion. If 6.53 kg per capita is an accurate statistic, then the annual consumption for the US should be 1.63 billion kg. At an average retail price of $3.63/kg this would confirm the $28.8 billion of retail sales. However, the US 'catch' for fish and shellfish for 1988 is quoted as 4.94 billion kg. There seems thus little justification for increasing prices based on claimed shortages (Hirsch, 1991). Despite the rise of the sale of all seafood, the packaging of fish has been one of the most neglected areas of packaging technology.

There have been few human health hazard problems with seafood products even though they are amongst the least stable of foods and spoil rapidly (Whittle, Hardy and Hobbs, 1990). If fish products are maintained at 0°C throughout handling and processing, there will be few bacterial health hazards unless there is gross contamination at some stage. Most bacteria that causes food poisoning are mesophilic and generally will not grow below 8°C. *Listeria* and *Campylobacteria* species have been highlighted in recent years as the cause of food poisoning in chilled foods. *Listeria monocytogenes* has become a major concern in recent years for the following reasons:

1. It is a psychrotopic pathogen and can therefore grow at chill temperatures.

2. It is widespread in the environment.
3. It appears to be more heat tolerant than many other vegetative microbes.
4. The disease it causes, listeriosis, is associated with a high mortality rate (an average of 35% in food outbreak situations).

Listeria are able to grow at 0°C, however, proper refrigeration, i.e near 0°C, and good hygienic practices, especially to avoid recontamination, minimize the risk of poisoning from these organisms (Farber, 1991). *Clostridium botulinum* has been an ongoing concern in food processing for many years. The incidence and levels of *C. botulinum* spores in raw and semi-preserved meats in North America are very low; however, the incidence in fish has been as high as 23.5%. *C. botulinum* type E is a particular concern in seafoods, since it is a natural seafood contaminant which can grow at temperatures as low as 3.3°C (Farber, 1991).

It has been indicated that a modern systems approach to optimize quality, yield, and financial return in the catching and subsequent handling, processing, packaging, chilling, distribution and retailing of fish stocks would result in a marked improvement in quality and acceptability of most seafood species. However, much more is needed requiring major changes in the entire food processing industry from catching to retailing as well as in government and regulatory agencies.

4.2 QUALITY AND NUTRITIONAL VALUE OF SEAFOOD

Loss of initial quality of fish is predominantly due to enzymatic and physical processes. The introduction of undesirable flavors appears to be caused predominantly by bacterial metabolism and to a much more minor extent by chemical processes. The most practical way to modify these deteriorative processes is to lower the temperature at which fish is stored. Ideally, fish could be stored just above their freezing point, about −1.5°C, with some possible advantage in minimizing deteriorative changes. Maintaining this temperature using present methods of refrigeration is difficult as either some fish would freeze or there would be an increase in temperatures. Gas packing combined with chilling is expected to result in considerable reduction in deteriorative processes for seafood.

Over recent years, there has been a considerable surge in the interest of the nutritional aspects of foods. Much of the population is aware of adverse effects of high levels of fat, calories, cholesterol and sodium, and the loss of nutritional value of harshly 'processed foods', and the possible health concerns of chemically preserved foods. The trend is towards eating foods that are 'natural', fresh and healthy. Seafood in particular has received considerable attention in recent years because of its nutritional benefits. The interest in unsaturated fats and ω-3 fatty acids has increased the popularity of consuming fish and fish products.

The current demand for convenient, healthy, restaurant-quality food has sparked an explosion of ready-to-eat and heat-and-eat types of frozen, refrigerated and shelf-stable foods. These include minimally processed entrées, meats, vegetables, salads, pastas, sauces and other side dishes. Many of them are designated as the 'new generation of refrigerated' or 'chilled' foods. These 'new generation' foods contain minimal amounts or no preservatives and many are packaged in a modified atmosphere to extend their refrigerated shelf-life. Depending upon the product, these refrigerated foods are expected to have a shelf-life of 40–60 days. Consumers prefer the refrigerated foods because of their better taste, texture and quality. They also perceive these foods to be relatively fresh, nutritious and 'close to natural', as compared with 'fast foods', and more convenient compared with frozen foods. These items can be purchased at grocery stores and do not need a special trip to fast-food outlets or restaurants. These new generation, refrigerated foods have gained significant popularity in many European countries and in Japan. Currently, in the US, the annual chilled-food market is about $500 million. This is expected to increase by more than 45% in the next 5 years if some technical problems are resolved. The market for microwavable frozen foods has grown considerably in the last decade due to the penetration of microwave ovens and changing demographics where people have less time to stay home and prepare a meal. This growth is not expected to continue since consumers are now demanding quality along with convenience. Sales have been forecasted to grow more slowly from $1.6 billion in 1988 to $2 billion in 1993 (Morris, 1989).

4.3 FOOD PRESERVATION METHODS

4.3.1 The problem with microorganisms

The most common way a food loses its acceptable quality is due to the growth of microorganisms or from non-microbial causes, e.g. lipid oxidation. Similarly, a food that seems acceptable can be unsafe for consumption because of the growth or presence of pathogenic microorganisms or due to the presence of toxic chemicals.

Microorganisms, which include bacteria, yeasts, moulds and viruses, gain access to food through various sources. Their level and types vary greatly depending upon their source, the effectiveness of sanitation used during all phases of handling, and the methods used in the processing and storage of a food. Different physical, chemical and biological preservation methods are applied with an intention to make a food safer and to extend its shelf-life by destroying and retarding or preventing the growth of pathogenic and spoilage microorganisms that might be present in the food. Physical methods include temperature treatment (both high and low), reduced water activity, controlled or modified atmospheres and radiation. Chemical methods include adding inorganic and organic chemicals, and smoking. Biological methods include allowing food-grade starter culture microorganisms to produce antimicrobial metabolites during natural and controlled fermentation (Light and Walker, 1990a).

The usual course of preservation methods and the action of some preservatives is to kill a portion of the population, injure some of them, and have no effect on others, i.e. they remain as normal cells or spores. Generally it is not the objective to kill all the microorganisms in food since this causes a significant decrease in quality. The relative proportions of cells in each of these three groups vary depending upon the type of food, microbial content, means of preservation and several other factors. The surviving population, which includes both injured and normal cells and spores can, under favorable conditions, multiply, reach a high population and make a food unsafe for consumption or make it lose its acceptable quality (Light and Walker, 1990a).

Preservation methods have made perishable products available year round instead of seasonally and has let us choose from a wide range of food products. They have also helped in the commercial processing of food in large volumes and in the distribution of food nationally and internationally. However, the basis of selecting any preservation method(s) for a food is dependent on several factors, including the type of food, the type of microorganisms, the extent of shelf-life desired and the cost.

4.3.2 The addition of hurdles to maintain food quality

In addition to adding preservatives to foods, typical food parameters can be adjusted to prevent growth by becoming hurdles to spoilage microorganisms. The higher the number of preservation methods or hurdles used, the greater is the microbial viability loss or growth retardation of survivors. Simultaneous application of more than one method increases the likelihood of producing microbially safe foods with potential for longer shelf-lives. The addition of hurdles can interfere either directly or synergistically and include pH, water activity (a_w), gas atmosphere composition and temperature. Generally, the lower the pH and a_w, the less the chance that microbes will grow and cause spoilage. In theory adjusting these parameters seems like a perfect solution to ensuring safety in products. However, since food systems are very complicated, these hurdles are very dependant on the food composition, packaging permeability and any changes within the package. For example, moisture may migrate from meat into a sauce thus having an effect on pH and a_w (Labuza, Fu and Taoukis, 1992). As will be discussed later in this chapter, gas atmosphere composition and temperature can have a drastic effect on the growth and activity of microorganisms. Minor food ingredients can also often have an effect in reducing the activity and growth of microorganisms. Spices, e.g. cinnamon, onion and garlic, have bacteriostatic effect. Organic acids, e.g. malic, lactic, acetic, benzoic, sorbic and propionic acids, can lower the pH of foods and have a bacteriostatic effect. Alternatively, the pH of some foods can be unintentionally increased by alkaline ingredients, e.g. calcium carbonate, calcium hydroxide and alkaline phosphates (O'Donnell, 1991).

4.3.3 Packaging as a means of preservation

(a) Role of packaging

Food packaging, along with transportation, refrigeration and warehousing systems, is crucial to our modern food distribution and marketing system. Without protective packaging, food spoilage and waste would increase tremendously, and consumers would pay much more for their food. The advent of modern packaging technologies and new or improved packing materials made possible the era of convenient food, stockpiling of manufactured foods in warehouses, and shipment of perishable products over thousands of miles. Sophisticated packaging that protects food from physical and environmental damage has enabled us to locate food processing facilities where they are most advantageous from the standpoint of raw material availability, labor supply and power. Without unit packaging, self-service grocery stores, supermarkets and food vending machines would not exist.

Today packaging of foods has many functions and is so complex that it has grown into a multibillion dollar industry since the days before frozen foods, overnight coast to coast deliveries, and before agriculture and food processing plants were so distant from major consuming areas. Almost half of all containers and packaging materials made in the US are for food. In the past, packaging emphasized the expectations of the producers and distributors. This emphasis has now shifted back towards the consumer since they are becoming more demanding and are more aware of the choices available (Piergiovanni, 1991). Packages must be developed to meet the requirement of each food product, including chemical and physical properties of the food, storage and distribution requirements, and requirements related to consumer acceptance.

A food package usually provides a number of functions in addition to being a protective container. One of these functions is to conveniently assemble food items. Putting similar items together or combining ingredients in handy sizes permits faster and easier warehousing, distributing, purchasing and handling of foods. Convenience packaging for consumers requires a shape and size of containers that allows for easy opening, pouring, serving, carrying, reclosing and storage. The package may also serve as a container for cooking, baking and serving, e.g. frozen dinners, boil-in-bag products and individual servings of jelly, margarine and other foods as used in food service establishments, and single portions of soups, puddings and cheese, etc., for lunch boxes and vending machines.

4.4 PACKAGING FOR SEAFOOD

4.4.1 Fish composition and initial bacterial spoilage

Fish spoilage is usually caused by enzymatic reactions within the fish and the metabolic activity of microorganisms. Fish muscle is an excellent substrate for bacterial growth since it is lower in undigestible connective tissue than the muscle of meat animals, has a high water activity and has neutral or near neutral pH (6.2 and above). Spoilage due to the fat, i.e. oxidative rancidity, varies considerably with the species and composition of the fish. The higher the fat content the faster spoilage may occur (Hobbs, 1987). Non-fatty fish, e.g. cod and haddock, have been characterized as having an average composition of about 80% water, 18% protein, 1% carbohydrate and less than 1% lipid. Fatty fish, e.g herring and mackerel, have about the same concentrations of carbohydrate and protein as that of non-fatty fish, but have a wide range of water and lipid content (1–30% lipid), which still totals about 80% (e.g. water 75% and lipid 5%). Fat content depends on the species and season, so a single species may contain different amounts of fat at different times of the year. Minor components of fish muscle include minerals, vitamins, sugars and low molecular weight non-protein nitrogenous compounds, and are readily utilized by spoilage organisms (Kraft, 1992).

The physical shape of a fish can also be a problem for spoilage since spoilage bacteria grow extensively on the surface of fish. Generally, fish have a high surface to volume ratio, which increases as processing proceeds from the whole fish form to the gutted form, then fillet form, so the rate of deterioration may also increase (Liston, 1980).

Generally, temperature is the most important factor that influences the composition and numbers of bacteria on fish. Fish from clean, cold waters (temperature of the sea bed of 10°C or less) have counts at the lower end of the range, particularly with storage of the fish in melting ice. The bacterial flora is primarily psychrotrophic or psychrophilic Gram-negative rods. Fish from tropical or subtropical areas harbor more mesophyllic bacteria which may be predominantly Gram-positive. In the latter case, changes in the flora during chilling and storage are more pronounced than that for fish from colder waters. Jay (1986) lists 19 genera of bacteria known to occur on fish and shellfish, with *Pseudomonas* most frequently observed. *Pseudomonas* are the most important spoilage types since they become dominant, leading to spoilage of fish at 0–6°C (Shaw and Shewan, 1968).

4.4.2 Packaging of fresh fish

Fresh fish is one of the most perishable of all foods. A characteristic of fresh fish is their moist and slippery surface, which makes handling difficult and causes rapid loss of quality. The rate of spoilage doubles for every 5.5 °C rise in temperature. They must be refrigerated or frozen immediately after harvest and kept this way until utilized. If fish are of poor initial quality, the flavor, color and texture loss will be more pronounced than if using fish of good quality. The best package cannot improve the quality of the contents so the fish must be of high quality prior to packaging (Fujita, 1990).

Seafood varies in shape and size and is required to be in ice or frozen for distribution. These facts make the packaging of seafood difficult. One solution to this problem is the new concept of container packaging that offers standardized handling conveniences and hygienic protection. Cultured fish, shellfish and some others are distributed as live fish. The fresh fish packages are required to offer easy disposal of drip, handling convenience and hygienic protection of the contents (Fujita, 1990).

(a) Packaging requirements

The rapid decline of retail seafood stores has been coupled with a change in consumer buying habits. More and more supermarkets now maintain fresh seafood counters as well as offering prepackaged fish. Extension of shelf-life by new packaging technologies has become an essential prerequisite for marketing success.

A suitable fresh fish package must: reduce fat oxidation, reduce dehydration, provide for less bacterial and chemical spoilage, eliminate drip and prevent odor permeation.

1. *Fat oxidation.* The highly unsaturated nature of fish oils and fats makes them very susceptible to rancidity. Spoilage also occurs from chemical, enzymatic or bacterial action. At room temperature and in the presence of oxygen the lipid components can rapidly become rancid. The use of oxygen barrier packages and cold storage drastically reduces this action.
2. *Dehydration.* Excess moisture loss from fresh fish leads to texture, flavor and color changes. Poor storage conditions or a poor barrier package will allow dehydration.

3. *Bacterial and chemical spoilage.* Enzymes and bacteria present in fish cause rapid spoilage. Powerful proteases in the digestive tract may, after death, perforate the intestinal wall and attack the surrounding flesh. This is why fish should be gutted washed or frozen immediately. The flesh of healthy fish is free of bacteria but the slime and digestive tracts contain many bacterial flora. Immediately after death these bacteria begin to grow and spoil the fish with development of off flavors and odors and pronounced textural changes. As with enzymes, immediate cleaning is the best means for preventing spoilage. Refrigeration alone is ineffective as many bacteria associated with fish are psychrophilic, that is they will grow at and below 0°C. Freezing will retard spoilage provided the temperature is held low enough to prevent further bacterial growth.
4. *Drip.* When a piece of fish is packaged, some juice may be lost from the fish and trapped in the package. The result is messy and unattractive. An absorbent pad may be placed in the package to soak up the drip. Another approach is to dip the fish in a polyphosphate solution long enough to swell the outer surface layer of cells. The rupturing of cell walls helps to prevent the escape of internal fluids.
5. *Odor permeation.* Odor from fish becomes stronger within several hours in permeable packages stored at 2°C. Although this odor may dissipate, a considerable amount of flavor volatiles will be lost. Foreign odors are easily picked up by fish. A good odor barrier is an essential part of a package.

(b) Bulk packaging

In order to transfer fresh fish to markets quickly, air transport is usually required. A lightweight and protective container becomes a necessity. Air shipment is an emerging trend in fresh seafood distribution. Formerly, wooden boxes with parchment liners were used for bulk packaging of fresh fish. In recent years, plastic containers, plastic corrugated cartons and metallic containers have been commonly used for fresh fish packaging. Fresh fish are often packed with ice in a foam polystyrene carton and are also packed in polyethylene bags, small wooden boxes and wax-coated paper (Fujita, 1990).

(c) Retail packaging

A shallow tray and transparent film overwrap is the most popular package for fresh refrigerated fish. Tray packs are overwrapped with semi-moisture-proof cellophane, polystyrene, oriented polypropylene or polyvinyl chloride. A significant disadvantage of such tray packs is that they tend to deteriorate if too much moisture is absorbed. Foam and clear polystyrene require the use an of absorbent blotter. In order to obtain improved rigidity for direct film overwrapped packages, a plastic or wax-coated paperboard insert is used. Sometimes, a vacuum pack is used to improve shelf-life. This requires higher barrier films, e.g. polyvinylidene chloride (PVDC) coated cellophanes, polyethylenes, polypropylenes or polyesters (Fujita, 1990).

(d) Frozen fish

Many marine products are frozen for longer shelf-life and can be distributed in frozen condition with the development of freezer chains. Seafood is processed and frozen on-vessel immediately after the catch or frozen on land at the factory after the cleaning, gutting, shell removing, filleting and cutting processes. They are frozen in freezing pans, in freezing pans with water, dry pack freezing or individual quick freezing by contact freezing, air-blast freezing or immersion freezing.

Large fish are typically frozen in the round. Usually ice glaze treatment is given to the surface followed by distribution without a further package. Medium-size fish are, in general, individually air-blast frozen, packed in a polyethylene sheet or bag and then placed in a water-resistant corrugated fibreboard box, plastic corrugated box or foam polystyrene box. For small-size seafood like fillets of consumer size, packaging almost identical to that of frozen foodstuffs is applied, including polyethylene bags, vacuum packaging using laminated film, shrink packaging, wrapped tray packaging, paper boxes, etc.

Frozen seafood is characterized by its hardness and irregularity of shape, especially with the protrusion of fins, spines and sharp edges. To maintain quality under these conditions, freezing resistance and antifreezing strength is required from the package. Seafood is also susceptible to freezer burn. In a cold and dry atmosphere, ice sublimes, the surface of frozen product is easily dehydrated and rancidity occurs through oxidation. Therefore, packages are required to protect the contents from such adverse influence. Moisture resistance, gas-barrier capability, waterproofing, oil resistance and air tightness are important requirements for the package (Fujita, 1990).

4.4.3 Processed fish

Processed fish, e.g. dried, salted, seasoned, pickled and smoked items, are distributed in normal, cool or freezing conditions. These products are packed in the following containers depending on the type of process: wooden boxes, paper containers, plastic containers, polyethylene bags, vacuum packaging, gas-replacement packages, packaging with deoxidative and desiccating agents, and others.

Various species of seafood are packed in cans and glass bottles with brine, oil, seasoning, etc. Tin plate is usually employed for canning with various coatings on the inner surface, like oleoresinous type C enamel and phenolic-type seafood-enamel coating. Aluminum and chrome-plated iron are also used. In canned crabmeat, chelating parchment paper is used inside for preventing struvite. Deep-draw two-piece cans and pull-top cans made of aluminum have recently been used. The flexible retort pouch and semi-rigid tray are also used for seafood products, e.g. a three-ply laminated bag composed of an outer layer of polyester film, a middle layer of aluminum foil and an inner layer of modified polypropylene (Fujita, 1990).

4.5 MODERN FOOD PACKAGING

A common misconception that many food processors make about food packaging is to think expensive is better. Choosing appropriate packaging for foods is very important but also very difficult. Foods are essentially biological systems and have certain requirements for atmosphere in order to maintain quality. It is especially difficult to find optimum packaging for mixed food products, e.g. meat and vegetables, since each has its own biological requirements. As well, some meat products require oxygen and an expensive oxygen impermeable film will not yield the best-quality product. It is also important to know the intended storage conditions of the product, i.e. temperature, relative humidity and expected shelf-life. This is one of the reasons why multilayered plastics are very popular (Muller, 1991).

4.5.1 Functions of food packaging

The basic function of food packaging is to protect the product from physical damage and contaminants, to delay microbial spoilage, to allow greater handling convenience, and to improve presentation. Some common film properties include: gas barriers, moisture barriers, strength, temperature resistance/sealability, cost and environmental effects.

1. *Gas barriers*. Gas barriers are important when dealing with MAP and *sous vide* products where prevention of gasses migrating in or out is important for determining and extending shelf-life. Gas barriers are directly proportional to the thickness of the packaging film. High gas barrier films include saran, PVDC and ethyl vinyl alcohol copolymer (EVOH).
2. *Moisture barriers*. Moisture barriers should maintain the same moisture level in the package throughout the shelf-life of the product. PVDC is a good moisture barrier while EVOH has poor moisture barrier properties.
3. *Strength*. Strength is required in packaging films to protect the pro-

duct during processing (if applicable), distribution and retail handling. Oriented films are generally stronger than unoriented films; however, oriented films cannot be used for thermoforming applications since they are not as extensible. Nylon and polyester are commonly used for oriented films.

4. *Temperature resistance/ heat sealability*. The temperature to which a container is subjected during filling, processing, storage or preparation limits the selection of materials. Heat sealability is usually provided by polyolefin polymers, e.g. as low density polyethylene (LDPE), linear low density polyethylene (LLDPE) and medium density polyethylene (MDPE). LDPE is used in a wide range of packaging materials due to its low cost and moderate strength. LLDPE and MDPE are often used in hot fill or boil-in applications due to their hot tack properties, puncture resistance and seal strength (Muller, 1991). Crystallized polyester (CPET) can tolerate temperatures up to 220°C and can be reheated in both microwave and convectional ovens. Polypropylene (PP) can tolerate temperatures of up to 130°C and can only be reheated in the microwave (Forshaw, 1990). Care must be given to choosing packaging that will be heated in microwave or conventional oven since there may be interaction with the food contained (Piergiovanni, 1991).
5. *Cost*. Since these packaging materials are used for convenience-type products, very few consumers will be motivated to find other uses for them. Therefore the packaging material must be inexpensive to produce. CPET is more rigid than PP so trays can often be made thinner. However, CPET is almost twice the cost of PP, partly due to the fact that it needs to be crystallized and that processing speeds are slower. The incorporation of barrier layers into PP can initiate considerable cost and waste (Forshaw, 1990).
6. *Effect on the environment*. Attention has been paid to the effect that food packaging has on the environment. There is a trend towards the elimination of chlorinated polymers (PVC) and PVDC. The replacement of the latter is a problem since no polymer has been found that is as effective as PVDC in terms of acting as a good barrier to gasses and vapors (Piergiovanni, 1991).

The various procedures that a product will undergo determines the materials, lidding system, barriers added and any special designs added to the product. Foods that will be frozen require trays that can withstand the low temperatures. Either PP or CPET can be used in these cases. For chilled products a less expensive grade of CPET can be used. Retort sterilization requires trays that are generally designed with thicker walls and top edges. Aluminum lids and foils are appropriate only when microwave heating will not be employed. Shelf-stable products require a hermetic seal to prevent microbial spoilage and impermeable films to prevent deterioration from the permeation of gasses and vapors. The latter is achieved by incorporating barrier structures in PP trays and developing high-barrier CPET alternatives. For products requiring retort sterilization, polyester, heat sealable films that can withstand high temperatures are used (Forshaw, 1990).

4.5.2 Active packaging

Active packaging is a term to describe a new technology trend whereby packaging takes on an active role in preserving the product. It can be defined as:

> any form of positive interaction between the packaging and the environment inside it, or between the packaging and the food itself, in order to extend its shelf-life or to improve the acceptability and/or convenience of the product (Piergiovanni, 1991).

Oxygen absorbers or scavengers can be added to control oxygen levels, e.g. reduced iron. Similarly a CO_2 generating system can be added. Some enzymatic systems have the dual benefit of being oxygen scavengers and producing antimicrobial by-products. Alcohol oxidase and glucose oxidase dips can be used to extend the shelf-life of fresh fish (Labuza, 1990). Temperature control is of major concern for minimally processed foods. Devices can be added to indicate the age of the product or show incorrect temperatures during storage. Active packaging also includes multifunctional packages, e.g. for marketing and reheating as well as protective purposes (Piergiovanni, 1991).

4.6 PROCESSED SEAFOOD AS A CHILLED CONVENIENCE PRODUCT

Seafood packs range from the most simplistic forms of overwraps or boxes suitable for wholesale marketing, to retail packs using complex material and sophisticated vacuum and gas packing techniques. The primary aim of most packaging is to prevent contamination and to increase marketing. However, there are important secondary benefits like maintaining hygiene and slowing down deteriorative processes, particularly those associated with microbiological, physical and chemical spoilage. Although all forms of packaging give advantages over unpackaged material, most attention from the viewpoint of quality maintenance has concentrated on two forms of packaging, i.e. vacuum packaging and gas packaging.

4.6.1 Vacuum packaging

Vacuum packing can considerably extend the viable shelf-life of many cooked foods. It should be stressed that vacuum packaging must be used under strict conditions of hygiene and control, and not as a means to forgo proper sanitation. This method could be applied to cook–chill food systems to increase food quality and microbiological assurance. Cook–chill is a hot-fill system where pumpable foods, e.g. soups, stews and gravies, are prepared and pasteurized in conventional steam kettles and pumped directly into barrier-plastic casings at the precise degree of 'doneness' while still above pasteurization temperature. The casing is then clipped and turned into cold water to arrest cooking and chill the product to approximately 4°C in about 1 hour. The chilling induces a vacuum (Morris, 1989). The use of vacuum packaging, in gas-impermeable and heat-stable materials, for hot, pasteurized foods or for pasteurizing foods within the vacuum pack has advantages which include:

1. No or low risk of post-pasteurization contamination.
2. Ease of handling.
3. Inhibition of growth of aerobic spoilage organisms.
4. In-pack chilling can include iced water bath chilling.
5. Inhibition or slowing of deleterious oxidative reactions in the food during storage due to oxygen barrier properties of the packaging material.

The disadvantages of the method include:

1. Added expense of equipment for vacuum packaging.
2. Growth of anaerobic organisms, normally restrained by competitive growth of aerobes, could be stimulated or allowed by the conditions within vacuumized food packs; the anaerobic environment may provide optimal conditions for the recovery of heat-injured micro-organisms.
3. Normal spoilage microorganisms (i.e. pseudomonades, yeasts and molds) are inhibited. These organisms give the consumer the tell-tale signs of spoilage, e.g. foul odors, sliminess and textural changes (Rhodehamel, 1992).
4. While in-pack, foods cannot be properly inspected for spoilage or unwanted changes without breaking the bag seal and thereby shortening the specific sample's shelf-life.
5. Without specifically designed devices it is difficult to measure the core temperature of vacuum packed foods.
6. Added training is necessary for food production staff.

There are a number of criteria required of the films used for vacuum packaging in large-scale production methods, especially those which require in-pack pasteurization (i.e. the pasteurization of the food after it has been packaged). These requirements include:

1. High durability, i.e. ability to withstand considerable mechanical stress during packaging, handling and transport.
2. Retention of flexibility even at low temperatures (−2 to 4°C) to enable satisfactory handling in the packaging and refrigeration rooms.
3. Ability to withstand heating to at least 100°C without structural damage, leaching of potentially toxic plastics or plasticizers or changes in any gas permeability.
4. Impermeability to liquids, including oils and fats, and macromolecules, e.g. proteins.
5. Impermeability to gases, in particular oxygen, so that oxidative deterioration of the packaged food-stuffs is limited or inhibited.
6. Manufactured from non-toxic, food-acceptable, odorless materials.
7. Must be able to create air-tight durable heat seals to close packs.

Many of these criteria have been met by a range of materials, mostly multi-laminated plastics. A wide range of materials are now manufactured by a number of companies throughout the world which are suitable or even specifically designed for use in large-scale vacuum packaging/ cook–chill operations.

4.6.2 MAP/CAP

Modifying the atmosphere in order to aid food preservation has gained a great deal of attention since it has the potential to meet consumers demands for fresh, high-quality convenience products. Modification of the atmosphere inside a particular packaging material can be achieved at three levels of sophistication (Jones, 1989):

1. *Vacuum packaging*. Predominantly used for retail packs; the original air atmosphere is evacuated; the atmosphere which develops during storage is mainly the result of biological activity.
2. *CAP*. Mainly used for bulk storage or transport; the gas composition, humidity and temperature can be constantly controlled to provide optimal conditions for long-term storage of meat and other foods.
3. *MAP*. Used for bulk and retail packs in which the packs are flushed and filled with nitrogen or gas mixtures; the gas composition may subsequently change as a consequence of pack permeability, and chemical and biological activities. Potential advantages of MAP include:

   potential shelf life increases of 50–400%; reduced economic loss; distribution of products over longer distances and with fewer deliveries, leading to decreased distribution costs; provides a high quality product; and easier separation of slices (e.g. vacuum-packaged versus MAP bacon).

Potential disadvantages of MAP include:

> visible added cost; necessity of temperature control; different gas formulation are required for each product type; and special equipment and training is required (Farber, 1991).

The choice of the type of packaging atmosphere is restricted by certain considerations, especially the effect on food, e.g. food color, presence of oxidative deterioration, etc. In certain sectors of the food industry, MAP is well established and the benefits of the system are proven. However, for seafood products the technology is still in progress, particularly in understanding the preservation mechanisms by the use of different gases, e.g. nitrogen, CO_2 and oxygen or mixtures of these. The obvious effect of nitrogen and CO_2 is to replace the oxygen, thereby delaying oxidative rancidity and inhibiting the growth of aerobic spoilage organisms. CO_2 is water and lipid soluble and is the gas mainly responsible for the bacteriostatic effect seen on microorganisms packaged under this environment. CO_2 seems to achieve a bacteriostatic effect by increasing the lag phase of growth and decreasing the growth rate during the logarithmic phase. The solubility of CO_2 decreases as the temperature increases, so for maximum antimicrobial effect from CO_2 the MAP product should be kept as cool as possible (Farber, 1991). It has been found that increased CO_2 levels can raise the temperature minimum for some pathogens. *L. monocytogenes*, which has a minimum growth temperature of about 2.5°C, can grow at temperatures as low as 0°C. In elevated CO_2 levels, the organism may not be able to grow below 10°C (Labuza, Fu and Taoukis, 1992). To be effective, CO_2 must dissolve in the liquid phase of the food and it must be applied at relatively high concentrations (40% or greater) to insure its availability for a longer time. The general atmosphere recommended for white fish is 40% CO_2, 30% N_2 and 30% O_2. For fatty fish it is 60% CO_2 and 40% N_2. Oxygen is excluded to minimize rancidity (O'Beirne, 1990). An example of atmospheres and ideal temperatures to store fish products is shown in Table 4.1 (Farber, 1991).

For fresh and cured meats, in which gas and vacuum packing has been very successful in extending shelf-lives, the safety of the system relies on the effectiveness of chill temperature control and on the growth of non-pathogenic spoilage flora compared with food poisoning organisms. Seafood is not like other muscle foods and is very susceptible to microbial spoilage and chemical deterioration. Therefore many food processors are hesitant to use MAP on individual consumer packages. It has been shown from studies that MAP products using CO_2 have improved stability over fresh fish. Because an atmosphere containing high levels of CO_2 tends to inhibit spoilage organisms (Hotchkiss, 1988), it is possible that during the extended shelf-life of a MAP product pathogen growth may occur or even be stimulated before spoilage is evident. This is particularly worrisome in the case of MAP fish products,

Table 4.1 MAP fish products

	Temperature (°C)	*Atmosphere*	*Shelf-life (days)*	
			air	*MAP*
Cod fillets	2	48–100% CO_2 remainder O_2 or N_2	–	–
Trout, golden croaker	4	100% CO_2	–	–
Cooked cray fish	4	80% CO_2:20% air	14	≥21
Trout	1.7	80% CO_2:20% N_2	12	~20
Spotted shrimp	0–2	50% CO_2:50% air	12–16	
Swordfish	3.5	100% CO_2 plus 40 and 70% CO_2 remainder O_2 or N_2	–	–
Brown shrimp	4	100% CO_2 66% CO_2:34% O_2	4–5	>14
Rockfish fillets	1.7	80% CO_2:20% air	≤6	≤13
Perch	1.1	100% CO_2	~14	≥25
Rock cod	4	80% CO_2:20% air	,	≥21

Adapted with permission from Farber (1991).

where nonproteolytic type E clostridia is capable of growing at low temperatures and producing toxin before organoleptic spoilage is evident (Post *et al.*, 1985). A CAP fish product has made its appearance in recent years. The package is displayed refrigerated, equipped with an oxygen permeable membrane permitting oxygen to circulate through the package. The oxygen level in the package prevents *C. botulinum* problems. Equivalent concerns are related to the growth of pathogen in chilled foods, e.g. *L. monocytogenes* and others. The extended refrigerated shelf-life of these products could allow for the growth of some of these psychrotrophic bacteria to very high numbers (Palumbo, 1986). The results from Farber *et al.* (1990) suggested that MAP products of good microbial quality can be produced, but it would not guarantee *Listeria*-free products. Because of the long shelf-life of MAP products, the potential for psychrotrophic pathogens to grow to high numbers is increased.

The role of competing flora may be important in restricting the germination and growth of Gram-negative spoilage bacteria in fresh fish packed under vacuum or MAP/CAP conditions. It has been found that harmless lactic acid bacteria can retard the growth of *C. botulinum*. Lactic acid bacteria have actually been employed to extend the shelf-life of shrimp. Since the lactic acid bacteria are facultative, they could be seeded in a CAP or MAP package to retard the growth of the harmful microbes. In an ideal situation there would be a higher amount of Gram-positive lactic acid bacteria in the starting sample compared with spoilage bacteria. The lactic acid bacteria would have a greater chance of multiplying throughout the storage period, retaining the background microflora, including any pathogenic microorganisms. The addition of lactic acid starter cultures as a hurdle to spoilage and pathogenic microorganisms may be beneficial in certain MAP products (Farber, 1991).

The use of high-quality starting materials for MAP products is of upmost importance. Fresh raw materials generally have a low bacterial load resulting in a longer bacterial lag phase and a lower final cell density. In addition, the presence of pathogens would be minimal and the lactic acid bacteria would be in a position to dominate more easily (Farber, 1991). Packaging technologies cannot be employed to salvage poor quality raw materials.

Information about the microbiological safety of MAP products appears to be lacking. Although there is great potential for abuse of MAP products, there have been no reports of illnesses linked to the consumption of these products. As with many traditional methods of food preservation, MAP technology has developed first and our understanding of the microbiology of these processes has come much later. With rapid advances being made in both packaging materials and methods, and improved chill distribution, there will be further opportunities for novel use of modified atmospheres. Confidence in the safety of new applications of modified atmospheres in food preservation will surely be greater if the scientific basis underlying these processes has been thoroughly researched.

4.6.3 *Sous vide* technology

The idea of cooking foods in sealed bags, pouches or other containers was successfully used in the 1960s in two large-scale systems (the Nacka and the AGS systems). However, neither of these systems achieved the kind of quality that might be associated with gourmet cooking or haute cuisine. In the late 1970s a French chef, George Pralus, set out to solve the problem of weight and juice loss in pâté de foie gras. He began, following the principle of cuisine en papillote, by wrapping the foie gras in a plastic layer and cooking. The result was not successful but he persevered and, eventually, by using three layers of plastic to enclose the foie gras during cooking, he was able to reduce weight loss dramatically from about 40% to around 5%. This remarkable result led to a collaboration with a plastics manufacturer and the evolution of a multilaminate plastic material that was both heat resistant and able to retain significant gas impermeability even after high temperature treatment.

Sous vide is comparable to a CAP/MAP method. The main difference is in the post-packaging pasteurization step which provides a significant advantage. A CAP/MAP product begins to deteriorate from the time it is packaged while a *sous vide* product has virtually the same quality after 21 days as it did on the first day (Lingle, 1992). Numerous developments and changes have led to the general acceptance of the method in France and the use of the method as an haute cuisine adjunct to standard cook–chill procedures. Claims made for the method in terms of shelf-life extension and quality retention may have obscured the more important aspect of microbiological safety and the vulnerability of the method to abuse.

(a) The sous vide *process*

The literal translation of *sous vide* is 'under-empty' and merely describes the process of placing food stuffs in evacuated containers. *Sous vide* cooking involves the pre-packaging of foods in plastic bags or pouches, forming a vacuum and cooking the food within the bags at controlled temperatures. A general flow chart of this process can be seen in Figure 4.1. There are several variations of the method which allow the pre-packaging of part-cooked foods, raw foods and the holding of raw foods at chill temperatures for some time before cooking. However, the system of *sous vide* cooking in its simplest form involves the following:

1. *Preparation*. This involves the preparation of the raw ingredients, e.g. washing, peeling, trimming, boning, seasoning, etc., prior to cooking. Traditional approaches are used for these methods of preparation of the food for cooking, as in standard cook–chill methods.
2. *Pre–cooking*. The food is heat processed inside a plastic container. As the food is not directly exposed to radiant heat, the browning effect is minimized. It is possible that some vegetables have to be blanched before packaging because of their strong flavors. For example, brussels sprouts have to be blanched to reduce their powerful flavor and aroma before being sealed in the plastic container or pouch.
3. *Vacuum packaging*. The prepared food is placed into special heat resistant, air impermeable plastic bags or pouches. These containers, produced from multi–laminate plastic materials, are the key to the *sous vide* process. Once the food is in the pack, a vacuum is created, using a specially designed vacuum cabinet, to remove all residual air from the bag. The pack is then heat sealed immediately before any loss of vacuum occurs. The use of a vacuum serves to inhibit the progress of oxidation reactions which generally lead to the deterioration of foods during storage either before or after cooking. Also, the use of a vacuum serves to pull down the plastic film tightly onto the food surface. This allows maximal heat transfer into the food through the plastic during cooking.
4. *Pasteurization*. The packaged food is pasteurized for a set time and temperature appropriate to the individual characteristics of the food to be cooked. Cooking can be carried out in a water bath, where the

Fresh, top-quality ingredients

↓

Basic preparation
(seasonings, addition of ingredients and sauces
under strict hygienic and quality-control conditions)

↓

Packaging
(ingredients weighed, placed in a plastic dish and covered
with a fine plastic seal, impermeable to air and contaminants)

↓

Air extraction and hermetic sealing
(air is removed with the aid of a machine specialized for
creating a vacuum and the product is hermetically sealed)

↓

Pasteurization
(slow heating in an autoclave with a
precise electronic regulator)

↓

Quick-chilling
(inside a quick-chill chamber)

↓

Storage in cold chamber
(between 0° and 3°C)

↓

Reheating
(for 10 to 15 minutes in boiling water or
4 to 5 minutes in the microwave)

↓

Service

Figure 4.1 Flow chart of *sous vide* processing.

vacuumized bags or pouches are immersed in water of a controlled temperature. Another popular method is the use of water immersion mode during retorting. Pasteurization is the most critical step in the *sous vide* process. As the method relies on the strict control of temperature to produce highly flavored, tender products, it is essential that the user is confident that the procedures adopted are sufficient to fully pasteurize all products.

5. *Rapid chilling*. After a proper pasteurization step, the packaged food is chilled at 1–3°C as rapidly as possible (i.e. within 90 min). Direct chilling in an iced water bath is made possible because the food is pre-packaged. This method, which is an option that is normally not possible with most standard cook–chill menu items, is both cheap and efficient.
6. *Chilled storage*. Storage at 0–3°C is required for the period of time prior to re-heating and service. Storage conditions must be as rigorous as those laid down for standard cook–chill menu items. Since the food is stored under vacuum in sealed containers this leads to certain advantages in slowing oxidation reactions and in inhibiting the growth of aerobic microorganisms.

7. *Regeneration.* Re-heating and service of the product may take place in the bag with the product still under vacuum, i.e. in a heated water bath, or after the contents of the bag have been emptied into a separate container for heating on top of the cooker or inside the oven in a heat-proof dish. Alternatively, after puncturing the bag, the product could be re-heated in a microwave oven. In both of the different methods of re-heating, the food must be re-heated to at least 70°C at its center as advised for standard cook–chill menu items.

The advantages of *sous vide* cooking over conventional cooking are the following:

1. The cellular structure of the food does not change during cooking and shrinkage is minimized.
2. Vacuum packaging before heating allows the product to cook in its own juices.
3. As a properly controlled cook–chill method it is extremely hygienic, with no post-cooking contamination by air, handling or refrigeration odors.
4. Excellent food cost control is gained.
5. Up-grading of menu items is made possible by preparing food in advance.
6. When used properly, the system can result in manpower savings.

The disadvantages of *sous vide* cooking are the following:

1. It has limited applications. On its own, *sous vide*/cook–chill, like classic cook–chill, cannot provide an entire menu range because many items do not suit this unique production method.
2. There is a need for extra equipment (e.g. vacuum packaging devices) and the need to package each item separately, thus incurring the additional costs of the packaging material.
3. The method is somewhat labor-intensive, involving extra steps not normally encountered in standard cook–chill operations (pre-cooking procedures, packaging and vacuumizing), so there must be a added cost in production.
4. The technique must be used within a controlled and hygienic environment, using high-quality raw materials, proper packaging materials and very careful control of cooking (pasteurization), chilling, chilled storage conditions and re-heating. A mistake in the monitoring or control of any of these parameters could lead to the growth of potentially lethal microorganisms within the bag.

The vacuum skin packaged tray offers several advantages for retail; it is in a form that consumers are familiar with, it provides high contact clarity so the foods' visual appeal is maximized and it avoids typical adverse connotations the consumers perceive with CAP/MAP products, e.g. condensation and moisture migration (Lingle, 1992).

The *sous vide* bag or pouch provides the ideal environment, given improper pasteurization, for the growth of anaerobic organisms capable of growth at chill temperatures, including *L. monocytogenes*, *C. botulinum* and *Escherichia coli*. Since products are only pasteurized it is important to keep them chilled to prevent the germination and growth of spores. If *sous vide* products are subject to even mild temperature abuse at any stage after pasteurization other foodborne pathogens, due to poor quality raw materials or poor handling practices, could become a potential problem (Rhodehamel, 1992).

(b) Studies performed on sous vide *cooking*

Two formulated *sous vide* products are shown in Figure 4.2, above is a mackerel/rice sample and below is a meat sauce/vermicelli sample. These are characteristic *sous vide* products and are typical of the ones that will be discussed below.

In light of the microbial concerns expressed about the *sous vide* process, most studies have been focused on this aspect. One such study investigated chicken, vegetable and fish products. The products were cooked then chilled at 1–3°C for up to 21 days. Microbial analysis performed showed that after 21 days storage at 1–3°C there was negligible microbial growth in the chicken and vegetable samples, whereas the fish sample had counts greater than 6×10^6. Due to the delicate nature of the fish product it was sealed under inadequate vacuum and cooked under less rigorous conditions than the other products. It is worthwhile to mention that the fish product was still considered acceptable after 14 days. The more rapid microbial spoilage of the fish product illustrates the importance of carefully following each stage of the *sous vide* process (Light *et al.*, 1988).

As mentioned earlier in this chapter, *sous vide* is performed under pasteurization conditions to enhance quality but that all microorganisms are not killed off. In order to have a safe product, there has to be a balance between heating food enough to kill microbes effectively but not to heat too much to have an adverse effect on quality (Light and Walker, 1990b). A pasteurization schedule, based on a time/temperature treatment specific to a particular product, is required to ensure safety of that product. Ghazala *et al.* (1993) indicated that the time/temperature treatments should be designed so that the coldest part of the product (geometric center of the food) would obtain the designated pasteurization values to affect a certain number of decimal reductions of the target microorganism. Pasteurization value (P_v) represents the minimum number of minutes at a specific temperature required for the product's coldest point to be held and is the required number (n) of decimal reductions of the target microorganism multiplied by its decimal reduction time (D) at a specific temperature (Smith *et al.*, 1990). The temperature sensitivity of the specified microorganism is the z value. The time/

temperature data collected are used to evaluate heat penetration data, including: come up time (CUT), heating rate index (f_h), cooling rate index (f_c), heating rate lag factor (j_{ch}) and cooling rate lag factor (j_{cc}). Process times can be evaluated using different established methods based on this heat penetration data (Ghazala *et al.*, 1993).

A study by Ghazala *et al.* (1991) established *sous vide* processing schedules for two different products based on the thermal destruction

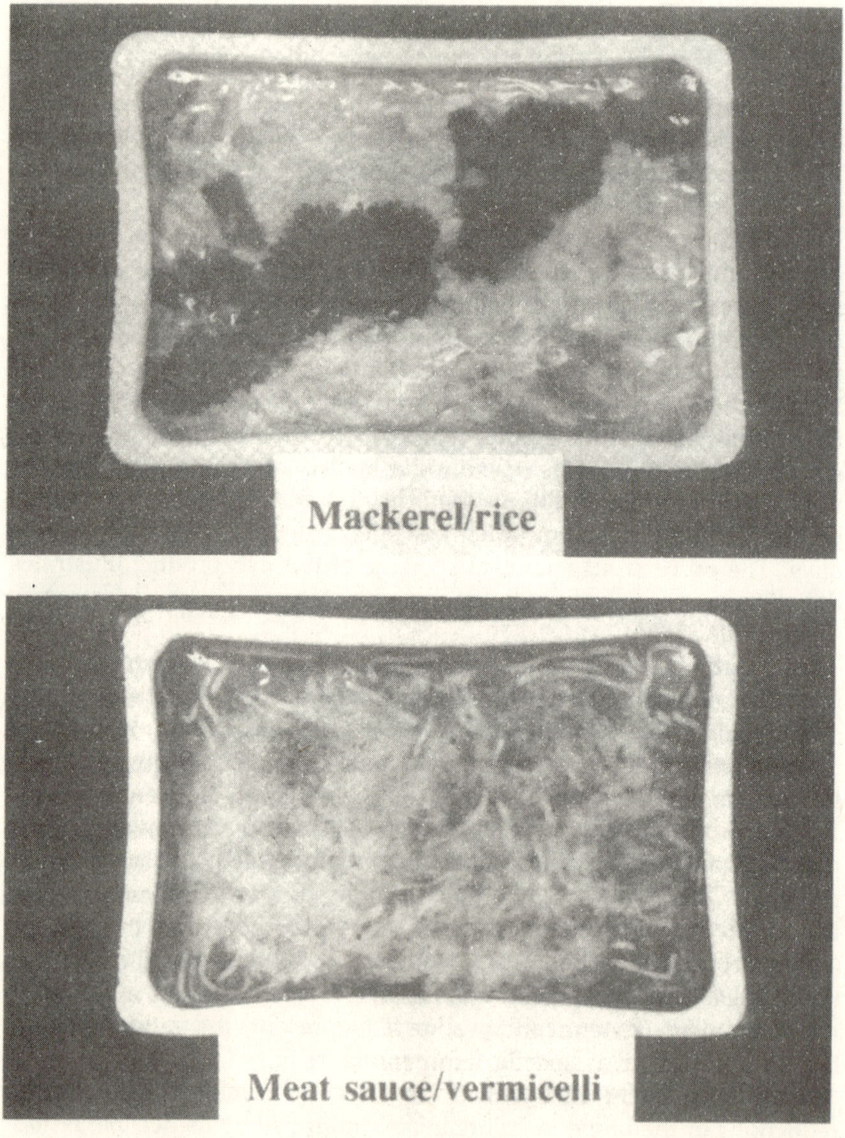

Figure 4.2 Photographs of two *sous vide* products.

Table 4.2 Comparison of process times evaluated by both the General and Stumbo's formula methods

Product	*Temperature* (°C)	P_v (*min*)	*Process time (min)* *Stumbo's method*	*General method*	*Error (%)*
Spaghetti/ meat sauce	60	235.0	289.5	295.3	−1.96
	65	71.0	104.2	105.7	−1.42
	70	15.5	49.8	50.1	−0.59
	75	7.0	39.1	39.3	−0.51
Rice/ salmon	60	312.7	347.3	351.1	−1.08
	65	72.7	109.6	110.1	−0.45
	70	18.9	56.0	56.2	−0.36
	75	7.8	42.5	42.6	−0.24

Reproduced by permission from the International Institute of Refrigeration, 177 boulevard Malesherbes, F-75017 Paris (Ghazala *et al.*, 1991).

of *Streptococcus faecium* based on a 13*D* ($n = 13$) decimal reduction. Using two different methods to analyze the heat penetration data, the P_v was determined for spaghetti/meat sauce and rice/salmon products as shown in Table 4.2. The P_v is equivalent to the minimum desired processing time at a particular temperature. The actual process time as determined by a particular method, i.e. Stumbo's method and the General method, can be adjusted to meet the minimum desired pasteurization value (P_v).

Pasteurization schedules are necessary for each *sous vide* product in order to ensure safety. As can be expected, this can be quite time consuming and costly. A computer model was developed to determine *sous vide* pasteurization schedules using a combination of thermal kinetics of *S. faecium* and the heat penetration history of the product (spaghetti/meat sauce and rice/salmon). It was found from this study that a computer model can predict and verify pasteurization schedules for different *sous vide* products (Ghazala *et al.*, 1993).

A complex process is involved in formulating a *sous vide* product. The pH, water activity, salt content, type of barrier film and details of ingredients have to be manipulated to make the product safe under vacuum (Perkins, 1990). Simpson *et al.* (1991) studied the effects of salt and pH on the thermal resistance of *S. faecium* in both nutrient broth and *sous vide* products (spaghetti/meat sauce and rice/salmon). In the study brain heart infusion broth was adjusted to pH 5, 6, 7 and 8 and the effect of salt concentration (2.5–10%) was evaluated at each pH value. The *D* and *z* values of *S. faecium* were determined at each of these conditions to determine thermal resistance. It was found that the *D* values at pH 5–6 ranged from 13 to 18 min compared with 23–24 min

Table 4.3 Effect of pH on *D* values at 60°C

pH	D *values (min)*	
	Streptococcus agar	*Enterococcus agar*
5	18.1	15.7
6	13.1	14.7
7	24.1	22.1
8	23.3	26.8

Reproduced by permission from the International Institute of Refrigeration, 177 boulevard Malesherbes, F-75017 Paris (Simpson *et al.*, 1991).

at pH 7–8, indicating that *S. faecium* experiences an increased thermal resistance at a neutral or a slightly alkaline pH (Table 4.3). High salt concentrations can also exert a protective effect on thermal resistance as shown in Table 4.4. The *D* values at pH 5 and 7 increased in the 7.5–10% salt concentration range. The combined effect of pH, salt and processing temperature can have a pronounced effect on thermal stability of microorganisms (Table 4.5). With an increase in salt content from 0.5 to 2.5% salt, the *z* value decreased from 9.5 to 8.6°C and 9.3 to 9.1°C and pH values of 7.0 and 5.0, respectively (Simpson *et al.*, 1991).

Most of the literature about *sous vide* cooking indicates that it is a remarkable new process that has much potential in the food industry. However, to date, very little research has been performed on *sous vide*. Most of the knowledge about *sous vide* has come from chefs and companies working with different products. As well, in the few scientific studies performed, very few food products were studied. The major points that have come out of these studies include:

1. Good extension of shelf-life in terms of microbiological quality is possible using *sous vide*, provided, and this is the crucial point, very high standards of raw materials, preparation, cooking, chilling and storage are adhered to.
2. In order to ensure full protection of products from aerobes during extended storage, packs should be properly evacuated. The use of inadequate vacuums to protect soft products could pose high risks of unchecked microbial growth.
3. The *sous vide* method cannot provide an entire menu range because many items do not suit this unusual production method. In addition, some products will have far better shelf-life, about 21–28 days, potential than others. However, those with a good sensory shelf-life appear to be extremely stable in terms of sensory quality (Light and Walker, 1990a).

Table 4.4 *D* values (in min) at different salt concentrations at 60°C

Treatment		*Salt*				
pH	*Agar*	*0.5%*	*2.5%*	*5.0%*	*7.5%*	*10%*
7.0	*Streptococcus*	24.1	26.1	20.0	26.0	32.6
	Enterococcus	22.1	27.6	20.0	28.3	35.3
5.0	*Streptococcus*	18.1	16.9	17.7	21.1	25.7
	Enterococcus	15.7	16.5	15.6	19.3	29.3

Reproduced by permission from the International Institute of Refrigeration, 177 boulevard Malesherbes, F-75017 Paris (Simpson *et al.*, 1991).

Table 4.5 *z* values (°C) for *S. faecium* in nutrient broth at different levels of pH and salt

Treatment	*Control (0.5% salt, pH 7.0)*	*2.5% Salt pH 7.0*	*Control (0.5% salt, pH 5.0)*	*2.5% Salt pH 5.0*
Streptococcus agar	9.5	8.6	9.3	9.1
Enterococcus agar	9.7	8.3	10.3	9.3

Reproduced by permission from International Institute of Refrigeration, 177 boulevard Malesherbes, F-75017 Paris (Simpson *et al.*, 1991).

In the light of some of above mentioned studies it would seem that, after using the appropriate techniques and complying with stringent working practices, the *sous vide* cooking method in combination with controlled chilled storage can be used successfully to preserve seafood and other food products for up to 28 days. However, there is insufficient evidence to provide conclusive proof of the system's commercial viability, for example in catering applications, particularly with regard to potential problems arising from the abuse of products in terms of temperature and time of heat processing and also chilling and chilled storage. Further, more research is needed to establish time and temperature tolerances, pasteurization levels, possible toxicological transfer of substances from the packing material during the slow pasteurization process and microbiological hazards.

4.7 CONCLUSIONS

There is a growing demand by consumers worldwide for high-quality foods particularly high-quality convenience meals that can be stored in the refrigerator for an extended period of time and quickly cooked in a microwave. Seafood is of great interest to people today because of its high nutritional content and health benefits. Previous technologies for preserving fish do little to enhance these qualities. New technologies that employ a packaging material coupled with more traditional technologies of modified or controlled gas atmosphere packaging (MAP/CAP) are being implemented to extend the shelf-life of fresh seafood and products that contain them. Further development of these useful technologies should be supported by scientific studies of the microbiological aspects. More studies are needed to evaluate gas composition (i.e. CO_2, O_2 and N_2 level), a_w, pH, temperature and permeability of packaging film/container on the shelf-life of vacuum packaged and MAP/CAP products. It will also be beneficial to determine how these parameters act as hurdles to spoilage microorganisms. Active packaging is an interesting packaging concept and many systems are being incorporated into the packaging material. These systems will help to extend the high-quality life of fresh and extended shelf-life foods and maintain the nutritional content.

It appears that, after using the appropriate techniques and complying with stringent working practices, the *sous vide* cooking method in combination with controlled chilled storage can be used successfully to extend seafood shelf-lives. As indicated in this chapter, initial studies using seafood have been encouraging, and with further research packaging under *sous vide* conditions will be common. Nevertheless, research is needed to establish time and temperature tolerances, pasteurization levels, and possible microbiological hazards associated with the slow pasteurization process of *sous vide* products.

ACKNOWLEDGMENTS

The author wishes to thank Ms Hilary Steele and Dr Thamir Al-Kanani for their help with the chapter, and her research team at McGill University and Memorial University of Newfoundland.

REFERENCES

Farber, J.M. (1991) Microbiological aspects of modified-atmosphere packaging technology – a review. *J. Food Protect.*, **54**, 58–70.

Farber, J.M., Warburton, D.W., Gour, L. and Milling M. (1990) Microbiological quality of foods packaged under modified atmospheres. *Food Microbiol.*, **7**, 327–34.

Forshaw, T. (1990) Prepared meals – choices for packaging in plastics. *Food Rev.*, **17**, 23–25.

Fujita, T. (1990) Seafood products, in *Food Packaging*, (ed. T. Kadoya), Academic Press, New York, pp. 335–48.

Ghazala, S., Ramaswamy, H.S., Smith, J.P. *et al.* (1991) Establishing *sous vide* processing schedules for spaghetti/meat sauce and rice/salmon based on thermal destruction of *Streptococcus faecium*. Paper no. 354 presented at the *XVIII*th *International Congress of Refrigeration*, Montréal, Québec, 10–17 August.

Ghazala, S., Ramaswamy, H.S., Smith, J.P. and Simpson, M.V. (1993) Thermal process simulations for *sous vide* processing of fish and meat foods. *Food Res. Int.*, submitted.

Hirsch, A. (1991) Which would be the most suitable package for seafood?, in *Flexible Food Packaging: Questions and Answers*, Van Nostrand Reinhold, New York, pp. 134–7.

Hobbs, G. (1987) Microbiology of fish, in *Essays in Agricultural and Food Microbiology*, (eds J.R. Norris and G.L. Pettipher), John Wiley, New York, pp. 199–226.

Hotchkiss, J.M. (1988) Experimental approaches to determining the safety of food packaged in modified atmospheres. *Food Technol.*, **42**, 55–64.

Jay, J.M. (1986) Spoilage of miscellaneous foods, in *Modern Food Microbiology*, 4th edn, (ed. J.M. Jay), Van Nostrand Reinhold, New York, pp. 234–48.

Jones, M.V. (1989) Modified atmospheres, in *Mechanisms of Action of Food Preservation Procedures*, (ed. G.W. Gould), Elsevier Applied Science, London, pp. 247–84.

Kraft, A.A. (1992) Spoilage of eggs and fish, in *Psychrotrophic Bacteria in Foods: Disease and Spoilage*, CRC Press, Boca Raton, FL, pp. 99–111.

Labuza, T.P. (1990) Active food packaging technologies. *Food Sci. Technol. Today*, **4**, 53–6.

Labuza, T.P., Fu, B. and Taoukis, P.S. (1992) Prediction for shelf-life and safety of minimally processed CAP/MAP chilled foods: a review. *J. Food Protect.*, **55**, 741–50.

Light, N. and Walker, A. (1990a) Food preservation: the microbiological quality of cook–chill foods, in *Cook-chill Catering: Technology and Management*, Elsevier Applied Science, New York, pp. 43–67.

Light, N. and Walker, A. (1990b) Cooking in vacuumized containers–*sous-vide* cuisine, in *Cook–Chill Catering: Technology and Management*, Elsevier Applied Science, New York, pp. 157–77.

Light, N., Hudson, P., Williams, R. *et al.* (1988) A pilot study on the use of vacuum cooking as a production system for high quality foods in catering and retail. *Int. J. Hospitality Management*, 7, 21–7.

Lingle, R. (1992) Why *sous-vide* makes sense. *Prepared Foods*, **18**, 98–100.

Liston, J. (1980) Fish and shellfish and their products, in *Microbial Ecology of Foods*, Food Commodities International, International Commission on the Microbiological specifications of food, Academic Press, New York, Vol. 2.

Morris, C.E. (1989) Convenience foods: the second revolution. *Food Engineering*, **September**, 69–82.

Muller, S. (1991) Packaging requirements of food. Presented at a *Food Packaging Workshop* sponsored by the Canadian Institute of Food Science and Technology, Montreal, Quebec, 16 June.

O'Beirne, D. (1990) Chilling combined with modified atmosphere packaging, in *Processing and Quality of Foods, Vol. 3, Chilled foods: The Revolution in Freshness*, (ed. P. Zeuthen, J. C. Cheftel, C. Eriksson *et al.*), Elsevier Applied Science, New York, pp. 3.190–203.

O'Donnell, C. (1991) MAP maps future. *Prepared Foods*, **September**, 58–9.

Ooraikul, B. and Stiles, M.E. (1991) Introduction: review of the development of modified atmosphere packaging, in *Modified Atmosphere Packaging of Food*, (eds B. Ooraikul and M.E. Stiles), Ellis Horwood Limited, Chichester, pp. 1–17.

Palumbo, S.A. (1986) Is refrigeration enough to restrain foodborne pathogens. *J. Food Protec*, **49**, 1003–9.

Perkins, C. (1990) *Sous vide*. *Seafood Business*, **January/February**, 108–18.

Piergiovanni, L. (1991) Food preservation and packaging, in *Eurofood Trends and Technology*, (ed. C. Lake), IBC Business Publishing, London, pp. 116–19.

Post, L.S., Lee, D., Solberg, M., Furgang, D., Specchio, J. and Graham, C. (1985) Development of botulinal toxin and sensory deterioration during storage of vacuum- and modified atmosphere packaged fish fillets. *J. Food Sci.*, **50**, 990–6.

Rhodehamel, E.J. (1992) FDA's concerns with *sous vide* processing. *Food Technol.*, **46**, 73–6.

Shaw, B.G. and Shewan, J.M. (1968) Psychrophilic spoilage bacteria of fish. *J. Appl. Bacteriol.*, **31**, 89.

Simpson, M.V., Smith, J.P., Ramaswamy, H.S. *et al.* (1991) Thermal kinetic studies on *Streptococcus faecium* in nutrient broth/*sous vide* products. Paper No. 353 presented at the *XVIIIth International Congress of Refrigeration*, Montréal, Québec, 10–17 August.

Smith, J.P., Toupin, C., Gagnon, B. *et al.* (1990) Hazard analysis critical control point approach (HACCP) to ensure the microbiological safety of *sous vide* processed meat/pasta product. *Food Microbiol.*, **7**, 177–98.

Whittle, K.J., Hardy, R. and Hobbs, G. (1990) Chilled fish and fishery products, in *Chilled Foods: The State of the Art*, (ed. T.R. Gormley), Elsevier Applied Science, London, pp. 87–116.

PART TWO

Fisheries Biomass Bioconversion

CHAPTER 5

Traditional fermented fish: fish sauce production

P. Saisithi

5.1 INTRODUCTION

Fish is the most important source of animal protein for the people in developing countries. Approximately 53% of the world's catch is from these countries and most of the catch is consumed within the countries. In the Southeast Asian region, people recieve 60–70% of their protein from fish.

Traditionally, preservation of fresh fish is commonly done by salting, followed by smoking and sun-drying. Salting and drying in a tropical climate can be prolonged due to high humidity and frequent rainfall. This prolonged salting allows fermentation to set in, and people gradually acquire the taste and the aroma of fermented fish. It is, at least, a change from ordinary salted fish which can become unpalatable. Fermentation is in great demand to enhance the flavor or even to mask the taste of tainted food. The demand for fermented food is so great that more varieties are sought. By partially supplementing ordinary salted fish with carbohydrate sources, such as palm sugar, roasted rice or cooked rice, traditional fermented fish products with different tastes and aromas are obtained.

Unlike highly concentrated and purified food products which are often too expensive for the people in developing countries, traditional fermented fish is comparatively cheap and well-known to them. Nutritionally, traditional fermented fish may not be as good as fresh fish, but fresh fish is highly perishable. If hygienic practice is not strictly observed, fresh fish can be highly hazardous to health. Traditional fermented fish is formed by biochemical changes similar to natural

Fisheries Processing: Biotechnological applications.
Edited by A.M. Martin.
Published in 1994 by Chapman & Hall, London. ISBN 0 412 58460 3

processes. The food value of fermented fish may be altered from the fresh fish but in many aspects it benefits the consumers. The advantages of traditional fermented fish products to the people in developing countries are: high acceptability, low cost, ease of preparation, safety, and improved digestability and absorbability. Since most traditional fermented fish is considered as a product of lactic fermentation, it also has therapeutic and prophylactic effects against certain types of gastrointestinal infection.

Protein calorie malnutrition is of concern to public health authorities in many developing countries. It has been estimated that several tens of millions of children of preschool age suffer from this condition to varying degrees (Anonymous, 1974). Traditional fermented fish is probably good food for patients with malabsorption. The reason for better absorption is the particle size effect. In general, in fermented foods the particle size is much smaller than in fresh food. The importance of the nutritive value of fermented fish products in the daily diet of the Southeast Asian population has been discussed to a certain extent by different investigators (Amano, 1962; van Veen, 1965; Areekul *et al.*, 1974: Garby and Areekul, 1974; Suwanik, 1979. Anonymous, 1973; Saisithi, 1981). Currently, more attention has been given to improving traditional fermented fish products in order to increase the availability of low cost protein food for local consumption.

5.2 TYPES OF TRADITIONAL FERMENTED FISH PRODUCTS

Due to favorable climatic conditions, there are several indigenous fermented foods in Asia. The importance of the products to the Asian people is so much that, in 1981, The Association for Science Cooperation in Asia (ASCA) chose the topic 'Traditional Food Fermentation as Industrial Resources in ASCA countries' for the 8th conference. In 1986, a concise handbook of Indigenous Fermented Foods in the ASCA Countries was published, which included a few traditional fermented fish products. Traditional fermented fish products are indigenous to Southeast Asian Countries such as Laos, Kampuchea, Vietnam, Thailand, Mayanmar, Malaysia, Indonesia and Philippines. In Thailand alone there are at least 16 traditional fermented fish products currently utilized (Sundhakul *et al.*, 1975). There are probably similar amounts in the other countries; however, there can be more varieties than appears because the process described as fermentation may actually form a series of well-defined products (Adam, Cook and Rattagool, 1985). Therefore, many attempts have been made to group fermented fish products according to presumed common characteristics. Rao (1967), for example, simply divided traditional fermented fish into three groups according to the final product appearance as follows:

1. Fermented fish in which the fish still retains its original form when the fermentation was started.
2. Fermented fish in which the fish is macerated into paste-like products.
3. Fermented fish from which the fish is completely hydrolyzed into liquid form.

Since Rao's classification is based on product appearance rather than processing methods, fermented fish products of the same group may have different types of fermentation. Amano (1962) divided fermented fish products into three categories according to processing methods as follows:

1. Traditional products mainly fermented by the action of enzymes normally present in fish flesh and entrails to which salt has been added.
2. Traditional products fermented by the combined effects of fish enzymes supplemented with microbial enzymes supplied in the form of starter cultures on fish flesh and entrails with added salt.
3. Non-traditional products manufactured by accelerated fermentation, acid ensilage and chemical hydrolysis.

According to Amano's classification most traditional fermented fish products fall into the first group. Only one or possibly two traditional products fall into the second category. Hence, the classification using only the sources of enzymes as the method to divide the fermented products is impractical for traditional fermented fish products.

Recently, Adam, Cook and Rattagool (1985) divided the traditional fermented fish products of Southeast Asia according to the substrates used in the fermentation processes as follows:

1. Products made from fish and salt.
2. Products made from fish, salt and carbohydrate.

Here again, considering the added ingredients as substrates does not represent a true classification of traditional fermented fish products. The addition of carbohydrates to a mixture of fish and salt is usually intended for different purposes. For example, roasted rice, which is added in the late fermentation period, not only provides flavor to the products but also absorbs excess moisture and prevents the fermented fish from sticking together. The roasted rice is sometimes replaced by rice bran. Sugar is often used to provide an energy source for lactic acid bacteria; thus hastening the bacterial growth rate. Cooked rice is sometimes used instead of sugar. In other cases cooked rice is used as a substrate for yeasts and molds. Saisithi (1987), therefore, proposed to classify traditional fermented fish products, based on both the type of substrate used and the source of the enzymes used in the fermentation process, into three groups as follows:

1. Traditional fermented fish in which the fish is fermented by the combined action of fish enzymes and bacterial enzymes normally present in the fish/salt mixture.
2. Traditional fermented fish in which the fish and carbohydrate are fermented mostly by bacterial enzymes normally present in the fish/salt/carbohydrate mixture.
3. Traditional fermented fish in which the fish is fermented mostly by fish tissue enzymes and the carbohydrate is fermented by yeasts and molds added in the form of a starter culture.

It should be noted here that the traditional fermented fish products mentioned by Saisithi (1987) were based on research performed on Thai traditional fermented fish products. The fermentation process of the products appears to involve the lactic acid bacteria in all three groups. The lactic acid bacteria used are not the same species for all products, they depend on the different types of carbohydrates being used, the amount of salt added and whether or not a starter culture is added. In order to give the reader a better understanding of how traditional fermented fish products are divided, each group of products is discussed separately in the following sections.

5.2.1 Traditional fermented fish products – Group I

The varieties of fermented fish products in this group consist of the two most important and well-known products of Southeast Asia, i.e. fish sauce and fish paste (Table 5.1).

Both fish sauce and fish paste are prepared from whole fish. Fish is the only available substrate in lactic acid fermentation. The addition of salt to fish reduces the water activity (a_w) and thus keeps the fish from spoilage microorganisms. The enzymes for the fermentation process come partly from the fish digestive system and partly from the bacteria

Table 5.1 Fish sauce and fish paste of Southeast Asia

Countries	*Local names*	
	fish sauce	*fish paste*
Burma	Ngapi	Ngapi
Indonesia	Ketjab-ikan	Trassi
Kampuchea	Nuoc-mam-gau-ca	Prahoc
Laos	Nam-pla	Pradec
Malaysia	Budu	Balachan
Philippines	Patis	Bagoong
Vietnam	Nuoc-mam	Mam-tom

naturally present in the fish and in the salt. The microorganisms which are responsible for the fermentation, therefore, have to be either halophilic or haloduric in nature. The only difference between the two products is the texture of the end products. Fermentation starts with the fish gut enzymes which gradually break down the large molecules of fish tissues into smaller ones. These tissue hydrolysis products are, in turn, used as substrates for lactic acid bacteria. Complete hydrolysis of fish tissues produces a liquid form known as fish sauce. In producing the fish paste, only partial fish gut enzyme hydrolysis is allowed. In doing so, most of the excess liquid formed during the salting process is drained off. This not only limits the fish enyzmes' activities but also the necessary water activity for bacterial growth. Full hydrolysis of fish tissues is further prevented by partial drying of the fish substrate. The end product, in contrast to fish sauce, remains in the solid form.

Another type of traditional fermented fish product which should belong to this group is Pla-ra (in Thai). This product exists in many countries of the Southeast Asian region (Table 5.2). The product is often placed in different groups from fish sauce and fish paste due to the fact that carbohydrate is added. Research shows that the carbohydrate added during the preparation of the product is not actually utilized as a substrate in the fermentation process. Lean fresh water fish must be used because fatty fish can spoil the product due to rancidity. During the preparation of the raw material, the fish is gutted to remove the gut enzymes, and scaled and kneaded to squeeze out fat from the fish tissues. The fish is then washed and drained before salting. The salted fish is packed tightly into a container. During this time protein hydrolysis and fermentation begin; however, without the fish gut enzymes, hydrolysis of fish tissues is far from completed. Before the fermented fish is fully mature, it is taken out of the container and ground roasted rice (or rice bran) is mixed in with the fish. The mixture of fermented

Table 5.2 Traditional fermented fish products of Group I (in which the fish retain their original form)

Countries	*Common names*	*Ingredients*	*References*
Indonesia	Wadi	fish/salt/roasted rice	Rao (1967)
Kampuchea	Mam-ca-sat	fish/salt/roasted rice	Amano (1962); Rao (1967)
Laos	Padec	fish/salt/roasted rice	Rao (1967)
Malaysia	Pekasam	fish/salt/roasted rice	Yeoh, Aziz and Adnan (1985)
Thailand	Pla-ra	fish/salt/roasted rice	Hongtongdaeng (1979)
Vietnam	Mam	fish/salt/roasted rice	Rao (1967)

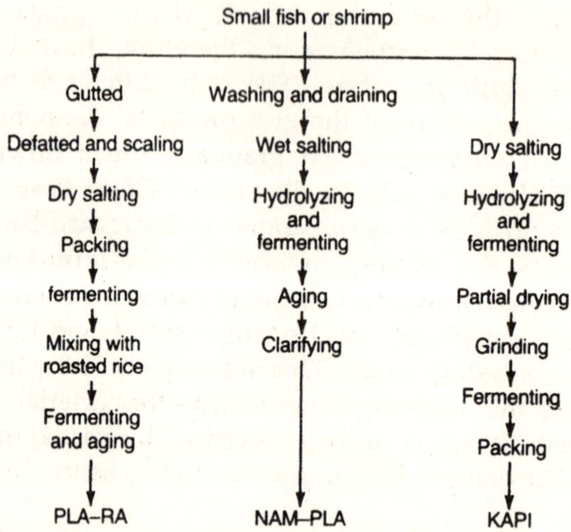

Figure 5.1 Processing methods of traditional fermented fish products – Group I.

fish and roasted rice is repacked for further aging. The main function of the roasted rice is to absorb the excess moisture produced during the aging period and to prevent the fish from sticking to each other. Rice bran is often used in some areas instead of roasted rice. The processing techniques of the traditional fermented fish products of group I can be summarized as shown in Figure 5.1. The only lactic acid bacterium responsible for the fermented fish products of this group is *Pediococcus halophilus*. The sole presence of *Pediococcus halophilus* is probably due to the high salt concentration in the products of Group I.

5.2.2 Traditional fermented fish products – Group II

There are more varieties of the fermented fish products in this group than in the previous one. However, the total production is much less and the consumption is strictly localized, as compare with the products of group I. Fish used in the preparation of the products can be either marine or fresh water fish. The fish can also be prepared in different forms, such as whole dressed fish, pieces and minced fish. Carbohydrates are usually added in the form of cooked rice, although palm sugar is sometimes used. The processing procedures of the traditional fermented fish in this group are summarized in Figure 5.2.

The fermented products of this group are prepared from either dressed fish or whole fish. The fish is dressed by removing the gut, the scales and the gills. If the fish is large, both sides of the body are slit to facilitate the salting process. The ratio of salt to fish is approximately 1:3 or 1:4. Salting and fermenting times vary between 1 day (for Filippino

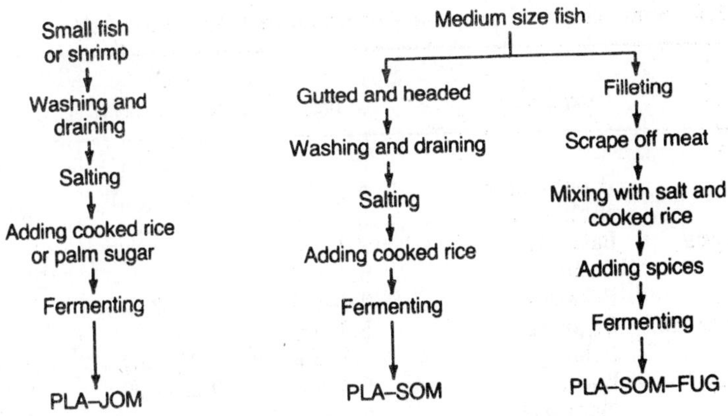

Figure 5.2 Processing methods of traditional fermented fish products – Group II

Burong-dalag) to 2–3 days (for Indonesian Makassar and Thai Pla-som), with the exception of Japanese funo-sushi which takes as long as 2 months. After preliminary salting and fermentation, cooked rice is mixed in and fermentation is allowed to take place again. The rate of the second fermentation is much higher with added carbohydrate as the substrate. The fermentation takes only 2–3 days. Coloring matter, such as Angkak (red rice), is sometimes added with the cooked rice. The traditional fermented fish products with a red color are Thai Pla-paeng-daeng, Filipino Bulong-dalag and Indonesian Makassar. Table 5.3 shows a few of the fermented fish products in Group II.

Palm sugar is often used when whole small shrimp or fish are used as the protein substrates. The products, in this case, have the appearance of paste-like products.

Table 5.3 Traditional fermented fish products of Group II in Southeast Asia

Countries	*Local names*	*Ingredients*	*References*
Kampuchea	Paak or Mam-chao	fish/salt/cooked rice	Amano (1962)
Indonesia	Makassar	fish/salt/boiled rice/Angkak/spice	Rao (1967)
Japan	Fung-sushi	fish/salt/cooked rice	Matsushita (1937)
Malaysia	Cincalok	acetes/salt/ cooked rice	Yeoh, Aziz and Adnan (1985)
Philippines	Bulong-dalag	fish/salt/cooked rice/angkak	Orillo and Paterson (1968)
Thailand	Pla-paeng-daeng	fish/salt/cooked rice/angkak	Neetiwatanapong (1979)
Thailand	Kung-som	shrimp/salt/palm sugar or cooked rice	Boonkohsrang (1973)

Table 5.4 Some chemical properties of traditional fermented fish

Countries	*Local names*	*pH*	*Acidity (% lactic acid)*	*Salt (%)*
Group I				
Japan	Ika-shiokara	6.0–7.0	–	8.0–15.0
	shottsuru	5.5–6.0	–	27.0–30.0
Malaysia	Balachan	7.2–7.8	–	20.0–25.0
	Budu	5.4–6.2	–	21.7–28.2
	Pekasam	4.1–5.4	1.4–2.6	9.8–10.5
Philippines	Bagoong	5.5–6.0	–	13.0–25.3
	Patis	5.5–5.9	1.02–1.04	20.0–25.0
Thailand	Budu	5.3–6.0	0.22–1.29	14.9–27.5
	Nam-pla	6.5–6.7	–	20.5–27.4
	Pla-ra	4.7–6.2	0.37–3.15	7.8–17.9
	Tai-pla	5.2–6.2	1.01–2.34	25.0
Group II				
Malaysia	Cincalok	4.8–5.9	up to 2.8	7.0–17.0
Philippines	Balao-balao	3.6–3.8	1.8 –2.00	–
	Burong-isda	3.5–3.8	–	1.33
Thailand	Pla-paeng-daeng	3.9–5.2	1.42–2.10	4.5–9.2
	Pla-som	4.0–4.6	2.12–4.01	2.25–5.9
	Som-fug	4.1–5.0	1.08–2.83	2.5–4.8

The main characteristics of the products in this group are:

1. Carbohydrate is the principle substrate.
2. The acid thus produced is the main preservative.

This is in contrast to the products in Group I, where the fish is the main substrate and the preservative action is the high salt concentration. However, the microorganisms that are responsible for the fermentation of the products in both groups are naturally occurring microorganisms. The chemical and bacteriological properties of the products in Group I compared with Group II are shown in Tables 5.4 and 5.5.

5.2.3 Traditional fermented fish products – Group III

The fermented fish products in this group are similar to the products of Group II, as far as the nature of substrates are concerned. The only difference is that the causative microorganisms are added as a starter culture. However, naturally occurring microorganisms are not inhibited. The product in this group is found probably only in Thailand and it is called *Pla-jao*. Pla-jao is the result of two types of fermentation. Salted fish is fermented by the naturally occurring microorganisms found with the fish and/or microorganisms present in the salt. The added steamed

Table 5.5 Microorganisms found in products groups I and II

Countries	*Local names*	*Microorganisms*
Group I		
Japan	Shottsuru	*Pediococcus homeri*, *Halobacterium* spp.
	Ika-shiokara	*Micrococcus* spp., *Staphylococcus* spp., *Debaryomyces* spp.
Philippines	Bagoong	*Pediococcus* spp., *Bacillus* spp.
	Patis	*P. halophilus*, *Micrococcus* spp., *Halobacterium* spp., *Halococcus* spp., *Bacillus* spp.
Thailand	Budu and Nampla	*P. halophilus*, *Staphylococcus* spp., *Halobacterium* spp., *Halococcus* spp.
	Pla-ra	*Pediococcus* spp., *P. halophilus*, *Staphylococcus* spp.
	Tai-pla	*Pediococcus* spp., *P. halophilus*
Group II		
Philippines	Balao-balao	*Leuconostoc* spp., *Lactobacillus* spp., *Pediococcus* spp., *Streptococcus* spp.
	Burong-isda	*Leuconostoc* spp., *Lactobacillus* spp., *Streptococcus* spp., *Micrococcus* spp.
Thailand	Pla-paeng-daeng	*Pediococcus* spp., *Staphylococcus* spp., *Micrococcus* spp., *P. halophilus*
	Pla-som and Som-fug	*Pediococcus* spp., *Lactobacilus* spp., *Staphylococcus* spp.
Thailand	Pla-chom and Kung-chom	*Pediococcus* spp., *P. halophilus*, *Lactobacilus* spp., *staphylococcus* spp.

rice, however, is fermented by the starter microorganisms. The traditional processing method is shown in Figure 5.3.

Medium to large size fresh water fish are commonly used for preparing the product. The dressing is done by scaling, evicerating and cutting off the head of the fish. The dressed fish is slit transversely on both sides to facilitate salt penetration. After the fish has been salted overnight, it is mixed with cooked glutinous rice previously inoculated with the starter culture. The mixture is packed tightly into a container and fermentation is allowed to take place for approximately 15 days. The fermented product is slightly sweet and needs to be cooked before being eaten. The chemical properties of the product are: pH, 4.0–5.3; acidity, 1.08–3.80% (calculated as percent of lactic acid); and salt, 4.4–9.5%. The microbiogical make-up consists of *Pediococcus cerevisiae*, *Staphylococcus* spp., *Micrococcus* pp., *Bacillus* spp., *Saccharomyces* spp. and *Endomycopsis* spp. The last two species are obviously from the starter culture.

In conclusion, the traditional fermented fish products of Southeast Asian countries consist of products of different forms, textures and flavor. They can be used as condiments in cooking, or as appetizers or

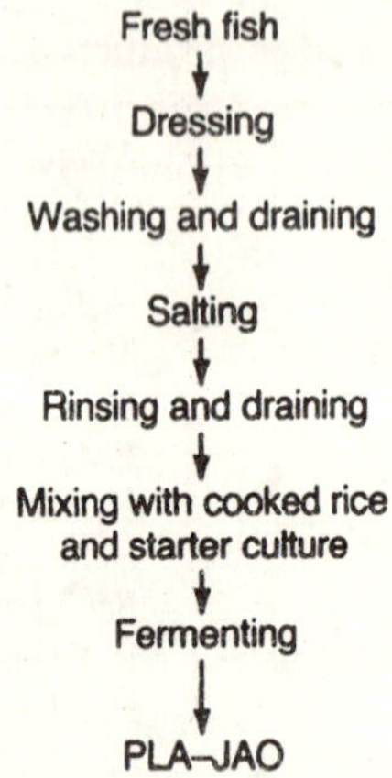

Figure 5.3 Processing methods of traditional fermented fish products – Group III.

main dishes. However, low salted products have a tendency to be hazardous to health due to poor sanitary practice. These products are, therefore, gradually disappearing from local markets. However, high salted products, especially products of Group I, have become internationally recognized (see Tables 5.5 and 5.6). Fish sauce, for example, is now fully industrialized and is an export commodity of

Table 5.6 Countries which produce fish sauce, local names and species of fish used for production

Countries	*Local names*	*Species of fish used*
Kampuchea	Tuuk-prahoc	no record
Indonesia	Ketjab-ikan	*Clupea* spp., *Stolephorus* spp., *Puntius* spp., *Osteochilus* spp., *Ctenops* spp.
Japan	Uwo-shoyu or Qunaga	*Loligo bleekeri*, *Katsuwonus pelamys*
Laos	no record	no record
Malaysia	Budu	no record
Philippines	Patis	*Stolephorus commersonii*, *S. indicus*, *Clupeoides lile*, *Sardinella perforata*, *S. frimbriata*, *S. longiceps*, *Decapterus macrosoma*
Thailand	Nampla and Budu	*Stolephorus* spp., *Rastrelliger* spp., *Clupea* spp., *Sardinella* spp., *Cirrhinus* spp.
Vietnam	Nuoc-mam	*Stolephorus* spp., *Clupeoides lile*, *Eagraulis mystax*, *Decapterus* spp.

many countries, including Thailand. Hence, the background, traditional processing methods, technological improvements and the status of fish sauce production are considered separately in the following sections.

5.3 FISH SAUCE PRODUCTION

Of all of the traditional fermented fish products, fish sauce is the most well-known product. It is not well-known only within the Southeast Asian region but also in Europe and North America. Fish sauce is produced from many countries of Southeast Asian region, as shown in Table 5.6. At the present, Thailand is the largest producer in this region. There are approximately 390 fish sauce factories in Thailand, which utilize about 64 000 metric tons of fish annually. There are many species of small fish, both marine and fresh water, that can be used as the raw material for fish sauce production (see also Table 5.6).

Fish sauce is used mainly as a condiment. The quality of fish sauce, therefore, is judged mainly by the bouquet of the sauce. However, fish sauce also provides a substantial amount of daily nitrogen intake. According to Amano (1962), fish sauce could supply as much as 7.5% of an individual's total nitrogen intake. The amount of organic nitrogen in fish sauce from different countries is shown in Table 5.7.

Table 5.7 Comparison of organic nitrogen contents in different fish sauces

Fish sauce	*Total nitrogen (%)*	*Amino nitrogen (g %)*	*Volatile nitrogen (g %)*	*Formol (g %)*	*Peptide (g %)*
Nam-pla	1.96	ND	0.21	ND	ND
Budu	1.77	1.17	0.12	1.85	0.48
Nuoc-man	1.90	1.05	0.40	1.45	0.45
Patis	0.02	0.98	0.18	0.17	0.84

ND, not determined.

Apart from organic nitrogen there are also vitamins and minerals in fish sauce. For example, vitamin B_{12} is found in large quantities (Amano, 1962) If a person consumes 25 ml of fish sauce daily, he or she would receive approximately 0.08–1.46 μg of vitamin B_{12} (Areekul *et al.*, 1974). This amount of vitamin B_{12} would be sufficient as a protective measure against megaloblastic anemia. The amounts of minerals found in fish sauce are: phosphorus, 0.266–0.566 g/l; calcium, 0.439–0.541 g/l; magnesium, 2208–2310 mg/l; organic sulfur, 0.54–1.16 g/l; and iron, 10–22 mg/l (Van Veen, 1965; Garby and Areekul, 1974; Suwanik, 1977).

5.4 PROCESSING OF FISH SAUCE

Traditional fish sauce fermentation is actually a combination of salting, enzyme hydrolysis and bacterial fermentation. Fish sauce processing was probably first recorded by Rosé (1918a,b) in Vietnam. Nuoc-mam (Vietnamese Fish Sauce) was formerly exported to Thailand and Malays (Gruvel, 1925), and there is no doubt that Vietnam was the largest producer of fish sauce of that time. Therefore, it is worth while to look into the Vietnamese method of fish sauce manufacture. Van Veen (1953) described the fish sauce production in Vietnam as

> The fresh but uncleaned fish are mixed with salt and dumped into the vats. The amount of salt varies according to the kind of fish; usually four to five parts of salt to six parts of fish are used (Rose, 1918a). For some species of fish proportionally more salt is added to the upper layer in the vat. The fish are piled above the top of the vat. A final layer of salt is placed on the pile of fish. After 3 days, the collected liquid (called nuoc-bai) which is turbid and bloody in appearance, is drained off. A part of it is returned to the vat; another portion is saved and used later. On exposure of the liquid to the air, the red color soon vanishes; the liquid also becomes somewhat clearer (Guillerm, 1930). Meanwhile the fish have settled below the top of the vat and the layer of the salt has almost disappeared. The fish are now packed down thoroughly. Some of the clear liquid is then poured back over the fish until they are covered to a depth of about 10 cm. The contents are then covered with large bamboo wicker work of split bamboo on top of which weights are placed.

Traditionally, fish sauce processing in Thailand is similar to Vietnam. Fish and salt ratios can be varied from 2:1 for marine fish to 3:1 for fresh water fish. Fresh and uncleaned fish is dumped on a concrete floor and the excess liquid allowed to drain off. One man throws salt on to the fish while another man mixes the fish and salt together. It is very important that the mixing is done thoroughly. The fish/salt mixture is then placed into fermentation tanks (or vats) which have already had salt spread on the bottom. The fish/salt mixture does not pile above the top of the tank but instead is placed about 50 cm below the edge of the tank. Another layer of salt is spread on top of the fish. After 1 week, the fish may float on top of the natural brine thus formed and rise to the top of the fermentation tank. They are weighted down until all of the fish is submersed under the liquid. It is necessary to keep the fish submersed at all times, otherwise they will become rancid. While the salting process is going on, the hydrolysis of fish tisues by fish gut enzymes provides the necessary nutrients for bacterial fermentation to begin. The length of time that is needed for the full flavor of fish sauce to develop varies

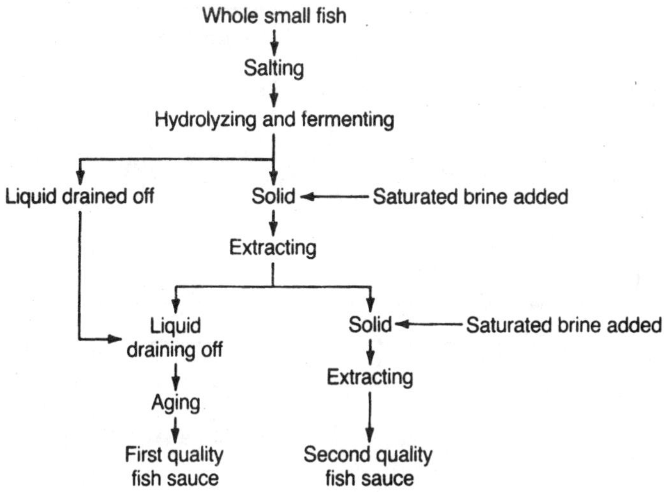

Figure 5.4 Sequential operation of Thai fish sauce.

from 8 to 18 months. After fermentation is completed, the liquid is drained off and saturated brine is added to the residue to extract the left-over soluble matters. Traditionally, the first and the second liquids are combined to produce first quality fish sauce. The procedures for making Thai traditional fish sauce are shown in Figure 5.4.

Salting techniques have been improved. Research reveals that salt is used only to prevent putrefactive microorganisms from spoiling the fish and the maximum amount to prevent the growth of putrefactive cocci is 15% salt. It is, therefore, recommended that the amount of salt used for the production of fish sauce should be between 18 and 20%. A salt content of more than 15% also inhibits the proteolytic activity of each of the alkaline and acid proteinases from the pyloric caeca and stomach of sardines. However, minimum inhibition was observed when the fish muscle protein was used as substrate (Noda *et al.*, 1981). It is also established that proteolysis of fish protein comes mainly from the fish gut enzymes (Chevey, 1931–2). The more enzymes present in the gut, the quicker the fish tissues become liquefied. Usually fish will have the highest amount of gut enzymes during the feeding season. It is also found that the tissue protein of active swimming fish will be hydrolyzed faster than that of slow moving fish (Fuji, Hirose and Era, 1951). To use too high salt concentration, therefore, could cause unnecessary delay in the fermentation of the fish sauce.

Salting by hand is being replaced by the machines. The types of machines often used in fish sauce factories are either of cascade type or drum mixer type (Figures 5.5 and 5.6). Salting by hand can be very slow due to the small amount of fish that can be salted at any given time in order to make sure that all fish is salted uniformly. Fish may be spoiled

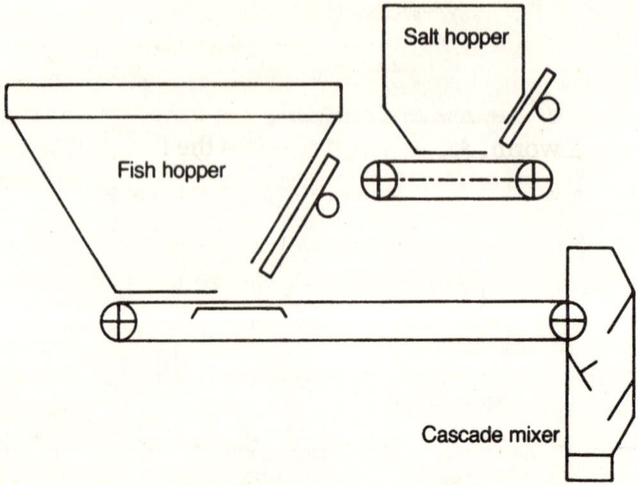

Figure 5.5 Diagram of fish-curing layout for small fish.

before salting is complete. The autolysis of fish tissues begins almost immediately after the fish is dead and bacterial spoilage will soon follow. The salting process, therefore, should be completed while the fish is still fresh. Salting and mixing machines can speed up the process.

The cascade-type mixer is illustrated in Figure 5.5. The small fish fall from the hopper and settle in a neat layer on a belt conveyor where they are spread with salt supplied by a second conveyor belt from the salt hopper. The thickness of the layers of fish and salt is controlled by the width of the slits at the outlets of the hoppers. The salt-strewn fish fall

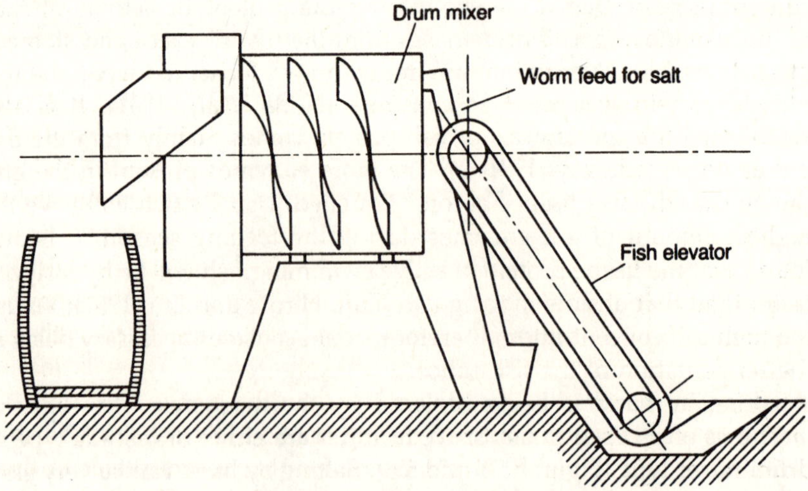

Figure 5.6 Diagram of herring curing installation

from the belt of the lower conveyor into the cascade mixer, and are then emptied into the tub and then to the fermentation tank.

The salt-curing machine illustrated in Figure 5.6 consists of a revolving drum mixer. The fish are carried to the mixer by elevator and the salt is fed by a worm. As the drum revolves the fish/salt mixture passes gradually to the other end, falling into a barrel or a fermentation tank.

Another salting technique used in the Philippines is to put salt and fish in the vat in alternative layers with layers of salt at the bottom and top (Hamm and Clague, 1950). This salting technique also helps to speed up the salting process and it is especially good with large fish. Mixing fish and salt directly is best for small fish such as sardines and anchovies. Since small fish have a large surface area, salt can attach to the surface to as much as 18% in contrast to 3–4% for larger fish. Small fish, usually, are salted in bulk. When the fish/salt mixture is poured into the vat, the proportion of salt is gradually increased from the bottom upwards in such a way that the top layers of fish have about 50% more salt than the bottom ones. The topmost layer is then covered by a layer of salt 1.5–2 cm thick.

New fermentation tanks equipped with a recycling system are now under testing at commercial scales. It is claimed that this new technique will cut the production time by more than half and that fish sauce can be produced under full hygienic control.

5.5 CHANGES DURING FERMENTATION

When the fish/salt mixture is transferred to a vat, tank or barrel, it has to be ensured that the salt remains evenly distributed around the fish. The proper distribution of salt over the fish is most important, because diffusion is slow and if one small part is deprived of salt, the fish will spoil.

As soon as salt comes into contact with the fresh fish natural brine will form fairly quickly. If the salting process is slow, the natural brine can be expected to cover the fish at the end of 1–2 days. In this case, fish in the top layer will start to spoil. Therefore, saturated brine should be added to the tank until all the fish is covered by the brine. The fish sauce manufacturers often claim that they have difficulties with this, mainly because the fish float up on top of the brine – this may happen when the brine is saturated or fish belly cavities are filled with gas. When fish are mixed with salt, it is unlikely that the fish will float up because the mixture has greater density than the brine.

The natural brine thus formed during the first few days is reddish in appearance and is highly turbid. Some manufacturers prefer to draw this liquid out and replace it with saturated brine. Others leave the liquid in the tank until it becomes ripe fish sauce, which usually takes about 1 year. The liquid which is drawn off after the first 2–3 days is kept for later use. The first batch of fish sauce traditionally consists of about 40–50% of the total amount of fish and salt. Undoubtedly, this could only be caused by the incomplete hydrolysis of fish tissues, due either to too high salt concentration or too little fish gut enzymes or both. The residue left after the first extraction of fish sauce still contains a high amount of undigested meat. Therefore, second, third and fourth extractions are performed by adding equal amounts of saturated brine as in the first extraction, leaving it for 3–4 months or less and then the liquid is again drawn off. The first and the second extractions are combined and sold as first quality fish sauce.

The proper use of salt and fish with strong enzymic activities helps to cut down the fermentation time appreciably, now the fermentation of fish sauce takes approximately only 4 months. Attempts to use commercial enzymes to replace fish gut enzymes proved to be of little value (Greig and Estrella, 1988). There are also indications that, apart from the fish gut enzymes, there are also proteolytic enzymes from halophilic bacteria normally present in the salt which could take part in the protein hydrolysis of fish tissues (Suntinanalert, 1979)

The red color of the initial sauce is gradually disappears and become clearer. The solid matter settles down to the bottom of the tank. It is at this stage that the actual fermentation begins through the specific group of lactic acid bacteria. The typical aroma of fish sauce is developed. There is also a change in the color of the sauce. The more protein and lipid hydrolyzed into low molecular weight compounds, the better the quality of the fish sauce. The apparent qualities of Thai fish sauce are the clear amber color and the typical aroma. Because of the great importance of Nam-pla, (the Thai name for fish sauce) to the economy of Thailand, an official standard was set up in 1970 by The Thai Industrial Standards Institute.

5.6 MICROBIOLOGY OF FISH SAUCE

It was found previously that typical fish sauce flavor was lacking when fish sauce was produced under aseptic conditions (Krempf, 1974). Boez and Guillerm (1930) have isolated an anaerobic, spore-bearing bacteria (a *Clostridium* sp.) that is said to produce the typical flavor of Nuoc-mam. It has been known for a long time that the maturation process of salted fish also depends on the activity of the microflora. Under the influence of the microflora, carbohydrates are fermented by the lactic acid bacteria and the intestinal bacilli, and it is the substances formed that give the salted fish its pleasant flavor and slightly sour flesh. Mees (1934) has isolated and identified the bier sarcina (*Pediococcus halophilus*) as the flavor producing bacteria in salt-pickled anchovies.

Table 5.8 Changes in bacterial numbers at different stages of Patis fermentation

Fermentation time	*Bacterial numbers/g*
Fresh fish	6.5×10^6
Immediately after salting	2.9×10^6
1 week	4.6×10^4
2 weeks	5.4×10^3
3 weeks	3.2×10^2

In the Philippines, Hamm and Clague (1950) began to study the bacterial quality of Patis (the local name for Filipino fish sauce). Bacterial numbers of Patis at different stages of fermentation were investigated and are shown in Table 5.8. The reduction in bacterial numbers is apparently due to the salting (most of the spoilage bacteria originally present on fish will die off quickly after contacting with salt), leaving only the high salt tolerant cocci to grow slowly in the medium. Kasemsarn (1963), therefore, believed that *Staphylococcus* spp. was responsible for the fermentation of Nam-pla. Saisithi (1967) was the first to isolate *Pediococcus halophilus* from Nam-pla.

In contrast to Hamm and Clague's findings, Amorntiparat (1979) found that there was an increase of the total viable numbers of bacteria after the third day of Budu fermentation (Thai and Malaysian name for fish sauce). At the 13th day of fermentation the highest number of bacteria, 9.7×10^5 cells/g, was obtained. The bacteria found at this stage are mostly *Pediococcus halophilus* and *Staphylococcus* spp. *Pediococcus halophilus* becomes predominant until the end of 3 months fermentation time. Itoh *et al*. (1985) identified acid producing bacteria in Thai fish sauce as *Pediococcus halophilus, Staphylococcus saprophyticus, Micrococcus varians* and *Paracoccus halodenitrificans*. They also isolated other halophilic lactic acid bacteria which grew well on media with a high salt concentration at alkaline pH and produced lactic acid that lowered the pH to about 5.0. The bacteria were identified as *Aerococcus haloveridans* and the heterofermentative lactic acid producing bacteria were identified as *Peptococcus anaerobius*.

5.7 CHEMISTRY OF FISH SAUCE

Fish sauce is the hydrolyzed product of the fish tissues which consist mainly of protein and lipid. The degradation pathways of the lipid component during the fermentation process is not fully understood. Saisithi (1967) found that the amount of total volatile acids (TVA) increased as the fermentation time increased and reached a maximum

Table 5.9 Volatile acids composition of fish sauce

Volatile acids	*Total amount (meq/l)*	*References*
Formic, acetic, propionic and iso-butyric acids	87	Saisithi (1967)
n-butyric/acetic 1/1 to 1/20	175	Nguyen An-Cu and Vialard-Goudou (1953)
(Acetic, 210–530 mg/100 ml; propionic, 11–47 mg/100 ml; eso-butyric, 1 mg/100 ml; *n*-butyric, 6–13 mg/100 ml; iso-valeric, 3–30 mg/100 ml)	78	Howard and Dougan (1974)

at approximately 9 months of fermentation. The volatile acids found in Nam-pla and Nuoc-mam are shown in Table 5.9. These volatile acids have been established as the flavor components of the fish sauce. It is believed that these volatile acids are the by-products of lactic acid fermentation by lactic acid bacteria present in the fish sauce (Saisithi, 1967).

Proteolysis of fish tissues by fish gut enzymes (Chevy, 1931–2; Fuji, Hirose and Era, 1951) and possibly by microbial enzymes (Suntinanalert, 1979; Mittranond and Thongthai, 1984) gives rise to amino acids and small molecular peptides. Howgate (1971) found that the amount of

Table 5.10 Amino acid composition of Nam-pla

Amino acid	*Quantity (mg/100 ml)*		
	undigested	*digested*	*fish flesh*
Taurine	167	211	–
Aspartic acid	1601	2009	1800
Threonine	819	873	834
Serine	683	861	924
Glutamic acid	1448	3280	2704
Glycine	552	1041	816
Alanine	1194	1272	1136
Valine	469	974	928
Methionine	331	335	432
Isoleucine	372	379	800
Leucine	434	431	1472
Tyrosine	99	63	656
Phenylalanine	519	541	752
Histidine	707	649	576
Lysine	1767	1982	1692
Arginine	4	34	1120
Citrulline	1300	894	–

amino acids during the fermentation of Nam-pla increased gradually from 2937 to 12 466 mg/100 ml at the end of 12 months of fermentation. The amino acid composition of Nam-pla is shown in Table 5.10. The total amino acid content in fish sauce is approximately the same as that found in the protein of fish tissues. Taurine, which is considered as a non-protein component but is found in Nam-pla, could be derived from the amino acids cysteine or cystine (Meister, 1965). This would account for the disappearance of cysteine and cystine from the fish sauce. The same is true for citrulline, which is a derivative of arginine. These amino acids, and others such as tyrosine, tryptophan, proline, OH-proline and gamma-amino butyric acid, are metabolized by the bacteria (Bremstedt, 1962), whereas the disappearance of some amino acids is due to non-enzymic reactions (van Klaveren and Legendre, 1965). Saisithi (1967) believed that the color of fish sauce resulted from the non-enzymic browning reaction between the amino acids in the fish sauce and the ribose – a product of post-mortem degradation of ATP (Partmann, 1965; Tarr, 1965).

REFERENCES

Adams, M.R., Cook, R.D. and Rattagool, P. (1985) Fermented fish products of Southeast Asia. *Trop. Sci.*, **25**, 61–73.

Amanoc, K. (1962) The influence of fermentation on the nutritive value of fish with special reference to fermented fish products of Southeast Asia, in *FAO International Symposium on Fish in Nutrition*, (eds E. Heen and R. Krauzer), Fishing News (Books) Ltd, London, pp. 180–200.

Amorntiparat, M. (1979) A microbiological study on the Thai traditional fermented food products: Budu. *MS Thesis*, Kasetsart University Bangkok, Thailand (in Thai).

Anonymous (1973) Tables of food components for Thai traditional fermented food products. *J. of Dept. Med. Sci.*, **15**(1/2), 47–49.

Anonymous (1974) World Food Conference. *Report E-CONF 65/20*, United Nations, New York.

Areekul, S., Thearawibul, R. and Matrakul, D. (1974) Vitamin B_{12} content in fermented fish sauce and soyabean sauce. *Southeast Asian J. Trop. Med. Public Health*, **5**, 461.

Boez, L. and Guillerm, J. (1930) Quoted by van Veen (1953).

Bremstedt, F. (1962) Amino acid composition of fresh fish and influence of storage and processing. *International Symposium on Fish in Nutrition*, (eds E. Heen and R. Krauzer), Fishing News (Books) Ltd, London, pp. 61–7.

Bremstedt, F. (1957). Geschmacke und Garuchstoffe in Fischfleishch Arch: *Fischereiwiss.*, **8**, 4–102.

Boonkohsrang, A. (1973) A microbiological study of the Thai traditional fermented Kung-som. *MSc Thesis*, Kasetsart University, Bangkok, Thailand (in Thai).

Chevey, P. (1931–2) *Rapport sur le fonctionnement de l'Institut Oceanographique de l'Indochine*, no. 21. (Quoted by van Veen, 1953.)

Fuji, M., Hirose, M. and Era, S. (1951) Studies of the proteases of fish. I. The distribution of proteases in the body of sardine. *Bull. Jpn. Soc. Sci. Fish.*, **16**, 553–555.

Garby, L. and Areekul S. (1974) Iron supplementation on Thai fish sauce. *Ann. Trop. Parasitol.*, **68**, 467–476.

Gruvel. A. (1925) L'Indochine. Ses richesses marine et fluviales. *Société d'editions geographiques, maritimes, et coloniales*, Paris. (Quoted by van Veen, 1953.)

Greig, R.I.W. and Estrella, D.C. (1988) A study of the acceleration of fish sauce production using enzymes in *Procedings of the Food Science and Technology in Industrial Development Conference*, Vol. 1, pp. 275–80.

Guillerm, J. (1930) *Trans. Congr. Far East Assoc. Trop. Med. 8th Congress (Bangkok)*, p. 1. (Quoted by van Veen, 1953.)

Hamm, W.S. and Clague, J.A. (1950) Temperature and salt purity effects on the manufacture of fish paste and sauce. *Fish and Wildlife Service, US Dept. Interior. Res. Rep.* **24**, 11.

Hongtongdaeng, P. (1979) A microbiological study on the Thai traditional fermented food products. *MS Thesis*, Kasetsart University, Bangkok, Thailand (in Thai).

Howard, G.E. and Dougan, J. (1974) Analytical indications of a more rational approach to accelerated fish sauce production, in *International Symposium on Fishery Products*, (ed. R. Kreuzer), Fishing News (Books) Ltd, London.

Itoh, H., Hadioetomo, R.S., Nikkuni, S. and Okada, N. (1985) Studies on lactic acid bacteria in fish sauce. I Chemical composition and microflora of fish sauce. *Rep. National Food Res. Inst.*, **47**, 23–30.

Itoh, H., Hadioetomo, R.S., Nikkuni, S. and Okada, N. (1985) Studies on lactic acid bacteria in fish sauce. II Identification of salt-tolerance and acid-producing bacteria from fish sauces. *Rep. National Food Res. Inst.*, **47**, 31–40.

Kasemsarn, B. (1963) Studies on fish sauce fermentation. *MS Thesis*, University of Washington, Seattle.

Krempf, A. (1974) La conservation et utilisation des produits de la pêche sous les tropiques. *Communications presentées par l'Institut Oceanographique de l'Indochine au Congrès Scientifique du Pacifique 1933*, Vancouver, no. **29**, pp. 31–7. (Quoted by can Veen, 1953.)

Matsushita, K. (1937) Studies on 'Funasushi' (fermented crucian carp). *Bull. Agri. Chem. Soc. Jpn.*, **13**, 629–38.

Mees, R.H. (1934) Undersuchingen uber die biersarcina. *PhD Thesis*, Technical University Delft, Holland.

Meister, A. (1965) *Biochemistry of Amino Acids*, 2nd edn. Academic Press, New York.

Mittranond, C. and Thongthai C. (1984) Experimental fermentation of fish sauce using facterial cultures. Paper presented at the *JSPS-NRCT seminar*, Osaka, Japan (unpublished).

Neetiwatanapong, C. (1979) A microbiological study on the Thai traditional fermented food products: Tai-pla and Pla-paeng-daeng. *MS Thesis*, Kasetsart University, Bangkok, Thailand (in Thai).

Nguyen-An-Cu and Vialard-Goudou A. (1953) Sur la nature delacidile volatile de la saumure Vietnamiene (Nuoc-mam). *Compt. Rend.*, **236**, 2128.

Noda, M., Thanh Vo Van, Kusakabe, I. and Murakami, K. (1981) Substrate specificity and salt inhibition of five proteinases isolated from the pyloric caeca and stomach of sardine. *J. Inst. Appl. Biochem.* 1565–9.

Noda, M. and Murakami, K. (1981) Studies on proteinases from the digestive organs of sardine. II. Purification and characterization of two acid proteinases from the stomach. *Biochim. Biophys. Acta*, **658**, 27–34.

Orillo, C.A. and Pederson, C.S. (1968) Lactic acid bacterial fermentation of Burong Dalag. *J. Appl. Microbiol.*, **16**, 1669–71.

Partmann, W. 1965. Changes in protein nucleotides and carbohydrates during rigor-mortes, in *FAO International Symposium on the Technology of Fish Utilization*, Fishing News (Books) Ltd, London.

Rao, J. (1967) Fish processing in the Indopacific area. Indo-Pacific Fisheries Council. Regional Studies No. 4, FAO Regional Office for Asia and the Far East, Bangkok, Thailand.

Rosé E. (1918) Quoted by van Veen (1953).

Saisithi, P. (1967) Studies on the origins and development of the typical flavor and aroma of Thai fish sauce. *PhD Thesis*, University Washington, Seattle.

Saisithi, P. (1981) Roles of fermented foods in solving malnutrition. *Microbial Utilization of Renewable Resource*, **2**, 200–10.

Saisithi, P., Sundhakul, M., Daengsubha, W. and Suyanandana, P. (1975) A brief description of Thailand traditional fermented food products. *Thai J. Agric. Sci.*, **8**, 205–219.

Sundahakul, M., Daengsubha, W. and Suyanandana, P. (1975) A brief description of Thailand traditional fermented food products. *Thai J. Agric. Sci.*, **8**, 205–19.

Suntinanalert, P. (1979) Roles of microorganisms in the fermentation of Nampla in Thailand: relationship of bacteria isolated from different geographical lacalities in Thailand. *MS Thesis*, Mahidol University Bangkok, Thailand.

Suwanik, R. (1979) Iron and iodine fortification of common salt and fish sauce. *Research Report*, Medical Association of Thailand.

Tarr, H.L.A. (1965) Enzymic degradation of glycogen and adenosine-5-phosphate in fish muscle. *FAO International Symposium on the Technology of Fish Utilization*. Fishing News (Books) Ltd, London.

van Veen, A.G. (1953) Fish preservation in Southeast Asia. *Adv. Food Res.*, **4**, 209–231.

van Veen, A.G. (1965) Fermented and dried seafood products in Southeast Asia. *Fish as Food*, **2**, 227–250.

van Klaveren, F.W. and Legendra, R. (1965) Salted cod. *Fish as Food*, **3**, 133–163.

Yeoh, Q.L., Aziz, A. and Adnan, N.A. Md. (1985) Fermented fish products of Malaysia, in *Proceedings of Food Conference '85*, Phillipines International Conference Center, 18–23 February, Manila, Phillipines.

CHAPTER 6

Proteases from aquatic organisms and their uses in the seafood industry

N.F. Haard and B.K. Simpson

6.1 INTRODUCTION

The term 'fish proteases' refers to hydrolytic enzymes from aquatic species that catalyze degradation of peptide bonds in protein molecules. Proteases play a vital role in biotechnology, food processing and other industries, as well as in a variety of physiological processes (Table 6.1). There is considerable demand for enzymes with the right combination of properties for specific applications in industry. Worldwide sales of industrial enzymes were estimated at about $1.5 billion for 1990 with approximately 48% of all these sales deriving from proteases (Knorr and Sinskey, 1985). Proteases used in industry are mainly derived from plant, animal and microbial sources, whereas their counterparts derived from marine and other aquatic sources have not been extensively used. Some of the reasons for the under-utilization of fish proteases include the following:

1. relatively few studies have been carried out on these enzymes, hence there is limited information about their potential;
2. the seasonal nature of the source material which precludes supply in a steady manner;
3. the unfavorable attitude by workers to the 'odors' associated with the source material (i.e. fish offal).

Although the microbial world offers a diverse source of enzymes, there are very few species of microorganisms actually used for industrial

Fisheries Processing: Biotechnological applications.
Edited by A.M. Martin.
Published in 1994 by Chapman & Hall, London. ISBN 0 412 58460 3

Table 6.1 Some processes facilitated by proteolytic enzymes

Industry/field	*Commodity and/or applications*
Food	Cereals, baked goods: Improves product handling characteristics; improves texture and loaf volume of bread; and decreases dough mixing time.
	Egg and Egg Products: Improves quality of dried products.
	Meats: Tenderization; protein recovery from bones; hydrolysis of blood proteins.
	Fish: Fish protein hydrolysates; viscosity reduction; skin removal, roe processing.
	Pulses: Tofu; soy sauce; protein hydrolysates; off-flavor removal soy milk.
	Dairy: Cheese curd formation; accelerate cheese aging; rennet puddings.
	Brewing: Fermentation and filtration aid; chill proofing.
	Wine: Clarification; decrease foaming, promote malolactic fermentation.
	Cocoa: Facilitates fermentation for chocolate production.
Biotechnology	Leather: Bating and dehairing.
	Textiles: Detergents.
Clinical	Reduction of tissue inflammation; dissolution of blood clots; promotion of wound healing; digestive aid.
Diagnostics	Diagnosis of candidosis.
Physiological processes	Cellular processes, e.g. digestion, translocation, activation of enzymes or hormones; activation of protoxins; protein turnover.

enzyme production at this time. This is largely because microorganisms intended for use as protease source must first be subjected to rigorous toxicological evaluation for safety, and this screening process is both tedious and costly. Furthermore, animal and plant proteases are traditionally preferred as food processing aids over their microbial counterparts because of adverse consumer perception of products derived from microorganisms. The marine environment contains a large pool of diversified species adapted to a variety of habitat conditions. These organisms include bony fish (Teleosti) which comprises of about 7000 fresh water and 13 000 salt water species known at this time. One would expect to find in these fish species the equivalent of any protease found in other terrestrial animals and plants, since they all carry out virtually the same metabolic processes. Genetic variation within species (Haard, Helbig and Feltham, 1981) together with adaptation to different environmental conditions have resulted in differences in the properties of fish proteases as compared with their counterparts derived from other sources. Some of these distinctive properties include higher catalytic efficiency at low temperature, lower thermal stability and substantial catalytic activity/stability at neutral to alkaline pH. The exploitation of the unique properties of fish proteases in the food processing industry is only at the beginning stage. Some enzymes from marine organisms are currently being used as process aids in the seafood industry (Almas, 1990; Raa, 1990; Stefansson and Steigrimsdottir, 1990; Wray, 1988) and this is expected to cascade with the recent advances in recombinant DNA technology which is enabling the cloning of proteases in 'generally recognized as safe' (GRAS) organisms. The purpose of this chapter is 2-fold:

1. To review the industrial uses of proteases in general.
2. To survey some of the sources and current uses / potential uses of fish proteases.

6.2 GENERAL PROPERTIES AND APPLICATIONS OF PROTEASES

Proteases make up about 48% of all industrial enzymes (Fox, Shannon and Bjarnason, 1991), and are mostly used in detergents, leather production and the food industry (Godfrey and Reichelt, 1983), as processing aids for many products including baked goods; beer and wine; cereals; milk and dairy products like cheese, chocolate; eggs and egg products; meat and fish products; legumes; and for the production of protein hydrolyzates and flavor extracts (Table 6.1). Several of these processes involve the breakdown of proteins, which may be achieved by chemical and/or physical means, i.e. through the use of chemicals at extreme pH and/or extreme temperature. However, there are several advantages with using proteases to achieve protein modification in foods,

1. They are more specific, thus obviating undesirable side reactions.
2. They are active under mild conditions of pH and temperature.
3. They are effective at low concentrations.
4. They can readily be inactivated after they have carried out the desired transformation in the food material.

Proteases differ in their catalytic and physical properties and whether or not a particular protease is suitable for a given application depends on several factors, such as substrate specificity, sensitivity to pH and temperature, presence or absence of activators and/or inhibitors, availability, technical service support, and cost.

6.3 CRITERIA FOR SELECTION OF A PROTEASE IN INDUSTRY

For a protease to be used as a processing aid, it must exhibit specificity for the substrate, i.e. it must be able to act on the intended substrate, otherwise there cannot be any transformation. Different proteases act on peptide bonds to varying degrees, and this feature should be considered in the selection of an enzyme for processing. In certain applications (e.g. cheese making), the hydrolysis of a limited number and type of peptide bond is of paramount importance. Thus, most proteolytic enzymes are unsuitable for cheese making because their broad specificity results in lower yield of curd and excessive proteolysis (Brewer, Helbig and Haard, 1984) to influence adversely the texture and flavor of the final product. The enzyme of choice in cheese making, chymosin, has a relatively narrow specificity for peptide bonds in milk protein, including the hydrolysis of a critical phenylalanine methionine bond in κ-casein. Likewise, a limitation of using a wide variety of currently available proteases for meat tenderization is their inability to exclusively hydrolyze critical bonds in both the extracellular and intracellular matrix, i.e. collagen and key proteins which affect dissolution of the myofibril z discs (Mohr, 1980). At the other extreme, production of protein hydrolyzates for use as flavors or bacterial growth media may require enzyme(s) with much broader specificity to achieve a high degree of hydrolysis for the formation of low molecular weight peptides and free amino acids.

In addition to the ability to act on the intended substrate, a protease must be capable of functioning appreciably well within the pH range of the food material, since it may not always be desirable to alter the pH of the food material to the optimum pH of the enzyme. For example, pH manipulation of the substrate may result in an excessive increase in salt content and may also change a food's taste. In addition to the catalytic activity of the enzyme, the stability of the enzyme is also an important consideration. For example, the instability of most pepsins at neutral pH limits their use in production of sweet curd cheese (Haard *et al.* 1983; Brewer, Helbig and Haard, 1984; Shamsuzzaman and Haard, 1984). Proteases active in the acid, neutral and alkaline pH range are commercially available.

The protease must also be appreciably active and stable within the temperature range at which processing is carried out. In this regard, thermostable enzymes are preferred in reactions that proceed at elevated temperatures such as the use of papain in barbecue sauce to the tenderization of meat during cooking (Metz, Boyce and Sigmiller, 1975). High operating temperature for short duration may reduce problems of microbial contamination. However, disadvantages to using heat stable enzymes at elevated process temperature include high energy costs, destruction of heat labile flavors and nutrients, etc., alteration in the physical integrity of the product, a relative increase in undesirable side reactions that have a higher Arrhenius energy of activation (E_a) than the enzyme reaction, and an inability to heat inactivate the enzyme and stop the reaction after the enzyme has done its job (Almas, 1990; Simpson and Haard, 1987a; Hultin, 1978). The latter problem is particularly important, otherwise problems of bitterness or excessive tissue deterioration may arise.

Certain compounds naturally present in certain food materials, act as stabilizers, activators or inhibitors, and should be taken into consideration before using a particular protease as a processing aid. For example, raw soybeans have high levels of trypsin inhibitors, and need to be pretreated to eliminate or inactivate these inhibitors before trypsin-like and/or chymotrypsin-like proteases can be used as effective processing aids. Proteases differ considerably in their response to specific inhibitors, i.e. thiol proteinase, metalloproteinase, aspartic acid proteinase, and serine proteinase. For all four groups of protease there are naturally occurring protein or peptide inhibitors, e.g. α-2-macroglobulin, pepstatin and Kunitz inhibitor (Laskowski, 1986; Lorier and Aitkens, 1991). Moreover, proteases may differ in their response to general protein reactive substances, e.g. salt, acid, alkali, urea, detergents, etc. In addition to variation in response to inhibitors based on different active sites, it is possible that homologous enzymes, having the same active site, may differ somewhat in their response to inhibitors or inactivators (Wiggs, 1974; Godfrey and Reichelt, 1983). For example, a thyroid protease from burbot (*Lota lota*) has different iodoacetate sensitivity when isolated from fish in Winter and Spring seasons.

6.4 CLASSIFICATION OF FISH PROTEASES

Fish proteases may be classified using the same criteria used for proteases from other animal, plant or microbial sources. For example, they can be classified on the basis of the following:

1. Their similarities to well-characterized proteases, as trypsin-like, chymotrypsin-like, chymosin-like, cathepsin-like, etc.
2. Their pH activity profiles as acid, neutral or alkaline proteases.
3. Their substrate specificity.
4. Their mechanism of catalysis.

The latter method of classification is the system used by the Enzyme Commission of the International Union of Applied Biochemists and, based on this approach, fish proteases may also be classified into four groups as serine proteases, acid or aspartate proteases, thiol or cysteine proteases and metalloproteases. Examples of fish proteases in the four categories that have been characterized from fish are summarized in Table 6.2. The enzymes in the different groups may be distinguished from each other on the basis of their response to various inhibitors.

Table 6.2 Classification of fish proteases based on mode of catalysis

Group	*Examples*	*Sources*	*Reference(s)*
Serine	trypsin	capelin	Hjelmeland and Raa (1982)
		catfish	Yoshinaka *et al*. (1984)
		cod	Simpson, Simpson and Haard (1989); Simpson and Haard (1984a,b); Overnell (1973)
		cunner	Simpson and Haard (1985a)
		goldfish	Jany (1975)
		sardine	Murakami and Noda (1981)
		starfish	Chen, Yau and Chen (1978)
	chymotrypsin	capelin	Kalac (1978)
		herring	Kalac (1978)
		dogfish	Ramakrishna, Hultin and Atallah (1987)
Aspartyl/ acid	pepsin	cod	Squires, Haard and Feltham (1986a,b); Arunchalam and Haard (1985)
		dogfish	Merrett, Bar-Eli and Van Vunakis (1969)
		hake	Sanchez-Chiang and Ponce (1981)
		salmon	Fruton and Bergmann (1940)
		trout	Owen and Wiggs (1971); Twinning *et al*. (1983)
		tuna	Tanji, Kageyama and Takahasi (1988)
	gastricin	hake	Sanchez-Chiang and Ponce (1981)
		seal	Shamsuzzaman and Haard (1984, 1983)
Cysteine/ thiol	cathepsin	carp	Hara, Suzumatsu and Ishihara (1988)
		cod	Schmitt and Siebert (1967)
		herring	Schmitt and Siebert (1967)
		sole	Schmitt and Siebert (1967)
		flounder	Schmitt and Siebert (1967)
		salmon	Ting, Mongomery and Anglemier (1968)
		squid	Hameed and Haard (1985)
		tilapia	Sherekar, Gore and Ninjoor (1988)
		mussel	Zeef and Dennison (1988)
Metallo	collagenase	crab	Grant, Sacchettini and Welgus (1987); Eisen and Jeffrey (1969)
		rockfish	Bracho and Haard (1991)
		lobster	Chen (1992)
	calpains	cod	Gill (1992)
		herring	Gill (1992)
		hake	Gill (1992)
		salmon	Gill (1992)

6.4.1 Serine proteases

(a) Trypsins

Trypsin has been isolated from a number of aquatic species including Greenland cod (Simpson and Haard, 1984b,c), Atlantic cod (Simpson, Simpson and Haard, 1989; Benediktsson, 1987; Asgeirsson and Bjarnason, 1988, 1989; Gildberg, 1993), capelin (Hjelmeland and Raa, 1982), sardine (Murakami and Noda, 1981), catfish (Yoshinaka *et al.*, 1984), starfish (Chen, You and Chen, 1978), crayfish (Pfleiderer, Zwilling. and Senneborn, 1967) and cunner (Simpson and Haard, 1985b, 1987b; Simpson, Simpson and Haard, 1989). Fish trypsins are similar to trypsins from warm blooded animals with respect to their molecular size, amino acid composition and sensitivity to serine protease inhibitors. Trypsin from fish tends to be stable in alkaline medium and very unstable in acidic conditions whereas trypsin from mammals is most stable in acidic medium. Compared with bovine trypsin, the turnover number of fish trypsins is about one order of magnitude larger for both amidase and esterase reactions. The Arrhenius energy of activation is considerably lower for fish trypsins than for mammalian trypsins. A lower E_a indicates a reaction less responsive to temperature change. The free energy of activation (ΔG^*) for esterase and amidase reactions is lower for fish trypsins, indicating the fish enzymes are better catalysts of these reactions (Simpson and Haard, 1987a). Further, the relative contributions of ΔS^* and ΔH^* to ΔG^* are different for the trypsins from cold water fish. The ΔS^* is more important than ΔH^* in lowering ΔG^* values for the fish enzymes, while in the case of bovine trypsin ΔH^* is more important than ΔS^* in lowering the ΔG^* barrier. Given the relatedness of the internal structure of protein molecules to ΔS^* (Low, Bada Somero, 1973), it would appear that the cold-adapted fish trypsins have relatively more flexible structures than the enzymes from warm blooded animals.

Trypsin and chymotrypsin from higher vertebrates do not efficiently hydrolyze native globular proteins (Pfleiderer, Zwilling and Sonneborn, 1967; Jany, 1976; Mihalyi, 1978), and it is generally assumed that acid denaturation and hydrolysis in the stomach is necessary for efficient protein digestion by intestinal enzymes. Studies have shown that trypsin and trypsin-like enzymes from stomachless fish including crayfish (Pfleiderer, Zwilling and Sonneborn, 1967), cunner (Simpson and Haard, 1987b), and mullet (Marshall, 1990) are able to hydrolyze native, globular proteins, unlike trypsin from animals with functional stomachs. Cunner pancreas contains a 24 000 Da trypsin and a 14 000 Da trypsin-like enzyme. After 4 h incubation with equimolar concentrations of enzyme, approximately 12% of RNAase activity was lost with bovine trypsin whereas cunner trypsin and trypsin-like enzymes caused 58 and 72% inactivation, respectively (Simpson and Haard, 1987b). Cunner trypsin is also more efficient in hydrolyzing native hemoglobin than Greenland cod and Atlantic cod trypsin (Simpson, Simpson and Haard, 1989).

(b) Chymotrypsin

Chymotrypsins have been recovered and characterized from marine species such as the spiny dogfish (Racicot, 1984; Ramakrishna, Hultin and Atallah, 1987), capelin (Kalac, 1978) and herring (Kalac, 1978). They are single polypeptide molecules with molecular size of approximately 25 kDa, and are stable and active at slightly alkaline pH (7.5 – 8.5). Fish chymotrypsins demonstrated a higher rate of reaction and hydrolyzed more peptide bonds in various protein substrates (casein, collagen and bovine serum albumin) than their mammalian counterpart from beef pancreas. The fish enzyme also had greater activity toward soy protein isolate at various reaction temperatures (5, 20 and 35°C) than bovine chymotrypsin (Ramakrishna, Hultin and Atallah, 1987), and achieved clotting of milk. Chymotrypsins from herring and capelin exhibited greater activities toward the synthetic substrate, *N*-benzozyl-L-tyrosine ethyl ester (BTEE), than their bovine counterpart (Kalac, 1978). These findings were similar to those made by Simpson and Haard (1987a) with Greenland cod trypsin versus bovine trypsin.

6.4.2 Aspartyl or acid proteases

(a) Pepsins

Fish pepsins have been isolated from the gastric glands of various aquatic animals, including capelin (Gildberg and Raa, 1983), American smelt (Haard, Helbig and Feltham, 1981), Polar cod (Arunchalam and Haard, 1985), Greenland cod (Squires, Haard and Feltham, 1986a,b), Atlantic cod (Almas, 1990; Brewer, Helbig and Haard, 1984; Haard, 1986), and dogfish (Merrett, Bar-Eli and van Vunakis, 1969). Pepsins from cold water fish exhibit a low Arrhenius activation energy, a high apparent Michaelis constant, a low temperature optimum, a low thermal stability and a high pH optimum in comparison to pepsins from cold-blooded animals with temperate or warm habitat and warm blooded animals. Fish pepsins are reported to be very resistant to autolysis at low pH (Raa, 1990). The thermal denaturation temperature of pepsins from cold water fish is about 20°C lower than that from mammals. Accordingly, it is possible to inactivate fish pepsins by relatively mild heat treatment, an advantageous property in cheese manufacture after the milk clotting reaction is complete (Haard *et al.*, 1982; Brewer, Helbig and Haard, 1984; Simpson and Haard, 1987a; Almas, 1990). There is a direct relationship between the habitat temperature of fish and Arrhenius energy of activation for proteolysis (Simpson and Haard, 1987a; Haard *et al.* 1982). E_a for hydrolysis of hemoglobin ranges from as low as 2.9 kcal/mol for Polar cod pepsin B to 11.2 kcal/mol for pepsin from warm blooded animals (Arunchalam and Haard, 1985).

(b) Chymosin (from marine mammals)

Chymosin is an acid protease with some distinct characteristics from other gastric proteases, e.g. the pH optimum of proteolysis, stability at pH near 7.0 and narrow substrate specificity. Chymosin generally occurs in milk-fed ruminants whereas pepsins occur in adult ruminants and in simple stomached animals. The occurrence of chymosin in non-ruminant animals has not been well established. Recently, it was shown that the gastric proteases of harp seal include two isoenzymes with properties like chymosin rather than pepsin (Haard *et al.*, 1983; Shamsuzzaman and Haard, 1983, 1984, 1985). The chymosin-like protease has properties which distinguish it from pepsin including a relatively high ratio of milk clotting to proteolytic activity, a pH optimum of 2.2–3.5 for hemoglobin hydrolysis, inability to inactivate ribonuclease, low activity on *N*-acetyl-L-phenylalanyl-3,5-diiodo-l-tyrosine, instability in 6 M urea and zymogen molecular weight of 33 800 Da. The ratio of chymosin-like isoenzyme to pepsin isoenzyme was greater in 1 week old animals than in adult animals. Chymosin, traditionally obtained from the abomasum of suckling calves, is a valuable enzyme for cheese making and was in short supply (de Koning, 1978) until the introduction of recombinant DNA technology.

6.4.3 Cysteine or thiol proteases

(a) Cathepsins

Cathepsins have been isolated from a number of aquatic species including carp (Hara, Suzumatsu and Ishihara, 1988), cod (Schmitt and Siebert, 1967), salmon (Ting, Montgomery and Anglemier, 1968), squid (Hameed and Haard, 1985) and sole (Geist and Crawford, 1974). They have been characterized as single polypeptide chains with molecular sizes ranging from 13.6 to 25 kDa. Cathepsins from different species display maximum activity over a broad pH range, from pH 3.5 in some, to pH 8.0 in others. They are activated by Cl^- ions, and generally require sulfhydryl and/or chelating agents for activity. They are invariably inhibited by iodoacetate and heavy metals. Fish muscle contains about 10 times more catheptic enzyme activity than does mammalian muscle (Siebert and Schmitt, 1965), and are believed to play a major role in fish flesh softening during postharvest handling and storage (Warrier, Ninjoor and Nadkarni, 1985).

6.4.4 Metalloproteases

(a) Collagenases

Collagenases have been recovered and characterized from the digestive, glands of various marine organisms such as crab, prawn and lobster (Eisen and Jeffrey, 1969; Eisen *et al.*, 1973; Grant, Sacchettini and Welgus, 1983; Baranowski, Nip and Moy, 1984; Nip, Moy and Zhang, 1985; Chen, 1992). Collagenases from marine organisms resemble their counterparts from mammalian sources in their action on native collagen. However, the digestive collagenases differ from mammalian tissue collagenases in exhibiting trypsin-like and chymotrypsin-like specificities. These proteases are most active within the pH range of 6.5–8.0 and are inactivated at pH values lower than 6.0. They are inhibited by diisopropylfluorophosphate and phenylmethysulfonyl fluoride. Collagenases are capable of degrading the polypeptide backbone of native collagen under conditions that do not denature the protein. Collagenases from hepatopancreas of marine invertebrates have been implicated in postharvest texture deteriorations in these animals during storage. A highly specific metalloproteinase with collagenase activity has been isolated from the muscle of rockfish (Bracho and Haard, 1991).

(b) Calpains

Also known as calcium-activated neutral proteases, calpains have been characterized in the skeletal muscle of various marine species including Atlantic cod, hake, herring and salmon (Gill, 1992). These enzymes are extracted in the active form from fish species several hours after death, unlike their counterparts from mammalian tissue. Furthermore, the fish calpains were active against myofibrillar proteins, including the myosin heavy chain. Calpains are believed to promote flesh softening in fish during postharvest storage. Thus, it would be expected that their continued activity in fish muscle during iced storage would be deleterious for products intended for use in making surimi-type products.

6.5 TRADITIONAL APPLICATIONS OF PROTEASES IN THE SEAFOOD INDUSTRY

Until recently, the use of enzymes in the seafood industry has been limited to very few applications, i.e. viscosity reduction of stickwater in the fish meal industry, preparation of fish protein hydrolyzates for food or animal feed (Mohr, 1980) and fermentation processes depending on endogenous proteases (Stefansson and Steingrimsdottir, 1990).

6.5.1 'Stick water' viscosity

In the fish meal industry, dissolved protein that is concentrated by evaporation can contribute to excessive viscosity and prevent efficient evaporator operation. The addition of neutral or alkaline bacterial proteinases, when the solids content is above 20% results in fuel savings and product improvement (Jacobsen and Lykke-Rasmussen, 1984). To achieve rapid viscosity reduction of a protein solution, an endo protease, which catalyzes hydrolysis of the polypeptide chain at random sites is preferable to an exo protease.

6.5.2 Fish protein concentrate (FPC)

Fish protein hydrolyzate produced by proteinase treatment is sometimes called 'biological fish protein concentrate'. The FPC process is a means to transform inexpensive pelagic fish and fish processing wastes to a protein concentrate which can be used in food and other applications rather than a lower value animal feed. The process normally involves limited hydrolysis of the ground material with protease(s), e.g. ficin or papain, inactivation of the enzyme, filtration of the hydrolysate, concentration and spray drying of the filtrate. Fish protein concentrates are soluble and have been used in milk replacers, food supplements, and broth flavorants. However, a problem with FPC is the formation of bitter tasting hydrophobic peptides (Mohr, 1980), and excessive degradation of protein to peptides and amino acids. Hydrolysis with proteases that have a relatively narrow specificity for peptide bonds can avoid some of these problems (Simpson and Haard, 1987a).

6.5.3 Fishery products obtained by autolytic proteolysis

A number of traditional fishery products are prepared by fermentation. Endogenous proteases, rather than the action of microorganisms, play the major role in transforming raw material to finished product. These products include fish sauce, cured fish products like matjes herring and fish silage (Stefansson and Steingrimsdottir, 1990).

Annual production of fish sauce is estimated at about 250 000 tons (Campbell-Platt, 1987). During the fermentation, endogenous enzymes slowly degrade the fish tissue in the presence of 20–30% salt to form a clear liquid with a high content of free amino acids and excellent flavor. The traditional fermentation is a slow process requiring 1 year or longer. Addition of exogenous proteases to the fermentation can accelerate the process; however, the flavor characteristics of the finished product are normally inferior to the traditional product. However, supplementation with a protease fraction from squid hepatopancreas accelerates the fermentation and yields a product having sensory properties superior to the product obtained by traditional methods (Raksakulthai, Lee and Haard, 1986; Raksakulthai and Haard, 1992).

Fish silage is a liquid protein hydrolyzate produced from fish processing waste and small pelagic fish under acidic conditions. Silage has been produced for many years in Northern Europe and is used as a protein supplement in animal feed (Raa and Gildberg, 1982; Gildberg and Almas, 1986). The raw material is acidified with inorganic or organic acids and autolysis to a liquid product depends on endogenous enzymes from the viscera as well as from the flesh (Haard *et al.*, 1985). Autolysis normally proceeds rapidly at ambient temperature and it does not appear that added enzymes substantially benefit the process. However, preparation of silage from certain raw materials may be aided by exogenous proteases. For use in ruminant feed, the degree of protein hydrolysis after liquefaction and lipid oxidation can be controlled by addition of formaldehyde (Haard *et al.*, 1985).

Less extensive protein hydrolysis occurs in products like cured herring (Stefansson and Steingrimsdottir, 1990), matjes (Simpson and Haard, 1984c) and fermented squid (Lee, Simpson and Haard, 1982). The whole fish or meat is brined and biochemical reactions occur that change the taste characteristics of the product without drastically breaking down the muscle structure. Ripening of these products appears to depend on endogenous digestive enzymes and the best products are obtained from herring harvested during the heavy feeding stage, when digestive protease activity is high. Addition of enzyme preparations from herring viscera (Eriksson, 1975; Opshaug, 1982), or trypsin from Greenland cod pyloric ceca (Simpson and Haard, 1984c) or bovine pancreas (Anonymous, 1975) has been used to prepare artificially ripened herring. Likewise, addition of fish trypsin, bovine cathepsin C or a squid hepatopancreas extract rich in cathepsin C to fermenting squid improves the formation of taste active amino acids (glutamate, alanine, leucine, serine, lysine, arginine and proline) and sweet delicious taste (Lee, Simpson and Haard, 1982). It would appear that the use of industrial enzyme supplementation in production of such products is not widely practiced, perhaps because appropriate enzyme preparations are not available.

6.6 RECENT APPLICATIONS OF FISH PROTEASES IN THE SEAFOOD INDUSTRY

In recent years, additional applications of proteases in the seafood industry have emerged. These include the selective removal of skin, hydrolysis of membranes and other supportive tissue that envelops roe, roe sacks and other tissues, and recovery of pigment and flavor extracts.

6.6.1 Skin removal

A number of inventions using enzymes to remove the skin from fish that are difficult to skin by mechanical means have been described in recent years. Processes have been described for deskinning of tuna using 60°C steam treatment, together with an ill-defined mixture of proteases and carbohydrases (Fehmerling, 1973); herring with the aid of fish pepsins at low pH (Joakimsson, 1984); squid using a weak salt solution containing papain (Hempl, 1983; Raa, 1990); and skate using heat pretreatment, and a combination of an unspecified protease and carbohydrase (Wray, 1988; Stefansson and Steingrimsdottir, 1990). Research is also being carried out to develop an enzymatic process to loosen shrimp shells from muscle tissue prior to peeling (Raa, 1990).

6.6.2 Scale removal

In Iceland, recent studies have emphasized use of enzymes to descale fish (Stefansson and Steingrimsdottir, 1990, 1989a,b). Scaling of species like ocean perch and haddock by mechanical methods is harsh and results in tearing of skin and lower fillet yield. Scales can be gently removed by spraying with water after incubating in an enzyme solution at 0°C.

6.6.3 Roe production

Proteases, including fish pepsins, have been used to hydrolyze the supportive tissue that envelops salmon and trout roe and roe sacks (Sugihara *et al.*, 1973; Gildberg and Almas, 1986; Gildberg, 1988, 1993; Raa, 1990). Pepsins apparently split the linkages that adhere the egg cells to the roe sack without affecting the eggs. In 1986, about 50 tons of caviar was produced in Canada and Scandinavia using this process. The yield of roe from rainbow trout is double that of the traditional riddle technique using a proteolytic enzyme supplied by NOVO Nordisk Bioindustries (Anonymous, 1990). Research has also been done to produce shrimp roe products using enzymes (Stefansson and Steingrimsdottir, 1990).

6.6.4 Membrane removal

Other applications of proteases include removal of the collagenous membrane surrounding cod liver prior to canning (Steingrimsdottir, 1987). The cod liver is sometimes infested with sealworm which are mainly contained in this membrane. The throughput of liver in canning plants can be increased by at least 20–30% using enzymes rather than the manual method (Stefansson and Steingrimsdottir, 1990). The black membrane surrounding cod swim bladder can also be removed from salted product by an enzymic process (Steingrimsdottir and Stefansson, 1988). The process is currently used by all swim bladder processors in Iceland and results in an 8-fold increase in throughput over mechanical methods. Proteases have also been used to loosen or remove the intestines from scallop muscle (Stefansson, Johannesson and Magnusdottir, 1987; Raa, 1990). By using enzymes from fish viscera, it is possible to selectively break the tissue which connects the soft tissue to the muscle and gonads (Raa, 1990).

6.6.5 Seafood flavorings

Seafood flavors are in high demand for use in products like artificial crab and fish sausage. Proteolytic enzymes can aid the extraction of flavor compounds from shells and other materials. The taste and volatile components of shrimp heads were recovered by enzymatic digestion using corolase N and koji or bacterial strains with high proteolysis rates followed by concentration and spray drying (Pan, 1990). The product contained 9–12% free amino acids, mainly taurine, arginine, glycine and proline and inosine monophosphate as the major nucleotide. Shrimp flavorant can be used as an additive in surimi-based products and cereal-based extrusion products like shrimp chips. Proteases are used in this process, although other types of enzymes may also be involved. A French company, ISNARD LYRAZ, has developed an enzymatic process to recover seafood flavoring materials from marine animals (In, 1990). Enzymes are used to liquefy the material which allows separation of bones and shells and facilitates concentration to 50–60% dry matter. The degree of protein hydrolysis is used to optimize the process with regard to yield and off-flavor formation. Enzymes have also been used to recover flavor from oysters.

6.6.6 Pigments

Various methods have been devised to recover carotenoid pigments from crustacean wastes for use in the feed of farmed fish and shellfish which depend on carotenoids for their bright red color. Extraction of shell waste with oil (Chen and Meyers, 1982) reduces ash and chitin levels and achieves a good recovery of pigment, but it suffers the disadvantage of yielding a product devoid of protein, thereby decreasing the stability of the carotenoid to oxidation and failing to recover a valuable nutrient. Since about one-third of the dry matter in crustacean shell waste is protein, an enzymatic process has been developed to recover the protein along with the carotenoid in its native carotenoprotein state. Proteolytic enzymes were used to recover carotenoprotein from shrimp (Simpson and Haard, 1985a; Cano-Lopez, Simpson and

Haard, 1987) and crab (Manu-Tawiah and Haard, 1987) shell waste. About 80% of the protein and 90% of the astaxanthin pigment from shrimp process waste can be recovered as an aqueous dispersion after trypsin hydrolysis. Lower yield of protein (60–70%) and pigment (35–50%) are recovered in the carotenoprotein fraction when protease with broad specificity is used rather than trypsin. Protein associated astaxanthin is more resistant to oxidation (Simpson and Haard, 1985a) and is deposited in the skin and flesh of rainbow trout more efficiently (Long and Haard, 1988) than the free pigment. Use of this process by industry will depend on a less costly source of trypsin than is currently available.

6.7 OTHER POTENTIAL APPLICATIONS OF FISH PROTEASES IN INDUSTRY

6.7.1 Fish trypsins

A number of experiments have demonstrated the benefit of using fish trypsin to exploit its high molecular activity at low reaction temperature and its relatively low heat stability (Simpson and Haard, 1984a–c, 1987a). Supplementing the fermentation (10°C) of matjes herring with equivalent amounts of pure bovine and Greenland cod trypsins resulted in increased protein solubilization and formation of free amino acids (Simpson and Haard, 1984a). The initial rate of protein solubilization with cod trypsin was almost double that observed with bovine trypsin. Similar results were observed when squid was fermented at 4°C, i.e. added cod trypsin increased the formation of free amino acids and sensory scores of the finished product.

It is known that treatment of milk with bovine trypsin can prevent formation of oxidized flavor (Lim and Shipe, 1972); however, the trypsin must be inactivated after an appropriate reaction time to avoid excessive hydrolysis of milk proteins. Bovine and cod trypsins were similarly effective in preventing oxidation of milk lipids at concentrations greater than 0.0013%; however, cod trypsin but not bovine trypsin was completely inactivated by the pasteurization process. This was expected since trypsin from Greenland cod is considerably more heat labile than bovine trypsin (Simpson and Haard, 1984c).

Cod trypsin has also been used to advantage in extracting carotenoprotein from shrimp waste, but it is necessary to purify the enzyme from other digestive proteases (Cano-Lopez, Simpson and Haard, 1987). Trypsin and chymotrypsin, purified from cod viscera, will soon become commercially available according to a recent report (Stefansson and Steingrimsdottir, 1990).

The ability of trypsin from stomachless fish to hydrolyze native proteins may find application in the food processing industry as well as other areas. For example, pectin methyl esterase causes a problem of juice instability where the serum and particulate matter separate. The heat resistance of citrus esterase is such that pasteurization causes deterioration of the products flavor. Research at the University of Florida has shown that mullet trypsin is effective in inactivating pectin methyl esterase in citrus juice (Marshall, 1990). Another example of a possible application of enzymic enzyme inactivation is the treatment of the cuticle of raw shellfish products like shrimp so as to inactivate polyphenol oxidase to preventpost harvest discoloration by the phenolase reaction.

6.7.2 Fish pepsins

The temperature coefficient (Q_{10}) for the milk clotting reaction is 1.2 for Polar cod pepsin compared with 2.0–2.2 for pepsins from mammals. Because of the high molecular activity of cold-adapted pepsins at low reaction temperatures, they are excellent candidates for cold renneting milk (Brewer, Helbig and Haard, 1984; Simpson and Haard, 1987a). There are, however, problems with use of cold-adapted pepsins for cheese making, namely somewhat lower yields of curd and development of bitter off-flavors during aging of Cheddar cheese (Brewer, Helbig and Haard, 1984). Atlantic cod pepsin is now produced commercially in Norway (Gildberg and Almas, 1986; Almas, 1990; Raa, 1990) and the product appears to be useful for enzymatic caviar production, skin removal, cheese production and as a digestive aid in fish feed.

6.7.3 Chymosin from marine mammals

Chymosin is the principal component of calf rennet, used to form the curd in cheese making. A large number of proteases from plant, microbial and animal sources have been screened as rennet substitutes but relatively few have shown promise. Two major problems with rennet substitutes are a lower yield of curd due to loss of protein and fat to the whey and the formation of texture and flavor defects during ripening of the cheese. Cheddar cheese prepared with a simple extract of seal gastric protease (Shamsuzzaman and Haard, 1983) or from the purified chymosin-like isoenzyme (Shamsuzzaman and Haard, 1985) gave significantly higher sensory scores than cheese made with calf rennet. In particular, Cheddar made with the crude extract from harp seal underwent accelerated aging (Shamsuzzaman and Haard, 1983) and this was not found with the purified chymosin-like enzyme (Shamsuzzaman and Haard, 1985). Moreover, the seal enzyme is able to clot milk at neutral pH because, like chymosin, it is more stable than pepsin at neutral to alkaline pH. The seal enzyme catalyzed the same amount of non-protein nitrogen formation in aging cheese as calf chymosin. Seal enzyme, like calf chymosin and unlike pepsin, has a relatively narrow specificity for hydrolysis of casein.

6.7.4 Fish collagenases

Industrial applications of specific collagenolytic enzymes may be sought in the food (meat tenderizing) and leather (bating and dehairing) industries. Traditional proteinases used as meat tenderizers are mainly of plant origin, i.e papain, ficin and bromelain (Bernholdt, 1975; Dransfield and Etherington, 1981). However, these proteinases currently used for meat tenderization attack both connective tissue and myofibrillar proteins, thus they can bring about over-tenderization, leading to mushiness and undesirable texture softening. The ideal enzyme for degrading insoluble collagen in a meat system must not only demonstrate specificity for collagen, but must also function at:

1. the relatively low pH of meat;
2. the low temperature at which meat is held during storage;
3. the higher temperatures reached during cooking.

Collagenases have been found to have the desired properties, and especially collagenases from marine sources have been shown to offer high quality meat and meat products, and their use is encouraged for commercial application (Cronlund and Woychik, 1987; Miller, Strange and Whiting, 1989).

6.7.5 Plasteins

The term 'plastein reaction' is used to refer to the re-synthesis of high molecular weight protein-like substances from an ill-defined highly concentrated protein hydrolyzate. Although the plastein reaction has been extensively studied, its exact mechanism still remains a subject of great controversy. Thus far, plastein formation has been facilitated with proteases like trypsin, chymotrypsin, papain and pepsin, and at temperatures between 37 and 50°C, where hydrophobic interactions are relatively more important. The predominantly hydrophobic products resulting from the reaction have limited solubility in aqueous systems. Polar amino acids may be incorporated into plasteins if they are first chemically modified (Aso *et al.*, 1985; Hajos, 1986), and after completion of the plastein reaction, the ethyl ester linkages may be hydrolyzed yielding hydrophilic plastein products. However, the inability to produce water-soluble plasteins directly poses a limitation to the use of this reaction in the food industry, because hydrophobic products have limited application in food products. Since digestive proteases have high catalytic activity at low reaction temperatures where hydrophilic interactions are favored, it would be expected that cold-adapted trypsins, chymotrypsins and pepsins from fish could facilitate the precipitation of water-soluble plasteins at low temperatures for incorporation into certain food products. The plastein reaction has great potential in the food industry, where it could be employed to:

1. improve the quality of protein isolates and hydrolysates by improving flavor, solubility and nutritive value;
2. prepare gel-like products with excellent viscoelastic properties;
3. prepare products with high or low levels of particular amino acids as a dietary supplements for certain target groups;
4. prepare special types of peptides with important flavor and other properties.

6.8 CONCLUSION

The information on the properties and mechanism of action of proteases from aquatic species, in particular marine organisms, is growing rapidly and this has permitted their utilization in a variety of processes. Based on the upsurge of interest in these enzymes, it is accurate to say that the era of marine enzyme technology has started. Thus far, proteases from aquatic species have been produced by classical means and it is predicted that the rapid advances in recombinant DNA technology (Dixon, 1990; Wasserman, 1990) will enhance their production from fish processing waste as well as from more exotic species on a commercial scale for use in hitherto unexploited reactions.

REFERENCES

Almas, K.A. (1990) Utilization of marine biomass for production of microbial growth media and biochemicals, in *Advances in Fisheries Technology and Biotechnology for Increased Profitability*, (eds M.N. Voigt and J.R. Botta), Technomic, Lancaster, PA, pp. 361–72.

Anonymous. (1975) Production process for salted herrings (Matjes). *British Patent 1 403 221*.

Anonymous. (1990) Enzyme breaks up trout roe. *Fish Farming Int.*, **August**, 71.

Arunchalam, K. and Haard, N.F. (1985) Isolation and characterization of pepsin from Polar cod, *Boregadus saida*. *Comp. Biochem. Physiol.*, **80B**, 467–73.

Asgeirson, B. and Bjarnason, J.B. (1988) Studies on an enzyme preparation from cod viscera. *Report RH-1288*, Science Institute, University of Iceland, Reykjavik, Iceland.

Asgeirson, B. and Bjarnason, J.B. (1989) Purification and characterization of trypsin from the poikilotherm *Gadus morhua*. *Eur. J. Biochem.*, **180**, 85–94

Aso, K., Kimura, H., Watanabe, M. and Arai, S. (1985) Chemical properties of enzymatically modified proteins produced from soy protein by covalent attachment of methionine. *Agric. Biol. Chem.*, **49**, 1649–54.

Baranowski, E.S., Nip, W.K. and Moy, J.H. (1984) Partial characterization of a crude enzyme extract from freshwater prawn, *Machrobranchium rosenbergii*. *J. Food Sci.*, **49**, 1494–95.

Benediktsson, B. (1987) Production of proteolytic enzymes from cod viscera. *Internal report*, Icelandic Fisheries Laboratories, Reykjavik, Iceland.

Bernholdt, H.F. (1975) Meat and other proteinaceous foods, in *Enzymes in Food Processing*, 2nd ed., (ed. G. Reed), Academic Press, New York, pp. 473–92.

Bracho, G.E. and Haard, N.F. (1991) Characterization of alkaline metalloproteinases with collagenase activity from the muscle of Pacific rockfish (*Sebastes* sp.), in *Proceedings of Joint Meeting of the Atlantic Fisheries Technologists and Tropical/Subtropical Fisheries Technologists*, (ed. S.W. Otwell).

Brewer, P., Helbig, N. and Haard, N.F. (1984) Atlantic cod pepsin. Characterization and use as a rennet substitute. *Can. Inst. Food Sci. Technol. J.*, **17**, 38–43.

Campbell-Platt, G. (1987) *Fermented Foods of the World – A Dictionary and Guide*, Butterworths, London.

Cano-Lopez, A., Simpson, B.K. and Haard, N.F. (1987) Extraction of carotenoprotein from shrimp process wastes with the aid of trypsin from Atlantic cod. *J. Food Sci.*, **52**, 503–504, 506.

Chen, C.-S., Yan, T.-R. and Chen, H.-Y. (1978) Purification and properties of trypsinlike enzymes and a carboxypeptidase A from *Euphasia superba*. *J. Food Biochem.*, **21**, 349–66.

Chen, H.-M. and Meyers, S.P. (1982) Extraction of astaxanthin pigment from crawfish waste using a soy oil process. *J. Food Sci.*, **47**, 892–900.

Chen, Y. (1992). Characterization of semi-purified collagenase fraction from lobster hepatopancreas. *MSc Thesis*, McGill University, Montreal, Canada.

Cronlund, A.L. and Woychik, J.H. (1987) Solubilization of collagen in restructured beef with collagenases and amylase. *J. Food Sci.*, **52**, 857–60.

de Koning, P.J. (1978) Coagulating enzymes in cheese making. *Dairy Ind. Int.* **43**, 7.

Dixon, B. (1990) Cod waste yields useful enzymes. *Biotechnology*, **8**, 791.

Dransfield, E. and Etherington, D. (1981) In *Enzymes and Food Processing*, (eds G.G. Birch, N. Blankebrough and K.J. Parker), Applied Science, London, pp. 177–178.

Eisen, A.Z. and Jeffrey, J.J. (1969) An extractable collagenase from crustacean hepatopancreas. *Biochim. Biophys. Acta*, **191**, 517–526.

Eisen, A.Z., Henderson, K.D., Jeffrey, J.J. and Bradshaw, R.A. (1973) A collagenolytic protease from the hepatopancreas of the fiddler crab, *Uca pugilator*. Purification and properties. *Biochemistry*, **12**, 1814–22.

Eriksson, C. (1975) Method for controlling the ripening process of herring. *Canadian Patent 969 419*.

Fehmerling, G.B. (1973) Separation of edible tissues from edible flesh of marine creatures. *US Patent 3 729 324*.

Fox, J.W., Shannon, J.D. and Bjarnason, J.B. (1991) Proteinases and their inhibitors in biotechnology, in *'Enzymes in Biomass Conversion,'* (eds G.F. Leatham and M.E. Himmel), ACS, Washington, DC.

Fruton, J.S. and Bergmann, M. (1940) The specificity of salmon pepsin. *J. Biol. Chem.*, **204**, 559–60.

Geist, G.M. and Crawford, D.L. (1974) Muscle cathepsins in three species of Pacific sole. *J. Food Sci.*, **39**, 548–51.

Gildberg, A. (1988) Aspartic proteinase in fishes and aquatic invertebrates. *Comp. Biochem. Physiol.*, **91B**, 425–35.

Gildberg, A. (1993) Enzymic processing of marine raw materials. *Process Brochem.*, **28**, 1–5.

Gildberg, A. and Almas, K.A. (1986). Utilization of fish viscera, in *Food Engineering and Process Applications*, (eds M. Le Maguer and P. Jelen), Applied Science, London, Vol. 2, p. 383.

Gildberg, A. and Raa, J. (1983) Purification and characterization of pepsins from Arctic fish capelin *(Mallotus villosus). Comp. Biochem. Physiol.*, **75A**, 337–42.

Gill, T.A. (1992) Unpublished data, Technical University of Nova Scotia, Halifax., NS, Canada.

Godfrey, T. and Reichelt, J. (1983) *Industrial Enzymology*, The Nature Press, New York.

Grant, G.A., Sacchettini, J.C. and Welgus, H.G. (1983) A collagenolytic serine protease with trypsinlike specificity from the fiddler crab, *Uca pugilator. Biochemistry*, **22**, 354–6.

Haard, N.F. (1986) Atlantic cod protease. I. Characterization with casein and milk substrate and influence of Sepharose immobilization on salt activation, temperature characteristics and milk clotting reactions. *J. Food Sci.*, **51**, 313–6, 326.

Haard, N.F., Helbig, N. and Feltham, L.A.W. (1981) The temperature characteristics of pepsin from two stocks of American smelt (*Osmerus mordax*), in *Proceedings Workshop of the Labrador Coasting Offshore Region*, Newfoundland Institute of Cold Ocean Science, St Johns, pp. 174–96.

Haard, N.F., Feltham, L.A.W., Helbig, N. and Squires, J. (1982) Modification of proteins with proteolytic enzymes from the marine environment, in *Modification of Proteins*, (eds R.L. Feeney and J.R. Whitaker), ACS, Washington, DC, pp. 223–4.

Haard, N.F., Shamsuzzaman, K., Brewer, P. and Arunchalam, K. (1983) Enzymes from marine organisms as rennet substitutes, in *Use of Enzymes in Food Technology*, (ed. P. Dupuy), Lavoisier, Paris, pp. 237–41.

Haard, N.F., Kariel, N., Herzberg, G. and Feltham, L.A.W. (1985) Stabilization of protein and oil in fish silage for use as ruminant feed supplement. *J. Sci. Food Agric.*, **36**, 229–41.

Hajos, G. (1986) Incorporation of essential amino acids into protein by enzymatic peptide modification (EPM). *Nahrung*, **30**, 418–22.

Hameed, K.S. and Haard, N.F. (1985) Isolation and characterization of cathepsin C from Atlantic shortfinned squid, *Illex illecebrosus. Comp. Biochem. Physiol.*, **82B**, 241–6.

Hara, K., Suzumatsu, A. and Ishihara, T. (1988) Purification and characterization of cathepsin B from carp ordinary muscle. *Nihon Suisan Gakkaishi.*, **54**, 1243–52.

Hempl, E. (1983) Taking a short-cut from the laboratory to industry-scale production. *Infofish Marketing Digest*, **4**, 30.

Hjelmeland, K. and Raa, J. (1982) Characteristics of two trypsin type isozymes isolated from the Arctic capelin (*Mallotus villosus*). *Comp. Biochem. Physiol.*, **71B**, 557–62.

Hultin, H.O. (1978) Enzymes from organisms acclimated to low temperatures, in *Enzymes. The interface Between Technology and Economics.*, (eds J.P. Danehy and B. Wolnak, eds), Marcel Dekker, New York, pp. 161–78.

In, T. (1990) Seafood flavourants produced by enzymatic hydrolysis, in *Advances in Fisheries Technology and Biotechnology for Increased Profitability*, (eds M.N. Voigt and J.R. Botta), Technomic, Lancaster, PA, pp. 425–36.

Jacobsen, F. and Lykke-Rasmussen, O. (1984), Energy-savings through enzymatic pretreatment of stickwater in the fish meal industry. *Proc. Biochem.*, **19**, 165–9.

Jany, K.-D. (1976) Studies on the digestive enzymes of the stomachless bonefish *Carassius auratus gibelio* (Bloch): endopeptidase. *Comp. Biochem. Physiol.*, **53B**, 31–38.

Joakimsson, K.G. (1984) Enzymatic deskinning of herring (*Clupea harengus*). *Thesis*, Institute of Fisheries, University of Tromso, Norway.

Kalac, J. (1978) Studies on herring (*Clupea harengus* L.) and capelin (*Mallotus *villosus*) pyloric ceca protease. III. Characterization of anionic fractions of chymotrypsins. *Biologia* (Bratslava), **33**, 939–45.

Knorr, D. and Sinskey, A.J. (1985) Biotechnology in food production and processing. *Science*, **229**, 1224–9.

Laskowski, M. (1986) Protein inhibitors of serine protease-mechanism and classification. *Adv. Exp. Med. Biol.*, **199**, 1–17.

Lee, Y.Z., Simpson, B.K. and Haard, N.F. (1982) Supplementation of squid fermentation with proteolytic enzymes. *J. Food Biochem.*, **6**, 127–34.

Lim, D. and Shipe, W.F. (1972) Proposed mechanism for the antioxidative action of trypsin in milk. *J. Dairy Sci.*, **55**, 753–8.

Long, A. and Haard, N.F. (1988) The effect of carotenoid protein association on pigmentation and growth rates of rainbow trout, *Salmo gairdneri*, in *Proceedings of the Aquaculture International Congress*, Vancouver, BC, pp. 99–101.

Lorier, M.A. and Aitken, B.L. (1991) Method for treating fish with alpha-2-macroglobulin. *US Patent 5 013 568*.

Low, P.S., Bada, J.L. and Somero, G.N. (1973) Temperature adaptation of enzymes: roles of free energy, the enthalpy, and the entropy of activation. *Proc. Natl. Acad. Sci. USA*, **70**, 430–2.

Manu-Tawiah, W. and Haard, N.F. (1987) Recovery of carotenoprotein from the exoskeleton of snow crab, *Chionectes opilio*. *Can. Inst. Food Sci. Technol. J.*, **20**, 31–3.

Marshall, M. (1990) Unpublished data. University of Florida, Gainesville.

Merrett, T.G., Bar-Eli, E. and Van Vunakis, H. (1969) Pepsinogens A, C, and D from the smooth dogfish. *Biochemistry*, **8**, 3696–702.

Metz, F.E., Boyce, F.O. and Sigmiller, J.L. (1975). Process for making barbecue sauce with a tenderizer. *US Patent 3 930 030*.

Mihalyi, E. (1978), *Application of Proteolytic Enzymes to Protein Structure Studies*, 2nd edn, CRC Press, Palm Beach, FL, Vol. 1, pp. 215–308.

Miller, A.J., Strange, E.D. and Whiting, R.C. (1989) Improved tenderness of restructured beef steaks by a microbial collagenase derived from *Vibrio* B-30. *J. Food Sci.*, **54**, 855–857.

Mohr, V. (1980). Enzyme technology in the meat and fish industries. *Process Biochem.*, **August/September**, 18–21, 32.

Murakami, K. and Noda, M. (1981) Studies on proteinases from the digestive organs of sardine – purification and characterization of three alkaline proteinases from the pyloric ceca. *Biochim. Biophys. Acta*, **658**, 17–26.

Nip, W.K., Moy, J.H. and Tzang, Y.Y. (1985) Effect of purging on quality changes of ice-chilled freshwater prawn, *M. rosenbergii*. *J. Food Technol.*, **20**, 9–15.

Opshaug, K.R. (1982) Method of enzyme maturing of fish. *PCT Patent No. 82/03533*.

Overnell, J. (1973) Digestive enzymes of the pyloric ceca and the associated mesentery in the cod (*Gadus morhua*). *Comp. Biochem. Physiol.*, **46B**, 519–31.

Owen, T.G. and Wiggs, A.J. (1971) Thermal compensation in the stomach of the brook trout (*Salvelinus fontinalis* Mitchill). *Comp. Biochem. Physiol.*, **40B**, 465–73.

Pan, B.S. (1990). Recovery of shrimp waste for flavourant, in *Advances in Fisheries Technology and Biotechnology for Increased Profitability*, (eds M.N. Voigt and J.R. Botta), Technomic, Lancaster, PA, pp. 437–52.

Pfleiderer, V.G., Zwilling, R. and Sonneborn, H.-H. (1967) Eine protease vom molekulargewicht 11,000 and eine trypsinahinche frackion aus *Astacus fluviatilis*. *Z. Physiol. Chem.*, **348**, 1319–31.

Raa, J. (1990). Biotechnology in aquaculture and the fish processing industry. A success story in Norway, in *Advances in Fisheries Technology and Biotechnology for Increased Profitability*, (eds M.N. Voigt and J.R. Botta), Technomic, Lancaster, PA, pp. 509–24.

Raa, J. and Gildberg, A. (1982). Fish silage: A review. *CRC Crit. Rev. Food Sci. Nutr.*, **16**, 383–419.

Racicot, W.F. (1984). A kinetic and thermodynamic comparison of bovine and dogfish chymotrypsins. *PhD thesis*, University of Massachusetts, Amherst, MA.

Raksakulthai, N. and Haard, N.F. (1992) Correlation between the concentration of peptides and amino acids and the flavor of fish sauce. *ASEAN Food J.*, **7** 86–90.

Raksakulthai, N., Lee, Y.Z. and Haard, N.F. (1986) Influence of mincing and fermentation aids on fish sauce prepared from male, inshore capelin, *Mallotus villosus*. *Can. Inst. Food Sci. Technol. J.*, **19**, 28–33.

Ramakrishan, M., Hultin, H.O. and Atallah, M.T. (1987) A comparison of dogfish and bovine chymotrypsins in relation to protein hydrolysis. *J. Food Sci.*, **52**, 1198–1202.

Sanchez-Chiang, L. and Ponce, O. (1981) Gastricinogens and gastricins from *Merluccius gayi* – purification and properties. *Comp. Biochem. Physiol.*, **68B**, 251–257.

Schmitt, A. and Siebert, G. (1967) Distinguishing aliphatic dipeptidases from cod muscle. *Z. fur Physiol. Chem.*, **348**, 1009–16.

Shamsuzzaman, K. and Haard, N.F. (1983) Evaluation of harp seal protease as a rennet substitute for Cheddar cheese. *J. Food Sci.*, **48**, 179–83.

Shamsuzzaman, K. and Haard, N.F. (1984) Characterization of a chymosin-like enzyme from the gastric mucosa of harp seal *Pagophilus groenlandicus*. *Can. J. Biochem. Cell Biol.*, **62**, 699–708.

Shamsuzzaman, K. and Haard, N.F. (1985) Milk clotting activity and cheese making properties of a chymosin-like enzyme from harp seal, *Pagophilus groenlandicus*. *J. Food Biochem.*, **9**, 173–92.

Sherekar, S.V., Gore, M.S. and Ninjoor, V. (1988) Purification and characterization of cathepsin B from the skeletal muscle of fresh water fish, *Tilapia mossambica*. *J. Food Sci.*, **53**, 1018–23.

Siebert, G. and Schmitt, A. (1965) Fish tissue enzymes and their role in determining changes in fish, in *The Technology of Fish Utilization*, (ed. R. Kreuzer), Fishery News (Books) Ltd, London, pp. 47–52.

Simpson, B.K. and Haard, N.F. (1984a) Trypsin from Greenland cod as a food processing aid. *J. Appl. Biochem.*, **6** 135–43.

Simpson, B.K. and Haard, N.F. (1984b) Trypsin from Greenland cod, *Gadus ogac*. Kinetic and thermodynamic characteristics. *Can. J. Biochem. Cell Biol.*, **62**, 894–900.

Simpson, B.K. and Haard, N.F. (1984c) Trypsin from Greenland cod, *Gadus ogac*. Isolation and comparative properties. *Comp. Biochem. Physiol.* **79B**, 613–22.

Simpson, B.K. and Haard, N.F. (1985a) Extraction of carotenoprotein from shrimp processing offal with the aid of trypsin. *J. Appl. Biochem.* **7**, 212–22.

Simpson, B.K. and Haard, N.F. (1985b) Characterization of the trypsin fraction from cunner, *Tautogolabrus adspersus*. *Comp. Biochem. Physiol.* **80B**, 475–80.

Simpson, B.K. and Haard, N.F. (1987a) Cold adapted enzymes from fish, in *Food Biotechnology*, (ed. D. Knorr), Marcel Dekker, New York, pp. 495–527.

Simpson, B.K. and Haard, N.F. (1987b). Trypsin and trypsinlike enzymes from the stomachless cunner. *J. Agric. Food Chem.*, **35**, 652–6.

Simpson, B.K., Simpson, M.V. and Haard, N.F. (1989) On the mechanism of enzyme action: digestive proteases from selected marine organisms. *Biotechnol. Appl. Biochem.*, **11**, 226–34.

Squires, E.J., Haard, N.F. and Feltham, L.A.W. (1986a) Pepsin isozymes from

Greenland cod, *Gadus ogac. 1. Purification and physical properties. Can. J. Biochem. Cell Biol.*, **64**, 205–14.

Squires, E.J., Haard, N.F. and Feltham, L.A.W. (1986b) Pepsin isozymes from Greenland cod. *Gadus ogac.* 2. Substrate specificity and kinetic properties. *Can. J. Biochem. Cell Biol.*, **64**, 215–22.

Stefansson, G. and Steingrimsdottir, U. (1989a). Descaling of redfish. *Internal Report*, Icelandic Fisheries Laboratories, Reykjavik, Iceland.

Stefansson, G. and Steingrimsdottir, U. (1989b). Enzymatic descaling of haddock. *Internal Report*, Icelandic Fisheries Laboratories, Reykjavik, Iceland.

Stefansson, G. and Steingrimsdottir, U. (1990). Application of enzymes for fish processing in Iceland – present and future aspects, in *Advances in Fisheries Technology and Biotechnology for Increased Profitability*, (eds M.N. Voigt and J.R. Botta), Technomic, Lancaster, PA, pp. 237–50.

Stefansson, G., Johannesson, J. and Magnusdottir, E. (1987) Removal of intestines from scallop muscle with enzymes. *Internal Report*, Icelandic Fisheries Laboratories, Reykjavik, Iceland.

Steingrimsdottir, U. (1987). Enzymatic removal of cod liver membrane. *Internal Report*, Icelandic Fisheries Laboratories, Reykjavik, Iceland.

Steingrimsdottir, U. and Stefansson, H. (1988). Enzymatic removal of swimbladder membrane. *Internal Report*, Icelandic Fisheries Laboratories, Reykjavik, Iceland.

Sugihara, T., Yashima, C., Tamura, H., Kawasaki, M. and Shimizy, S. (1973) Process for preparation of Ikuna (salmon egg). *US Patent 3 759 718*.

Tanji, M., Kageyama, T. and Takahashi, K. (1988) Tuna pepsinogens and pepsins. Purification, characterization and amino terminal sequences. *Eur. J. Biochem.*, **177**, 251–9.

Ting, C.-Y., Montgomery, M. and Anglemier, A.F. (1968) Partial purification of salmon muscle cathepsins. *J. Food Sci.*, **33**, 617–21.

Twining, S.S., Alexander, P.A., Huibregtse, K. and Glick. D.M. (1983) A pepsinogen from rainbow trout (*Salmo gaidneri*). *Comp. Biochem. Physiol.*, **75B**, 109–12.

Warrier, S.B.K., Ninjoor, V. and Nadkarni, G.B. (1985) Involvement of hydrolytic enzymes in the spoilage of Bombay duck (*Harpodon nehereus*): technical report, in *Harvest and Postharvest Technology of Fish*, Society of Fisheries Technology, India, pp. 470–72.

Wasserman, B.P. (1990) Evolution of enzyme technology: progress and prospects. *Food Technol.*, **44**, 118–22.

Wiggs, A.J. (1974) Seasonal changes in the thyroid proteinase of a teleost fish, the burbot, *Lota lota*, L. *Can. J. Zool.*, **52**, 1071–8.

Wray, T. (1988) Fish processing: new uses for enzymes. *Food Manufacture*, **63**(7) 48.

Yoshinaka, R., Sato, M., Suzuki, T. and Ikeda, S. (1984). Enzymatic characterization of anionic trypsin of the catfish *Parasilurus asotus*. *Comp. Biochem. Physiol.*, **77B**, 1–6.

Zeef, A.H. and Dennison, C. (1988) A novel cathepsin from the mussel (*Perna perna* Linne). *Comp. Biochem. Physiol.*, **90B**, 204–210.

CHAPTER 7

Bioprocessing of chitin and chitosan

B.K. Simpson, N. Gagne and M.V. Simpson

7.1 INTRODUCTION

Chitin is a non-toxic, biodegradable, high molecular weight, naturally occurring polymer of *N*-acetylglucosamine and glucosamine residues, which is widely distributed in nature. The constituent monosaccharide units in chitin are linked together by β(1–4) glycosidic bonds. It is the second most abundant natural polymer after cellulose (Ruiz-Herrera, 1978) with which it bears a structural resemblance, except that the some of the –OH groups bonding the second carbon position of cellulose are acetylated in chitin (Tsugita, 1991). Chitin is found associated with protein in the exoskeletons of marine invertebrates, insects and arachnids (Austin *et al.*, 1981), and in the cell walls of various fungi and algae. Chitosan is the *N*-deacetylated form of chitin and is derived from chitin by deacetylation, i.e. the acetamide groups in chitin are substituted into amino groups in chitosan. Chitosan forms the body wall of most fungi, molds and yeasts (Ruiz-Herrera, 1978), and like chitin, chitosan is non-toxic and biodegradable. Chitin and chitosan are produced by fresh water invertebrates (e.g. arthropods, bryozoans and zooplanktons) in eutrophic stagnant waters. In the fresh water environment, arthropods are the main chitin producers, with annual chitin production of about 51 g/m^2). In the salt water environment, the major chitin producers include marine haloplanktonic crustaceans (e.g. copepods, cladecera and euphausiaceae); and marine benthic and pelagic species (including various crustaceans, hydrozoans and bryozoans). Some terrestrial animals also possess chitinous integuments (e.g. insects, crustacea, annelids and mollusks), while fungi, molds and yeasts also possess chitinous cell walls (Ruiz-Herrera, 1978). In animals, these chitinous polymers

Fisheries Processing: Biotechnological applications.
Edited by A.M. Martin.
Published in 1994 by Chapman & Hall, London. ISBN 0 412 58460 3

carry out a protective and/or mechanical function, serving as 'envelopes' for eggs and other latent forms of life such as cysts; they contribute to the formation of locomotor appendages and masticatory organs such as jaws and radula; and also serve a protective role in the intestinal mucosa of some arthropods and annelids (Muzzarelli, 1977).

Shellfish processing waste (i.e. derived from shrimp, crab, lobster, prawn, squid and crawfish) contain 14–35% chitin on a dry weight basis, and it is estimated that more than 100 billion tons of chitin are produced annually by these animals and microorganisms. However, this abundant resource material has been put to relatively little use thus far, and these source materials for chitin are becoming a growing waste disposal problem for the fish and shellfish processing industries (Revah-Moiseev and Carroad, 1981). The amount of annually accessible chitin is currently estimated at about 150 000 tons (Allan, Fox and Kong, 1987), and the total sales of chitin/chitosan are projected to reach almost US$ 2 billion during the next 10 years (Knorr, 1991).

Japan is the leading producer and user of chitin and chitosan. According to Hirano (1988), industrial chitosan was first produced in the world by Kyowa Yushi, Inc., (Chiba, Japan) and, since then, several other Japanese companies have been engaged in the commercial production of chitin or chitosan. Commercial chitin and chitosan are also produced to a lesser extent in the US and other countries by, for example,:

1. Polypate of Calcutta (India), which makes the products from prawn;
2. Kypro Co. (Seattle, Washington, USA), which produces the products from the Dungeness crab;
3. the Fisheries Central Board in Szczecin (Poland), which fabricates the products from Antarctic krill and squid pens (Muzzarelli *et al.*, 1981; Brzeski, 1988);
4. Bioshell Products (Albany, Oregon, USA);
5. NovaChem (Halifax, Canada).

Commercial chitin and chitosan are supplied in the form of powders, flakes, fibers, beads and sheets, and show variation in quality with each of these companies. The variation in quality of commercial chitin and chitosan is due to the fact that current methods for producing chitin/chitosan polymers cause some modifications such as depolymerization and deacetylation of the native molecule, resulting in products with inconsistent physicochemical characteristics, because of the rather non-specific nature of the action of these chemicals (Knorr and Klein, 1986). For the purpose of preparing chitin/chitosan of less modified structures, milder treatments are preferable. The biological approaches recover products with better/improved consistent properties; however, the yields are rather low. This chapter reviews various approaches used to recover chitinous polymers from their source materials, with more emphasis on the biological methods. Some properties and applications of these natural polymers are also discussed.

7.2 GENERAL PROPERTIES OF CHITIN AND CHITOSAN

A number of studies on chitin chemistry, structure and application indicate that these polymers have an unusual combination of properties, including high tensile strength, bioactivity and biodegradability, which make them attractive speciality products. They are natural, non-toxic, high molecular weight, water-insoluble, biodegradable compounds with high water holding capacities and a capacity to form films or coatings. Furthermore, chitin is crystalline and insoluble in ordinary solvents (Tracey, 1957; Muzzarelli, 1977; Austin *et al.*, 1981). However, it is soluble in strong mineral acids and in anhydrous formic acid, but insoluble in alkali. Its high degree of crystallinity enables chitin to be cast into films or membranes. The film or membrane forming properties of chitin may be demonstrated by dissolving it in a suitable solvent, and then coating the resulting solution on a particular surface to produce a film (Muzzarelli, 1977).

Pyrolysis of chitin at high temperatures (about 900°C) produces a number of flavor compounds (Knorr, 1984). The pyrolyzate fraction can be divided into four groups:

1. the acidic fraction, made up of short chain fatty acids (e.g. butyric acid, with the characteristic flavor of butter);
2. the phenolic fraction, comprising phenol, xylenol and cresol;
3. the neutral fraction made up of naphthalene and xylene;
4. the basic fraction made up of picoline, nicotine and pyrazines.

The pyrazines from the basic fraction are a group of heterocyclic nitro-genous compounds which contribute significantly to the characte-ristic flavor of toasted and roasted foods.

The deacetylated form of chitin, chitosan, is also remarkably stable to most reagents, including aqueous alkali (Muzzarelli, 1977) and, according to Von-Furth and Russo (1906), three out of four acetyl groups can be removed from chitin during deacetylation. According to Muzzarelli *et al.* (1981), commercial chitosans derived from shellfish chitin exhibit acetyl contents varying from 22 to 43%. The characteristic difference between chitin and chitosan is their solubility. For example, chitosan, unlike chitin, is soluble in dilute acid solutions; however, both are insoluble in alkaline solutions and organic solvents. Because chitosan is soluble in dilute acids, e.g. formic and acetic acids, and its salt is soluble in water, it finds wider applications in industry than chitin.

At acidic pH, chitosan is a linear polyelectrolyte with a high charge density of $-NH_3^+$ (i.e. one charge per glucosamine unit), making it a high positively-charged polyelectrolyte, and it is this feature about chitosan that makes it a more versatile and valuable product than chitin. Since many materials carry negative charges (e.g. proteins, anionic polysaccharides, nucleic acids, etc.), they can interact strongly with the positive charge of chitosan to produce electric neutrality. Thus, chitosan adheres readily to natural polymers such as hair and skin, which are composed of negatively-charged mucopolysaccharides and proteins. Chitosan can also form complexes with metal ions and for these reasons, it has been used:

1. in the treatment of water;
2. for biomass recovery from industrial waste water;
3. as an active ingredient in hair fixing products.

Chitosan facilitates wound healing, reduces blood serum cholesterol (Nagyvany *et al.*, 1979) and stimulates the immune system (Nishimura *et al.*, 1985). When seeds are coated with chitosan, the latter induces a protective response in the germinating plant which results in increased crop yields (Hadwiger, 1984; Sandford, 1989). Chitin and chitosan are also capable of interacting with transition metals by chelation, sorption, ion exchange or coordinate covalent binding to produce various colored compounds.

7.3 METHODS FOR PREPARING CHITIN

Even though chitin and chitosan are widely distributed in nature, the most practical source for preparing these polymers commercially is from shellfish processing wastes. As mentioned earlier, chitin in shellfish waste is tightly associated with proteins, lipids, pigments and calcium deposits. Therefore these source materials have to be pretreated to remove these components. Current procedures for preparing chitin from shellfish waste consist of two main steps, i.e. $CaCO_3$ separation or demineralization with dilute acid or chelating agents (Tsugita, 1991), followed by protein separation or deproteinization with dilute alkali or proteolytic enzymes. Alternatively, the deproteinization process may precede the demineralization step (No, Meyers and Lee, 1989; Sandford, 1989). Subsequent to the demineralization and deproteinization steps, the product may be decolorized with acetone and/or hydrogen peroxide. Decoloration has also been achieved with NaOCl (No, Meyers and Lee, 1989), although Austin (1988) recommended peroxide bleaches over their chlorine counterparts, since the latter could form *N*-chloro compounds with chitin. The general scheme for preparing chitin from shellfish waste is summarized in Figure 7.1.

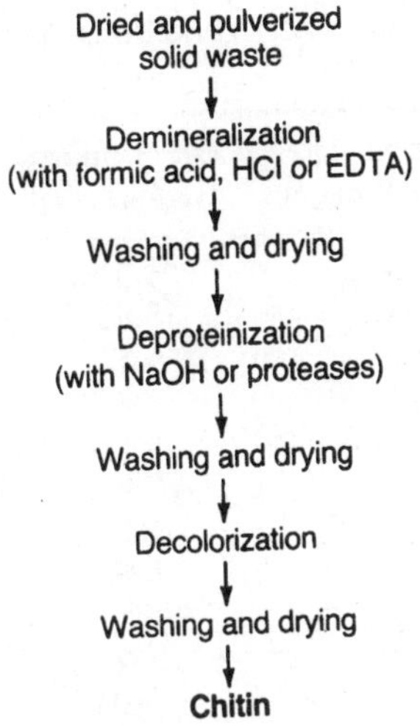

Figure 7.1 General scheme for preparation of chitin.

7.3.1 Chemical methods

(a) Demineralization

Demineralization may be achieved by treating the shells with a dilute aqueous solution of HCl, EDTA (pH 7.7) or formic acid. The HCl method was first described by Hackman (1954) and it is still the most widely used. Demineralization with HCl involves extraction of the shells with the acid at ambient temperature for 3 h. Hackman (1954), Whistler and BeMiller (1962) and Broussignac (1968) all used HCl for shellfish waste demineralization. In this reaction, $CaCO_3$ is converted into soluble $CaCl_2$ with the evolution of CO_2 gas. These studies show that the degree of demineralization attainable depends and/or increases with the concentration of the acid. Some of the disadvantages with HCl as the demineralization agent include the following: use of HCl at concentrations above 1.25 N adversely affects the viscosity of the final product due to depolymerization (Muzzarelli, 1977). For example, Madhavan and Ramachandran (1974) gave a comparative account of demineralization conditions on the viscosity of chitosan solutions (Table 7.1).

To control the deacetylation (and also depolymerization), Lusena and Rose (1953) suggested the use of HCl at pH not lower than 3. Furthermore, EDTA, H_2SO_3 or formic acid may be used instead of HCl for demineralization (Foster and Hackman, 1957; Horowitz, Roseman and Blumethal, 1957). The use of CH_3COOH and H_2SO_3 for decalcifying shellfish waste during chitin production evolved based on the need to control chitin degradation caused by HCl (Madhavan and Ramachandran, 1974; Peniston and Johnson, 1978), and involved extraction of the

Table 7.1 Effect of demineralization conditions on the viscosity of chitosan solutions

HCl concentration (M)	*Treament time (min)*	*Ash in chitin (%)*	*Viscosity of 1% solution (cP)*
1.25	30	24.34	106.85
	60	18.82	97.07
	120	6.33	58.05
	180	2.97	46.44
1.50	30	15.34	49.28
	60	7.90	43.95
	120	3.14	40.06
	180	1.46	38.84
2.00	30	2.72	37.66
	60	1.76	31.52
	120	1.03	26.94
	180	0.65	17.79

From Madhavan and Ramachandran (1974).

shells with 1.75 N CH_3COOH or 1.67 N H_2SO_3 for 12 h at ambient temperature. Alternatively, supersaturated EDTA solution in phosphate-buffered formalin may be used to demineralize the shells (Muzzarelli, 1977). In studies carried out in our laboratory (Gagne, 1992), demineralization with formic acid was milder and recovered a better product.

(b) Deproteinization

The chemical methods that have been used to deproteinize shellfish waste for chitin production include digestion with NaOH (Hackman, 1954), refluxing of shells with dimethylformamide (Takeda and Katsuura, 1964) or extraction with mild H_2SO_3 solution at 25°C for 3 h (Brine and Austin, 1981). By far, the most widely used method for deproteinization involves digestion with alkali. Hackman (1954), Whistler and BeMiller (1962) and Horowitz, Roseman and Blumenthal (1957) all used dilute aqueous solution of NaOH for removal of protein, with constant stirring and for prolonged hours in an inert atmosphere. The limitations with this approach include the following.

1. Requirement of large amounts of alkali which results in deacetylation.
2. A decrease in the molecular size of the product (Muzzarelli, 1977).

To prevent such modifications, milder treatments are preferable.

(c) Decoloration of chitin

Decoloration of chitin may be achieved by use of organic solvents such as acetone or ethanol. Other ways of removing these pigments include the use of cold formic acid on the carapace, or mixtures of $(NH_4)_2SO_4$ and H_2SO_4 Whistler and BeMiller (1962) used a combination of 95% ethanol, acetone and ether to decolorize chitin, but this could not remove all the pigments. For commercial chitin preparation, further treatment with bleaching reagents is required. These may include sodium hypochlorite (Blumberg *et al.*, 1951), 3% hydrogen peroxide (Brine and Austin, 1981) and ethyl acetate (Brzeski, 1982). According to No, Meyers and Lee (1989), the treatment that gives the most acceptable commercial product is extraction of the pigments by refluxing with absolute acetone, followed by treatment with 75% acetone and then bleaching with 0.315% cold sodium hypochlorite (Blumberg *et al.*, 1951). However, more recent studies suggest that hydrogen peroxide is preferable as a decoloring agent (Austin, 1988; Gagne, 1993).

7.3.2 Biological methods

The desire for producing chitins with more consistent physicochemical properties necessitates the use of milder treatments for removing some of the components associated with the shell, such as proteins. Some investigators have attempted to deproteinize crustacean waste intended for chitin/chitosan production by enzymatic digestion (Hackman, 1960; Takeda and Abe, 1962; Takeda and Katsuura, 1964); however, complete removal of protein was not attained. Shimahara, Ohkouchi and Ikeda (1982) described a procedure for deproteinizing crustacean carapace by growing cells of proteolytic bacteria (*Pseudomonas maltophilia* LC 102) on the raw material, and found this method of deproteinization to be preferable for the purpose of preventing deacetylation.

In our own laboratory, various proteolytic enzymes (i.e. chymotrypsin, papain and bacterial protease) were used to deproteinize crustacean shells (Gagne, 1993). In these studies, the response surface methodology (RSM) approach was used to optimize conditions for achieving maximum deproteinization with these proteases. Of these three proteases tested, chymotrypsin was found to be the most effective, achieving a degree of deproteinization comparable with the chemical approach using NaOH. The optimum conditions established with the RSM approach using chymotrypsin were: pH 8 for 72 h at 40°C and an enzyme to substrate ratio of 7:1000 (w/w).

7.4 METHODS FOR PREPARING CHITOSAN

Chitin prepared from shellfish waste may be processed further into chitosan by deacetylation. Deacetylation may also be achieved by either chemical means (i.e. with hot concentrated NaOH solution) or by biological means using chitin deacetylase or chitin deacetylase producing microorganisms like *Mucor rouxii*, *Phycomyces blakesleeanus* and *Asper-*

Chitin

↓

Deacetylation with NaOH
or chitin deacetylase

↓

Chitosan

Figure 7.2 General scheme for the preparation of chitosan from chitin.

gillus niger). The general scheme for the preparation of chitosan from chitin is shown in Figure 7.2.

7.4.1 Chemical methods

Some of the limitations with the chemical approach for preparing chitosan from chitin include the following.

1. It requires high temperature and therefore entails high energy cost.
2. It involves the use of high volumes of concentrated NaOH, which is caustic.
3. There is simultaneous production of liquid waste containing protein and other non-protein nitrogenous compounds (Shimahara *et al.*, 1989).

It has been suggested that deacetylation proceeds to about 70% within the first hour of treatment with 50% NaOH solution at 100°C, but it progresses only gradually after this, reaching about 80% in 5 h (Wu and Bough, 1978). Further, deacetylation can rarely extend beyond 80%

progresses only gradually after this, reaching about 80% in 5 h (Wu and Bough, 1978). Further, deacetylation can rarely extend beyond 80%
unless the alkali fusion procedure is applied along with fractionation (Horton and Lineback, 1965); Muzzarelli, 1977). Extended treatments with hot concentrated NaOH solution could result in an almost completely *N*-deacetylated product (i.e. about 90% deacetylation), however, such a treatment is also accompanied by degradation of the molecular chain. Since treatment with alkali leads to depolymerization of the material, deacetylation may be carried out in an inert atmosphere to alleviate the depolymerization problem and thereby recover chitosans of higher viscosity (Muzzarelli, 1977). In this regard, Mima *et al.* (1982) have described a method for preparing chitosan having a degree of deacetylation of up to 100% by alkali treatment of chitin with repeated washing of the intermediate products with water.

7.4.2 Biological methods

erson, 1962). Thus, preparation of chitosan from chitin by biological methods may be accomplished either:

1. by fermentation, whereby cultures of chitin deacetylase producing microorganisms are inoculated in media containing chitinous substrate, as demonstrated with *A. niger* mycelia; or
2. by a direct treatment of chitin polymers with extracts of microbial chitin deacetylase enzyme (Knorr and Klein, 1986).

McGahren *et al.* (1984) prepared chitosan by fermentation of *Absidia coerulea* in a medium containing glucose or molasses as carbon source (pH 4.5), together with a high concentration of ammonium salt, trace quantities of mineral salt and yeast extract. Other microorganisms that have been shown to produce chitin deacetylase include *A. niger* (Knorr and Klein, 1986), *Mucor meehei* (Kauss, Jeblick and Young, 1982/1983), *Absidia butleri, Ab. glauca, Ab. orchidis, Circinella mucoroides, C. muscae, Phycomycetes nitens, Rhizopus acetoinus, R. chinensis, R. circanans, R. microsporus, Zygorhyncus californiensis* and *Z. exponens* (Shimahara *et al.*, 1989). Of all the mucoraceae strains listed, *Ab. butleri* was found to be the most efficient in producing chitosan from chitin. Some problems associated with the production of chitosan by fermentation include low yields and variation with age of culture. On the other hand, chitosan production by fermentation has the advantage of recovering a product with a very low degree of acetylation (5–10%, White, Farina and Fulton, 1979), compared with chitosans from shellfish chitin with acetyl content ranging from near 22 to 43% (Muzzarelli *et al.*, 1981). Various researchers have also attempted to produce chitosan by fungal fermentation, involving the enzyme chitin deacetylase (Araki and Ito, 1975; McGahren *et al.*, 1984), but these have shown inconsistent trends in the degree of deacetylation, low yields (Knorr and Klein, 1986) as well as variation in molecular weight (Arcidiacono, Lombardi and Kaplan, 1988). However, it is to be expected that appropriate manipulation of processing procedures and controls over biosynthesis could alleviate some of the problems with yield, degree of acetylation and wide variations in molecular weight. Thus, the added advantage with the use of microorganisms in chitosan production is the potential that exists for genetic manipulation of the organism.

The enzyme, chitin deacetylase, was first recovered by Araki and Ito (1974) from *Mucor rouxii* extracts by growing the organism in a peptone (1%), yeast extract (0.2%) and glucose (2%) medium (pH 4.5). The mycelia were harvested by filtration through a sintered glass funnel, frozen and powdered in a mortar, then extracted with 5 mM CH_3COONa buffer (pH 5.2). The crude extract thus obtained was further purified by $(NH_4)_2SO_4$ precipitation, ion exchange chromatography (IEX) on CM–cellulose and DEAE–cellulose. The enzyme has also been recovered from the plant pathogen, *Colletotrichum linde-*

muthianum (Kauss and Bauch, 1988), by inoculating the spores in a malt extract (5%)/yeast extract (2%)/glucose (2%) medium at 22–25°C with vigorous rotation for about 5 days and then filtering to obtain the filtrate as the crude enzyme. This crude extract was purified further by $(NH_4)_2SO_4$ fractionation and IEX on DEAE–Sephacel. The recovered enzyme fraction may then be applied to chitin for deacetylation to chitosan.

Since shellfish wastes are rich in various other useful nutrients and biochemicals, e.g. protein, carotenoid pigments, minerals and fat (Simpson and Haard, 1985), apart from chitin and chitosan, a recovery scheme based primarily on enzymatic methods and mild chemical treatments used in our laboratory, to ensure maximum economic use of this abundant waste material, is suggested in Figure 7.3.

Recovery of carotenoprotein by a trypsin-aided process for
use as feed supplement for cultured salmonids
(Simpson and Hoard, 1985)

↓

Demineralization
with 1.75 N CH_3COOH for 12 h at ambient temperature

↓

Further deproteinization with proteases (e.g. chymotrypsin)

↓

Decoloration with H_2O_2

↓

Washing and drying

↓

Chitin

↓

Deacetylation with chitin deacetylase

↓

Chitosan

Figure 7.3 Suggested scheme for production of chitin and chitosan by biological means for maximum economic use of shellfish waste.

7.5 ADVANTAGES OF BIOLOGICAL VERSUS CHEMICAL METHODS FOR PREPARING CHITIN/CHITOSAN

As noted earlier in this chapter, the chemical procedures currently used to recover chitin/chitosan result in commercial products of inconsistent physicochemical characteristics, because of the rather non-specific nature of the action of the chemicals (Knorr and Klein, 1986). Also, chitin content varies with the kind of species as well as the part of the species used as source material. For example, the limbs of red crab have more chitin than the carapace (Hirano, 1988). Chitosans of different origins and from different production plants also do not have the same properties. This might also be due to variations in the source material – natural species variation of chitin have been demonstrated (Muzzarelli, 1977) – as well as differences in the isolation procedures employed for chitin production and subsequent processing to chitosan. For example, the viscosity and degree of acetylation of commercial chitosan fabricated by Kyowa Tokyo were 263 cP (at 15°C) and 26% respectively, compared with the product made by Kurita Kurifax whose viscosity and acetyl content were 189 cP (at 15°C) and 41% respectively. Further, the viscosity (198 cP at 15°C) and degree of acetylation (26%) of the products derived from Chesapeake differed from the two products manufactured by Kyowa Tokyo and Kurita Kurifax described above (Muzzarelli *et al.*, 1981). The essence of recovering chitin/chitosan with consistent properties lies in the following factors: it is useful in defining the quality of product, and ensures reproducible results.

7.6 APPLICATIONS OF CHITIN AND CHITOSAN

Chitin and chitosan are already part of our food supply (Knorr, 1984), and numerous applications for these natural polymers in the food and pharmaceutical, as well as cosmetic, paint and textile, industries have been identified. The advantages of using chitinous polymers as food and other processing aids include the following: they are natural, non-toxic components of foods; are biodegradable and have been approved for potable water application, and as a seed coating by the US Environmental Protection Agency (EPA). They are also approved by both the Food and Drugs Administration and EPA for use as precipitating agents of proteinaceous material in food processing, and as a feed supplement at a level of incorporation not exceeding 0.1% (Sandford, 1989). Chitin and chitosan are also used as a clarifying agent for wines and drinking water in the UK and other European countries (Ornum, 1991). Chitinous polymers also have antimicrobial properties and other health-related benefits, such as the capacity to tie up bile salts, as well as their ability to lower blood serum cholesterol levels.

In the food, pharmaceutical and cosmetic industries, there has always been a need for inexpensive suspensions of polymers for use as thickening, suspending, stabilizing, gelling and viscofying agents. Water soluble polysaccharides such as alginates, cellulose, guar gum, starch and xanthan gum and their derivatives are commonly used for these purposes (Muzzarelli, 1977). Since chemical modification of chitin could produce water soluble forms of chitin, interest has arisen in the utilization of these products in food applications. The film-forming ability of chitin has been the subject of several studies, some of which have led to industrial uses in various fields such as photographic films (Ryan and Yankowski, 1961), as reverse osmosis membranes (Yang and Zall, 1984) and in cosmetics (Gross, Konrad and Mger, 1983; Lang and Clausen, 1988).

7.6.1 Current applications

The principal use of chitosan has been as a 'flocculating agent' for clarification of industrial effluent and other waste products, e.g. sludge from breweries, sewage, stick water from fish processing plants, slaughterhouses, etc. (No and Meyers, 1989), and as a chelator of environmental contaminants (e.g. PCBs, heavy and/or radioactive metals). Chitosan is able to carry out this function because of its highly positively-charged nature, which enables it to combine with negatively-charged biomolecules or form complexes with metal ions, which then flocculate and are separated by filtration, centrifugation or pressing. The capacity of chitin and chitosan to form complexes with metal ions has been exploited in Japan for the treatment of water.

Chitosan also reacts with aldehydes under suitable conditions to form gels, and this reaction is also used as a basis for entrapment of cells and enzymes. Chitosan is put to a number of uses in the biotechnological industry, such as the manufacture of chitosan beads by Fujibo, Inc. (Shizuoka, Japan) for use in bioreactors to immobilize whole cells, enzymes like glucose isomerase or for chelating metals. Products based on this capacity to form gels are commercially available from Seikagaku Kogyo, Inc. (Tokyo, Japan). Chitosan is also used in the food industry as a food quality enhancer. For example, Nihon Kayaku, Inc. (Tokyo, Japan) and Tanmi Foods, Inc., (Nagoya, Japan) produce dietary cookies and noodles enriched with chitosan because of its hypocholesterolemic value. Furthermore, vinegar products containing chitosan are made and sold by Nakano, Inc. (Handa, Japan), also because of the cholesterol lowering ability of chitosan (Hirano, 1989).

Chitosan is an excellent moisturizer because of its high water holding capacity, and this property has been exploited in the making of cosmetic (hair and skin) products in Japan (Hirano, 1989). The water holding capacity, as well as the good film-forming and cationic properties of chitosan have enabled the extensive use of this material in skin and hair applications. Products for fixing hair (i.e. shampoo, conditioner, setting lotion, etc.), usually consist of film-forming synthetic polymers (e.g. polyvinylpyrolidone, polyvinylacetate or polyacrylic compounds, etc.; Gross, Konrad and Mger, 1983). The concentration of chitosan salts (about 1%) required for fixing the hair is comparatively lower than that of conventional resins (about 3%) to fix the same hair (Gross, Konrad and Mger 1983). Yanagida (1985) described the use of powdered or granular chitin or chitosan as an abrasive for skin cleansing. In this example, chitin or chitosan is added to the usual cosmetic emulsion and it acts mechanically on the skin. Bernadet (1960) has also described a

Table 7.2 Some other applications of chitin/chitosan polymers

Industry	*Application*
Food	Biomass recovery as in cheese processing; sludge treatment as in brewing; as packaging films or edible food wraps; in flavor production; and as a food thickener and/or stabilizer.
Paper and pulp	Papermaking additive for surface strength improvement.
Water purification	Removal of color; in waste water treatment; and for removal of heavy metals (e.g. Hg, Cd, Pb, Ag and Ni), radioactive metals (e.g. U and Co), and environmental contaminants like PCBs.
Plant protection	As defence against pathogens; inhibition of fungal growth; and the suppression of plant parasites.
Animal nutrition	As feed supplement; dietary fiber; digestive aid; and for the promotion of growth of bifidobacteria.
Chemical industry	As oil refinery waste water decontaminant; as molecular sieves; as flame retardant; in reverse osmosis; as an adhesive; for enzyme/cell immobilization; in chromatography.
Cosmetics	In dry shampoos; hair conditioners and hair setting lotions.
Pharmaceuticals	As diluent for drug and other tablet manufacture.
Biomedicals	As bile salt sequesterant; anticoagulant; vascular grafts; sutures; aggregation of cells; artificial kidney membrane.

patented procedure for incorporating chitin in lotions, creams and gels for the treatment of cellulitis.

The use of chitosan as coating material for seeds such as wheat, to enhance crop yield (Sandford, 1989) has been approved by the US Environmental Protection Agency (EPA) and the Association of Animal Feed Control Officials, Inc. (AAFCO). When used as a seed coat, chitosan enhances the ability of the plant to protect itself from its natural predators, such as pathogenic fungi (Sandford, 1989). Other applications of chitin/chitosan polymers include those summarized in Table 7.2.

7.6.2 Other potential applications

Chitin and chitosan have potential for use as dietary fiber. It has been found that addition of up to 10% of chitin to the diet of chickens resulted in normal growth and vigor of the animals, and also increased growth of *Bifidobacterium* in the gut (Austin *et al.*, 1981). The bacteria block the growth of other microorganisms in the gut and generate lactase required for the digestion of milk. This could be significant for both humans and animals with lactose intolerance. It has also been shown that when chitin is added to bread whose gluten has been partly replaced by other protein sources, there is prevention of depression of loaf volume (Knorr and Betschart, 1978, 1981). When used as dietary fiber, chitin exhibits a potential of concurrent reduction in cholesterol levels in humans (Knorr, 1986).

The potential for the use of chitosan powder in a biosensor for monitoring lipid oxidation in muscle foods such as fish/shellfish, was recently demonstrated by Weist and Karel (1992). The rationale for this application is based on the ability of the primary amino groups of chitosan to form a stable fluorosphere with volatile aldehydes, e.g. malonaldehyde from the breakdown of fats (Weist and Karel, 1992).

The sequestering ability of chitinous polymers for transition metal ions produces antimicrobial action. The exact mechanism of the antimicrobial action of chitin/chitosan is still uncertain; however, a few plausible speculations have been put forward which suggest that chitin/chitosan act by:

1. depriving some of these microorganisms of vital ions, e.g. Cu. Thus, the microorganism is killed by the disturbance in the equilibria involving metal ions (Muzzarelli, 1977);
2. blocking or destroying membrane function (Papineau *et al.*, 1991);
3. causing leakage of intracellular constituents (Papineau *et al.*, 1991);
4. causing the formation of polyelectrolyte complexes with acidic polymers at bacterial cell surfaces (Muzzarelli *et al.*, 1990). This property of chitinous polymers has promoted their incorporation in the manufacture of personal hygiene products like toothpaste and mouthwashes (Muzzarelli, 1977).

In addition, chitosan is able to chelate many harmful metals (e.g. Hg, Pb, Cd, Ag, Zn, Cu, U, etc.). Thus, there is interest in the use of this product for the detoxification of hazardous waste materials.

Chitinous polymers also have potential for use as non-absorbable carriers of food additives. The rationale underlying this concept of non-absorbable food additives stems from the fact that functional ingredients which are absorbed intact or metabolized may interact with target organs or tissues, and constitute a potential risk. The use of chitin/chitosan as non-absorbable food carriers for highly concentrated food ingredients was investigated in animal feeding trials with the food dye FD&C Red No. 40 and these studies showed that when the dye is attached to the chitinous polymer, absorption of the dye by the animal was reduced (Watkins and Knorr, 1983). Investigations on the potential use of chitin and chitosan in controlled drug delivery systems have been carried out by many authors (Muzzarelli, 1983; Nagai, Sawayanagi and Nambu, 1984; Baba *et al.*, 1988). Its polymeric character along with its possession of potentially reactive functional group have given it unique possibilities for use in controlled release technology.

The capacity of chitosan effectively to aid the coagulation of protein from food process wastes and thereby reduce suspended solids (Fujita, 1972; Bough, 1976) is of interest for exploitation in biomass recovery. Over the past two decades, increasing attention has been given to the use of polyelectrolytic coagulants of natural origin in aiding the coagulation of colloidal and suspended solids from food processing wastes.

Processing of clarified fruit juices commonly involves the use of clarifying agents including gelatin, bentonite, silica sol, tannins, polyvinylpyrolidone or various combinations of these compounds. Chitosan salts, which carry a strong positive charge, have been shown to be equally effective in reducing juice turbidity. Soto-Peralta, Miller and Knorr (1989) found that both acid soluble and water soluble chitosan salts are effective as fining agents. Comparison of the effectiveness of an acid soluble chitosan salt or the water soluble form with the conventional silica sol/gelatin/bentonite treatment for the clarification of fruit juices revealed that chitosan is effective at low dose application levels (i.e. 0.6–0.7 kg/m^3); and there was little effect of temperature on the removal of colloidal and dispersed particles in apple juices by chitosan (Soto-Peralta, Miller and Knorr, 1989). Chitosan has also been shown to retard enzymatic browning in apple and pear juices (Sapers, 1992).

The film-forming properties of chitin has led to the recommendation for its use as oven and other food wraps. The use of *N,O*-carboxymethylchitin films to preserve fruits over long periods has been approved in both Canada and the US (Davies, Elson and Hayes, 1988). Application of chitin and chitosan as wound healing accelerators, as artificial skin and fibers for absorbable sutures have been tested (Muzzarelli, 1977; Malette, Quigley and Adickes, 1982). The use of contact lenses made of chitosan has also been reported (Markey, Bowman and Bergamini, 1988).

7.7 CONCLUSION

Current methods for preparing commercial chitin and chitosan result in products with inconsistent properties. Biological methods for preparing these products are milder and thus relatively more capable of recovering products with more consistent properties. With the advances in biotechnology, it is to be expected that the biological approaches would be relatively more amenable to manipulation to result in improvements in yield, as well as control variation in properties in the products recovered from various sources. At the present time, there seems to be a fair balance between production and utilization of these polymers. However, the large quantities of shellfish waste generated annually; the need to convert food processing waste to value-added products to alleviate environmental pollution concerns; and the ever growing interest in novel applications of chitin and chitosan, makes it predictable that production of these products will increase considerably in the foreseeable future.

REFERENCES

Allan, G.G., Fox, J.R. and Kong, N. (1978) A critical evaluation of the potential sources of chitin and chitosan, in *Proceedings of the 1st International Conference on Chitin/Chitosan*, (eds R.A.A. Muzzarelli and E.R. Parisen), MIT Sea Grant Program, Cambridge, MA.

Araki, Y. and Ito, E. (1974) A pathway of chitosan formation in *Mucor rouxii*. Enzymatic deacetylation of chitin. *Biochem. Biophys. Res. Commun.*, **56**, 669–75.

Araki, Y. and Ito, E. (1975) A pathway of chitosan formation in *Mucor rouxii*. Enzymatic deacetylation of chitin. *Eur. J. Biochem.*, **55**, 71–5.

Arcidiacono, S., Lombardi, S.J. and Kaplan, D.L. (1988) Fermentation, processing and enzyme characterization for chitosan biosynthesis by *Mucor rouxii*, in *Chitin and Chitosan*, (eds G. Skjak-Braek, T. Anthonsen and P. Sanford), Elsevier Applied Science, New York, p. 319.

Austin, P.R. (1988) Chitin solutions and purification of chitin. *Methods Enzymol.*, **161**, 403–7.

Austin, P.R., Brine, C.J., Castle, J.E. and Zikakis, J.P. (1981) Chitin: new facet of research. *Science*, **212**, 749–53.

Baba, S., Uraki, Y., Miura, Y. and Tokura, S. (1988) Controlled release and hydrolysis of prodrug using carboxymethyl-chitin as a drug carrier, in *Chitin and Chitosan*, (eds G. Skjak-Braek, T. Anthonsen and P. Sanford), Elsevier Applied Science, New York, p. 702.

Bartnicke-Garcia, S. and Nickerson, W.J. (1962) Isolation, composition, and structure of cell walls of filamentous and yeast-like forms of *Mucor rouxii*. *Biochim. Biophys. Acta*, **58**, 102–19.

Bernadet, M.F. (1969) Skin cosmetics from chitin. *French Patent FR 1552026*.

Blumberg, R., Southall, C., Van Rensberg, N.J. and Volkmann, O.B. (1951) South African fish products. XXXII. The rock lobster: A study of chitin production from processing wastes. *J. Sci. Food Agric.*, **2**, 571–6.

Bough, W.A. (1976) Chitosan – a polymer from seafood wastes, for use in treatment of food process wastes and activated sludge. *Process Biochem.*, **11**, 13–16.

Brine, C.J. and Austin, P.R. (1981) Chitin variability with species and method of preparation. *Comp. Biochem. Physiol.*, **69B**, 283–6.

Broussignac, P. (1968) Chitosan, a natural polymer not well known by the industry. *Chim. Ind. Genie. Chim.*, **99**, 1241–6.

Brzeski, M.M. (1982) Concepts of chitin and chitosan isolation from Antarctic krill (*Euphasia superba dana*) shells on a technical scale, in *Proceedings of the 2nd International Conference of the Japanese Society of Chitin and Chitosan*, (eds S. Hirano and S. Tokura), Sapporo, Japan.

Brzeski, M.M. (1988) Production and application of chitin and chitosan in Poland, in *Chitin and Chitosan*, (eds G. Skjak-Braek, T. Anthonsen and P. Sanford), Elsevier Applied Science, New York, p. 161.

Davies, D.H., Elson, C.M. and Hayes, E.R. (1988) *N,O*-carboxymethylchitosan, a new water soluble chitin derivative, in *Chitin and Chitosan*, (eds G. Skjak-Braek, T. Anthonsen and P. Sanford). Elsevier Applied Science, New York, p. 467.

Foster, A.B. and Hackman, R.H. (1957) Application of ethylenediaminetetraacetic acid in the isolation of crustacean chitin. *Nature*, **180**, 40–41.

Fujita, T. (1972) Recovery of proteins. *Japanese Patent 01 633*.

Gagne, N. (1993) Optimization studies on chitin/chitosan from crustacean waste. *MSc Thesis*. McGill University.

Gross, P., Konrad, E. and Mger, H. (1983) Investigation on chitosan as a natural film-forming ingredient in hair cosmetic products under the consideration of ecological aspects. *Parfuem Kosmetique*, **64**(7), 71–9.

Hackman, R.H. (1954) Studies on chitin. 1. Enzymatic degradation of chitin and chitosan esters. *Aust. J. Biol. Sci.*, **7**, 168–78.

Hackman, R.H. (1960) The occurrence of complexes in which chitin and protein are covalently linked. *Aust. J. Biol. Sci.*, **13**, 568–77.

Hadwiger, L.A. (1984). Chitosan, a natural regulator in plant–fungal pathogen interactions increases crop yields, in *Chitin, Chitosan and Related Enzymes*, (ed. J.P. Zikakis), Academic Press, San Diego, p. 291.

Hirano, S. (1988) Production and application of chitin and chitosan in Japan, in *Chitin and Chitosan*, (eds G. Skjak–Braek, T. Anthonsen and P. Sanford), Elsevier Applied Science, New York, p. 37.

Horowitz, S.T., Roseman, S. and Blumenthal, H.J. (1957) Preparation of glucosamine oligosaccharides. 1. Separation. *J. Am. Chem. Soc.*, **79**, 5046–9.

Kauss, H. and Bauch, B. (1988) Chitin deacetylase from *Colletotrichum lindermuthianum*. *Methods Enzymol.*, **161**, 518–23.

Kauss, H., Jeblick, W. and Young, D.H. (1982/83) Chitin deacetylation from the plant pathogen *Colletotrichum lindemuthianum*. *Plant Sci. Lett.*, **28**, 231–36.

Knorr, D. (1984). Use of chitinous polymers in food. *Food Technol.*, **38**, 85–97.

Knorr, D. and Betschart, A.A. (1978) The relative effect of an inert substance and protein concentrates on loaf volume of breads. *Food Sci. Technol.*, **11**, 198–203.

Knorr, D. and Betschart, A.A. (1981) Water absorption and loaf volume of protein fortified breads. *Food Sci. Technol.*, **14**, 306–12.

Knorr, D. and Klein, J. (1986) Production and conversion of chitosan with cultures of *Mucor rouxii* or *Phycomyces blakesleeanus*. *Biotechnol. Letts.*, **8**, 691–4.

Knorr, D. (1991) Recovery and utilization of chitin and chitosan in food processing waste management. *Food Technol.*, **44**, 114–22.

Kreger, D.R. (1954) Observation of cell walls of yeast and some other fungi by X-ray diffraction and solubility tests. *Biochim. Biophys. Acta*, **13**, 1–9.

Lang, G. and Clausen, T. (1988) The use of chitosan in cosmetics, in *Chitin and Chitosan*, (eds G. Skjak-Braek, T. Anthonsen and P. Sanford), Elsevier Applied Science, New York, p. 139.

Lusena, C.V. and Rose, R.C. (1953). Preparation and viscosity of chitosan. *J. Fish Res. Board Can.*, **10**, 521–2.

Madhavan, P. and Ramachandran, N.K.G. (1974) Utilization of prawn waste: isolation of chitin and its conversation to chitosan. *Fish. Technol.*, **11**, 50–53.

Malette, W.G., Quigley, H.J. and Adickes, E.P. (1982) *Chitosan Effect in Vascular Technology*, (eds R.A.A. Muzzarelli, C. Jeuniaux and G.W. Gooday), Plenum Press, New York, p. 435.

Markey, M.L., Bowman, L.M. and Bergamini, M.V.W. (1988) Contact lenses made of chitosan, in *Chitin and Chitosan*, (eds G. Skjak-Braek, T. Anthonsen and P. Sanford), Elsevier Applied Science, New York, p. 713.

McGahren, W.J., Perkinson, G.A., Growich, J.A. *et al.* (1984) Chitosan by fermentation. *Process Biochem.*, **19**, 88–90.

Mima, S., Miya, M., Iwamoto, R. and Yoshikawa, S. (1982) Highly deacetylated chitin and its properties, in *Proceedings of the 2nd International Conference of the Japanese Society of Chitin and Chitosan*, (eds S. Hirano and S. Tokura), Sapporo, Japan, p. 21.

Muzzarelli, R.A.A. (1977) *Chitin*, Pergamon Press, Oxford.

Muzzarelli, R.A.A. (1983) Chitin and its derivatives: new trends of applied research. *Carbohydr. Polym.*, **3**, 53–8.

Muzzarelli, R.A.A., Tanfani, F., Emanuelli, M. *et al.* (1981) The production of chitosans of superior quality. *J. Appl. Biochem.*, **3**, 316–321.

Muzzarelli, R., Tarsi, R., Fillipini, O. *et al.* (1990) Antimicrobial properties of *N*-carboxymethyl chitosan. *Antimicrob. Agents Chemother.*, **34**, 2019–23.

Nagai, T., Sawayanagi, Y. and Nambu, N. (1984) Application of chitin and chitosan to pharmaceutical applications, in *Chitin, Chitosan and Related Enzymes*, (ed. J.P. Zikakis), Academic Press, New York, p. 21.

Nagyvany, J.J., Falk, J.D., Hill, M.L. *et al.* (1979) The hypolipidemic activity of chitosan and other polysaccharides in rats. *Nutr. Rep. Int.*, **20**, 677–84.

Nishimura, K.S., Nishimura, N., Nishi, N. *et al.* (1985) Immunological activity of chitin derivatives, in *Chitin in Nature and Technology*, (eds R.A.A. Muzzarelli, C. Jeuniaux and G.W. Gooday), Plenum Press, New York, p. 469.

No, H.K. and Meyers, S.P. (1989) Crawfish chitosan as a coagulant in recovery of organic compounds from seafood processing streams. *J. Agric. Food Chem.*, **37**, 580–3.

No, H.K., Meyers, S.P. and Lee, K.S. (1989) Isolation and characterization of chitin from crawfish shell waste. *J. Agric. Food Chem.*, **37**, 575–9.

Ornum, J.V. (1991) Chitosan, in *Proceedings of the 34th Annual Conference of the Canadian Institute of Food Science Technology*, Montreal, Québec, June 16–19.

Papineau, A.M., Hoover, D.G., Knorr, D. and Farkas, D.F. (1991). Antimicrobial effect of water soluble chitins with high hydrostatic pressure. *Food Biotechnol.*, **5**, 47–57.

Peniston, Q.P. and Johnson, E.L. (1978) Process for determination of chitin in crustacean shells. *US Patent 4066 735*.

Revah-Moiseev, S. and Carroad, P.A. (1981) Conversion of the enzymatic hydrolysate of shellfish waste chitin to single cell proteins. *Biotechnol. Bioeng.*, **23**, 1067–78.

Ruiz-Herrara, J. (1978) The distribution and quantitative importance of chitin in fungi, in *Proceedings of the 1st International Conference on Chitin/Chitosan*, (eds R.A.A. Muzzarelli and E.R. Parisen), MIT Sea Grant Program, Cambridge, MA.

Ryan, W.H. and Yankowski, E.L. (1961) Photographic image-receiving material. *German Patent 1 116 969*.

Sandford, P.A. (1989) Chitosan: commercial uses and potential applications, in *Chitin and Chitosan – Sources, Chemistry, Biochemistry, Physical Properties and Applications*, (eds G. Skjack-Braek, T. Anthonsen and P. Sandford), Elsevier Applied Science, London, pp. 51–69.

Sapers, G.M. (1992) Chitosan enhances control of enzymatic browning in apple and pear juices by filtration. *J. Food Sci.*, **57**, 1192–3.

Simpson, B.K. and Haard, N.F. (1985) Extraction of carotenoprotein from shrimp processing offal with the aid of trypsin. *J. Appl. Biochem.*, **7**, 212–22.

Shimahara, K., Ohkouchi, K. and Ikeda, M. (1982) A new isolation method of crustacean chitin using proteolytic bacterium, *Pseudomonas maltophilia*, in *Proceedings of the 2nd International Conference of the Japanese Society of Chitin and Chitosan*, (eds S. Hirano and S. Tokura), Sapporo, Japan, p. 10.

Shimahara, K., Takiguchi, Y., Kobayashi, T. *et al.* (1989) Screening of mucoraceae strains suitable for chitosan production, in *Chitin and Chitosan: Sources, Chemistry, Biochemistry; Physical Properties and Applications*, (eds G. Skjak-

Braek, T. Anthonsen and P. Sandford), Elsevier Applied Science, London, pp. 171–8.

Soto-Peralta, N.V., Miller, H. and Knorr, D. (1989) Effects of chitosan treatment on the clarity and colour of apple juice. *J. Food Sci.*, **54**, 495–6.

Takeda, M. and Abe, E. (1962) Isolation of crustacean chitin. Decalcification by disodium ethylenediaminetetraacetate and enzymic hydrolysis of incidental proteins. *Norisho Suisan Koshusho Kenkyu Hokohu*, **11**, 339–45.

Takeda, M. and Katsuura, H. (1964) Purification of king crab chitin. *Suisan Daigaku Kenkyu Hokoku*, **13**, 109–116.

Tracey, M.V. (1957) Chitin. *Rev. Pur. Appl. Chem.*, **7**, 1–19.

Tsugita, T. (1990) Chitin/chitosan and their application, in *Advances in Fisheries Technology and Biotechnology for Increased Profitability*, (ed. M.N. Voigt and J.R. Botta), Technomic, Lanaster, PA, pp. 287–298.

Von-Furth, O. and Russo, M. (1906) Veber Kristallinische Chitosanverbendungen aus Sepienshulpen. Ein Beitrag zur Kenntnis des Chitins. *Beitr. Bhem. Physiol. Path.*, **8**, 163–8.

Watkins, T.R. and Knorr, D. (1983) *In vivo* dye binding of chitin and its effects on gerbil growth and gut function. *Nutr. Rep. Int.*, **27**, 189–97.

Weist, J.L. and Karel, M. (1992) Development of a fluorescence sensor to monitor lipid oxidation. 1. Fluorescence spectra of chitosan powder and polyamide powder after exposure to volatile lipid oxidation products. *J. Agric. Food Chem.*, **40**, 1158–1162.

Whistler, R.L. and BeMiller, J.N. (1962) Chitin. *J. Org. Chem.*, **27**, 1161–5.

White, S.A., Farina, P.R. and Fulton, I. (1979) Production and isolation of chitosan from *Mucor rouxii*. *Appl. Environ. Microbiol.*, **38**, 323–8.

Wu, A.C.M., and Bough, W.A. (1978) Coagulation of cheese whey solids, in *Proceeding of 1st International Conference on Chitin/Chitosan*, (eds R.A.A. Muzzarelli and E.R. Parisen), MIT Sea Grant Program, Cambridge, Mass, p. 88.

Yanagida, T. (1985) Application of chitin and chitosan in Japan. *Kokai Yokkyo Koho. JP 61/210014 A2 (80/210014).*

Yang, T. and Zall, R.R. (1984) Chitosan membranes for reverse osmosis application. *J. Food Sci.*, **49**, 91–93.

CHAPTER 8

Applications of crustacean wastes in biotechnology

M.E. Hansen and A. Illanes

8.1 INTRODUCTION

Over the past two decades the shellfish industry has experienced a significant expansion, making crustacean waste materials available in concentrated areas and in much larger quantities. This is a consequence of the impressive growth of the frozen food and canned food industry. Three to four decades ago, shellfish processing was oriented around fresh products and, consequently, markets were local and processing factories were small.

The situation is now different. Shellfish are processed in large facilities all over the world, each variety with production rates in the order of several thousand tons per year. Crustacea are mainly processed into frozen or canned products, frozen shrimp representing the largest volume in the frozen food market today (Johnson and Peniston, 1978).

This expansion and concentration of the shellfish industry, together with environmental and energy restrictions, have forced the nation's seafood processors to consider alternatives to the inland dumping or hauling out to sea of the waste materials (Carroad and Tom, 1978).

The most commercially harvested crustacean species are crab, shrimp, prawn, Antartic krill and crayfish. In the crustacean meal industries, waste material may represent more than 80% of the landing. Some of these industries have been established without any consideration for enviromental protection and by-product reclamation. Shellfish, on a live weight basis, contain about 25% solids, 20–25% edible meat and 50–60% recoverable waste, the rest being losses in the form of solubles and processing fines (Johnson and Peniston, 1978; Knorr, 1991).

Fisheries Processing: Biotechnological applications.
Edited by A.M. Martin.
Published in 1994 by Chapman & Hall, London. ISBN 0 412 58460 3

In 1978, shellfish processing waste was estimated at 154 000 dry tons, krill waste being estimated at 800 000 dry tons per year (Allan, Fox and Kong, 1978). However, krill waste is essentially unavailable, since processing is mainly done by freezing on board fishing vessels far away from the coast, wastes being dumped into the sea.

Disposal in the future will have to cope with more stringent ecological standards, making the recovery of valuable by-products mandatory. Integral utilization of renewable resources is certainly a goal worth pursuing and can be envisaged as new reclamation technologies are developed. This certainly applies to shellfish processing wastes and by-products.

The chapter outlines the current options for the utilization of shellfish wastes, with particular emphasis on biotechnological applications.

8.2 QUALITY AND COMPOSITION OF CRUSTACEAN WASTE

Shellfish wastes represent a substantial portion of the raw material. These materials, mainly shells, viscera, heads and adhered meat, are partly used today in the production of fishmeal for animal consumption, the rest being wasted even though it is environmentally offensive.

The relative amount of these materials varies with the species and the technology of processing. Usual ranges are presented in Table 8.1.

Table 8.1 Composition range for crustacean waste

	Percent
Solids	30–40
Chitin: dry basis	14–30
Protein: dry basis	15–40
$CaCO_3 + Ca_3(PO_4)_2$	35–55
Lipids	0–10

Chitin content varies with the species, as seen in Table 8.2 for leading commercial species in the Southern Pacific (González, 1966). Protein in waste comes from unrecovered meat but also from residues adhered to the shell matrix. Its content varies widely, depending on species and type of primary shellfish processing. For example, Chesapeake Bay blue crab is commonly processed by hand, leaving considerable amounts of unrecovered meat in the waste material. Shrimp in Atlantic and Gulf fisheries are beheaded at sea, losing 40% of the live weight and close to 55% of the total waste. In other areas, crab butchering and extraction are separate operations, so that butchering wastes are not recovered

Table 8.2 Chitin content in crustaceans

Crustacean	*Chitin (% dry basis)*
Crab (*Cancer* spp. and *Homolaspis plana*)	15–30
King crab (*Lithodes antarticus*)	15–30
Yellow prawn (*Pleuroncodes monodon*)	14
Red prawn (*Cervimunida johni*)	14
Nylon shrimp (*Heterocarpus reedi*)	17
Krill (*Euphasia superba*)	25

(Johnson and Peniston, 1978). Peeling is a key operation in terms of yield, losses being higher for small species like krill. Mechanical peeling uses large amounts of water which extracts adhered protein and thus reduces yield. Water leaching, if properly done, can be beneficial in terms of recovery of valuable products from the waste water.

The nutritional value of shellfish waste protein is high, making recovery attractive. Amino acid profiles are presented for crustacean waste protein and compared with standard proteins in Table 8.3 (Burger, 1971). The profile for crustacean waste protein is excellent, the content of essential amino acids being significantly higher than the

Table 8.3 Amino acid profile of different proteins (g amino acid/16 g nitrogen)

Amino acid	*Crustacean*	*Soybean meal*	*Fishmeal*
Lys[a]	8.34	2.88	5.07
His	2.97	1.12	1.59
Arg	8.06	3.24	3.81
Asp	8.63	–	–
Thr[a]	3.91	1.89	2.82
Ser	4.69	2.52	2.52
Glu	17.80	–	–
Pro	4.54	–	–
Gly	7.52	1.89	3.68
Ala	7.14	–	–
Cis + Met[a]	2.60	2.25	3.46
Val	5.62	2.25	3.46
Ile[a]	5.17	2.20	3.06
Leu[a]	8.14	3.42	4.98
Tyr[a]	3.61	1.58	2.22
Phe[a]	5.05	2.20	2.75
Trp	0.73	0.58	0.78
Met	–	0.63	1.95

[a]Essential amino acid.

standards, with sulfur amino acids higher than in soybean meal, although lower than in fishmeal.

According to the species, the protein may contain significant amounts of the important natural pigment astaxanthin. In fact, crustacean waste is an interesting source of natural pigments, especially astaxanthin and astaxanthin ester. However, these pigments are highly unstable to temperature and chemicals, which makes extraction a very critical operation (Perceval, 1978; Simpson, 1978).

Minerals in shellfish waste material usually consist of 90% or more calcium carbonate, the rest being calcium phosphate. The high amount of calcium is a problem for recovery and disposal.

Fats are largely derived from viscera and vary from 0% in clean crab-leg shells to perhaps 5% or more in some shrimp and lobster species (Johnson and Peniston, 1978) and 10% or more in krill wastes (Anderson de Pablo and Romo, 1978).

The proteolytic activity in shellfish waste is high, which lowers the yield of protein recovery. This in turn has led to some studies to evaluate the recovery of other valuable products like proteases and amino acids from waste waters.

The recovery of some of these products requires the development of a strategy for fractionation, the technology to use being defined by the main product recovered.

8.3 RECOVERY OF CRUSTACEAN WASTE COMPONENTS

8.3.1 Preliminary considerations

Shellfish waste is a highly perishable material; therefore, preservation and transport to the processing plant is a major problem for the recovery of their waste components. In a matter of a few hours, bacterial and/or enzymic degradation of protein and chitin in the shell matrix begins, affecting extraction yield and quality of product. Location of the recovery plant at or close to the primary shellfish processing facility is therefore highly recommended.

Seasonal fluctuations in the quantity and quality of shellfish waste is a major problem, the process having to be flexible enough to accomodate these fluctuations. Physical differences among species have also to be considered. For example, some crab species have highly calcified and dense claw tips, which are resistant to extraction and demineralization. Shrimp shell is more flexible and light, which hinders filtration and centrifugation. Flexibility in the design of the processing plant is thus highly desirable to allow proper handling of different species (Johnson and Peniston, 1982).

8.3.2 Chitin/chitosan and protein recovery

Major components in shellfish waste are chitin and residual protein. Chitin is the predominant molecule in crustacean structure. It is a polymer of *N*-acetylglucosamine units linked β(1–4). The monomer is 2-acetoamide-2-deoxy-β-D-glucose, whose structural formula is ($C_8H_{13}NO_5$). The amino groups in chitin are acetylated. Deacetylated chitin is called chitosan, which is a β(1–4) polymer of the 2-amino-2-deoxy-β-D-glucose monomer.

In its natural form, however, a broad range of partially deacetylated polymers exists, the difference between chitin and chitosan being rather arbitrary since completely acetylated and completely deacetylated polymers do not occur in nature nor as end processsing products (Montenegro, 1980). In practice a polymer where most residues are acetylated is called chitin while the opposite is true for chitosan.

Several processes have been proposed for the fractionation of shellfish waste for the production of chitin and chitosan. Most processes for chitin production include alkaline extraction of organics and acid solubilization of minerals, the protein being recovered in the former stage. During such extractions the chitin molecule also suffers some structural changes, including a moderate degree of deacetylation. For the production of chitosan, substantial deacetylation is obtained by alkaline treatment of chitin under severe conditions.

(a) Protein extraction

During the first stage the protein from adhered tissue and viscera is leached from the shells; chemically bound protein is then released more slowly (De la Cerda, 1980).

(b) Demineralization

Once protein has been extracted, $CaCO_3$ is removed from the chitin–carbonate matrix. Strong mineral acids are used, solubilizing calcium ions and producing CO_2. Demineralization has to be complete in order to ensure the quality of the chitin so produced.

(c) Deacetylation

When chitosan is to be produced, extensive deacetylation is necessary. The removal of the *N*-acetyl groups is difficult and very strong conditions are required. Concentrated NaOH or KOH solutions are used at high temperature and pressure, the quality of chitosan depending on the degree of deacetylation obtained (Muzzarelli, 1977; Wu and Bough, 1978; De la Cerda, 1980).

(d) Recovery of extracted protein

Following extraction, protein is recovered by solid–liquid separation, concentration and drying. Isoelectric precipitation is a convenient method for concentration. By adjusting the pH to the isoelectric point, protein solubility will be minimal and the precipitate will be easily recovered by settling or centrifugation and then washed to produce a protein isolate (Simpson, 1978; De la Cerda, 1980).

(e) Process alternatives

Process alternatives for the production of chitin/chitosan with protein recovery are presented as a block diagram in Figure 8.1. Four alternatives arise from this diagram.

- Alternative I represents the most simple and conventional process which includes washing, milling, deproteinization, demineralization and deacetylation. Both adhered and chemically bound protein are extracted.
- Alternative II includes storage, washing, dense phase separation, deproteinization, demineralization and deacetylation. In this scheme,

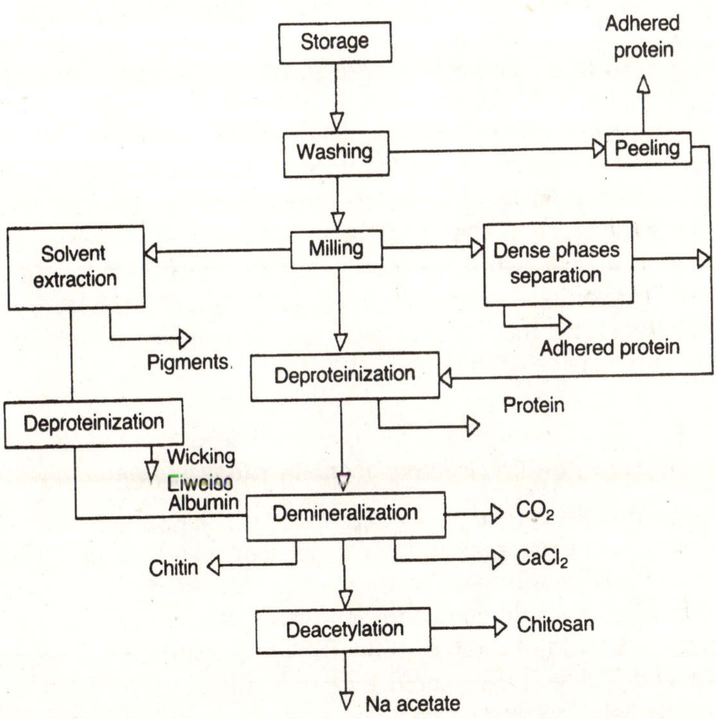

Figure 8.1 Process alternatives for shellfish wastes fractionation.

adhered protein is removed by salting-in, producing a protein with a high salt content whose use as food is doubtful.

- Alternative III considers storage, washing, milling, solvent extraction system, deproteinization, demineralization and deacetylation. Lipids and pigments (oleoastacin) are extracted in this process; therefore, the protein obtained is a high quality product called Wicking–Eiweiss albumin (Marín, 1977). Solvent extraction is complex and cost intensive, and the recovery of pigments requires antioxidants and low temperatures for drying, which might make this process unattractive.
- Alternative IV includes storage, washing, deboning, deproteinization, demineralization and deacetylation. As in alternative II, adhered meat is removed, but the quality is higher and it can be used as a food ingredient for sauces, cold cuts and patties. Deboning is an additional operation that can be worthwhile to include, depending on the quality of the food protein recovered (Ugarte, 1966; Marín, 1977; De la Cerda, 1980).

The most relevant operations for the production of chitin and chitosan are deproteinization, demineralization and deacetylation. Different methodologies have been reported, most of which are based on laboratory procedures, industrial know-how being severely protected.

Methods such as those described by Hackman (1954), Whistler and BeMiller (1962), Horowitz, Poseman and Blumenthal (1957) consider prolonged extraction with NaOH solutions, producing not only deproteinization, but also some extent of deacetylation, which contaminates the solubilized protein. These methods are therefore inadequate if the protein is to be recovered efficiently. The method as described by Foster and Hackman (1957) requires expensive chemicals and prolonged extractions, its industrial application being doubtful. The method proposed by Takeda and Abe (1962) and Takeda and Katsuura (1964) includes enzymes which are expensive, and which require rigorous control and prolonged operations. Its cost at the industrial level might be prohibitive. The method reported by Broussignac is simple and relatively inexpensive (Broussignac, 1968).

8.3.3 Product recovery of crustacean waste water

Valuable products can be recovered from the liquid effluents of a shellfish processing plant, although a major disadvantage lies in the fact that these are highly diluted. Protein may be recovered by precipitation with coagulants. Precipitation conditions have to be carefully controlled and frequent addition of antioxidants during precipitation is required. Together with protein, carotenoid pigments are co-precipitated in the form of a protein–carotenoid complex whose nature has been studied for shrimp waste water, protein exerting a stabilizing effect on the pigment (Cano-López *et al.*, 1987).

Lipids and pigments have been recovered from waste water by solvent extraction. Proteolytic enzymes have been found in substantial amounts in waste waters. Protease content in Antartic krill has been reported by Kimoto *et al.* (1986), Kimoto, Fukamizu and Murakami (1986) and Osnes and Mohr (1986), trypsin being the prevalent enzyme (Osnes, Ellingsen and Mohr, 1986). Although proteolytic action is highly undesirable in shellfish processing, these proteases are remarkably active at low temperatures (Seki, Ozawa and Arai, 1975), which can be interesting for some industrial applications.

The recovery of some of these products from waste waters will depend on economic factors at individual plant locations and environmental requirements from local pollution control authorities.

8.4 APPLICATIONS OF PRODUCTS RECOVERED FROM CRUSTACEAN WASTES

Chitosan is a very interesting compound and an impressive number of potential uses have been proposed (Skjak-Braek, Anthonsen and Sandford, 1989), although only biotechnological applications, or applications which relate to the area of biotechnology will be considered in this chapter. These are: enzyme immobilization, food and feed, recovery of biomolecules (downstream processing), agriculture, effluent treatment, medicine and cosmetics, effluent treatment, and others such as mammalian cell culture. Particular emphasis will be given to enzyme immobilization.

8.4.1 Chitin and chitosan as carriers for enzyme immobilization

Although this is a minor field of application in quantitative terms, chitin and chitosan have been frequently reported as suitable carriers for enzyme immobilization.

An immobilized enzyme is a biocatalyst in which the enzyme has been attached to, or contained within, an inert, frequently solid, matrix. Immobilization widens the scope of enzyme applications for industrial purposes (Hultin, 1983; Poulsen, 1984), in clinical medicine (Klein and Langer, 1986) and in analysis (Bowers and Carr, 1980). For industrial purposes, immobilized enzymes offer the advantage of being robust catalysts, able to withstand the rigorous conditions found in a bioreactor, and can be used in continuous operation for long periods of time without catalyst recharge. For clinical use, immobilized enzymes are preferred because of enhanced stability, reduced immunogenicity and flexibility of mode of addition. For analytical purposes, immobilized enzymes have the advantage of enhanced stability and flexibility of configuration, which allows them to be incorporated as part of measuring devices (e.g. in enzyme electrodes).

Much of the success for an immobilized enzyme relies in the characteristics of the carrier. Desirable properties for a material to be used as an enzyme carrier are: large surface area per unit volume; permeability; hydrophilicity; insolubility; chemical, mechanical, thermal and biological stability; rigidity; regenerability; flexibility for shaping; low cost; and compatibility with product specifications.

Chitin and chitosan satisfy most of these properties. They are hydrophilic, insoluble, and chemically and biologically stable under a wide range of conditions, mechanically resistant, regenerable after use, readily shaped in different forms, inexpensive, and easily activated by chemical reagents. Although chitosan has been used as an adsorbent, when compared with other organic carriers used for enzyme immobilization its protein binding capacity is modest, which may be a consequence of the lower surface area per unit volume. On the other hand, since most of the enzyme is immobilized on the external surface of the carrier particle, catalysts produced from chitosan are essentially free of diffusional restrictions, as will be shown later.

When dealing with immobilized enzymes it is difficult to establish a clear-cut difference between chitin and chitosan, since the extent of acylation is seldom reported. Enzyme carriers are prepared from natural sources and different methods, such as those illustrated earlier, can produce polymers with varying degrees of free amino groups. The term chitin is sometimes used to denote the raw material from which chitosan is made, while elsewhere it simply refers to a material with a low level of deacylation, chitosan being the opposite. In the rest of this section we will use the term chitosan, because any method used to extract chitin from its natural source will produce some extent of deacylation.

Methods for enzyme immobilization include those in which the enzyme is attached to a solid matrix by adsorption or covalent bonding and those in which the enzyme is entrapped in membranes or gels. Enzymes have been immobilized in chitosan by adsorption, covalently through bifunctional reagents such as glutaraldehyde, and also by entrapment.

The immobilization process can be evaluated in terms of immobilization yield (Y) and specific activity of the immobilized enzyme (A) which can be defined as:

$$Y = \frac{E_I}{E} = \frac{E_I}{E_I + E_R + E_L}$$

$$A = \frac{E_I}{W}$$

where E_I represents the immobilized enzyme activity, E the total enzyme activity contacted with the carrier, E_R the residual enzyme activity in solution after immobilization, E_L the enzyme activity lost and W the unit weight (volume) of carrier. Activity is expressed in international units, one international unit (IU) being the amount of enzyme that converts 1 μmol of substrate per minute at saturation conditions. Y represents the fraction of enzyme activity that is expressed by the immobilized catalyst, while A represents the catalytic capacity of the enzyme particle. Both parameters are sometimes in compromise and a pre-established criterion of optimization is therefore needed. We have defined optimum conditions as those which maximize A at a value of Y over a preset minimum, since A will determine the enzyme reactor size. However, for costly enzymes, Y may have a higher incidence and therefore may be considered as the parameter to maximize.

Another relevant property of the immobilized enzyme is increased stability, when compared to its soluble counterpart which can be assessed in terms of enzyme half-life. It is expected that immobilization will enhance enzyme stability to a considerable extent.

A considerable amount of information has been reported on enzyme immobilization using chitosan. An outline will be presented aiming to evaluate chitosan as a carrier for enzyme immobilization.

Stanley *et al.* (1975) immobilized fungal lactase by covalent linkage to chitosan through glutaraldehyde and also by simultaneous adsorption and cross-linkage with glutaraldehyde. Better results were obtained with the former system, in which 60 IU/g of carrier were obtained at a yield of 60%. The charge capacity of the carrier was 5 mg protein/g carrier. The apparent Michaelis constant (K_m) of the immobilized enzyme was 86 mM, which is almost equal to the Michaelis constant of the soluble enzyme, 85 mM, indicating that the catalyst particle is free of diffusional restrictions. Illanes *et al.* (1988), with a similar enzyme (fungal lactase, 70 IU/mg of protein), obtained specific activities over 400 IU/g at yields of 35% when immobilizing the enzyme to glutaraldehyde-activated chitosan at optimized conditions. The apparent K_m of the immobilized enzyme was 90 mM as compared with 87 mM for the soluble counterpart, which reflects an enzyme catalyst free of difussional restrictions (Illanes, Zúñiga and Ruiz, 1990). The catalyst, in the form of flakes, was used for the operation of a continuous packed bed reactor. Mechanical behavior was excellent, no attrition or elution being observed, and the flow pattern was homogeneous, without channeling or significant void spaces appearing within the catalyst bed. Enzyme operation half-life at 40°C was 120 days as compared with 10 days for the soluble enzyme (Illanes, Ruiz and Zúñiga, 1990). As seen in Figure 8.2, chitosan-immobilized lactase performed well when compared with leading commercial catalysts.

Pifferi, Tramontini and Malacarne (1989) immobilized fungal endopolygalacturonase (35 IU/mg protein) on monomaleyl and trimaleyl chitosan by simple adsorption and adsorption assisted by cross-linkage

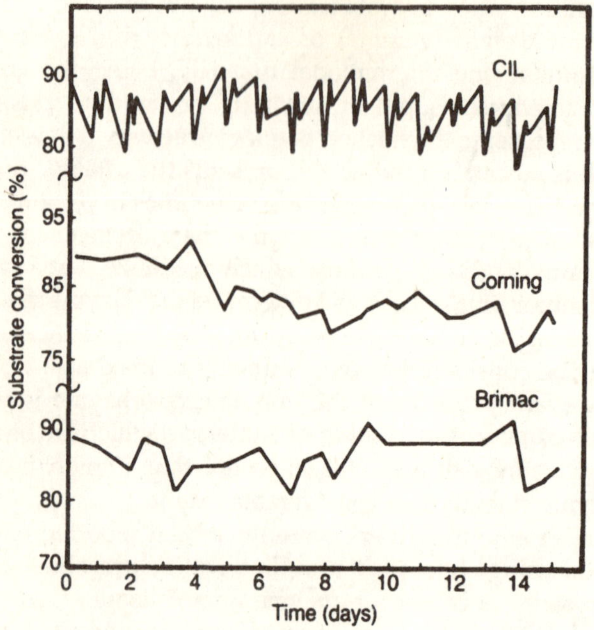

Figure 8.2

with glutaraldehyde. Specific activities from 1800 to 3100 IU/g of carrier, corresponding to 52–90 mg protein/g of carrier, were obtained at yields ranging from 19 to 32%. Best results were obtained when immobilizing the enzyme to cross-linked trimaleyl chitosan. The catalyst half-life was 21 days at 25°C, twice the value for the soluble enzyme. The apparent K_m was 0.24 mM, while for the soluble enzyme the K_m was 0.06 mM. Thus in this case diffusional retrictions are observed as a result of the molecular size of the substrate (polygalacturonic acid, 30 000 Da).

Working with a purified α-galactosidase (Mitsutomi and Ohtakara, 1984), Ohtakara, Mitsutomi and Uchida (1986) immobilized the enzyme on glutaraldehyde cross-linked colloidal chitosan at different enzyme to chitosan ratios. The enzyme specific activity of the carrier ranged from 35 to 343 IU/g at yields ranging from 70 to 46%. Residual enzyme increased from 13 to 39%, suggesting that the carrier was saturated with protein at the higher enzyme loads, which is surprising since the highest protein load was only 0.52 mg/g carrier.

Illanes, Chamy and Zúñiga (1986) immobilized crude invertase produced from autolyzed bakers' yeast cells (31 IU/mg protein) (Illanes and Gorgollón, 1985) on glutaraldehyde cross-linked chitin, obtaining a catalyst with 74 IU/g of carrier at a yield of 62%. The protein charge capacity was 3.3 mg/g of carrier and the specific activity increased only to corresponding protein loads of 5 mg/g, the carrier being saturated with protein beyond that point. Enzyme half-life in the absence of substrate at 53°C was 25 days, which is one order of magnitude higher than the half-life for the soluble enzyme. Long-term packed bed reactor operation with immobilized invertase at the same temperature produced no elution of enzyme and the operational half-life was estimated as 180 days. The carrier was regenerated three times with a loss in immobilization capacity of about 3% per cycle. All this demonstrates the feasibility of using chitosan as an industrial enzyme carrier (Illanes *et al.*, 1988). Chitosan was far superior to other carriers tested for invertase immobilization and diffusional restrictions were milder. The apparent K_m was 68.5 mM as compared with 39.2 mM for the soluble enzyme. Invertase immobilized on activated charcoal, however, has an apparent K_m of 185 mM (Illanes, Chamy and Zúñiga, 1986). Optimum pH was displaced from 4.5 to 4.3 as a result of immobilization, which is in agreement with the net positive charge exhibited by the carrier over this pH range.

Synowiecki, Sikorski and Naczk (1981a) immobilized purified Fluka invertase (aproximately 200 IU/mg protein) on krill chitosan by adsorption and covalent binding with glutaraldehyde, obtaining specific activities of 1500 and 1200 IU/g of carrier respectively, which correspond to 7.5 and 6 mg protein/g carrier. The reported immobilization yield for the former was 45%. Although results were slightly higher for simple adsorption, immobilization was in this case strongly dependent on ionic strength, suggesting that binding is mainly of ionic nature. In fact, at ionic strengths higher than 0.01 M the yield started to decrease, reaching 16% at an ionic strength of 0.25 M.

Synowiecki, Sikorski and Sikorski-Siondalska (1987) have reported an extensive study on enzyme immobilization on chitosan by adsorption, which is in their opinion a better system. Adsorption is produced mainly by ionic attraction between the amino groups of chitosan and the carboxyl groups of aspartic and glutamic acid residues, and also by hydrogen bonding, van der Waals' forces and hydrophobic interactions. Although the system is simple and little denaturation ensues, attention should be paid to catalyst stability which is generally poor. These authors have tried to improve catalyst performance through activation with carbon sulfide, which interacts with amino groups in chitosan forming dithiocarbamino groups. The carrier then has both positive and negative charges, and is able to interact ionically with different amino acid residues in the enzyme molecule. Diastase immobilized by this scheme was almost as stable as a catalyst produced by covalent linkage through glutaraldehyde, but was obtained with a better immobilization yield, making carbon sulfide activation an interesting alternative for developing chitosan-based enzyme carriers. Special precautions have to be taken with ionic strength to avoid enzyme desorption.

The same authors studied the effect of deproteinization and demineralization on the immobilization of different enzymes on krill chitosan

(Synowiecki, Sikorski and Naczk, 1981b). Results prove that the level of residual protein and minerals hinders enzyme immobilization. However, in certain cases immobilizations were improved when deproteinization was conducted under milder conditions. This can be attributed to the different degree of deacetylation produced by the alkaline conditions for protein extraction which will affect covalent linkage through free amino groups in the chitosan molecule. In principle, one can assume that higher specific activities should be obtained when the number of free amino groups is higher although, in the case of invertase or amyloglucosidase, this is not always evident. Working with lactase we have obtained consistently better results when working with chitosan of lower degree of deacetylation (Illanes *et al.*, 1990b). This, although contrary to expectation, can be attributed to an excess of linkages which could produce improper folding of the enzyme molecule and thus lead to steric hindrances.

Kimura *et al.* (1989) tested different carriers for the immobilization of bacterial pullulanase, better results being obtained with chitosan beads. They reported specific activities of 85 and 135 IU/g of carrier at yields of 47 and 27%, respectively, corresponding to a protein load of 3.6 mg/g for the former. Kusano, Shiraishi and Takahashi (1989) immobilized a partially purified bacterial pullulanse (88 IU/mg protein) on chitosan by adsorption and covalent linkage through bifunctional reagents like glutaraldehyde, hexamethylene diisocyanate, toluene 2,4-diisocyanate and 4,4-diphenylmethane diisocyanate. Results are summarized in Table 8.4.

Yields were only moderate and enzyme losses were significant for the case of toluene 2,4-diisocyanate and 4,4-diphenylmethane diisocyanate. The apparent K_m for pullulanase immobilized on glutaraldehyde treated chitosan was 1.8 g/l, which compares favorably with values for pullulanase immobilized on porous glass (K_m = 6.4 g/l) and on amberlite IRC-50 (K_m = 6.8 g/l). The K_m for the soluble enzyme was 0.4 g/l, meaning that the immobilized enzyme is subjected to diffusional restrictions, which are milder in the case of chitosan and quite insignificant if one considers the polymeric nature of the substrate. Chitosan-immobilized pullulanase remained fully stable for 30 days at 60°C, which is a very good performance. The same enzyme immobilized on amberlite IRC-50 lost half of its activity after that period of time.

Leuba and Widmer (1979) immobilized different proteinases on Fluka chitosan (3.7 μmol-NH_2/mg). The support was prepared by reprecipitation and cross-linkage with epichlorhydrin. The support was then activated with glutaraldehyde prior to enzyme immobilization. Immobilization yields were 15 and 17% for subtilisin and pronase, and specific activities were 14 and 12.5 U/g of carrier, respectively. The unit of activity (U) was defined in this case as the amount of enzyme that produces an increase in absorbance at 280 nm of 1/min using casein as substrate.

Table 8.4 Immobilization of pullulanase on chitosan[a]

Cross-linking agent	E *(IU/g)*	RE *(IU/g)*	A *(IU/g)*	Y *(%)*	P *(mg/g)*	RP *(mg/g)*	IP *(mg/g)*	Y_P *(%)*
None	2000		412	21	22.7	13.6	9.1	40.1
Glutaraldehyde	2060	920	334	16	23.4	10.5	12.9	55.5
Hexamethylene diisocyanate	2380	1320	604	25	27.1	15.0	12.1	44.6
Toluene 2-4 diisocyanate	2060	420	430	21	23.4	4.8	18.6	79.6
4-4 diphenylmethane diisocyanate	2060	420	426	21	23.4	4.8	18.6	79.6

[a]P, protein loaded; RP, residual protein in solution after immobilization; IP, immobilized protein; Y_P, yield for protein immobilization (IP/P).

Protein charge capacity was of the order of 30 mg/g, which is rather high for chitosan; however, the catalytic efficiency of the immobilized enzymes was low, which is attributed to diffusional restrictions due to the high molecular weight of the substrate. Thermal stability, although modest, was significantly improved by immobilization.

Kim and Rhee (1989) immobilized inulinase on glutaraldehyde-activated chitosan and obtained, at optimal conditions, an immobilization yield of 23% with a specific activity of 77 IU/g of carrier. No data were presented to assess protein load. Specific activity was significantly higher for this carrier, as compared with others such as aminoethyl cellulose and diethylaminoethyl cellulose. Enzyme stability was also higher (estimated half-life was 24 days at 40° C) as was reactor productivity for the conversion of Jerusalem artichoke inulin into fructose.

Braun *et al.* (1989) have reported the immobilization of penicillin acylase from *Escherichia coli* on reprecipitated chitosan shaped into powder, particles and beads. The enzyme was a crude preparation of 0.25 IU/mg of protein. Two immobilization schemes were devised: one in which the enzyme was immobilized by adsorption–reticulation with glutaraldehyde as cross-linking agent (method 1) and another in which chitosan was activated with glutaraldehyde prior to covalent enzyme linkage (method 2). Results are summarized in Table 8.5.

Yields and specific activities were markedly dependent on particle size, very good results being obtained for particles less than 0.5 mm in diameter. Although results appear to be better for method 1, enzyme desorption was significant and therefore method 2 should be the choice for reactor operation. The effect of protein load was studied for penicillin acylace immobilized on chitosan powder according to method 2. Results are summarized in Table 8.6.

Results are rather surprising in terms of protein loads. Higher values exceed 1 g protein/g of carrier, which is more than one order of magnitude higher than all others published. The authors declare that their results are hard to compare with others, which is apparent from our analysis. Even though these results are spectacular in terms of protein load, enzyme yields and specific activities were high but not

Table 8.5 Immobilization of penicillin acylase on chitosan[a]

Method		A *(IU/g)*	Y *(%)*	*IP* *(mg/g)*	Y_P *(%)*
1	beads	33	15	380	38
1	particles	94	54	460	57
1	powder	150	85	660	82
2	powder	37	84	195	97

[a]See Table 8.4.

Table 8.6 Effect of protein load on enzyme immobilization[a]

P (mg/g)	IP (mg/g)	Y_P (%)	E (ui/g)	A (ui/g)	Y (%)	E_R (%)	L (%)
46.7	41.3	88.4	10.3	6	58.3	11.6	30.1
100	94.7	94.7	22	13	59.1	5.5	35.5
200	184.0	92.0	44	34	77.2	8.0	14.8
400	373.4	93.3	88	62	70.5	6.7	22.8
800	666.7	83.3	176	127	72.2	16.6	11.2
1200	1093.4	91.1	264	194	55.1	31.7	13.2

[a]See Table 8.4.

unusual, since the specific activity of the enzyme preparation was very small. It is interesting that enzyme losses (*L* in Table 8.6) were higher at smaller loads, which is hard to explain; however, enzyme yield decreased noticeably at loads higher than 800 mg protein/g carrier, the sharp increase in residual activity indicating that the carrier was saturated beyond that point.

From the published information on immobilization of enzymes on chitosan-based matrices, some general statements can be made:

- Chitosan is a very flexible material which can immobilize enzymes by simple adsorption, by reticulation assisted adsorption, by covalent linkage through bifunctional and other activating agents, and even by gel entrapment.
- Ionic interactions between amino groups in the chitosan molecule and negative charges in the amino acid residues of the enzyme molecule are prevalent in adsorption, although van der Waals' forces and hydrophobic interactions also play a part.
- Immobilization yields tend to be higher in adsorption than in covalent linkage due to the milder conditions required in the former. Also specific activities are higher but catalyst stability is rather poor and quite sensitive to ionic force and pH, a serious inconvenience for reactor operation. Immobilization on activated chitosan through covalent linkage is therefore the method of choice despite these disadvantages.
- When compared with that of other carriers, chitosan load capacity is rather low and can be estimated as between 10 and 30 mg/g. Only one report presents protein loads of several hundreds mg/g, well over this range. Despite this modest capacity, enzyme yield is good, indicating that the catalytic expression of the immobilized protein is quite high. In fact, immobilization is mainly superficial and diffusional restrictions are low even with high molecular weight susbtrates; this is a definite advantage of chitosan as an enzyme carrier.
- Chitosan is quite stable and can be used in prolonged operation and also regenerated after enzyme decay. When shaped properly it is

mechanically strong and can withstand harsh conditions such as those encountered in industrial reactor operation.

- When compared with other carriers, chitosan is almost always selected because of higher yields, low diffusional restrictions, moderate load capacities, ease of handling and flexibility of configuration.
- The effect of impurities and the level of deacetylation affects significantly the immobilization capacity of chitosan. Residual protein and minerals decrease immobilization capacity but the level of deacetylation does not correlate with enzyme yield as expected, possibly because of improper enzyme folding and steric hindrances. A thorough study on the effect of chitosan structure and purity on enzyme immobilization remains to be done.

8.4.2 Food biotechnology

Protein, chitin and pigments are valuable constituents in crustacean wastes from the standpoint of the food and feed industry.

(a) Animal feed

Chitin, residual protein and internal organs are the basis for shellfish meal, whose protein content is about 45%. Shellfish meal, high in chitin, is used as a supplement in diets for chicken and fish. Nutritional studies with chicken have shown a stimulatory growth effect on *Bifidobacteria* which favors lactose adsorption, a major problem in feeding these animals because of the high lactose content in usual diets (Spreen, Zikakis and Austin, 1984; Knorr, 1984). Diarrhea has been effectively controlled in milk whey-fed chickens when the diet is supplemented with shellfish meal (Austin *et al.*, 1981). The latter can also be used as a food supplement for lactose intolerants (Knorr, 1984).

Chitin digestibility has been tested in several animals (Patton and Chandler, 1975; Sugano and Fujikawa, 1980). Kono *et al.* (1990) studied the effect of chitin and chitosan as a dietary supplement for fishes and found higher growth rates at levels of 10% in all species tested. All species were able to digest chitin because of the presence of chitinase in their digestive tract (Kono, Matsui and Shimizu, 1987). Chitinase activity has also been found in the digestive tract of Atlantic salmon (*Salmo salar*) (Värum, Rosenhund and Smidsrod, 1989) and a variety of other species, which is very important for aquaculture (Knorr, 1991).

Hirano *et al.* (1990) studied chitin digestibility in rabbits and hens. No adverse effects were detected after 239 days of feeding rabbits and hens at rates lower than 0.8 and 1.4 g chitosan/kg body weight/day respectively. At levels of 3.6–4.2 g chitosan/kg body weight/day severe physiological disorders were observed in hens. No growth inhibition was observed in rats at levels as high as 8.5% of chitin and 10% of chitosan in their diets (Sugano and Fujikawa, 1980). Other reports with rats have shown no differences in growth rate, internal organs and blood serum composition at levels of 5% chitosan in their diet (Knorr, 1984; Chavasit and Torres, 1990).

An LD_{50} value of 16 g chitosan/kg was reported for mice by Arai Kinumari and Fujita (1968). However, Chavasit and Torres (1990) reported that according to US regulations only up to 0.1% chitosan is allowed in livestock feed since 1985. On the other hand, chitosan for direct use in foods has already been approved in Japan (Sandford, 1989), where it is a regular ingredient in several sale products, including noodles and breadsticks.

With regard to protein, shellfish waste contains a substantial amount which can be processed into a protein concentrate for animal feed. The amino acid profile was shown to be better than in most feed proteins (see Table 8.3). Shrimp meal amino acid profile was proved to be similar to casein (Simpson, 1978), the same being true for protein from shrimp and crab waste, although in this case the content of sulfur amino acids was a little bit lower (Johnson and Peniston, 1982) and methionine supplement increased its nutritional value.

However, there are certain organoleptic restraints in terms of color and residual odor, which make the product unsuitable for human consumption, unless purified at a high cost.

Amino acid requirements for poultry, swine and salmon can be met with crustacean derived protein (Burger, 1971).

Perhaps the most promising alternative for shellfish waste protein is fish feeding, especially salmon, which is a very relevant economic activity in Chile. For the case of salmon, pigmented protein is highly desirable and there is no need to decolorize or deodorize the product.

Feed requirements for salmon are 2 kg/kg salmon, with 1 mg of commercial pigment/kg eviscerated salmon. A synthetic pigment that contains 8% astaxanthin is available on the market, with a sale price of US$ 200/kg. The world market in 1990 was estimated at 32.8 ton, representing sales of US$ 66 millions. Adequate pigmentation can be obtained with 10–25% of shellfish meat in their diet, the adsorption of astaxanthin being more efficient in this case than in the case of synthetic pigment. In general, pigment retention is more efficient for astaxanthin from natural sources, but results vary with the species, the best reported value for synthetic astaxanthin being over 5% (Torrissen, 1989; Torrissen *et al.*, 1989).

The astaxanthin content in shrimp meal varied from 76 μg/g when vacuum dried to almost nil when drying was conducted according to usual industrial practice. Salmon fed with 20% vacuum dried crab meal had a distinctive appealing pink color, while standards remained white (Simpson, 1982).

Protein requirements for salmon diets are high and about 60% of good quality protein must be supplemented, usually in the form of premium fishmeal (Torrissen, Hardy and Shearer, 1989). However, Simpson (1978) made a complete study of salmon feeding with protein isolates from crab and shrimp, reporting excellent results in terms of pigmentation, low calcium content and low bacterial counting when compared with a commercial protein standard.

Shellfish meal for salmon feeding can hardly compete with fishmeal at the moment because of economic constraints derived from the high technology of isolation and drying required if the pigment and quality of protein is to be retained. As the price of synthetic astaxanthin remains high and that from natural sources like flowers or microalgae does not share a significant fraction of the market, the production of pigmented shellfish protein isolates is a promising alternative in the near future.

Bioconversion of chitin to single-cell protein for animal feed is an interesting alternative of shellfish waste utilization (Carroad and Tom, 1978; Cosio, Fisher and Carroad, 1982). The scheme for chitin bioconversion is quite similar to that proposed by Wilke *et al.* (1976) for cellulosic residues.

Carroad and Tom (1978) studied the bioconversion of chitinous waste to single-cell protein as animal feed. Chitin is degraded by chitinases produced by *Serratia marcescens* QMB1466. Chitin is degraded extracellularly to simple sugars which are metabolized to produce the protein enriched microbial product. Reid and Ogrydziak (1981) obtained mutant stains of the same organism that overproduced chitinase.

The yeast *Pichia kudriavzevii* grew well in a chitin hydrolyzate produced by treatment with chitinases. The amino acid profile of the protein in the yeast cell mass obtained was acceptable (Revah-Moiseev and Carroad, 1981).

As in cellulose, fermentation can be optimized if suitable pretreatment of the raw material is considered. Pretreatment of chitinous waste by deproteinization and demineralization allowed 80% degradation of chitin with chitinase (Cosio, Fisher and Carroad, 1982). A high activity of chitinase has been reported in cultures of the fungus *Beauveria bassiana* when grown in medium containing ammonium sulphate (0.25%), peptone (0.5%) or lactic acid (0.2%) and demineralized crab shell as inducer (Xiuli, Danmei and Qian, 1989). This type of organism may also be considered for single-cell protein production from chitinous waste.

(b) *Human food and food processing*

If lipids and pigments are extracted from shellfish meal, a high quality protein can be obtained for human consumption (Marín, 1977). This type of protein is called Wiking–Eiweiss albumin and is produced by alternative III as previously described (see p. 180). The product is extremely clean, white, tasteless and 100% digestible, being composed almost entirely of polypeptides and free amino acids. The product, with

94% protein (Ugarte, 1966), is a good egg white substitute, 1 kg of it substituting 330 eggs (Marín, 1977). Wiking–Eiweiss albumin has been tested for the production of ice cream, mayonnaise and confectionery. On the other hand, waste adhered muscle and viscera could be recovered to produce shellfish food snacks for human consumption.

According to Chavasit and Torres (1990), one of the most relevant future applications for chitin and chitosan is as a functional ingredient in foods.

Chitin and chitosan can be used as food additives. In fact, polysaccharides have been used traditionally by the food industry as thickeners, adhesives, extenders, and gel-forming and crystal controlling agents for the production of sauces, baking goods, ice cream, confectionery, cold cuts, snacks, fruit juices, jams, jellies, cheese and several others (Knorr, 1984)

Chitin and chitosan can be used as functional ingredients for texture control in foods, through increasing interactions among functional groups (Knorr, 1982, 1984). They can also be used as emulsifiers because of their remarkable ability to bind water (230–440% w/w) and fat (170–315% w/w), crystalline chitin being the best emulsifier (Knorr, 1982).

The chelating properties of chitosans prove advantageous in the removal of heavy metals, dyes, pigments, acids and organic solids such as proteins. This has resulted in utilizing about 50% of the currently produced chitosans in clarification, coagulation and flocculation processes in food processing waste and waste water treatment, as well as in downstream processing for the removal or recovery of microorganisms (Knorr, Beaumont and Pandya, 1989).

Perhaps the most relevant application is as a flocculating agent as suggested by Hirano (1989). According to him, there were 15 companies in Japan in 1986 which produced 1270 tons/year of chitin, from which 1170 tons were used for chitosan production, 70% of which was used as a flocculant.

Fruit juice production requires clarifying agents to reduce turbidity. Chitosan has proved to be very effective in the removal of colloids and dispersed particles from fruit concentrates because of its net positive charge, without affecting color (Chavasit and Torres, 1990; Soto-Peralta, Müller and Knorr, 1989).

Addition of water-soluble and acid-soluble chitosan to fruit juice processing has been as effective as treatment with silica sol/gelatine/bentonite, obtaining 2.9 FTU (formazin turbidity units). Acid-soluble chitosan is more effective at low concentrations, maximum clarification being obtained at 0.7 g/l; however, water-soluble chitosan is preferred because it is easier to manipulate and does not affect aroma (Imeri and Knorr, 1988; Soto-Peralta, Müller and Knorr, 1989).

Chitosan can be used as food preservative, being effective against *Escherichia coli* and *Fusarium oxysporium*. At concentrations higher than

0.02% *Escherichia coli* growth is completely inhibited (Hirano, 1989). Chitosan derivatives were more effective as bactericidal agents (Muzzarelli, 1989). Contamination by microorganisms has been prevented or suppresed in tissue cultures by placing chitosan or its derivates in contact with the cell culture (Malette, 1985)

8.4.3 Downstream processing applications

Chitosan is a biopolymer related to cellulose. Therefore, membranes for reverse osmosis and ultrafiltration can be made from chitosan with the additional advantage of being resistant to alkaline conditions and inorganic solvents.

Reverse osmosis membranes are made by dispersing acetic acid 2% chitosan solutions over a glass surface. With these membranes a flux of 1.67×10^{-3} $cm^3/cm^2/s$ was obtained at 680 p.s.i. with a separation capacity of 78.8% of NaCl in a solution of 2 g/l. Membrane permeability was improved by adding 40% poliethylene glycol and durability was increased by using 10% chloromethyl oxirane (Yang and Zall, 1984).

Chitin–chitosan membranes have been tested for the separation of amino acids and polypeptides, amino acids being selectively removed. A chitosan gel membrane made by controlled cross-linking, allowed permeation of a 141 000 molecular weight yeast alcohol dehydrogenase under a concentration gradient. This protein did not permeate through an uncross-linked chitosan film (Pellegrino *et al.*, 1990).

Microporous spray-dried particles made from water, chitosan and ethylenediamine have been used as chromatographic matrices for the immobilization of biologically active substances or as controlling agents for food ingredients (Rodriguez-Sánchez and Rha, 1981).

Chitosan is a suitable matrix for affinity chromatography. A 6-O-CM-chitin prepared from queen crab chitin has been reported to be an ion exchanger (Tokura, Nishimura and Nishi, 1983). Uraki *et al.* (1989) studied the adsorption behavior of amino acid side chains on the surface on 6-O-CM-chitin. An extremely high affinity of phenylalanine toward Ca-6-O-CM-chitin complex was found as in other neutral amino acids. This also occurs for a variety of phenylalanine containing peptides.

Senstad and Mattiasson (1989a) proposed to use chitosan as a ligand carrier in affinity precipitation. In this system affinity interaction is performed in solution, being therefore easier to scale-up than conventional affinity chromatography. The ligand carrier should be easily activated for subsequent coupling of the ligand and then easily precipitated by changes in the surrounding enviroment. The chitosan ligand carrier is soluble at pH lower than 6.5 and precipitates at higher pH values. These authors used this system for wheat germ aglutinnin (WGA) (Senstad and Mattiason, 1989b) and trypsin (Senstad and Mattiason, 1989a). For WGA purification they used natural chitin and obtained a yield of pure WGA of 70%. For trypsin, soybean trypsin inhibitor (STI) was covalently coupled to chitosan with glutaraldehyde and the purified trypsin–STI–chitosan complex was dissociated with glycine–HCl buffer. Protein yield was 72% and yield of trypsin activity was 93%.

Magnetic affinity adsorbents, based on chitin, agar and agarose, have been reported for lysozyme isolation. The highest capacity for enzyme retention was obtained with magnetic chitin; however, it also showed stronger binding of contaminant proteins than magnetic agar or agarose, thorough washing with water being required to remove the accompanying proteins from the chitin surface. The author concludes that magnetic chitin may have large-scale applications (Safarik, 1991).

Chitosan has also been tested for protein purification by adsorption. Cross-linked chitosan, Chitopearl 3510, was studied for potential use in the large-scale chromatographic separation of serum albumin (BSA). Adsorption of cattle serum albumin on cross-linked chitosan appears technically feasible for salt-containing solutions (Yoshida and Kataoka, 1989).

Chitosan has been used in cells and cells debris recovery. It has also been applied in *Candida utilis* for single-cell protein recovery, helping flocs development in yeast suspensions (Holland, 1989). Under optimun conditions of pH, flocculant dosage and cell concentration recoveries of 98% were obtained.

Agerkvist, Eriksson and Enfors (1989) studied cell debris flocculation in an *Escherichia coli* homogenate; they used selective flocculation in a cell disintegrate to obtain high β-galactosidase recovery. Under optimal conditions chitosan removes 98% of the cell debris; 85% of the nucleic acids and 50% of the proteins are precipitated. Selectivity in protein flocculation depends on pH and other protein properties.

8.4.4 Applications in agriculture

This area of application is one of the most recent and most promising. Mitchell and Alexander (1961) reported a marked reduction in root-rot in beans and vascular wilt in radishes, both caused by *Fusarium* spp., after chitin was added to soil. These early findings are the basis for current applications of chitin and chitosan in agriculture. Antifungal properties of chitosan have been shown in several cases (Allan and Hadwiger, 1979; Struszczyk, Pospieszny and Kotlinski, 1989) and different mechanisms have been proposed to explain its action. One states that chitin stimulates the growth of actinomycetes that excrete antibiotic and other lytic substances. According to this, the antifungal mechanism of chitin would be indirect (Leuba and Stossel, 1986). Hadwiger and Beckman (1980) found two regulatory functions for chitosan in plant–parasite interaction. Chitosan, which is a cell wall component of many fungi, activated specific genes in plants and in similar concentrations repressed fungal RNA synthesis completely (Hadwiger *et al.*, 1989). Interaction between positively charged free amino groups in chitosan and negatively charged phosphate residues in nucleic acids is a possible explanation for this double action (Kendra and Hadwiger, 1986).

The action of chitosan in peas and beans is well reported, producing changes in chromatin structure, which make the plants resistant to pathogens. On the other hand, chitosan produced a clear inhibition of growth in *Fusarium solani*, the mechanism of which has been compared to that in plants. According to this mechanism chitosan enters the plant nucleus, which is high in chromatin, producing alterations in specific regions which alter transcription; however, in fungal cells the effect is so drastic that DNA transcription into RNA is repressed completely (Hadwiger *et al.*, 1986)

Many fungal plant pathogens have cell walls rich in chitin. It has been suggested that the production of chitinases by some higher plants would have a direct effect on chitin present in the cell walls of these fungal pathogens (Boller, 1986). Chitinase synthesis would be triggered by the presence of chitin in the medium. In fact, Hirano *et al.* (1989) studied the induction of chitinase in seeds of Japanese radish, soybean, rice, hulled rice and black pine. By coating the seeds with a thin layer of chitosan, chitinase activity increased 1.5 times with respect to uncoated seeds. Coated seeds are generally resistant against the attack of pathogens and insects, probably by preventing their adhesion on plant tissues.

In general terms, Sandford (1989) stated that chitosan triggers a response in the seed, signaling the plant to protect itself from natural predators, such as pathogenic fungi, most of which have chitin in their cell walls.

Antifungal effects of chitosan have been clearly proved in beans and peas infected with *Fusarium solani, Aphanomyces euteiches, Phytophthora megasperma, Phytium paroecandrum, Fusarium oxysporum, Rhizoctonia solani* and *Thanateporus cucumeris* (Hadwiger *et al.*, 1989).

Recent results proved that a mixture of chitosan and liquid fertilizers sprayed over fruit trees produced beneficial effects by retarding the superficial drying of leaves, decreasing phytotoxic effects and improving fertilizer retention. Chitosan at a 1% level in liquid fertilizer increased the adherence of the active components to the plant surface, preventing washout by forming a polymeric biodegradable film. It also increases the stability and dispersion capacity, increases nutrient assimilation, and prevents water evaporation and product crystallization (Struszczyk, Pospieszny and Kotlinski, 1989).

Chitosan in the form of powder, flakes or gel, dissolved in an organic acid, has been used in the formulation of a fertilizer called 'Florograma', which has been succesfully used in cereals (Struszczyk, Pospieszny and Kotlinski, 1989).

Another recent agricultural application of chitosan is as an incrustating agent. Chitosan acetate mixed with other chemicals is used to protect seeds during storage and after sowing. Chitosan offers some distinctive advantages as an incrustating agent, producing a very persistent dressing from treatment to germination. It also helps in the germination process, reduces adhered dust and has additional bioactive actions (Struszczyk, Pospreszny and Kotlinski, 1989).

Agricultural application of chitosan is a fast-growing business in Chile. A chitosan-based preparation is being produced with the brand name Biorend by the company Bioagro SA. Biorend has been tested successfully in sugar beet, rapeseed, corn, wheat, rice, rye, beans, sunflower and chickpeas. In all cases significant yield increases have been obtained on test fields. In forestry, Biorend has been applied to eucalyptus and *Pinus radiata*. The product is formulated with 2.5% chitosan acetate and acts as biostimulating agent, producing roots 10–20% longer and stronger stalks.

Chitosan appears to have multiple physiological effects. Apart from being an antifungal agent, it has been proposed as an inducer of lignification (Pearce and Ride, 1982) and as a protease inhibitor. Preliminary experiments tend to prove that chitosan can also assist in healing virus produced lesions in plants (Struszczyk, Pospieszny and Kotlinski, 1989).

As shown, agriculture is a very promising field for chitosan application. Chitosan can act as a fungicide, virucide, growth enhancer, nutrient carrier and protective agent for plants and trees. A market already exists and it should increase significantly in the future.

8.4.5 Effluent treatment

Chitosan complexes very easily with metal ions present in polluted water, being able to remove them efficiently (Moore, Johnson and Sistrunk, 1987). In this sense, chitosan has been recognized to be superior to other polyelectrolytes.

Chitosan has been used in the purification of municipal and industrial waste waters, in the removal of pesticides, in the recovery of proteins from food plant effluents, and in the adsorption of tannins (Bough, 1975, 1976; Austin *et al.*, 1981; Jeuniaux, 1986)

In addition, chitosan has been used for harvesting microalgae, as a potential animal feed source, from fresh water and waste water (Lavoie and de la Noue, 1983).

The use of chitosan as a coagulant in waste water treatment is very promising. Chitosan can be used in the recovery of valuable products like proteins, which can then be used as animal feed without further fractionation (Landes and Bough, 1976). Moreover, environmental regulations on water effluents are increasingly stringent in terms of organic load (BOD), heavy metals and other toxics for which chitosan is a powerful adsorbent.

The most important application of chitosan in waste water treatment refers to secondary treatment of municipal waste waters. The addition of chitosan to activated sludge is now a practice in certain districts of Japan and the US. Chitosan is superior to other polymers in terms of biodegradability and sludge removal and has displaced inorganic conditioners in Japan in recent years (Hirano, 1989).

Chitosan has been used for the coagulation of solids in cheese whey. An optimal chitosan to solids ratio was between 2–2.5% at pH 6, at which 90% solids removal was obtained. The level of minerals and the molecular weight of chitosan correlated inversely with removal efficiency. Even though promising, chitosan coagulation of cheese whey has not been adopted as an industrial operation (Wu *et al.*, 1978).

8.4.6 Applications in medicine and cosmetics

Chitosan is a biocompatible, biodegradable and non-antigenic material which can be shaped in fibers, particles, films, gels and tablets, and can be administered orally or intravenously.

Medical uses for chitosan which have been tried include tissue regeneration, vascular surgery (Malette, Quigley and Adickes, 1989), as a hemostatic agent (Fradet *et al.*, 1986), as an inhibitor of thrombin hydrolytic activity (Okei *et al.*, 1986), and as a blood anticoagulant (Hirano *et al.*, 1985; Muzzarelli *et al.*, 1986).

Chitin and chitosan have been administered orally to control appetite and prevent gastritis (Ferro, 1986; Olsen *et al.*, 1989). Chitosan can also been tested in oral preparations to prevent bacterial plaques, where it effectively controls growth of *Candida albicans* (Sapelli *et al.*, 1989; Susuki *et al.*, 1986).

Several applications for chitosan in cosmetics are presented by Lang and Clausen (1989). Mixtures of chitosan with glycerol and antiseptic agents are stable preparations for hair and skin care. A number of chitin containing cosmetics are on sale from Wella Inc. (Hirano, 1989).

8.4.7 Other biotechnological applications

The film-forming ability of chitosan has been the object of many studies, some of which have led to industrial uses in various fields. Membranes of chitosan have been used in mammalian cell culture technology (Hwang, Rha and Sinskey, 1986; Kim and Rha, 1989; Izume *et al.*, 1989; Shoiya and Rha, 1989). Chitosan improved the adherence of chick embryo fibroblast cells to collagen–chitosan-coated dishes in the early stages of growth and the cell growth rate was not affected by the presence of chitosan (Izume *et al.*, 1989).

When mammalian cells are encapsulated, the chitosan capsule membrane provides additional surface area for cell growth, protects cells from shear and mechanical impact, and allows for compartmentalization that leads to higher cell density. Such high cell density systems may simulate natural tissue conditions and reduce serum usage. Encapsulation offers further advantages because it can serve, at the same time, as a separation aid by providing controlled porosity of the semipermeable capsule membrane. Hwang, Rha and Sinskey (1989) studied the properties of chitosan encapsulated proteins of various molecular weights. The permeability of the chitosan membrane could be controlled, therefore controlling the chitosan retention capacity for proteins of differents molecular weights.

Microcapsules for drugs and food additives have been prepared from different polymers, such as carageenan, agar, alginic acid derivates, gums, pectins and chitosan. Chitosan mixed with alginate has been used in the preparation of membranes for the encapsulation of tissue and bacterial cells. The biological material remained viable and protected by a membrane fully permeable to nutrients and metabolic products (Rodriguez-Sanchez and Rha, 1981).

REFERENCES

Agerkvist, I., Eriksson, L.B. and Enfors, S.O. (1989) Selective flocculation with chitosan in *E. coli* cell homogenates, in *Chitin and Chitosan. Sources, Chemistry, Biochemistry, Physical Properties and Applications*, Elsevier Applied Science, New York, pp. 543–558.

Allan, C.R. and Hadwiger, L.A.(1979) The fungicidal effect of chitosan on fungi of varying cell wall composition. *Exp. Mycol.*, **3**, 285–7

Allan, G.G., Fox, J.R. and Kong, N. (1978) A critical evaluation of the potential sources of chitin and chitosan; in *Proceedings of The First International Conference on Chitin/Chitosan*, (eds R.A.A. Muzzarelli and E.R. Pariser), MIT Sea Grant Report 78–7, Cambridge, MA, pp. 64–78.

Anderson, C.G., de Pablo, N. and Romo, C. (1978) Antartic krill (*Euphausia superba*) as a source of chitin and chitosan, in *Proceedings of The First International Conference on Chitin/Chitosan*, (eds R.A.A. Muzzarelli and E.R. Pariser), MIT Sea Grant Report 78–7, Cambridge, MA, pp. 54–63.

Arai, K., Kinumari, T. and Fujita, T. (1968) On the toxicity of chitosan. *Bull. Tokai Reg. Fish Lab.*, **56**, 89–95.

Austin, P.R., Brine, C.J., Castle J.E. and Zikakis, J.P. (1981) Chitin: new facets on research. *Science*, **212**, 749–53.

Boller, T. (1986) Chitinase, a defense of higher plant against pathogens, in *Chitin in Nature and Technology*, (eds Muzzarelli, R.A.A., Jeuniaux, C. and Goody, G.W.), Plenum Press, New York, pp. 223–30.

Bough, W.A. (1975) Reduction of suspended solids in vegetable canning waste effuent by coagulation with chitosan. *J. Food Sci*, **40**, 297–301.

Bough, W.A. (1976) Chitosan – a polymer from seafood waste, for use in treatment of food processing wastes and activated sludge. *Process. Biochem.*, **11**, 13–16.

Bowers, L. and Carr, P. (1980) Immobilized enzymes in analytical chemistry. *Adv. Biochem. Eng.*, **15**, 89–112.

Braun, J., Le Chanu, P. and Le Goffic, F. (1989) The immobilization of penicillin G acylase on chitosan. *Biotechnol. Bioeng.*, **33**, 242–6.

Broussignac, P. (1968) Un haut polymere peu connu dans l'industrie: le chitosan. *Chim. Ind. Genie Chim.*, **99**, 1241–7

Burger, G.H. (1971) *Pescado y las Industrias Derivadas de la Pesca*, 1st edn, Acribia Ediciones, Zaragoza, p. 15.

Cano-López, A., Simpson, B.K. and Haard, N.F. (1987) Extraction of carotenoprotein from process wastes with the aid of trypsin from Atlantic cod. *J. Food Sci.*, **52**, 503–6.

Carroad, P. and Tom, R. (1978) Bioconversion of shellfish wastes: process conception and selection of microorganisms. *J. Food Sci.*, **43**, 1158–61.

Chavasit, V. and Torres, J. (1990) Chitosan–poly (acrylic acid): mechanism of complex formation and potential industrial applications. *Biotechnol. Prog.*, **6**, 2–6.

Cosio, I., Fisher, R. and Carroad, P. (1982) Bioconversion of shellfish chitin waste: waste pretreatment, enzyme production, process design, and economic analysis. *J. Food Sci.*, **47**, 901–5.

De la Cerda, M. (1980) Estudio de Factibilidad TécnicoEconómica del Aprovechamiento de los Desechos de la Zona Central. *Thesis*, Universidad Técnica Federico Santa María, Valparaíso, Chile.

Ferro, A. (1986) Appetite moderating and antigastritis composition. *European Patent Application 219 156*.

Foster, A.B. and Hackman R.H. (1957). Applications of ethylenediamino tetraacetic acid in the isolation of crustacean chitin. *Nature*, **180**, 40–1.

Fradet, G., Brister, S., Mulder, D.S. *et al.* (1986) Evaluation of chitosan as a new hemostatic agent: *in vitro* and *in vivo* experiment, in *Chitin in Nature and Technology*, (eds R.A.A. Muzzarelli, C. Jeuniaux and G.W. Goody), Plenum Press, New York, pp. 485–91.

González, L.E. (1966) Revaloración de Residuos de Camarones y Langostinos: Obtención de Quitina y Quitosano. *Thesis*, Universidad Católica de Valparaíso, Valparaíso, Chile

Hackman, R.H. (1954) Chitin. I. Enzymatic degradation of chitin and chitin esters. *Austr. J. Biol. Sci.*, **7**, 168–178.

Hadwiger, L.A. and Beckman, J.M. (1980) Chitosan as a component of pea–*Fusarium solani* interactions. *Plant Physiol.*, **66**, 205–11.

Hadwiger, L.A., Kendra, B.W., Fristensky, B.W. and Wagoner, W. (1986) Chitosan both activates genes in plants and inhibits RNA synthesis in fungi, in *Chitin in Nature and Technology*, (eds R.A.A. Muzzarelli, C. Jeuniaux and G.W. Goody), Plenum Press, New York, pp. 209–14.

Hadwiger, L.A., Chiang, S., Victory, S. and Horovitz, D. (1989) The molecular biology of chitosan in plant–pathogen interaction and its application in agriculture, in *Chitin and Chitosan. Sources, Chemistry, Biochemistry, Physical Properties and Applications*, Elsevier Applied Science, New York, pp. 119–38.

Hirano, S. (1989) Production and application of chitin and chitosan in Japan, in *Chitin and Chitosan. Sources, Chemistry, Biochemistry, Physical Properties and Applications*, Elsevier Applied Science, New York, pp. 37–44.

Hirano, S., Tanaka, Y., Hasegawa, M. *et al.* (1985) Effect of sulfated derivates of chitosan on some blood coagulant factors. *Carb. Res.*, **137**, 205–15.

Hirano, S., Hayasahi, M., Nishida, T. and Yamamoto, T. (1989) Chitinase activity of some seeds during their germination process, and its induction by treating with chitosan and derivates, in *Chitin and Chitosan. Sources, Chemistry, Biochemistry, Physical Properties and Applications*, Elsevier Applied Science, New York, pp. 743–48.

Hirano, S., Hakura, C., Seino, H. *et al.* (1990) Chitosan as an ingredient for domestic animal feed. *J. Agric. Food Chem.*, **38**, 1214–8.

Holland, C.R. (1989) Recovery of single cell protein by chitosan in a batch dissolved air flotation system, in *Chitin and Chitosan. Sources, Chemistry, Biochemistry, Physical Properties and Applications*, Elsevier Applied Science, New York, pp. 559–66.

Horowitz, S., Poseman, S. and Blumenthal, H.J. (1957). The preparation of glucosamine oligosaccharides. I. Separation. *J. Am. Chem. Soc.*, **79**, 5046–9.

Hultin, O. (1983) Current and potential uses of immobilized enzymes. *Food Technol.*, **October**, 66–82.

Hwang, C., Rha, C.K. and Sinskey, A.J. (1986) Encapsulation with chitosan: trans-membrane diffusion of proteins in capsules, in *Chitin in Nature and Technology*, (eds R.A.A. Muzzarelli, C. Jeuniaux and G.W. Goody), Plenum Press, New York, pp. 389–96.

Illanes, A. and Gorgollón, Y. (1986) Kinetics of extraction of invertase from autolysed baker's yeast cells. *Enzyme Microb. Technol.*, **7**, 510–14.

Illanes, A., Chamy, R. and Zúñiga, M.E. (1986) Immobilization of invertase on crosslinked chitin, in *Chitin in Nature and Technology*, (eds R. Muzzareli, C. Jeuniaux and G. Gooday), Plenum Press, New York, pp. 411–15.

Illanes, A., Zúñiga, M.E., Chamy, R. and Marchese, M.P. (1988) Immobilization of lactase and invertase on crosslinked chitin, in *Bioreactor Immobilized Enzymes and Cells*, (ed. M. Moo-Young), Elsevier Applied Science, London, pp. 233–49.

Illanes, A., Zúñiga, M.E. and Ruiz, A. (1990). Inmovilización de Lactasa Microbiana. *Arch. Biol. Med. Exp.*, **23**, 159–164.

Illanes, A., Ruiz, A., Zúñiga, M.E. *et al.* (1990). Immobilization of lactase for the continuous hydrolysis of whey permeate. *Bioprocess Eng.*, **5**, 257–262.

Imeri, A.G. and Knorr, D. (1988) Effects of chitosan on yield and compositional data of carrot and apple juice. *J. Food Sci.* **53**, 1707–1712.

Izume, M., Taira, T., Kimura, T. and Miyata, T. (1989) A novell culture matrix composed of chitosan and collagen complex, in *Chitin and Chitosan. Sources, Chemistry, Biochemistry, Physical Properties and Applications*, Elsevier Applied Science, New York, pp. 653–6.

Jeuniaux, C. (1986) Chitosan as a tool for the purification of waters, in *Chitin in Nature and Technology*, (ed. R.A.A. Muzzarelli, C. Jeniaux and G.W. Goody), Plenum Press, New York, pp. 551–69.

Johnson, E.L. and Peniston, Q.P. (1978) The production of chitin and chitosan, in *Proceedings of The First International Conference on Chitin/Chitosan*, (eds R.A.A. Muzzarelli and E.R. Pariser), MIT Sea Grant Report 78–7, Cambridge, MA, pp. 80–87.

Johnson, E.L. and Peniston, Q.P. (1982) Utilization of shellfish waste for chitin and chitosan production, in *Chemistry and Biochemistry of Marine Food Products*, (eds R.E. Martin, C.C. Flick, C.E. Hebard and D.R. Ward), Avi Publishing, Westport, CT, pp. 415–428

Kendra, D.F. and Hadwiger, L.A. (1984) Characterization of the smallest chitosan oligomer that is maximally antifungal to *Fusarium solani* and elicits pisatin formation in *Pisum sativum*. *Exp. Mycol.*, **8**, 276–81

Kim, S.K. and Rha, C. (1989) Chitosan for encapsulation of mammalian cell culture, in *Chitin and Chitosan. Sources, Chemistry, Biochemistry, Physical Properties and Applications*, Elsevier Applied Science, New York, pp. 617–26.

Kim, C. and Rhee, S. (1989) Fructose production from Jerusalem artichoke by inulinase immobilized on chitin. *Biotechnol. Lett.*, **11**, 201–6.

Kimoto, K., Thanh, V. and Murakami, K. (1981) Acid proteinases from Antarctic krill *Euphausia superba*, partial purification and some properties. *J. Food Sci.*, **46**, 1881–4.

Kimoto, K., Fukamizu, A. and Murakami, K. (1986) Partial purification and characterization of proteinases from abdomen part muscle of Antarctic krill. *Nippon Suisan Gakkaishi*, **52**, 745–49.

Kimura, T., Yoshida, M., Oishi, K. *et al.* (1989) Immobilization of exo-maltotetrahydrolase and pullulanase. *Agric. Biol. Chem.*, **7**, 1843–1848.

Klein, M. and Langer, R. (1986) Immobilized enzymes in clinical medicine: an emerging approach to new drug therapies. *Trends Biotechnol.*, **4**, 179–86.

Knorr, D. (1982) Functional properties of chitin and chitosan. *J. Food Sci.*, **47**, 593–95.

Knorr, D. (1984) Use of chitinous polymers in food. *Food Technol.*, **38**, 85–97.

Knorr, D. (1991) Recovery and utilization of chitin and chitosan in food processing waste management. *Food Technol.*, **45**, 114–22.

Knorr, D., Beaumont, M.D. Pandya, Y. (1989) Potential of acid soluble and water soluble chitosans in biotechnology, in *Chitin and Chitosan. Sources, Chemistry, Biochemistry, Physical Properties and Applications*. Elsevier Applied Science, New York, pp. 101–118.

Kono, M., Matsui, T. and Shimizu, C. (1987) Effect of chitin, chitosan and cellulose as diet supplements on the growth of cultured fish. *Agric. Bio. Chem.*, **51**, 520–7.

Kono, M., Matsui, T., Shimizu, C. and Koga, D. (1990) Purification and some properties of chitinase from the stomach of Japanese eel. *Agric. Biol. Chem.*, **54**, pp. 973–6.

Kusano, S., Shiraishi, T., Takahashi, S. *et al.* (1989) Immobilization of *Bacillus acidopullulyticus* pullulanase and properties of the immobilized pullulanases. *J. Ferment. Technol.*, **68**, 233–37.

Landes, D.R. and Bough W.A. (1976) Effects of chitosan, a coagulator for food processing wastes, in the diets of rats on growth and liver and blood composition. *Bull. Environ. Contam. Toxicol.*, **15**, 555–63.

Lang, G. and Clausen, T. (1989) The use of chitosan in cosmetics, in *Chitin and Chitosan. Sources, Chemistry, Biochemistry, Physical Properties and Applications*, Elsevier Applied Science, New York, pp. 139–48.

Lavoie, A. and de la Noue, J. (1983) Harvesting microalgae with chitosan. *J. World Maricul. Soc.*, **14**, 685–94.

Leuba, J. and Stossel, P. (1986) Chitosan and other polyamines: antifungal activity and interaction with biological membranes, in *Chitin in Nature and Technology*, (eds R.A.A. Muzzarelli, C. Jeuniaux and G.W. Goody), Plenum Press, New York. pp. 215–20.

Leuba, J. and Widmer, F. (1979) Immobilization of proteinases on chitosan. *Biotechnol. Lett.*, **1**, 109–13.

Malette, W.G. (1985) Growth development of tissue cultures in non protein matrix. *GB Patent 2 145 114*.

Malette, W.G., Quigley, H.J. Jr and Adickes, E.D. (1989) Chitosan effect in vascular surgery, tissue culture and tissue regeneration, in *Chitin in Nature and Technology*, (eds R.A.A. Muzzarelli, C. Jeuniaux and G.W. Goody) Plenum Press, New York, pp. 435–41.

Marín, B. (1977) Estudio Técnico Económico Destinado a la Elaboración de Productos para Consumo Humano a Partir de Residuos de Camarones y Langostinos. *Thesis*, Universidad Catolica de Valparaíso, Valparaíso, Chile.

Mitchell, R. and Alexander M. (1961) Chitin and biological control of *Fusarium* diseases. *Plant Disease Rep.*, **45**, 487–90.

Mitsutomi, M. and Ohtakara, A. (1984) A simplified procedure for purification and crystallization of thermostable α-galactosidase from *Pycnoporus cinnabarinus*. *Agric. Biol. Chem.*. **12**, 3153–5.

Montenegro, P. (1980) Producción de Quitina a partir de Krill. *Thesis*, Universidad de Santiago, Santiago, Chile.

Moore, K.J., Johnson, M.G. and Sistrunk, W.A. (1987) Effect of polyelectrolyte treatments on waste strength of snap and dry bran waste waters. *J. Food Sci.*, **52**, 491–494.

Muzzarelli, R.A.A. (1977) *Chitin*. Pergamon Press, New York, pp. 89–92.

Muzzarelli, R.A.A. (1989) Amphoteric derivates of chitosan and their biological significance, in *Chitin and Chitosan. Sources, Chemistry, Biochemistry, Physical Properties and Applications*, Elsevier Applied Science, New York, pp. 87–100.

Muzzarelli, R.A.A., Tanfani, F., Emanuelli, M. *et al.* (1986) Sulfated *N*- carboxymethyl chitosans as blood anticoagulants, in *Chitin in Nature and Technology*, (eds R.A.A., Muzzarelli, C. Jeuniaux, and G.W. Goody), Plenum Press, New York, pp. 469–75.

Ohtakara, A., Mitsutomi, M. and Uchida, Y. (1986) Immobilization of thermostable alpha-galactosidase on chitin and its applications, in *Chitin in Nature and Technology*, (eds R.A.A. Muzzareli, C. Jeuniaux, and G. Goody), Plenum Press, New York, pp. 409–11.

Okei, W., Nishimura, O., Somorin, N. *et al.* (1986) Inhibitory action of sulphated chitin derivatives on the hydrolytic activity of thrombin, in *Chitin in Nature and Technology*, (eds R.A.A. Muzzarelli, C. Jeuniaux, and G.W. Goody), Plenum Press, New York, pp. 453–59.

Olsen, R., Schwartzmiller, D., Weppner, W. and Winandy, R. (1989) Biomedical application of chitin and its derivates, in *Chitin and Chitosan. Sources, Chemistry, Biochemistry, Physical Properties and Applications*, Elsevier Applied Science, New York, pp. 813–27.

Osnes, K.K. and Mohr, V. (1986) On the purification and characterization of exopeptidases from Antarctic krill *Euphausia superba*. *Comp. Biochem. Physiol. B: Comp. Biochem.*, **83B**, 445–58.

Osnes, K.K., Ellingsen, T.E. and Mohr, V. (1986) Hydrolysis of proteins by peptide hydrolases of Antarctic krill *Euphausia superba*. *Comp. Biochem. Physiol. B: Comp. Biochem.*, **83B**, 801–5.

Patton, R.S. and Chandler, P.T. (1975) *In vivo* digestibility: evaluation of chitinous materials. *J. Dairy Sci.*, **58**, 397–402.

Pearce, R.B. and Ride J.P. (1982) Chitin and related compounds as elicitors of the lignification response in wounded wheat leaves. *Physiol. Plant Pathol.*, **20**, 119–123.

Pellegrino, J.J., Geer, S., Maegley K. *et al.* (1990) Chitin chitosan membranes: separations of amino acids and polypeptides. *Ann. NY Acad. Sci.*, **589**, 229–44.

Perceval, P. (1978). The economics of chitin recovery and production, in *Proceedings of The First International Conference on Chitin/Chitosan*, (eds R.A.A. Muzzarelli and E.R. Pariser), MIT Sea Grant Report 78–7, Cambridge, MA, pp. 45–53.

Pifferi, G., Tramontini, M. and Malacarne. A. (1989) Immobilization of endopolygalacturonase from *Aspergillus niger* on various types of macromolecular supports. *Biotechnol. Bioeng.*, **33**, 1258–66.

Poulsen, P. (1984) Current applications of immobilized enzymes for manufacturing purposes. *Genetic Eng. Rev.*, **1**, 121–39.

Reid, J.D. and Ogrydziak, D.M. (1981) Chitinase-overproducing mutant of *Serratia marcescens*. *Appl. Environ. Microbiol.*, **41**, 664–9.

Revah-Moiseev, S. and Carroad, P. (1981) Conversion of the enzymatic hydrolysate of shellfish waste chitin to single-cell protein. *Biotech. Bioeng.*, **23**, 1067–74.

Rodriguez-Sánchez, D. and Rha, C.K. (1981) Chitosan globules. *J. Food Technol.*, **16**, 469–479.

Safarik, I. (1991) Magnetic biospecific affinity adsorbents for lysozyme isolation. *Biotechnol. Tech.*, **5**, 111–14.

Sandford, P.A. (1989) Chitosan: commercial uses and potential applications, in *Chitin and Chitosan. Sources, Chemistry, Biochemistry, Physical Properties and Applications*, Elsevier Applied Science, New York, pp. 51–70.

Sapelli, P.L., Baldassarre, V., Muzzarelli, R.A.A. and Emanuelli, M. (1986). Chitosan in dentistry, in *Chitin in Nature and Technology*, (eds R.A.A. Muzzarelli, C. Jeuniaux and G.W. Goody), Plenum Press, New York, pp. 507–13.

Seki, N., Ozawa, R. and Arai, K. (1975) Muscle proteins of Antarctic krill. *Nippon Suisan Gakkaishi*, **41**, 1287–92.

Senstad, C. and Mattiasson, B. (1989a). Affinity-precipitation using chitosan as ligand carrier. *Biotech. Bioeng.*, **33**, 216–20.

Senstad, C. and Mattiasson, B. (1989b) Purification of wheat germ agglutinin using affinity flocculation with chitosan and a subsequent centrifugation or flotation step. *Biotech. Bioeng.*, **34**, 387–93.

Shioya, T. and Rha, C. (1989) Transmembrane permeability of chitosan–carboxymethyl–cellulose capsule, in *Chitin and Chitosan. Sources, Chemistry, Biochemistry, Physical Properties and Applications*, Elsevier Applied Science, New York, pp. 627–34.

Simpson K.L. (1978) The recovery of protein and pigments from shrimp and crab meals and their use in salmonid pigmentation, in *Proceedings of The First International Conference on Chitin/Chitosan*, (eds R.A.A. Muzzarelli and E.R. Pariser), MIT Sea Grant Report 78–7, Cambridge, MA, pp. 253–62.

Simpson K.L. (1982) Carotenoid pigments in seafood, in *Chemistry and Biochemistry of Marine Food Products*, (eds R.E. Martin, G.J. Flick, C.E. Hebard and D.R. Ward), Avi Publishing, Westport, CT, pp. 121–36.

Skajk-Braek, G., Anthonsen T. and Sandford, P. (1989) *Chitin and Chitosan. Sources, Chemistry, Biochemistry, Physical Properties and Applications*, Elsevier Applied Science, New York.

Soto-Peralta, N.V., Müller, H. and Knorr, D. (1989). Effects of chitosan treatments on the clarity and color of apple juice. *J. Food Sci.*, **54**, 495–496.

Spreen, K.A., Zikakis, J.P. and Austin, P.R. (1984) The effect of chitinous materials on the intestinal microflora and the utilization of whey in monogastric animals, in *Chitin, Chitosan and Related Enzymes*, (ed. J.P. Zikakis), Academic Press, Orlando, FL, pp. 57–75.

Stanley, W., Watters, G., Chan, B. and Mercer, J. (1975) Lactase and other enzymes bound to chitin with glutaraldehyde. *Biotechnol. Bioeng.*, **23**, 231–3.

Struszczyk, H., Pospieszny, H. and Kotlinski, S. (1989) Some new applications of chitosan in agriculture, in *Chitin and Chitosan. Sources, Chemistry, Biochemistry, Physical Properties and Applications*, Elsevier Applied Science, New York, pp. 733–42.

Sugano, M. and Fujikawa, T. (1980) A novel use of chitosan as a hypocholesterolemic agent in rats. *Am. J. Clin. Nutr.*, **33**, 787–93.

Suzuki, S., Susuki, A., Tokoro, A. *et al.* (1986). Immunopotentiation effect of *N*-acetyl-chito-oligosaccharides, in *Chitin in Nature and Technology*, (eds R.A.A. Muzzarelli, C. Jeuniaux, and G.W. Goody), Plenum Press, New York, pp. 485–91.

Synowiecki, J., Sikorski, Z. and Naczk, M. (1981a) Immobilization of invertase on krill chitin. *Biotechnol. Bioeng.*, **23**, 231–33.

Synowiecki, J., Sikorski, Z. and Naczk, M. (1981b). The activity of immobilized enzymes on different krill chitin preparations. *Biotechnol. Bioeng.*, **23**, 2211–15.

Synowiecki, J., Sikorska-Siondalska, A. and El-Bedawey, A. (1987). Adsorption of enzymes on krill chitin modified with carbon disulfide. *Biotechnol. Bioeng.*, **29**, 352–4.

Takeda, M. and Abe, E. (1962) Isolation of crustacean chitin. Decalcification by disodium ethylene-diamino tetracetate and enzymic hydrolysis of incidental proteins. *Norihsho Suisan Koshusho Kenkyu Hokoku*, **11**, 339–406.

Takeda, M. and Katsuura, H. (1964) Purification of king crab chitin. *Suisan Dagaku Kenkyu Hokoku*, **13**, 109–16.

Tokura, S., Nishimura, S. and Nishi, N. (1983) Studies on Chitin X. Specific binding of calcium ions by carboxymethyl-chitin. *Polym. J.*, **15**, 597–602.

Torrissen, O.J. (1989) Pigmentation of salmonids: interaction of astaxanthin and cantaxanthin on pigment deposition in rainbow trout. *Aquaculture*, **79**, 363–74.

Torrissen, O.J., Hardy, R.W. and Shearer, K.D. (1989) Pigmentation of salmonids – carotenoid deposition and metabolism. *Rev. Aquat. Sci*, **1**, 205–229.

Ugarte, L. (1966) Posibilidades Técnico-Económicas de la Utilización de los Desechos de Camarón y Langostino en la Zona de Valparaíso. *Thesis*, Universidad Católica de Valparaíso, Valparaíso, Chile.

Uraki, Y., Nishi, N., Nishimura, S. and Tokura, S. (1989) Adsorption properties of calcium carboxymethylated chitin complex, in *Chitin and Chitosan. Sources, Chemistry, Biochemistry, Physical Properties and Applications*, Elsevier Applied Science, New York, pp. 537–40.

Värum, K.M., Rosenlund, G. and Smidsrod, O. (1989) Enzymatic degradation of chitosan in Atlantic salmon (*Salmo salar*), in *Chitin and Chitosan. Sources, Chemistry, Biochemistry, Physical Properties and Applications*, Elsevier Applied Science, New York, pp. 299–308.

Whistler, R. and Miller, B. (1962) Chitin. *J. Organic Chem.*, **27**, 1161–3.

Wilke, C.R., Cysewski, G.R., Yang, R.D. and von Stockar, U. (1976) Utilization of cellulosic materials through enzymatic hydrolysis. II. Preliminary assessment of an integrated processing scheme. *Biotech. Bioeng.* **18**, 1315–20.

Wu, A.C. and Bough, W.A. (1978) A study of variables in the chitosan manufacturing process in relation to molecular weight distribution, chemical characteristics and waste-treatment effectiveness, in *Proceedings of The First International Conference on Chitin/Chitosan*, (eds R.A.A. Muzzarelli and E.R. Pariser), MIT Sea Grant Report 78-7, Cambridge, MA, pp. 88–102.

Wu, A.C.M., Bough, W.A., Holmes, M.R. and Perkins, B.E. (1978) Influence of manufacturing variables on the characteristics and effectiveness of chitosan products. III. Coagulation of cheese whey solids. *Biotech. Bioeng.*, **20**, 1957–66.

Xiuli, H., Danmei, A. and Qian, R. (1989) Study on chitinase in cultures of *Beauveria bassiana*, in *Chitin and Chitosan. Sources, Chemistry, Biochemistry, Physical Properties and Applications*, Elsevier Applied Science, New York, pp. 309–18.

Yang, T. and Zall, R.R. (1984) Chitosan membranes for reverse osmosis application. *J. Food Sci.*, **49**, 91–93.

Yoshida H. and Kataoka T. (1989) Adsorption of BSA on cross-linked chitosan: the equilibrium isotherm. *Chem. Eng. J.*, **41**, B11–B15.

CHAPTER 9

Production of fish protein concentrates

E. Bárzana and M. García-Garibay

9.1 INTRODUCTION

Fish flesh is a source of highly nutritive protein consumed world-wide. Practically all species of fish have the same nutritional quality in terms of proteins; however, only a few of them are usually consumed, due to cultural factors, habits and other aspects such as poor appearance, unusual size and unappealing taste. Many species which are caught lack commercial value due to these considerations, but are equally as nutritive as the preferred ones. To gain commercialization they are often used for the production of fish meal, fish protein concentrate or fish protein hydrolyzates.

Table 9.1 shows the composition of fish flesh from several different species. Figure 9.1 shows the essential amino acid content of fish, as an average of several commercial kinds, according to Paul and Southgate (1978) and fish protein products reported by several authors, compared with the FAO/WHO 1973 standard.

Another source of fish proteins is present in the waste produced during the processing of commercial fish species to obtain fillets or during canning operations.

Seeking strategies to fill the protein gap between poor and well nourished people, a great interest in the utilization of fish proteins from non-commercial species and/or processing wastes arose during the 1950s. During the 1960s food technologists dedicated great research efforts on fish protein concentrates, basically with the aim of extending their use to developing countries. Nevertheless, as has been widely demonstrated, and fish protein is a good example, protein sources are not consumed for their nutritive value but rather by their organoleptic

Fisheries Processing: Biotechnological applications.
Edited by A.M. Martin.
Published in 1994 by Chapman & Hall, London. ISBN 0 412 58460 3

Table 9.1 Protein and fat content of some raw commercial fish species on a wet basis (g/100 g)

Species	*Protein (N* × 6.25)*	*Fat*
White fish		
cod	17.4	0.7
haddock	16.8	0.6
halibut	17.7	2.4
sole	17.1	1.4
Fatty fish		
herring	16.8	18.5
mackerel	19.0	16.3
trout	23.5	4.5
tuna	22.8	22.0

Source: Paul and Southgate (1978).
*N is nitrogen content in g/100 g.

properties and, unfortunately, in reduced amounts. If proteins are used as additives to be incorporated in a food or dish, the most important factors to be considered are their functional properties.

Food habits in any part of the world are serious barriers to the introduction of a new food as part of the everyday diet; thus, the

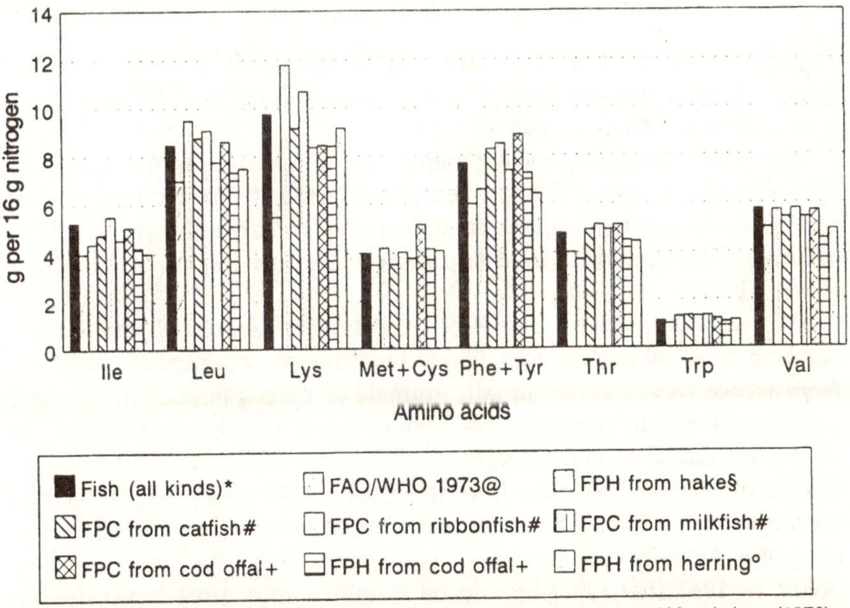

* Data from Paul and Southgate (1978); § Data from Yañez, Ballester and Monckeberg (1976); # Data from Sikka et al (1979); + Data from Lalasidis, Boström and Sjöberg (1978); @ Data adapted from Harper (1977); ° Data from Lalasidis and Sjöberg (1978).

Figure 9.1 Essential amino acid content of fishes and other food.

potential of introducing a new source of protein, beyond the economic determinants, depends on its incorporation in pre-existenting foods, making every effort to make its addition non-evident, or as a substitute of some products with adequate appeal. To do that, the protein must be neutral or inert in terms of sensory characteristics and should have adequate functional properties. These facts were very often ignored when new sources of protein were proposed two or three decades ago as the panacea to alleviate hunger in the world. Consequently, the use of fish protein concentrates (FPCs) had limited success when they were tried as a source of proteins for human consumption. However, fish proteins are very well utilized for animal feed which finally contributes to the human food supply, either supporting the growth of animal species for food or releasing foodstuffs for human consumption.

The use of proteolytic enzymes to produce fish protein hydrolyzates (FPHs) as a means of utilizing non-commercial species or wastes from the fish industry opened a new and attractive avenue to the sectors dealing with fish protein concentrates. Currently, there is an expanding demand for protein sources with adequate functional properties as food additives; in this context, more research is necessary to improve fish protein products.

9.2 FISHMEAL AND FISH PROTEIN CONCENTRATE

9.2.1 Definitions

Fishmeal is the crude flour obtained after milling and drying fish flesh. Due to its poor sensory quality and its lack of functional properties, its use is almost limited to animal feed.

The first technical developments for fishmeal production were achieved at the end of the last century (Mackie, 1983). One of the first reports of its use for humans came from South Africa during the 1930s where it was used as a food supplement for low income populations (Finch, 1977).

FPC was defined by the Protein Advisory Group of the United Nations as a stable product suitable for human consumption prepared from whole fish or other aquatic animals or parts thereof. The protein concentration is increased by removal of water and, in certain cases, of oil, bones and other materials (Finch, 1977). According to this definition, the simple removal of water from fish flesh produces a product that can be considered a FPC. Therefore, fishmeal should be considered in this category. Actually, the Food and Agricultural Organization (FAO) proposed in 1961 that FPCs should be classified into three types (Finch, 1977; Mackie, 1983):

- Type A: tasteless, colorless and odorless product containing a minimum of 67.5% of crude protein and no more than 0.75% of fat.

- Type B: fishmeal containing no more than 3% fat.
- Type C: crude fishmeal.

Clearly, the FPC concept includes fishmeal within its definition, with the understanding that the latter is a less refined product which contains obvious odor and taste related to its origin that limits its use in food product formulations. However, for the purpose of this chapter the term FPC will be reserved for the more refined products (types A and B), while the other product will be referred to as fishmeal.

Historically and from the functional point of view, such division of concepts is justified; while fishmeal has been used for animal feed for decades, the incentive to produce a more refined FPC was to obtain a safe product for human consumption with good organoleptic characteristics in order to be incorporated into other foods.

9.2.2 Raw materials

The world fish catch in 1989 was estimated as 100 millions tonnes (Venugopal, 1992). Since the 1970s it has been calculated that around 30% of the total catch is of low commercial value and diverted to the production of fishmeal and oil (Mackie, 1974, 1983; Sikorski and Naczk, 1981; Venugopal, 1992). These kinds of fish are mainly pelagic species (the shoaling, surface swimming varieties) such as anchovies, herring, jurel, mackerel, menhaden and sardines (del Valle and Aguilera, 1990). Pelagic species are fishes with reddish muscles and substantial fat reserves under the skin and within the flesh that imparts their characteristic oily appearance (Mackie, 1983).

The landed low commercial fish species come either from the intentional catch of pelagic fishes or as the by-catch of important commercial species such as shrimp. These two groups are the main source of raw material for the production of fishmeal and FPC.

The fish flesh from low cost species used as raw material may be handled either as ground whole fish, bone-free or fat-free fillets, or deboned fish mince (Mackie, 1983; Venugopal, 1992). Using minced fish technology instead of filleting, a yield increase of up to 50% in flesh and protein recovery can be reached (Venugopal, 1992). The use of lean species simplifies the production of fishmeal and FPC but these kinds of fish have a high demand as human food and are considerably higher in cost; however, several fatty fish species are landed in large amounts and their cost is low (Finch, 1977).

The high fat content of the fish flesh can become rancid very easily due to its unsaturated nature and represents a great disadvantage; even in lean or white species any content of fat exceeding 1.0% of the total weight must either be stabilized with antioxidants or removed to prevent rancidity. The amount of fat in the flesh of fatty fishes varies seasonally within a wide range, e.g. mackerel shows levels as high as 35% and as low as 1% (Mackie, 1982).

The other important source of raw material is the processing waste from commercial fish species; these include those with white lean flesh, such as cod, pollock, hake, flounder, etc., which contain 1–3% fat (Spinelli, Koury and Miller, 1975). The processing waste is the material left over after fillets are removed or in canning operations; this can be up to 64% of the weight of the fish, containing 10–13% protein (Mackie, 1974; Nair and Gopakumar, 1982). The parts that end up as waste include offals, fins, skin, heads, backbones, etc. (Mackie, 1983).

9.2.3 Processes

Fishmeal is produced from either whole fish, deboned fish flesh or fish processing wastes. The most widely used technique is the wet reduction process, which is operated continuously and requires large amounts of raw material. The fish is steam cooked, pressed, the press fluid centrifuged and the press cake is then dried. As most fishmeal is produced from fatty fish, oil is recovered after pressing mixed with stick water (Stansby, 1974). A diagram of the process is presented in Figure 9.2. The drying of the press cake can be achieved in rotary dryers using either steam or flame-heated air. For smaller factories a batch dry reduction process may be used, in which the fish is cooked and dried as a single operation in a steam-jacketed dryer (Stansby, 1974).

During the process waste waters are produced at various stages, as shown in Figure 9.2. The bail water is the water used to transport fish by pumping, i.e. it is the carrying fluid in operations such as unloading the catch from the ships, and it contains an average of 2.8% total solids and 1% crude protein. The blood water is the liquid exudate produced in the reduction plant during bulk storage of raw fish, and it contains 4.6–7.3% total solids and 2.2–4.4% crude protein. Finally, the stick water, which is produced during the process itself, is the residual watery phase from the centrifugation of the press liquor, and it contains an average of 9.4% total solids and 7.1% crude protein (del Valle and Aguilera, 1990). The stick water is recovered together with the fish oil and the two phases are separated by centrifugation; the crude fish oil is then freed of moisture and suspended solids by high speed centrifugation. The stick water can be concentrated in multiple effect evaporators to recover fish solubles either to be incorporated to the fishmeal stream process before drying or to be used separately for animal feed (Stansby, 1974).

The other type of product, FPC, is obtained after removing both oil and water from the flesh by solvent extraction. The most used solvents

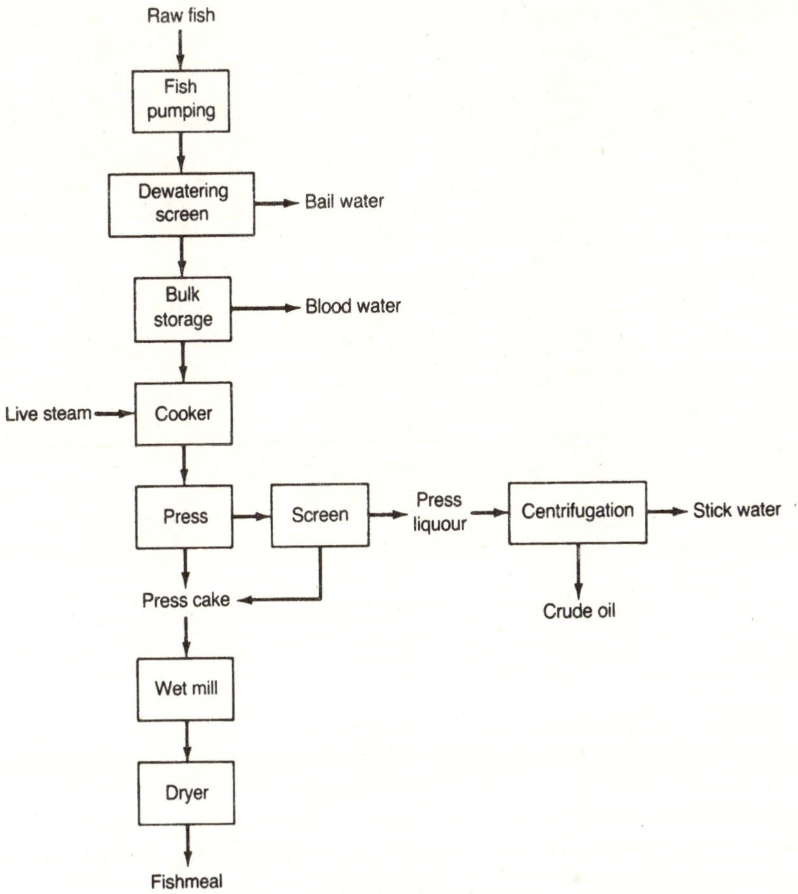

Figure 9.2 Fishmeal process.

are isopropanol, methanol, ethanol and 1,2-dichloroethane. Conditions for extractions, such as processing times and temperatures, repetitive extractions, etc., change from one processor to another. After removal of fat and water, the residual solvents are eliminated by air drying and the remaining pieces of bones by sieving (Mackie, 1983). Figure 9.3 shows a flow diagram for obtaining FPC.

After solvent extraction, the liquid phase is separated either by centrifugation or filtration. The solvents are then recovered from the liquid phase by distillation. Due to the complex nature of the lipids present in fish, the extraction efficiency depends on the solvent; if a mixture of two or more solvents is used, a higher yield of fat extraction can be attained. However, extraction processes with a single solvent are simpler and cheaper. Hence, several adequate procedures have been developed using only one solvent (Suzuki, 1981).

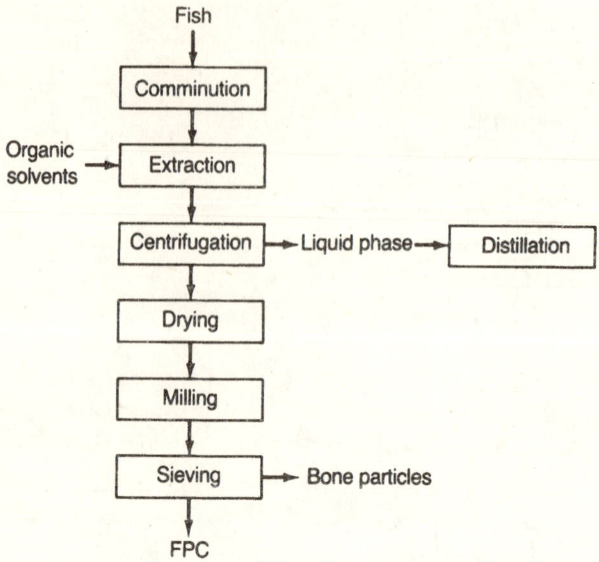

Figure 9.3 Flow diagram for FPC production.

9.3 FISH PROTEIN HYDROLYZATE

9.3.1 Definitions

The potential of FPC as a food additive is limited mostly because it lacks solubility and has poor functional properties (Spinelli, Koury and Miller, 1972). To solubilize FPC it is necessary to break down the protein into smaller sized peptides. Such treatment yields a mixture of proteinaceous fragments known as FPH.

The depolymerization of high molecular weight proteins occurs by the cleavage of peptide bonds through the addition of water molecules. In aqueous media the excess of water favors this lytic process and it is therefore known as 'hydrolysis'. Many substances can act as a catalyst for hydrolysis, e.g. acids, bases and enzymes. The latter are classified as proteases based on their substrate specificity for proteins and are assigned the numbers 3.4 (peptide hydrolases) by the Enzyme Commission. The fact that protease degradation yields fewer by-products compared with non-biological catalysts justifies its selection for use in industrial applications for foods and feeds.

Enzyme proteolysis has been used extensively in a wide range of processes, from hide dehairing and bating to the action of household detergents. In food products, enzyme modification of edible proteins permits the improvement of palatability and storage stability of available protein resources in a given region. Examples of such know-how date back to the origins of civilization, as in the case of fish autolyzates (McIver *et al.*, 1982) or fermented soya products in Southeast Asia.

An important parameter to grade proteolytic processes is the degree of hydrolysis (DH) since the functional properties of the final product, including the presence of bitterness, are associated with the size of the resulting peptides and hence to the extent of the reaction. Those aspects are discussed in more detail later in the chapter.

DH reflects the number of peptide bonds cleaved and is defined by the following relationship (Adler-Nissen, 1986):

$$\mathrm{DH} = (h/h_{\mathrm{tot}}) \times 100 \tag{1}$$

where h is the number of free amino groups at the end of the reaction and h_{tot} is the total number of peptide bonds in the original protein (total amino acids minus one). The free amino groups can be determined by reaction with trinitrobenzene sulfonic acid (TNBS).

It is obvious that DH does not gives any indication about the size of the peptides obtained. In fact, the dispersion of sizes will depend on the conditions of the reaction, the nature of the substrate and the catalyst used. For practical purposes an average peptide chain length (PCL) can be related to DH according to the following expression:

$$\mathrm{DH}/100 = (1/\mathrm{PCL}) - (1/\mathrm{PCL_o}) \tag{2}$$

where $\mathrm{PCL_o}$ is the number of amino acids in the substrate (the initial protein). For a typical protein with molecular weight of 30 000 or greater, equation (2) can be simplified to:

$$\mathrm{PCL} = 100/\mathrm{DH}. \tag{3}$$

9.3.2 Processes

Two conceptually different processes are used depending on the form of the protease added to the reaction vessel. The most straight forward method is the direct addition of the enzyme obtained from commercial sources. This approach is known as an enzymatic method and will be the focus of the forthcoming discussion. However, it is also possible to add a live culture of microorganisms in addition to appropriate nutrients to support its growth. The microbial strains should be protease producers. As a result, the proteases are synthesized and permitted to act on the protein simultaneously. This is a fermentative method and is covered in detail in other chapters of this book.

Interest in establishing commercially attractive processes for FPH dates back to the 1960s. An excellent review on historical developments has been published by Adler-Nissen (1986). Early reports described reaction conditions and enhancement of protein solubility (Sen *et al.*, 1962). In this case the final product was a fish peptone for use in fermentation media. Several patents have also been issued on the subject. One of them deals with a continuous process (Keyes and Meinke, 1966) for poultry feed. Also, a process is described in which precise conditions for limited hydrolysis of fish proteins are cited to obtain a stable emulsion (Rutman, 1971). In this case, the emulsion can be spray dried and easily rehydrated into milk-like dispersions with excellent nutritional properties. According to Mackie (1982), milk replacers constitute the main market for FPH. In fact, such products are used as basic ingredients in commercial milk substitutes in France (Baca, Peña-Vera and Díaz-Castañeda, 1991).

Other experimental processes include the use of membrane reactors (Cheftel, 1972). These types of systems have gained great attention for biotechnological applications. They allow enzyme reuse, an efficient separation of products and, in many cases, continuous operation (Nakajima, Shoji and Nabetani, 1992). Membrane reactors are particularly important in cases where the enzyme is inhibited by the product(s).

Using a membrane reactor of 100 cm^3 designed from data obtained in batch and semibatch experiments, 76% of total nitrogen was recovered in the permeate after 72 h of operation in continuous flow (Bhumiratana, Hill and Amundson, 1977). Total permeate was 71.3 times the reactor volume and the enzyme was added at intervals to compensate for the drop in catalytic activity. However, the flux of permeate declined steadily due to the build up of insoluble compounds along the membrane wall.

In a recent study, hydrolyzation of a reaction mixture with high solids content (fish ground with no addition of water) was evaluated (Baca, Peña-Vera and Díaz-Castañeda, 1991). The objective was to reduce energy consumption on the basis that spray drying of the solubilized protein is generally the most energy intensive step of the whole process. Yields of 15% in terms of whole fish were obtained at the pilot plant scale with the product showing excellent protein content and solubility values.

9.3.3 Sources of proteases

Considering that the hydrolytic process is aimed essentially at increasing protein solubility, proteases with broad specificity that render them capable of peptide cleavage at random, i.e. independent of amino acid composition, are used in industrial operations. In fact, many studies report the use of pronase, a preparation containing a mixture of proteases, with a very broad specificity and high rate of hydrolysis under adequate conditions (Hevia, Whitaker and Olcott, 1976). Proteases can also be used at a variety of conditions, ranging, in terms of pH optima, from pH 2 for pepsin to pH 8.5 for alcalase from *Bacillus* spp.

Commercial enzymes are obtained from animal, plant and microbial sources. Plant proteases include mostly bromelain, papain and ficin (Hevia, Whitaker and Olcott, 1976; Beddows and Ardeshir, 1979; Mackie, 1982; Quaglia and Orban, 1987a). Pepsin is the animal enzyme most widely used (Lin and Pigott, 1981), though chymotrypsin has also been tested (Montecalvo, Constantinides and Young, 1984). However, as a result of the advances in fermentation technology and bioseparations during the last decade, a variety of commercial microbial proteases, mostly of bacterial origin, is now available at low prices for use in the production of FPC (Mackie, 1982).

9.4 CHARACTERISTICS AND FUNCTIONAL PROPERTIES OF FISH PRODUCTS

9.4.1 Nutritional value

The biological value of proteins in FPC and FPH is generally as good as that of the fish from which it is derived (Mackie, 1983; Sikorski and Naczk, 1981). Table 9.2 shows the nutritional protein parameters of some products reported by several authors. The content of essential amino acids in fish products is very good and in most cases is superior to the FAO/WHO 1973 pattern, as can be seen in Figure 9.1. Particularly, fish muscle has an outstanding content of lysine, making fish products suitable for cereal fortification (Yañez, Ballester and Monckeberg, 1976).

The unit operations used during the production of FPC do not cause any significant loss or decrease of availability in amino acids. However,

Table 9.2 Nutritional parameters (%) of some fish products compared to casein

Product	*Protein efficiency ratio (PER)*	*Net protein utilization (NPU)*	*Biological value (BV)*	*Reference*
FPC catfish	99.7	–	–	Sikka *et al.* (1979)
FPC picked ribbonfish	112.7	–	–	Sikka *et al.* (1979)
FPC milkfish	85.3	–	–	Sikka *et al.* (1979)
FPC cod offal	–	100.1	100.6	Lalasidis, Boström and Sjöberg (1978)
FPH cod offal	101.2	94.1	93.4	Lalasidis, Boström and Sjöberg (1978)
FPH hake	115	–	–	Yañez, Ballester and Monckeberg (1976)
FPH threadfin bream	64.3	–	–	Nair and Gopakumar (1982)

the production of hydrolyzates can bring some undesirable changes. For instance, comparing a FPC from cod offal with its FPH obtained by alcalase and pancreatine treatment, a reduction in the content of valine, was observed, with a corresponding slight reduction on NPU and biological value (see Figure 9.1 and Table 9.2) (Lalasidis, Boström and Sjöberg, 1978). Other FPHs obtained by the same authors from fresh herring and commercial FPC also showed a limiting content of valine (Lalasidis and Sjöberg, 1978). Other authors (Yañez, Ballester and Monckeberg, 1976) reported that FPH from hake had a lower score of threonine.

The use of alkaline hydrolysis, which has been employed in some processes, can lead to the formation of toxic compounds such as lysinoalanine and reduces considerably the biological value of the product (Sikorski and Naczk, 1981; Mackie, 1983); nevertheless, under carefully controlled conditions, the nutritional quality of the protein can be maintained at adequate levels (Mackie, 1983).

9.4.2 Sensory quality

The important attributes responsible for the sensory quality of a protein aimed at use as a food additive are color, odor and flavor. In the case of FPC and FPH the color depends on the flesh pigments present in the particular species used as raw material and on products of non-enzymic browning formed during the process. Some products, such as dry FPH, are light brown powders (Sikorski and Naczk, 1981). A liquid FPH produced by a Japanese company was pink when it was obtained from red salmon, while the product was white when the raw material was pollack and beige if it was obtained from snow crab (Anonymous, 1987).

In some cases, e.g. when the fish protein is used as a meat extender, the color can be easily masked by the meat pigments. In other cases, such as in pasta fortification, the color may be a critical factor (Sikorski and Naczk, 1981).

The specific odor in fish products is mainly due to the oxidation of residual lipids. The composition and concentration of those lipids depend on the raw material, the polarity of the solvent or solvents used during the extraction and the conditions of extraction. If residual lipids are present in concentrations at around 0.1% the shelf-life of the product will be quite acceptable; however, only with concentrations of lipids of 0.05% or less can a suitable stability can be obtained (Sikorski and Naczk, 1981). Clearly, the concentration of lipids needed to yield products with high quality are very much lower than the level specified for a FPC type A.

Another important factor in terms of odor is the fishy note, which is given by amines and sulfur-containing compounds (Sikorski and Naczk,

1981). These compounds are volatile and can be removed by deodorization operations. However, if residual organic solvents are entrapped at sufficient concentration, a 'chemical note' will be present in the product (Sikorski and Naczk, 1981).

Concerning flavor, it is obviously influenced by the intensity of the odor notes. Specifically, FPH acquires bitter and umami (glutamate-like) tastes due to the hydrolysis reactions (Noguchi *et al.*, 1975; Hevia and Olcott, 1977; Sikorski and Naczk, 1981; Mackie, 1983). While the first taste is a drawback, the second one can be a desired attribute (having a flavor potentiating activity), particularly if the product is used as a meat extender or added to soups or gravies.

The intensity of bitterness depends on the degree of hydrolysis and the specificity of the protease, i.e. the size of peptides and the kind of peptidic bonds that were cleaved. Bitterness is due to hydrophobic peptides released during hydrolysis. For instance, pronase hydrolysates were less bitter than those FPHs produced with bromelain or ficin; the former had a higher proportion of low molecular peptides than the others, which also had more basic peptides (Hevia and Olcott, 1977). The proteolysis by endopeptidase results in a bitter off-flavor, while a more extensive hydrolysis using exopeptidases liberates a considerable amount of free amino acids with negligible bitterness (Lalasidis, Boström and Sjöberg, 1978). Debittering can also be accomplished by extracting the hydrophobic peptides with alcoholic solutions (Lalasidis and Sjöberg, 1978; Chakrabarti, 1983), by hydrophobic interaction chromatography (Lalasidis and Sjöberg, 1978) and by the plastein reaction (Lalasidis and Sjöberg, 1978; Mackie, 1983).

The umami flavor in FPH is due to oligopeptides containing a high molar ratio of glutamic acid (Noguchi, Arai and Yamashita, 1975).

9.4.3 Functional properties

While fishmeal and FPC show poor dispersibility, solubility, water-holding capacity and emulsifying properties, FPHs have superior functional characteristics. FPHs are generally readily dispersed in water (Sikorski and Naczk, 1981). The solubility of the FPH varies with the degree of hydrolysis. Quaglia and Orban (1987b) reported very good solubility of FPH from sardines in the pH range from 5 to 9 independent of the degree of hydrolysis; however, at pH between 5 and 2 the solubility decreased, being more pronounced in those products with a lower degree of hydrolysis. On the other hand, FPC is generally insoluble in water and has negligible water-holding capacity. The lack of functional properties in FPC is due to the use to heat and solvents that severely denature fish proteins. To improve the functionality some process parameters need to be modified or chemical modifications to the protein must be introduced (Sikorski and Naczk, 1981; Mackie, 1983).

For instance, by lowering the extraction temperatures up to 20°C, a significant increase of the water-holding capacity may result; however, as the efficiency of the process is reduced, repeated extractions must be conducted (Sikorski and Naczk, 1981). Important improvement of this functional property has also been obtained after addition of NaCl and pH reduction of the raw fish to 2.5 prior to the extractive operation (Sikorski and Naczk, 1981).

Acylation of FPC with acetic or succinic anhydride results in a product with high water-holding capacity and good dispersability in water to form viscous suspensions which do not coagulate even at 100°C (Sikorski and Naczk, 1981; Mackie, 1983). With high levels of acylation, proteins become more soluble; however, the nutritive value is reduced because essential amino acids such as lysine and sulfur-containing amino acids are acylated. A combination of acylation and proteolysis can be used to obtain more amenable products (Spinelli *et al.*, 1975; Groninger and Miller, 1979; Mackie, 1983). Spinelli and coworkers developed a method forming protein–phosphate complex to obtain functional fish protein isolates; these isolates were able to form colloidal dispersions in water (Spinelli *et al.*, 1975).

The capacity of FPC to emulsify fat is very weak, while FPH has excellent emulsification properties (Sikorski and Naczk, 1981; Mackie, 1983). The emulsifying properties of FPH depend on the degree of hydrolysis; FPH from sardines showed that the higher the degree of hydrolysis the lower the emulsifying capacity; the products with a low degree of hydrolysis had emulsifying properties higher than sodium caseinate (Quaglia and Orban, 1990).

By means of acylation, phosphate complex formation and the modifications of the process described above, the emulsifying properties of FPC can be improved (Spinelli *et al.*, 1975; Sikorski and Naczk, 1981; Mackie, 1983).

The capacity to form gels and stable foams are lacking in FPC, while FPH has suitable foaming properties (Sikorski and Naczk, 1981; Mackie, 1983). This property is greatly improved by acylation or by the combination of acylation and hydrolysis. Succinylated FPH showed whipping properties similar to those of egg albumin (Spinelli *et al.*, 1975).

9.5 UTILIZATION OF FISH PRODUCTS

Fishmeal is widely used as animal feed and, due to its lack of functional properties and to its flavor, is hardly used as human food. Concerning FPC, which was designed for human consumption, there is now little interest in its production on an industrial scale. The main problems with FPC were:

1. Economical factors. The economy of the process depends very much on the cost of fish. Additionally, the process is capital intensive and the production costs apart from raw material are high (Finch, 1977; Mackie, 1983).
2. The lack of clearly identified producers and consumers. FPC was intended as a means to alleviate the shortage of protein in poor countries; however, the market is not well defined in most cases. The lack of functional properties of FPC is the main reason that the products have not been commercialized. On the other hand, the development and use of this kind of product by governmental agencies or international organizations requires funding and political will, which are not always available. Even so, the protein scarcity was exaggerated; the real hunger problem in underdeveloped countries can be attributed to protein–caloric deficiencies due to inadequate food intake.

Nevertheless, intensive research has been done to incorporate FPC in human food, in addition to the utilization of fishmeal in animal feed.

The use of FPC in the human diet has been mainly as a cereal supplement in bakery commodities, breakfast cereals, and macaroni, in vegetable dishes or soups, and as a meat extender.

The Swedish company 'Astra' is possibly the only case of a private industry that successfully developed a type A FPC product. Such a product was introduced as a milk substitute for calves and a protein concentrate to supplement cereal flours for human consumption (Finch, 1977).

In Japan a product marketed with the trade name of Marinbeef was developed up to pilot scale as a meat extender for products such as hamburgers and meat loaves (Mackie, 1983). According to Vareltzis *et al.* (1990) it was possible to utilize FPC as an extender of minced beef to produce acceptable hamburger-type products; however, the fishy flavor was consistently identified even at a 20% level of replacement.

The use of type B FPC and fishmeal has shown a relative acceptability when used for food supplementation in countries that consume traditional fish products. These products can be used as a supplement in fish soups. Nordsildmel, a type B FPC developed by a Norwegian company, has been used to conduct acceptability tests in 14 countries, mainly in Africa, showing that it was generally accepted where fish consumption was high, but rejected where it was low (Finch, 1977).

In South Africa, a breakfast cereal was developed using type B FPC. In this country, other cereal products such as bread were supplemented under a broad governmental program (Finch, 1977; Mackie, 1983).

A liquid FPH called creamy fish protein, obtained either from red salmon, pollack or snow crab, was introduced in the US market in 1987 by a Japanese company. The product was used previously in Japan in combination with soya for the Japanese school lunch program. Suggested applications include high-protein beverages, soups, snacks, desserts, sauces, meat extenders and dairy analogs. The product can also be powdered by spray drying (Anonymous, 1987).

A spray-dried FPH has been incorporated into crackers, obtaining an adequate product in terms of sensory characteristics and overall acceptability when 10% FPH was added (Yu and Tan, 1990).

As succinylated FPH had interesting whipping properties, this form has been used successfully in desert toppings, a soufflé and a whipped gelatin dessert (Spinelli *et al.*, 1975).

Phosphate–protein complex isolates have been tested as egg albumin substitutes in angel cake formulations, as meat extenders in frankfurters and for the preparations of milk-like beverages (Spinelli *et al.*, 1975).

The above cases exemplify the challenges and interests that the food chemist has to face in order to make the massive use of fish processed products a true commercial reality. Certainly, much effort and financial resources are oriented to this goal all around the world.

REFERENCES

Adler-Nissen, J. (1986) *Enzymic Hydrolysis of Food Proteins*, Elsevier, London.

Anonymous (1987) Enzyme process creates creamy fish protein. *Food Eng.*, **59**(9), 44–5.

Baca, D.R., Peña-Vera, M.T. and Díaz-Castañeda, M. (1991) Production of fish protein hydrolysates with bacterial proteases; yield and nutritional value. *J. Food Sci.*, **56**(2), 309–14.

Beddows, C.G. and Ardeshir, A.G. (1979) The production of soluble fish protein solution for use in fish sauce manufacture. I. The use of added enzymes. *J. Food Technol.*, **14**, 603–12.

Bhumiratana, S., Hill, C.G. and Amundson, C.H. (1977) Enzymatic solubilization of fish protein concentrate in membrane reactors. *J. Food Sci.*, **42**(4), 1016–21.

Chakrabarti, R. (1983) A method of debittering fish protein hydrolysate. *J. Food Sci. Technol.*, **20**(4), 154–6.

Cheftel, C. (1972) Solubilisation enzymatique continue du concentré protéique de poisson. Essai de recyclage des enzymes. *Ann. Technol. Agric.*, **21**, 423–33.

del Valle, J.M. and Aguilera, J.M. (1990) Recovery of liquid by-products from fish meal factories: a review. *Process Biochem.*, **25**(4), 122–30.

Finch, R. (1977) Whatever happened to fish protein concentrate? *Food Technol.*, **31**(5), 44–53.

Groninger, H.S. and Miller, R. (1979) Some chemical and nutritional properties of acylated fish protein. *J. Agric. Food Chem.*, **27**(5), 949–955.

Harper, A.E. (1977) Human aminoacid and nitrogen requirements as the basis for the evaluation of nutritional quality of proteins, in *Food Proteins*, (eds J.R. Whitaker and S.R. Tannenbaum), AVI Publishing, Westport, CT, pp. 363–86.

Hevia, P., Whitaker, J.R. and Olcott, H.S. (1976) Solubilization of a fish protein concentrate with proteolytic enzymes. *J. Agric. Food Chem.*, **24**(2), 383–85.

Hevia, P. and Olcott, H.S. (1977) Flavor of enzyme-solubilized fish protein concentrate fractions. *J. Agric. Food Chem.*, **25**(4), 772–5.

Keyes, C.W. and Meinke, W.W. (1966) Method of processing fish. *US Patent 3 249 442.*

Lalasidis, G. and Sjöberg, L.B. (1978) Two new methods of debittering protein hydrolysates and a fraction of hydrolysates with exceptionally high content of essential amino acids. *J. Agric. Food Chem.*, **26**(3), 742–9.

Lalasidis, G., Boström, S. and Sjöberg, L.B. (1978) Low molecular weight enzymatic fish protein hydrolysates: chemical composition and nutritive value. *J. Agric. Food Chem.*, **26**(3), 751–6.

Lin, L.L. and Pigott, G.M. (1981) Preparation and use of inexpensive crude pepsin for enzyme hydrolysis of fish. *J. Food Sci.*, **46**, 1569–72.

Mackie, I.M. (1974) Proteolytic enzymes in recovery of proteins from fish waste. *Process Biochem.*, **9**(10), 12–14.

Mackie, I.M. (1982) Fish protein hydrolysates. *Process Biochem.*, **17**(1), 26–31.

Mackie, I.M. (1983) New approaches in the use of fish proteins, in *Developments in Food Proteins* – 2, (ed. B.J.F. Hudson), Applied Science, London, pp. 215–62.

McIver, R.C., Brooks, R.I. and Reineccius, G.A. (1982) Flavor of fermented fish sauce. *J.Agric. Food Chem.*, **30**, 1017–20.

Nair, A.L. and Gopakumar, K. (1982) Soluble protein isolates from low cost fish and fish wastes. *Fishery Technol.*, **19**(2), 101–3.

Montecalvo, J., Constantinides, S.M. and Yang, C.S.T. (1984) Enzymatic modification of fish frame protein isolate. *J. Food Sci.*, **49**, 1305–60.

Nakajima, M., Shoji, T. and Nabetani, H. (1992) Protease hydrolysis of water soluble fish proteins using a free enzyme membrane reactor. *Process Biochem.*, **27**, 155–60.

Noguchi, M., Arai, S., Yamashita, M., *et al.* (1975) Isolation and identification of acidic oligopeptides occurring in a flavor potentiating fraction from a fish protein hydrolysate. *J. Agric. Food Chem.*, **23**(1), 49–53.

Paul, A.A. and Southgate, D.A.T. (1978) *McCance and Widdowson's The Composition of Foods*. Elsevier/North-Holland Biomedical Press, London.

Quaglia, G.B. and Orban, E. (1987a) Enzymic solubilisation of proteins of sardine (*Sardina pilchardus*) by commercial proteases. *J. Sci. Food Agric.*) **38**, 263–9.

Quaglia, G.B. and Orban, E. (1987b) Influence of the degree of hydrolysis on the solubility of the protein hydrolysates from sardine (*Sardina pilchardus*). *J. Sci. Food Agric.*, **38**, 271–6.

Quaglia, G.B. and Orban, E. (1990) Influence of the enzymatic hydrolysis on structure and emulsifying properties of sardine (*Sardina pilchardus*) protein hydrolysates. *J. Food Sci.*, **55**(6), 1571–3, 1619.

Rutman, M. (1971) Process for preparing high-energy fish protein concentrate. *US Patent* 3 561 973.

Sen, D.P., Sripathy, N.V., Lahiry, N.L. *et al.* (1962) Fish Hydrolysates. I. Rate of hydrolysis of fish flesh with papain. *Food Technol.*, **16**(5), 138–41.

Sikka, K.C., Singh, R., Gupta, D.P. and Duggal, S.K. (1979) Comparative nutritive value of fish protein concentrate (FPC) from different species of fishes. *J. Agric. Food Chem.*, **27**(5), 946–9.

Sikorski, Z.E. and Naczk, M. (1981) Modification of technological properties of fish protein concentrates. *CRC Critical Rev. Food Sci. Nutr.*, **14**(3), 201–30.

Spinelli, J., Koury, B. and Miller, R. (1972) Approaches to the utilization of fish for the preparation of protein isolates. Enzymic modifications of myofibrillar fish protein. *J. Food Sci.*, **37**, 604–8.

Spinelli, J., Groninger, H., Koury, B. and Miller R. (1975) Functional protein isolates and derivatives from fish muscle. *Process Biochem.*, **10**(10), 31–6, 42.

Stansby, M.E. (1974) Fish and fishery products, in *Encyclopedia of Food Technology*, (eds A.H. Johnson and M.S. Peterson), AVI Publishing, Westport, CT, pp. 398–407.

Suzuki, T. (1981) *Fish and Krill Protein. Processing technology*, Applied Science, London.

Vareltzis K., Soultos N., Zetou, F. and Tsiaras, I. (1990) Proximate composition and quality of a hamburger type product made from minced beef and fish protein concentrate. *Lebensm. Wiss. Technol.*, **23**, 112–16.

Venugopal, V. (1992) Mince from low-cost fish species. *Trends Food Sci. Technol.*, **3**(1), 2–5.

Yañez, E., Ballester, D. and Monckeberg, F. (1976) Enzymatic fish protein hydrolyzate: chemical composition, nutritive value and use as a supplement to cereal protein. *J. Food. Sci.* **41**(6), 1289–92.

Yu, S.Y. and Tan, L.K. (1990) Acceptability of crackers ('Keropok') with fish protein hydrolysate. *Int. J. Food Sci. Technol.*, **25**, 204–8.

CHAPTER 10

Production of fish protein hydrolyzates by microorganisms

V. Venugopal

10.1 INTRODUCTION

Of the current world fish landing of around 100 million tonnes, consumer preference is restricted to only a few selected items. A significant proportion of total available fish forms the by-catch of shrimp trawling, and includes non-conventional species which are not commercially exploited. Nevertheless, these fish are sources of nutritive proteins which could be utilized for satisfying human protein requirements. The necessity of exploitation of under-utilized fish species as well as reduction of postharvest fishery losses have been pointed out by several authors (James, 1986; Morrissey, 1988; Burt, Hardy and Whittle, 1990; Sikorski, 1990).

10.1.1 Mince from under-utilized fish

Under-utilized fish could be used to develop several value-added products. The first step in utilization invariably consists of hygienic collection of the meat by a mechanical deboning operation. The meat mince obtained could be used as a raw material for the development of a number of products such as fermented items, fabricated foods and surimi-based products (Grantham, 1981; Regenstein, 1986; Venugopal, 1992; Venugopal, Ghadi and Nair, 1992). Fish protein hydrolyzate (FPH) is one product, currently used as a protein supplement for animal foods, which also has potential as a human food item (Kilara, 1985). At present FPHs are produced by digestion of fish meat by proteolytic enzymes

Fisheries Processing: Biotechnological applications.
Edited by A.M. Martin.
Published in 1994 by Chapman & Hall, London. ISBN 0 412 58460 3

from vegetable as well as animal sources. This article will discuss the scope of applying biotechnological methods for the production of FPH. At the outset, some aspects on the properties of fish mince, microbial extracellular proteases and immobilization of microorganisms will be discussed.

10.1.2 Properties of the mince

Fish mince, as compared with intact fish muscle, is more unstable and is prone to rapid deteriorative changes during storage (Grantham, 1981). Mechanical deboning causes disruption of the tissue and exposure of the meat to air which accelerates several deteriorative processes, such as lipid oxidation, bacterial proliferation and resulting protein changes. The higher temperature associated with deboning also affects the stability of the mince. The mince contains about 20% soluble components belonging to the sarcoplasmic fraction and about 2–3% connective tissue (although this may be as high as 10% in the case of elasmobranch species). The rest is composed of myofibrillar proteins. Washing of the mince in chilled water removes soluble compounds, significant amounts of lipids, blood and pigments to give concentrated myofibrillar proteins, having a bland taste (Lee, 1986). Fatty fish like mackerel or sardine may require washing with bicarbonate to remove oil (Putro, 1989). The mince may be stored frozen in the presence of cryoprotectants, e.g. polyphosphates, which can control protein denaturation during storage.

The washed myofibrillar proteins from most fish species are capable of turning into gel either on prolonged incubation at low temperature or by heating to 40–50°C (Lanier, 1986). Gelation results from cross-linking of randomly dispersed polymer chains in solution to give matrices which immobilize water in the interstitial structure. The process occurs in two stages consisting of initial unfolding of the protein chains followed by protein–protein interactions which help entrapment of the water. The stability of the gel depends upon the extent of such interactions. The presence of salts, low pH and mild heat can bring about unfolding of the proteins leading to gelation (Lanier, 1986; Foegiding, 1988; Zeigler and Foegeding, 1990). The washed mince could be the ideal raw material for the development of several foods (Grantham 1981; Lee, 1986). After gelation, the washed mince could be the ideal starting material for the preparation of hydrolyzates, as will be discussed later.

10.2 SOLUBILIZATION OF FISH MINCE

10.2.1 Conventional methods to make FPH

FPHs are distinguished from fermented foods by their high solubility in water, low fat and ash contents as well as high protein contents. FPHs

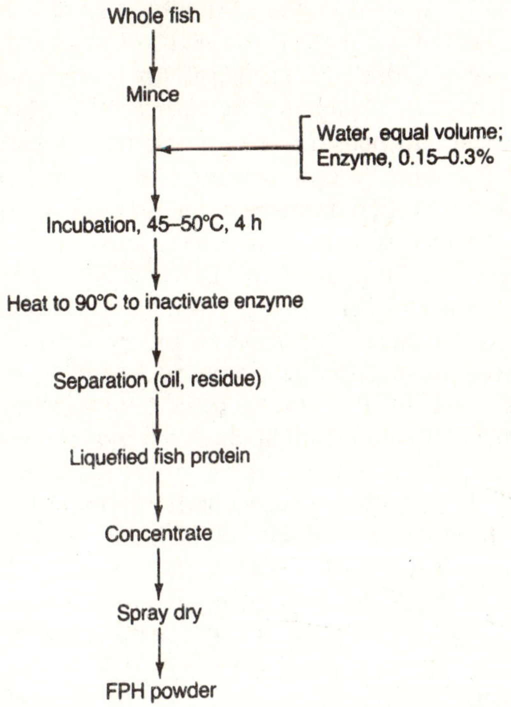

Figure 10.1 Flow diagram for production of FPH.

are produced by conventional methods by proteolytic digestion of fish and fish waste at optimal conditions of pH and temperature required by the enzymes. Liquefaction of the proteins takes place within a few hours. The hydrolyzate is then decanted and centrifuged to remove scales and bones. The soluble fraction is concentrated by evaporation and is suitably dried. The production of FPH has been described by several authors (Adler-Nissen, 1977, 1985; Kinumaki, 1978; Mackie, 1982a, b; Kilara, 1985; Qaglia and Orban, 1987a, b). The process is depicted in Figure 10.1.

By using different fish species, enzymes and digestion conditions, a wide range of hydrolyzates can be produced. The enzymes can be from plant (papain, ficin, etc.), animal (trypsin, pancreatin, etc.) or microbial (pronase, alcalase, etc.) sources. Uncontrolled or prolonged proteolysis results in the formation of smaller and highly soluble peptides completely lacking the functional properties of native proteins, such as water-holding capacity, emulsification capacity and foaming ability. By careful control of hydrolysis, it is possible to suitably modify the functional properties which are useful in food formulations (Mackie, 1982a). The degree of hydrolysis, which is expressed as the percentage of α-amino nitrogen in the soluble fraction (Monti and Jost, 1979), is important in optimizing the process parameters in the product development. Quaglia and Orban (1987b) showed higher solubility of sardine meat accompanied by a decrease in the molecular weight of protein fragments with increasing degree of hydrolysis. The feasibility of employing 23 different proteolytic enzymes was examined by Hale (1969), who observed that pancreatin, papain and pepsin were suitable for the process. Sen *et al.* (1962) defined digestion conditions using papain and found that at pH 7.0 maximum solubilization occurred in the first few hours of treatment. The authors advocated 40°C instead of 65°C for obtaining longer peptides during digestion. Venugopal and Lewis (1981) reported maximum solubilization of low cost fish using pepsin. Hevia, Whitaker and Olcott (1976) examined fish meat hydrolysis using bromelain, ficin and pronase, and showed that pronase gave a product with less bitterness. Lean fish species are ideal for the process since fatty species give products containing significant amounts of lipids. Such hydrolyzates may require centrifugation in order to remove the oil. Antioxidants may also have to be used to control rancidity development during hydrolysis. Ritchie and Mackie (1982) hydrolyzed several fish including both fatty and lean varieties using papain added at 5% of the weight of fish meat. The authors faced fewer problems in the preparation of spray dried FPH from lean fish varieties as compared with fatty species. The dry powder had 92% protein, 1.7% lipid and 6.4% ash. The yield of the powder varied from 8 to 16% of the initial raw material (Merritt, 1982). A process to make FPH from a low cost fish, i.e. dhoma, consisted of treatment of the mechanically deboned meat with papain at 55°C for 2 h. The powder obtained after spray drying of the hydrolyzate had a creamy color and 90% protein content. Polyphosphate improved the functional properties of the powder (Ghadi, Warrier and Ninjoor, 1987). Partial deodorization of the mince prior to enzyme treatment could be done by boiling it in water for a few minutes followed by screw pressing of the cooked meat (Venugopal and Lewis, 1981). Preparation of FPH from Antarctic krill, one of the major sources of non-conventional fish, has been described by Sikorski and Naczk (1981). The process consisted of initial low speed centrifugation to remove the gut of the fish followed by mechanical deboning. The mince obtained was treated with either neutrase or alcalase at 50°C for 2 h. The enzyme was inactivated by steaming and the hydrolyzate was spray dried.

10.2.2 Properties and uses of FPH

The protein hydrolyzates have better functional properties as compared with whole fish meat. However, bitterness is a commom problem of protein hydrolyzates, which has been attributed to the formation of peptides containing bulky hydrophobic groups towards the C-terminal end (Roy, 1992; Kilara, 1985). Bitterness is more intense if extensive hydrolysis of the protein has occurred. Several method have been suggested to mask the bitterness which include incorporation in the FPH of glutamic acid or glutamyl-rich peptides, polyphosphates, gelatin or glycine (Roy, 1992; Kilara, 1985). Another method is through the plastein reaction, which is brought about by the addition of certain proteases to the concentrated hydrolyzates, resulting in the synthesis of a new polypeptide. Raghunath and McCurdy (1991) observed significant amounts of plastein synthesis in autolyzed fish waste by pepsin at pH values ranging from 5 to 8. The plastein reaction also improves the nutritional quality and physical properties of the hydrolyzates (Satterlee, 1981). A protease derived from *Aspergillus* has been recognized to be beneficial for the purpose (Kilara, 1985). Biotechnologically, the bitterness score of a tryptic hydrolyzate of casein was decreased when the product was incubated with immobilized cells of *Erwinia ananas* (Watanabe, Shimizu and Arai, 1990).

The commercial aspects of FPH production have been discussed by Kinumaki (1978), Ritchie and Mackie (1982) and Merritt (1982). Piggott, Bucone and Ostrander (1978) studied the engineering aspects of a plant for FPH production. The spray drying costs could be reduced if the fish meat is directly treated with the enzyme, without addition of water, which gave higher solid contents in the hydrolyzate (Rebecca, Peña-Vera and Diaz-Castaneda, 1991). However, the production cost of FPH, based on a yield of 14% of the raw material, was more than that of fish meal. The economic advantage of FPH as a milk replacement has to be considered in this respect (Merritt, 1982).

Protein hydrolyzates are generally used for the modification of functional properties of foods and in dietetic foods as a source of small peptides and amino acids. As a result of its high solubility and amino acid balance, FPH has obvious advantages over dried products like fish protein concentrate or even human grade fishmeal, in these respects. Its high dispersability makes it suitable as a replacement for milk proteins. Although, at present, it is used as a milk replacer for calves in France and the US (Kilara, 1985), its potential has been pointed out as a protein supplement to cereal foods (Yanez, Ballester and Monckeberg, 1976), soups and bread (Kinumaki, 1978), and crackers (Yu and Tan, 1990). The development of products involving fish hydrolysis has been reviewed by Owens and Mendoza (1985).

10.3 MICROBIAL PROTEASES

10.3.1 Characteristics of microbial extracellular proteases

Microbial proteases can be either extracellular or intracellular. Extracellular enzymes are those secreted by the organisms and comprise mostly hydrolytic enzymes including proteases. Several microorganisms secrete a variety of proteases, the yields of which depend upon the nutritional composition of the growth media and the growth conditions. For practical applications, proteases are produced from high-yielding microbial strains by fermentations under controlled conditions. These enzymes act on proteins at specific sites and, depending upon the site of cleavage, they can be divided into exoproteases or endoproteases. Exoproteases cleave the protein substrate at the N-terminal (aminopeptidases) or at the C-terminal (carboxy peptidases) end. On the other hand, endoproteases (or proteinases) act on the protein molecule at sites away from the terminal amino acid residues. All proteinases which are of industrial importance are endoproteases rather than exoproteases. Microbial proteases have broad specificities as indicated by the requirements for specific amino acid residues near the splitting point. Apart from their sites of action on the protein, the enzymes are classified as acidic, neutral or alkaline, depending upon pH requirements for optimal activities. Acid proteases are widely distributed in molds and yeasts, but are seldom found in bacteria, which generally secrete neutral or alkaline proteases. Hartley (1960) classified the proteases based on their mechanism of action and sensitivities to various types of inhibitors. These classes include serine, thiol, metallic and acid (aspartic) proteases, which have serine, thiol, metal ion or aspartic acid, respectively, in their active sites. Several authors have summarized the information on microbial proteases relating to their production, extraction, classification, properties and uses (Glenn, 1976; Priest, 1983; Ward, 1985; Loffler, 1986). The role of microbial extracellular proteases in the spoilage of fish has been discussed by Venugopal (1990).

Serine proteinases have both serine and histidine in their active sites. They include trypsin-like proteinases and alkaline proteinases (Morihara, 1974). Trypsin-like proteinases have an optimum pH of 8 and may also possess esterase and amidase activities. Serine alkaline proteinases are produced by a number of bacteria, yeast and molds including *Aspergillus orizae*. These enzymes are commercially important since they are used in detergents. Thiol proteinases are activated by reducing compounds like cysteine and are inhibited by oxidizing agents. They are sensitive to thiol reagents such as parachloro mercurybenzoate. Acid proteinases show specificity towards aromatic or bulky amino acid residues at both sides. They have an optimal pH of 3–4, and are insensitive to inhibitors like difluorophosphate, parachloro mercurybenzoate and ethylenediamine tetraacetic acid. These proteinases find several industrial applications. The metalloproteinases, which involve a metal ion in their catalytic mechanisms, can be either neutral or alkaline. Several bacterial and fungal proteinases belong to this category. Some of the prominent microbial proteinases in terms of their structural or catalytic nature are grouped as follows:

Serine proteinases:
- *Bacillus* subtilisin
- *Pseudomonas* serine proteinase
- *Pseudomonas* elastase
- *Aspergillus* alkaline protease
- *Streptomyces* serine proteinase
- *Staphylococcus* serine proteinase
- *Saccharomyces* alkaline proteinase

Thiol proteinases:
- *Clostridium histolyticum* (clostripain)
- Yeast proteinase B
- Streptococcal proteinase
- Staphylococcal thiol proteinase

Acid proteinases:
- *Aspergillus oryzae* carboxy proteinase
- *Aspergillus saitoi* carboxy proteinase
- *Aspergillus niger* carboxy proteinase
- *Mucor pusillus* carboxy proteinase
- *Rhizopus* carboxy proteinase
- *Saccharomyces cerevisiae* carboxy proteinase

Most of these enzymes have molecular weights ranging from 20 to 50 kDa.

Synthesis of proteinases by microorganisms occurs at minimal levels during exponential growth. Enzyme secretion increases with a decrease in nutrients and reaches a maximum towards the late exponential or early stationary phase (Eveleigh and Montencourt, 1979; Glenn, 1976). Several regulatory mechanisms, including catabolite repression, end product inhibition and induction, operate in the syntheses and activities of proteases. Although certain amino acids and other low molecular weight compounds could inhibit enzyme synthesis, peptides and proteins are capable of inducing protease production. Keil-Dlouhia, Murshain and Keil (1976) reported that collagenase and a neutral protease of *Achromobacter iophagus* were induced by high molecular weight substrates such as collagen and β-casein. Similarly, trypsinized casein, neopeptone, gelatin and certain peptides were found to be potent inducers of some *Aeromonas* proteases (Litchfield and Prescott, 1970), while peptone induced collagenase synthesis in *Vibrio alginolyticus* (Reid, Woods and Rebb, 1980). Protease synthesis by *Serratia marcescens* was enhanced by bovine serum albumin and peptone (Braun and Schmitz, 1980). Washed cells of *Aeromonas hydrophila* (Alur, Nerkar and Venugopal, 1988) and *Bacillus subtilis* (May and Elliott, 1968) secreted proteases in the presence of myosin or casein hydrolysate, respectively. Similarly, protease secretion by *Erwinia chrysanthemi* was enhanced by hydrolyzed casein and not by whole casein (Wandersman, Ando and Bertheau, 1986). These data suggest the requirement of a certain short chain peptide to trigger the synthesis of the enzyme at the optimal level. It is now accepted that a basal level of constitutive exoenzyme degrades to the exogenous protein substrate and the resultant low molecular weight peptides enter the cells and induce further protease synthesis. Whooley, O'Callaghan and McLoughlin (1983) demonstrated growth-linked production of exoprotease by *Pseudomonas aeruginosa* in complex substrates such as proteins. They observed that the ability to synthesize protease provides the organism with a mechanism to survive in muscle foods.

The mechanism of enzyme secretion has been explained through the signal hypothesis (Blobel and Dobberstein, 1975). According to this, enzyme synthesis proceeds on the membrane-bound ribosomes of the organism. The primary precursors of the enzymes contain a hydrophobic amino acid sequence at the membrane receptors facilitating secretion of the growing peptides through the membrane. The signal sequence is cleaved proteolytically on the external side of the membrane and the nascent peptides acquires a stable conformation and activity.

10.3.2 Uses of microbial proteases in the food industry

As compared with animal and plant enzymes, microbial proteases have general advantages in food applications. These include their comparatively lower cost of production, less time required for enzyme isolation, possibilities for mass production and amenability of the organisms to genetic manipulations to improve the enzyme yields. The applications include production of oriental foods, bread-making, cheese ripening, meat tenderization and chill-proofing of beer. A number of microbial proteases also find applications in the detergent industry as well as for diagnostic purposes. Some of the organisms which are sources of commercial proteases include *Aspergillus oryzae, Aspergillus miehi, Mucor pusillus* and *Streptomyces griseus* (Taylor and Richardson, 1979; Ward, 1985; Loffler, 1986). Many of the proteases from these sources have been affirmed by the US Food and Drug Administration as 'generally recognized as safe' (GRAS) (Loffler, 1986).

10.3.3 Uses of microbial proteases in fish hydrolysis

A number of microbial enzymes have been found to be superior to plant enzymes in their ability to cause solubilization of proteins, including those from fish. For example, bacterial neutral and alkaline proteases were about 140% more efficient as compared with papain (Kilara, 1985). Hale (1969) showed that hydrolysis of raw fish with *Bacillus subtilis* protease at pH 8.5 gave a high yield of soluble product having an excellent balance of amino acids. Similarly, pronase, a bacterial protease, had twice the activity at 50°C as compared with that of ficin at 40°C or bromelain at 50°C. The product obtained using pronase was also less bitter (Hevia, Whitaker and Olcott, 1976). Liquefied fish protein having negligible bitterness could also be obtained using protease from *Aspergillus niger* (Kinumaki, 1978). Rebecca, Pena-Vera and Diaz-Castaneda (1991) examined the efficacies of several microbial proteases for the production of FPH from mullet (*Mugil cephalus*). The fish protein was hydrolyzed as such without the addition of water. An alkaline protease from *Bacillus subtilis* (Pescalase 560) was found to be more efficient than two neutral proteases from the same bacterium, giving more than 80% nitrogen solubilization. The bacterial proteases showed higher activities than fungal (*Aspergillus oryzae*) enzymes. The FPH had 70–85% available lysine and could be used for fortification of bakery products, as a milk substitute for calves as well as in fermentations as a highly concentrated soluble nitrogen source. Archer *et al.* (1973) examined the kinetics of solubilization of fish protein concentrate by *Bacillus subtilis* protease (monozyme). The authors showed that the enzyme was adsorbed on the surface of the substrate and the initial rate of reaction was proportional to the surface area of the substrate exposed to the aqueous phase. The overall kinetics were described by a sequence of two first-order processes – an initial fast reaction in which loosely bound polypeptide chains were cleaved from an insoluble protein particle and a second, slower reaction in which more compact core protein was digested. These results suggest that successful solubilization of fish meat could be achieved by employing certain microbial proteases. It is, however, essential that such microorganisms should be safe for food applications.

10.4 BIOTECHNOLOGICAL APPROACHES TO FISH MEAT SOLUBILIZATION

10.4.1 Immobilized enzymes and microbial cells

Recent advances in biotechnology have shown potential for a variety of food processing operations. Applications of microorganisms and microbial enzymes are some major areas of biotechnology (Nga and Lee, 1990). Immobilized microbial enzymes and microorganisms offer scope for the modification of food ingredients. This section will deal briefly with some of these techniques and their applications in food processing.

Biocatalysts (enzymes and whole cells) whose mobilities are restricted by chemical or physical means are called immobilized biocatalysts and can be used for bioconversions. The different techniques of immobilization have been described by a number of authors (D'Souza and Nadkarni, 1981; Klibanov, 1983; Klein and Wagner, 1983; Klein and Vorlop,

Table 10.1 Some examples of biocatalysts immobilized on different carriers

Biocatalyst	*Method of immobilization*	*Carrier*	*Reference*
Acetobacter aceti	adsorption	cellulose	Bar, Gainer and Kirwan (1986)
Clostridium thermocellum	adsorption	cellulose	Weigel and Dykatra (1986)
Pseudomonas spp.	adsorption	activated charcoal	Eharhardt and Rehm (1984)
Saccharomyces cerevisiae	adsorption	glass	Haecht, Bolipombo and Rouxhet (1985)
Saccharomyces cerevisiae	adsorption	polyster	Black *et al*. (1984)
Azotobacter vinclamdii	ionic binding	cellulose	Diluccio and Kirwan (1984)
Saccharomyces cerevisiae	covalent binding	kieselguhr	Navarro and Durand (1977)
Escherichia coli	entrapment	carrageenan	Umemura *et al*. (1984)
Saccharomyces cerevisiae	entrapment	calcium–alginate	Quereshi and Tamhane (1985)

1985; Hartmeier, 1986; Brodelius and Vandamme, 1987; Scott, 1987). Applications of immobilized cells have also been discussed by Chibata and Tosa, (1981), Linko and Linko, (1984) and Tanaka and Nakajima (1990).

Immobilization can be achieved by binding biocatalysts to one another or to carrier substances, by entrapping in the network of a polymer matrix or by means of membrane confinement. Binding to carriers could be through ionic adsorption, covalent binding or cross-linking. In adsorption, biocatalysts are held to the surface of a carrier by physical forces. The adsorbents could be selected from a wide variety of inorganic as well as organic substances. A disadvantage of the technique is the relative weakness of adsorptive binding which causes desorption by temperature fluctuation, ionic concentration, etc. Ionic binding is based on electrostatic attraction between oppositely charged groups of ion exchange resins such as DEAE–cellulose. The binding is, however, weaker than covalent binding. Covalent binding is generally employed for coupling enzymes but not whole cells. In cross-linking, the individual biocatalysts are formed with the help of multi-functional reagents such as glutaraldehyde. However, since the particles obtained are gelatinous and access to substrate becomes limited, they are not ideal for packed bed reactors.

Immobilization through matrix entrapment is widely used for whole cell systems. The network gives good cell retention and porosity for substrate and product transport. Hydrocolloids such as alginate and carrageenan are frequently used for the purpose. The network has to be formed in the presence of the cells to be entrapped and therefore the network forming reactions have to be adapted to the physiological requirements of stability and activity of the enzymes or cells. The process is referred to as ionotropic gelation since the water-soluble polyelectrolyte (alginate or carrageenan) is mixed with appropriate counter ions when solidification by polysalt formation occurs. The biocatalysts are formed by forcing the mixture of the hydrocolloid and the microorganism into an aqueous solution of the gelling material. The droplets falling into the solution are converted into beads upon gelation which contain the entrapped organism. The stability of the beads may be improved by drying (Scott, 1987) The well-known example is the calcium alginate gel which is obtained by gelation of a sodium alginate solution in $CaCl_2$ solution (Klein and Vorlop, 1985).

Table 10.1 gives some of the examples of the different techniques used for the immobilization of whole microbial cells.

10.4.2 Immobilised proteolytic enzymes

Immobilization of proteolytic enzymes has been reported by several authors. Cross-linking of pepsin and trypsin using glutaraldehyde has

Table 10.2 Some applications of immobilized proteases and proteolytic microorganisms

Process	*Bioreactor*	*Reference*
Casein hydrolysis	protease bioreactor	Haque and Mozaffar (1992)
Soy protein hydrolysis	pepsin on glass	Mason *et al*. (1975)
Milk coagulation	pepsin/calf rennet on glass	Cheryan *et al*. (1975)
Fish hydrolysis	*Bacillus megaterium* in calcium–alginate	Venugopal, Alur and Nerkar (1989)
Protein hydrolysis	protease on fibre	Deeslie and Cheryan (1981)
Protein hydrolysis	immobilized pronase	Lasch, Koelsch and Krelschmer (1987)
Meat fermentation	lactic acid bacteria in calcium alginate	Kearney, Upreon and McLoughlin (1990)
Debittering of casein hydrolysate	*Erwinia ananas* in calcium alginate	Watanabe, Shimizu and Aral (1990)

been developed by Khan and Siddiqui (1985). Proteolytic enzymes have also been immobilized on polyethylene terephthalate (Blassberger, Freeman and Goldstein, 1978), glass (Mason, Detar and Weetal 1975) and chitosan (Leuba and Widmer, 1979). Such systems have been proposed for food applications (Taylor and Richardson, 1979; Hultin, 1983). Lasch, Koelsch and Krelschmer (1987) demonstrated continuous hydrolysis of several proteins using immobilized protease reactors such as the stirred tank, packed bed, fluidized bed and bubbled columns. The last mentioned reactor gave maximum efficiency. Other applications include milk coagulation and cheese ripening (Table 10.2) (Deeslie and Cheryan, 1981; Loffler, 1986; Haque and Mozaffar, 1992). Commercial applications of these processes, however, have not yet been achieved. Some of the problems that may be encountered in these systems include autodigestion of immobilized proteases, problems relating to mass transfer and diffusion when working with viscous liquids (see Table 10.2).

10.4.3 Advantages of immobilized whole cells

As compared with immobilized enzymes, immobilized whole cells offer several advantages. These include the following:

1. processes for extraction and purification of enzymes are not necessary;
2. the yield of enzyme activity from immobilized cells is generally high;
3. the operational stability of immobilized cells is high;
4. the volume required for continuous reactors is low as compared with batch reactors or conventional fermentations (for review, see Dervakos and Webb, 1991).

Immobilization of proteolytic microorganisms could result in enhanced protease secretion by entrapped cells as compared with free cells (Klein and Vorlop, 1985; Hartmeier, 1986; Scott, 1987).

Klein and Vorlop (1985) pointed out the need for consideration of three factors in the development of an immobilized cell reactor. These are activity, efficiency and stability. The relative activity of a reactor is the ratio of activity of the immobilized system to that of free cells; while absolute activity is the rate of reaction based on the unit weight or volume of whole cells. High relative activity can be obtained using natural polymers such as alginate or carrageenan for cell entrapment. Efficiency is defined as the ratio of the actual reaction rate to the maximum possible reaction rate. Loss of efficiency is caused by transport limitation of a rate controlling substrate. Stability is the parameter used to characterize the kinetic performance of immobilized systems. Stability relates to both operational as well as storage stability of the system. Operational stability should be considerably higher for immobilized systems than for whole cells.

Several types of reactors have been designed for the use of immobilized whole cells. These consist of both batch-type and continuous reactors. In continuous reactors the efficiency is higher since problems of enzyme inhibition by products formed are minimal. Most bioreactor systems now being used for immobilized cells are of the continuous type, such as packed bed or fluidized bed systems (Scott, 1987). The reader is also referred to information provided by Hartmeier (1986). According to Scott (1987), immobilization of cells is well developed and, therefore, the emphasis should now be on process developments, including pilot plants, for the increased use of such systems in industry.

The efficiency of proteolysis using immobilized organisms could be improved using recombinant DNA technology, as pointed out by Loffler (1986). Such techniques would result not only in enhanced rates of enzyme secretion by the cells but also in improvement of pH as well as thermal stabilities of the secreted enzymes.

10.5 SOLUBILIZATION OF FISH MEAT BY IMMOBILIZED MICROBIAL CELLS

Only limited studies have been reported on the solubilization of fish meat using immobilized microbial cells. The work done in the author's laboratory (Venugopal, Alur and Nerkar, 1989) examined the efficacy of three immobilized bacteria in causing hydrolysis of meat obtained from a typical low cost fish, i.e. *Johnius dissumeri*. The mechanically deboned fish meat was partially deodorized by boiling in excess water for 15 min at low pH. The boiled meat was pressed in a screw press to remove water and the odor-bearing compounds. The fish meat (4 g) was suspended in 50 ml of water (pH adjusted to 8.0) and incubated with calcium alginate beads containing about 10^{11} cells/g. The extent of solubilization was determined by measuring tyrosine as well as the nitrogen content in the supernatants. Figure 10.2 shows the extent of solubilization of the fish meat by some immobilized microbial cells. It can be seen that the degree of solubilization, measured in terms of total nitrogen in the supernatant, was proportional to the incubation time and reached a maximum at 24 h. The other two bacterial cells were less efficient in causing solubilization as compared with *Bacillus megaterium*. The optimal pH for solubilization was about 9.0, although significant solubilization was observed in the pH range 7–9. Since the pH of the suspension decreased due to the liberation of acidic amino acid residues, pH was intermittently monitored using a pH wand. The calcium alginate beads containing *Bacillus megaterium* cells had a storage life of about 1 month. A bead-holding assembly was used for treatment of the fish

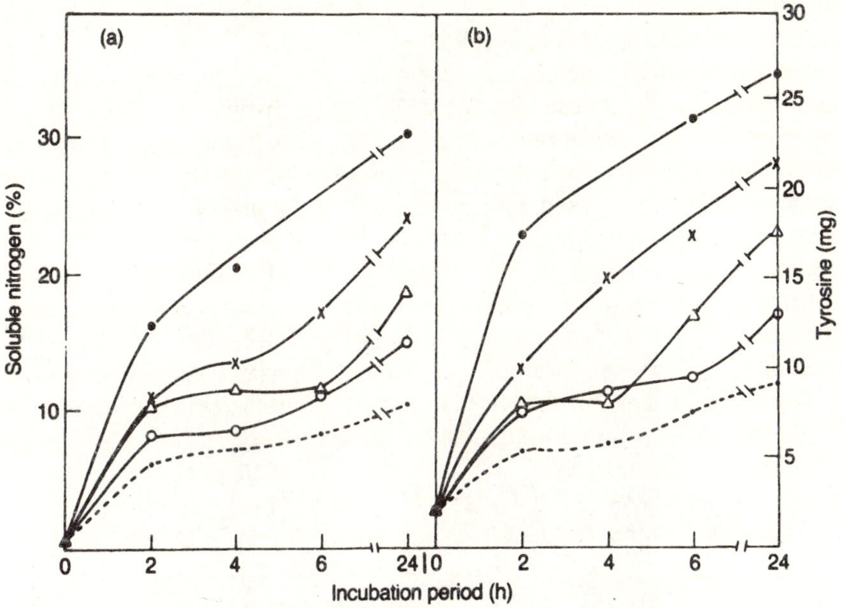

Figure 10.2 Reaction progress curves for solubilization of fish meat by immobilized bacterial cells. The fish meat (4 g) in 50 ml water was incubated with 4 g of immobilized bacterial cells. The degree of solubilization was expressed as percent of total nitrogen in the supernatant. ●, *B. megaterium;* △, *A. hydrophila;* ○, *P. marinoglutinosa;* ×, mixture of beads of above cells. From Venugopal *et al.* (1989). Copyright 1989, reprinted by permission of John Wiley & Sons, Inc.

meat in batch operations (Figure 10.3). The bead holder consisted of a perforated acrylic cylinder, 5 cm in length and 3.5 cm in diameter. The beads were distributed on disks inside the cylinder. The perforations on the cylinder helped in the contact of fish slurry with the beads. Movement of the slurry was attained by slow rotation of the assembly through its attachment to a motor. An acrylic piece attached to the bottom of the assembly facilitated stirring of the slurry during treatment. The equipment helped in the recovery of the beads after use.

10.6 FUTURE PROSPECTS

The foregoing sections were intended mainly to point out the possibilities of fish meat solubilization through biotechnological means. Although only limited studies have been carried out in this respect, scope exists for the use of microorganisms for the preparation of FPHs. Since the raw material is non-commercial fishery resources, the method for solubilization should be economically viable. Therefore, proteolyti-

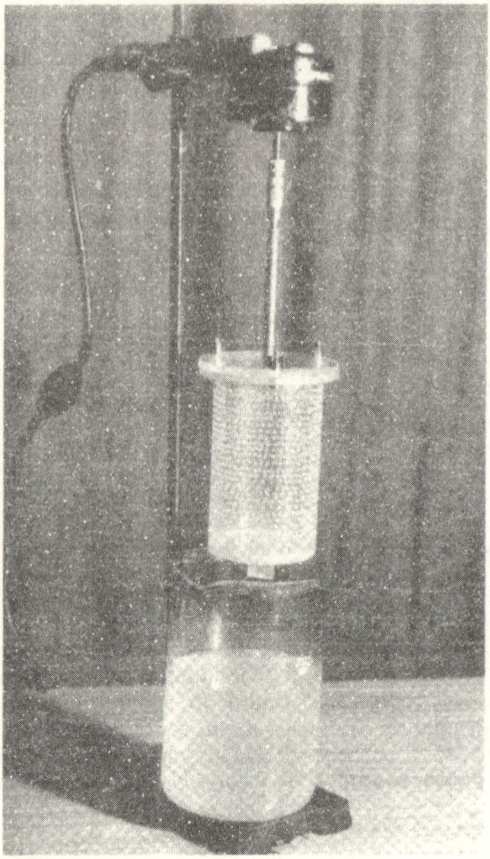

Figure 10.3 Bead-holding assembly for the solubilization of meat. From Venugopal *et al.* (1989). Copyright 1989, reprinted by permission of John Wiley & Sons, Inc.

cally active immobilized microbial cells appear to be ideal for the purpose.

Instead of the batch process (Venugopal *et al.*, 1989), the possibility exists for development of continuous reactors for meat solubilization. Continuous reactors for food processing involving hydrolysis have been recently reported. Sims and Cheryan (1992) observed the feasibility of using a membrane reactor for the hydrolysis of liquefied corn starch. Enzyme bioreactors for continuous production of casein hydrolyzate have been reported by Haque and Mozaffar (1992) and Mason (1975). Development of a continuous system for fish solubilization necessitates that the meat should be in a free-flowing state, so that the substrate could be run through a reactor containing immobilized microbial cells. Recent studies have shown that washed meat mince from low cost fish, such as threadfin bream, Atlantic mackerel, capelin and shark, could be readily converted into a gel (Venugopal and Shahidi, 1994; Venugopal, Doke and Nair, 1994). The washing step, which is accomplished by repeated treatment of the mince with chilled water, removes soluble nitrogenous compounds and pigments as well as significant proportions of lipids, and helps gelation (Lanier, 1986). The free-flowing characteristic of the gel helps continuous hydrolysis in a suitable reactor. The washed mince is highly deodorized and can give protein hydrolysates having significantly less odor. The treated proteins could be concentrated by appropriate means such as spray drying. Studies in our laboratory have already shown the possibility of preparing highly functional, spray dried powders from protein gels of some low cost fish items.

Immobilized microbial cells having proteolytic activity could also be used to improve the drying properties of fish stick water, a by-product from the fishmeal industry. Drying of the stick water often poses problems due to a viscosity increase accompanying the concentration of proteins. Conventionally, viscosity during drying operations is reduced by the addition of proteases such as alcalase (Jacobsen and Rasmussen, 1984). It is likely that the stick water could be treated with immobilized proteolytic microorganisms prior to dehydration. A third possible area of application is for the reduction in the fermentation time required for fish sauce production, which is usually prepared by the action of autolytic enzymes on fish muscle (Greig and Estrella, 1988). As observed recently by Dervakos and Webb (1991), while several advantages could be claimed for viable cell immobilization, the technique may also introduce certain problems relating to cell over-growth and leakage as well as mechanical stability of the matrix, depending upon individual application. Successful industrial applications of the process for fish protein processing would therefore depend upon careful evaluation of the advantages of the process for protein hydrolysis.

ACKNOWLEDGMENT

I thank Dr M.D. Alur for his valuable help in the experimental work described in this chapter.

REFERENCES

Adler-Nissen, J. (1977) Enzymatic hydrolysis of fish proteins. *Process Biochem.*, **12**, 18–23.

Adler-Nissen, J. (1985) *Enzymatic Hydrolysis of Food Proteins*, Elsevier, New York.

Alur, M.D., Nerkar, D.P. and Venugopal, V. (1988) Growth and protease secretion by spoilage bacteria: Influence of nitrogen fraction of proteinaceous foods on *Aeromonas hydrophila*. *J. Food Sci.*, **53**, 243–6.

Archer, M.C., Ragnarsson, J.O., Tannenbaum, S.R. and Wang, D.I.C. (1973). Enzymatic solubilization of an insoluble substrate, fish protein concentrate: process and kinetic considerations. *Biotechnol. Bioeng.*, **15**, 181–96.

Bar, R., Gainer, J.L. and Kirwan, D.J. (1986) Immobilization of *Acetobacter aceti* cells on cellulose ion exchangers. Adsorption isotherms. *Biotechnol. Bioeng.*, **28**, 1166–71.

Black, G.M., Webb, C., Mathews, T.M. and Atkinson, B. (1984) Practical reactor systems for yeast cell immobilization using biomass support particles. *Biotechnol. Bioeng.*, **26**, 134–41.

Blassberger, D., Freeman, A. and Goldstein, A. (1978) Chemically modified polyesters as supports for enzyme immobilization. Isocyanide, acylhydrazine and aminoaryl derivatives of polyethylene terephthalate. *Biotechnol. Bioeng.*, **20**, 309–15.

Blobel, G. and Dobberstein, B. (1975) Transfer of proteins across membranes. I and II. *J. Cell Biol.*, **67**, 835–51 and 852–62.

Braun, V. and Schmitz, G. (1980) Excretion of a protease by *Serratia marcescens*. *Arch. Microbiol.*, **124**, 55–61.

Brodelius, P. and Vandamme, E.J. (1987). Immobilized cell systems, in *Biotechnology*, (ed. J.F. Kennedy), VCS Verlagsgeselleschaft, Germany, Vol. 7a, pp. 405–64.

Burt, J.R., Hardy, R. and Whittle, K.L. (eds) (1990) *Pelagic Fish. The Resource and its Exploitation*, Fishing News (Books) Ltd/Blackwell Scientific Publications, Oxford.

Cheryan, M., Van Wyk, P.J., Olson, M.F. and Richardson, T. (1975) Continuous coagulation of milk using immobilized enzyme in a fluidized bed reactor. *Biotechnol. Bioeng.*, **17**, 585–598.

Chibata, I. and Tosa, T. (1981) Use of immobilized cells. *Annu. Rev. Biophys. Bioeng.*, **10**, 192–216.

Deeslie, W.D. and Cheryan, M. (1981) A CSTR-hollow fiber system for continuous hydrolysis of proteins. Performance and kinetics. *Biotechnol. Bioeng.*, **23**, 2257–71.

Dervakos G.A. and Webb, C. (1991) On the merits of viable-cell immobilisation. *Biotech. Adv.*, **9**, 559–612.

Diluccio, R.C. and Kirwan, D.J. (1984) Effect of dissolved oxygen on nitrogen fixation by *A. vinelandii*. II. Ionically adsorbed cells. *Biotechnol. Bioeng.*, **26**, 87–91.

D'Souza, S.F. and Nadkarni, G.B. (1981) Hen egg white. A novel support for the immobilisation of enzymes. *Biotechnol. Bioeng.*, **23**, 431–6.

Ehrhardt, H.M. and Rehm, H.J. (1984) Phenol degradation by microorganisms adsorbed on activated carbon. *Appl. Microbiol. Biotechnol.*, **21**, 32–6.

Eveleigh, D.E. and Montencourt, B.S. (1979) Increasing yields of extracellular enzymes, in *Advances in Applied Microbiology*, (ed. D. Perlman), Academic Press, New York, pp. 7–35.

Foegeding, E.A. (1988) Thermally induced changes in muscle proteins. *Food Technol.*, **42**(6), 58–61.

Ghadi, S.V., Warrier, S.B. and Ninjoor, V. (1987) Preparation and properties of fish protein hydrolysate from Dhoma, an under-utilized fish, in *Proceedings of the Symposium on Diversification of Post-harvest Technology for Low Cost Fish*, Society of Fishery Technologists, India, pp. 231–42.

Glenn, A.R. (1976) Production of extracellular proteins by bacteria. *Ann. Rev. Microbiol.*, **30**, 41–62.

Grantham, G.J. (1981) Minced fish technology. A review. *FAO Fish Technical Paper No. 216*. FAO, Rome.

Greig, R.I.W. and Estrella D.C. (1988) A study of acceleration of fish sauce production using enzymes, in *Food Science and Technology in Industrial Development: Proceedings of the Conference*, Indo-Pacific Fishery Commission, Thailand, pp. 275–81.

Hale, M.B. (1969) Relative activities of commercially available enzymes in the hydrolysis of fish proteins. *Food Technol.*, **23**(1), 107–10.

Haque, Z.U. and Mozaffar, Z. (1992) Casein hydrolysate continuous production using enzyme bioreactors. *Food Hydrocolloids*, **5**, 549–58.

Hartmeier, W. (1986) *Immobilized Biocatalysts. An Introduction*, Springer-Verlag, New York.

Hartley, B.S. (1960) Proteolytic enzymes. *Ann. Rev. Biochem.*, **29**, 45–72.

Haecht van, J.L., Bolipombo, M. and Rouxhet, P.B. (1985) Immobilization of *Saccharomyces cerevisiae* by adhesion. Treatment of the cells by Al ions. *Biotechnol. Bioeng.*, **27**, 217–24.

Hevia, P., Whitaker, J.R. and Olcott, H.S. (1976) Solubilization of fish protein concentrate with proteolytic enzymes. *J. Agri. Food Chem.*, **24**, 383–5.

Hultin, H.O. (1983) Current and potential uses of immobilized enzymes. *Food Technol.*, **37**(10), 66–82.

Jacobsen, F. and Rasmussen, O.L. (1984) Energy savings through enzymatic treatment of stickwater in fish meal industry. *Process Biochem.*, **19**(5), 65–169.

James, D.G. (1986) The prospects for fish for the malnourished. *Food and Nutrition (FAO)*, **12**(6), 20–30.

Kearney, L., Upton, M. and McLoughlin, A. (1990) Meat fermentations with immobilized lactic acid bacteria. *Appl. Microbiol. Biotechnol.*, **33**, 648–51.

Keil-Dlouhia, V., Murshain, R. and Keil, B. (1976) The induction of collagenase and a neutral protease by their high molecular weight substrates in *Achromobacter iophagus*. *J. Mol. Biol.*, **107**, 293–97.

Kennedy, J.F. (1978) Microbial cells immobilized and hung on solid supports and their applications to fermentation processes. *Enzyme Eng.*, **4**, 323–8.

Khan, S.S. and Siddiqui, A.M. (1985) Studies on chemically aggregated pepsin using glutaraldehyde. *Biotechnol. Bioeng.*, **27**, 415–19.

Kilara, A. (1985) Enzyme-modified protein food ingredients. *Process Biochem.*, **20**(5), 149–58.

Kinumaki, T. (1978) Possibilities for the production and application of liquefied fish protein concentrate in the IPFC area, in *Indo-Pacific Fishery Commission: Proceedings on the Symposium on Fish Utilization Technology*, Manila, Philippines, 8–17 May, pp. 492–506.

Klein, J. and Vorlop, K.-D. (1985) Immobilization techniques – cells, in *Comprehensive Biotechnology*, (ed. M. Moo-Young), Pergamon Press, New York, Vol. **2**, pp. 203–24.

Klein, J. and Wagner, F. (1983) Methods for the immobilization of microbial cells. *Appl. Biochem. Bioeng.*, **4**, 11–51.

Klibanov, A.M. (1983) Immobilized enzymes and cells as practical catalysts. *Science*, **219**, 722–7.

Lasch, J., Koelsch, R. and Krelschmer, K. (1987) Continuous production of protein hydrolysates in immobilized enzyme reactors. *Acta Biol.*, **7**, 227–235.

Lanier, T.C. (1986) Functional properties of surimi. *Food Technol.*, **40** (3), 107–14.

Lee, C.M. (1986) Surimi manufacturing and fabrication of surimi based products. *Food Technol.*, **40** (3), 115–24.

Leuba J.-L. and Widmer, F. (1979) Immobilization of proteinases on chitosan. *Biotechnol. Lett.*, **6**, 109–14.

Linko, P. and Linko, Y.-Y. (1984) Industrial applications of immobilized cells. *CRC Crit. Rev. Biotechnol.*, **1**, 289–338.

Litchfield, C.D. and Prescott, J.M. (1970). Regulation of proteolytic enzyme production by *Aeromonas proteolytica* 1. Extracellular endopeptidase. *Can. J. Microbiol.*, **16**, 17–22.

Loffler, A. (1986) Proteolytic enzymes: sources and applications. *Food Technol.*, **40** (1), 63–70.

Mackie, I.M. (1982a) General review of fish proteins hydrolysates. *Animal Feed Sci. Technol.* **7**, 113–24.

Mackie, I.M. (1982b) Fish protein hydrolysates. *Process Biochem.*, **17**, 26–32.

Mason, R.D., Detar, C.C. and Weetall, H.H. (1975) Protease covalently coupled to porous glass: preparation and characterization. *Biotechnol. Bioeng.*, **17**, 1019–27.

May, B.K. and Elliott, W.H. (1968) Characteristics of extracellular protease formation by *Bacillus subtilis. Biochem. Biophys. Acta*, **157**, 607–15.

Merritt, J.H. (1982) Assessment of the production costs of fish protein hydrolysates. *Animal Feed Sci. Technol.* **7**, 147–51.

Monti, J. and Jost, R. (1979) Papain-catalyzed synthesis of methionine-enriched soy plasteins. Average chain lengths of the plastein peptides. *J. Agri. Food Chem.*, **27**, 1281–5.

Morihara, K. (1974) Comparative specificities of microbial proteinases. *Adv. Enzymol.*, **41**, 179–243.

Morrissey, M.T. (ed.) (1988) *Post Harvest Fishery Losses. Proceedings of International Workshop*, The International Center for Marine Resource Development, University of Rhode Island, Kingston, RI.

Navarro, J.M. and Durrand, G. (1977) Modification of yeast metabolism by immobilization on to porous glass. *Appl. Microbiol. Biotechnol.*, **4**, 243–54.

Nga, B.H. and Lee, Y.K. (eds) (1990) *Microbiology Applications In Food Biotechnology*, Elsevier, New York.

Owens, J.D. and Mendoza, L.S. (1985) Enzymatically hydrolysed and bacterially fermented fishery products. *J. Food Technol.*, **20**, 273–93.

Piggott, G.M., Bucone, G.O. and Ostrander, J.G. (1978) Engineering a plant for enzymatic production of supplemental fish protein. *J. Food Proc. Preserv.*, **4**, 33–54.

Priest, F.G. (1983) Enzyme synthesis by microorganisms, in *Microbial Enzymes and Biotechnology*, (ed. W.M. Fogarty), Elsevier, London, pp. 319–66.

Putro, S. (1989) Surimi prospects in developing countries. *Infofish Int.*, **5**, 29–32.

Quaglia, G.B. and Orban, E. (1987a) Enzymatic solubilisation of proteins of sardine by commercial proteases. *J. Sci. Food Agri.*, **38**, 263–9.

Quaglia, G.B. and Orban, E. (1987b) Influence of degree of hydrolysis on the solubility of protein hydrolysates from sardine. *J. Sci. Food Agri.*, **38**, 271–6.

Quereshi, M. and Tamhane, D.V. (1985) Production of mead by immobilized whole cells of *Saccharomyces cerevisiae. Appl. Microbiol. Biotechnol.*, **21**, 280–1.

Raghunath, M.R. and McCurdy, A.R. (1991) Synthesis of plasteins from fish silage. *J. Sci. Food Agri.*, **54**, 655–658.

Rebecca, B.D., Pena-Vera, M.T. and Diaz-Castaneda, M. (1991) Production of fish protein hydrolysates with bacterial proteases. Yield and nutritional value. *J. Food Sci.*, **56**, 309–14.

Regenstein, J.M. (1986) The potential for minced fish. *Food Technol.*, **40**(3), 101–6.

Reid, G.C., Woods, D.R. and Rebb, F.T. (1980). Peptone induction of rifampicin insensitive collagenase production by *Vibrio alginolyticus. J. Gen. Microbiol.*, **142**, 447–54.

Ritchie, A.H. and Mackie, I.M. (1982) Preparation of fish protein hydrolysates. *Animal Feed Sci. Technol.*, 7, 125–33.

Roy, G. (1992) Bitterness: reduction and inhibition. *Trends Food Sci. Technol.*, **3**, 85–91.

Satterlee, L.D. (1981) Proteins for use in foods. *Food Technol.*, **35**(6), 53–70.

Scott, C.D. (1987) Immobilized cells. A review of recent literature. *Enzyme Microbiol. Technol.* **9**, 66–73.

Sen, D.P., Sripathy, N.V., Lahiry, N.L. and Sreenivasan, A. (1962) Fish hydrolysates: 1. Rate of hydrolysis of fish flesh with papain. *Food Technol.*, **16**(5), 138–41.

Sikorski, Z.E. (1990) *Seafood: Resources, Nutritional Composition and Preservation*, CRC Press, Boca Raton, FL.

Sikorski, Z.E. and Naczk, M. (1981) Modification of technological properties of fish protein concentrates. *CRC Crit. Rev. Food Sci. Nutr.*, **14**, 201–30.

Sims, K.A. and Cheryan, M. (1992) Hydrolysis of liquefied corn starch in a membrane reactor. *Biotechnol. Bioeng.* **39**, 960–967.

Tanaka, A. and Nakajima, H. (1990) Application of immobilised growing cells. *Adv. Biochem. Eng./Biotechnol.*, **42**, 97–131.

Taylor, M.J. and Richardson, T. (1979) Applications of microbial enzymes in food systems and biotechnology. *Adv. Appl. Microbiol.*, **25**, 7–19.

Umeura, I., Takamatsu, S., Sato, T. *et al.* (1984) Improvement of production of L-aspartic acid using immobilized microbial cells. *Appl. Microbiol. Biotechnol.*, **20**, 291–5.

Venugopal, V. (1990) Extracellular proteases of contaminant bacteria in fish spoilage. *J. Food Prot.*, **53**, 341–350.

Venugopal, V. (1992) Mince from low cost fish species. *Trends Food Sci. Technol.*, **3**, 2–5.

Venugopal, V. and Lewis, N.F. (1981) Isolation of proteins from low priced fish. *Fleischwirtschaft*, **61**, 1368–70.

Venugopal, V., Alur, M.D. and Nerkar, D.P. (1989) Solubilization of fish proteins using immobilized microbial cells. *Biotechnol. Bioeng.*, **33**, 1098–103.

Venugopal, V., Ghadi, S.V. and Nair, P.M. (1992) Value added products from low cost fish mince. *Asian Food J.*, **7**, 3–12.

Venugopal, V., Doke, S.N. and Nair, P.M. (1994) Gelation of shark proteins by weak organic acids. *Food Chemistry*, in press.

Venugopal, V., Martin, A. and Patel, T.R. (1994) Extractability and stability of washed capelin muscle in water. *Food Hydrocolloids*, in press.

Venugopal, V. and Shahidi, F. (1994) Thermostable dispersion from washed muscle of Atlantic mackerel. *J. Food. Sci.*, in press.

Wandersman, C., Ando, T. and Bertheau, Y. (1986) Extracullular proteases in *Erwinia chrysanthemi. J. Gen. Microbiol.*, **182**, 899–906.

Ward, O.P. (1985) Proteolytic enzymes, in *Comprehensive Biotechnology*, (ed. M. Moo-Young), Pergamon Press, Oxford, Vol. 3, pp. 789–818.

Watanabe, M. Shimizu, J. and Arai, S. (1990) Debittering of a tryptic hydrolysate of casein by incubating with ice nucleation-active bacterium *Erwinia ananas* and its aminopeptidase at low temperature. *Agri. Biol. Chem.*, **54**, 3351–3.

Whooley, M., O'Callaghan, J.A. and McLoughin, A.J. (1983) Effect of substrate on the regulation of exoprotease production by *Pseudomonas aeruginosa* ATCC 10145. *J. Gen. Microbiol.*, **129**, 981–8.

Wiegel, J. and Dykatra M. (1984) *Clostridium thermocellum* adhesion and sporulation while adhered to cellulose and hemicellulose. *Appl. Microbiol. Biotechnol.*, **20**, 59–65.

Yanez, E., Ballester, D. and Monckeberg, F. (1976) Enzymatic fish protein hydrolysate. Chemical composition and nutritive value and use as a supplement to cereal protein. *J. Food Sci.*, **41**, 1289–92.

Yu, S.Y. and Tan, L.K. (1990) Acceptability of crackers ('Kerpok') with fish protein hydrolysate. *Int. J. Food Sci. Technol.*, **25**, 204–8.

Ziegler, G.R. and Foegeding, E.A. (1990) The gelation of proteins. *Adv. Food Res.*, **34**, 203–98.

CHAPTER 11

Production of fish silage

S. Arason

11.1 INTRODUCTION

Fish material that is not for human consumption is used for the production of fishmeal which has a world market of considerable size. However, new ways of utilizing pelagic fish, fish waste and fish viscera are continuously being looked for.

The main emphasis has been on developing processes with low investment costs that can be used in fishing vessels and in small isolated places where fishmeal plants cannot be operated economically. One of the processes that have been considered is chemical preservation by acid or bases, either with or without hydrolysis (Arnesen, Arason and Jonsson, 1981; Raa and Gildberg, 1982). The product from this process is called fish silage.

Fish silage may be described as a liquid product made from fish or parts of fish and acid or more seldom base (e.g. sodium hydroxide). Liquefaction is caused by the action of enzymes naturally present in the fish, and is accelerated by the acid which creates the right conditions for the enzymes to break down the tissues and limits the growth of spoilage bacteria. Organic acids are most commonly used for silage production from fish.

Fish silage is used commercially in Scandinavia and Poland. The annual production is about 120 000 tons produced, mainly by the use of formic acid, acetic acid and mineral acids. The product is generally used as feed for fish, fur animals, pigs, etc.

11.2 BACKGROUND

Silage production is not a recent development. It was first introduced in Finland in 1920 by A.I. Virtanen (Raa and Gildberg, 1982). He treated

Fisheries Processing: Biotechnological applications.
Edited by A.M. Martin.
Published in 1994 by Chapman & Hall, London. ISBN 0 412 58460 3

green fodder with a mixture of sulfuric and hydrochloric acids. This method was adopted by Edin in the 1930s to preserve and liquefy different types of fish and fish waste (Edin, 1940).

Production of silage on an industrial scale started in Denmark in 1948 and 3 years later its production was about 15 000 tons per annum in 14 companies (Petersen, 1951).

In the beginning, silage production only made use of inorganic acids, such as hydrochloric and sulfuric acid; however, although these acids were relatively inexpensive, they were not very convenient because their preservative action first comes into effect when the pH value is down to about 2 (Arnesen, Arason and Jonsson, 1981). The feedstuff, therefore, had to be deacidified before it was fed to the animals (Petersen, 1953). The preservation effect of many organic acids, such as formic acid, becomes active at a higher pH level (pH 4). In recent years, formic acid has, therefore, increasingly been used in silage production.

In the former Yugoslavia, tests were made on the use of different types of acids and acid mixtures in silage and for the preservation of sardine offal. The conclusions were that least expensive was a 3% mixture of sulfuric acid and formic acid in the ratio of 3:1 (Lisac, 1961).

Norwegian scientists launched a large-scale research program in 1974 on the production of silage from fish viscera and offal. They found the nutritional value of silage deteriorates through storage, mainly because of break down of the amino acid tryptophan in acid conditions. This is usually not a problem when the silage is mixed with green fodder, because plants usually are low in tryptophan. If silage is mixed with dry green fodder, such as grass meal, and air can circulate freely around the blend, more acid is required to prevent the growth of mold, or at least 3% of formic acid. It is, however, sufficient to use a 1.5% mixture of formic and propionic acid (1:1), and that mixture is less expensive than pure formic acid (Gildberg and Raa, 1977; Ström *et al.*, 1980; Jangaard, 1991). In recent years Norway has had an annual production of 40 000 – 50 000 tons and in 1992 production of silage was about 60 000 tons from fish and fish waste (Jangaard, 1991; Stormo, 1993).

Silage can be produced by adding sugar or sugary grains along with lactobacilli to fish and fish waste. Lactobacilli convert sugar into lactic acid which preserves the fish and creates favorable conditions for the silage (Petersen, 1953; Nilsson and Rydin, 1968). In addition to acid, some types of lactobacilli produce other substances (antibiotics) which increase their preservation effect (Lindgren and Clevström, 1978; Schrøder *et al.*, 1980). These bacilli are also considered to prevent oxidation of fats (Raa and Gildberg, 1982). Silage produced by lactobacilli has not yet reached production. Feeding experiments with chicken have shown that fermented silage is by no means inferior as feed to silage produced in the conventional manner (Wirahadikusumah, 1969; Kompiang, Darwanto and Arifuddin, 1979). It is also conceivable that silage which is preserved through the use of lactobacilli is less expensive to produce than silage preserved through organic means, in particular where waste rich in carbohydrates is produced, such as molasses, whey, etc. (Arnesen, Arason and Jonsson, 1981).

Although acid silage is the most common, there are several other methods known. In Canada, an attempt was made to preserve viscera by nitrate; however, its preservation effects were short and it was, therefore, not very useful for long-term storage (Freeman and Hoogland, 1956). Some large-scale research was carried out in Iceland on silage production from viscera under basic conditions, using ammonia. This method was also used in India (Ramanathan and Moorjani, 1975). Feeding experiments were done in Iceland on meal produced from basic silage from viscera, indicating that the nutritional value was similar to that of good quality capelin meal (Dagbjartsson *et al.*, 1976). On the other hand, it has also been pointed out that in basic silage there is some danger of damage to important amino acids, which could result in a lower nutritional level and, to a certain extent, danger from the formation of lysinoalanin – an amino acid which can cause poisoning in animals (Anonymous, 1976; Gildberg and Raa, 1977; Raa and Gildberg, 1982).

In France and in the US animal feeds are produced from waste fish by methods that are similar to the production of silage. In this case enzymes are added to the raw material to break down the fish protein. The break down of the proteins takes place in a few hours, after which the bones, skin and fat are removed. The protein solution is then distilled into a concentrate or dried into meal. The product (hydrolyzate) is used mainly for feeding young animals to replace milk or milk powder (Tatterson and Windsor, 1976; Anonymous, 1977). Compared with the production of silage, the production of hydrolyzates is rather complicated and requires expensive equipment and exact control.

11.3 RAW MATERIALS

The raw material for silage production can be divided into the following categories: pelagic fish, trash fish, viscera, by-products and other.

In commercial industrial processing of fish for human consumption, yields are about 50% for direct human consumption. The other 50% consists of by-products from the processes, e.g. heads, skin, bones and viscera. In some countries the by-products are processed to fishmeal and some are converted to silage, but a large part is still wasted.

Fishmeal factories are usually placed at the main fishing ports where it is convenient to use the bulk of waste material for fishmeal. In small and isolated ports and fishing communities supplies of fish waste may be small and/or irregular. It is therefore generally not economical to produce fishmeal from waste in these places, but production of fish silage is a feasible option.

Fish silage offers an opportunity to make use of by-catch, viscera and by-products from the fishing vessels and the factory trawlers. By-catch is often thrown overboard because of its low price.

Pelagic fish may be caught periodically in quantities exceeding the local fishmeal processing or freezing capacities. Preserving and storage as silage is a convenient way of utilizing this resource.

11.4 COMPOSITION AND CHEMICAL CHANGES

The composition of fish silage is similar to that of the raw material used. If oil separation is carried out, the concentration of other components in the deoiled silage will increase in proportion to the removed oil. The approximate composition of certain materials is shown in Tables 11.1 and 11.2 (Gudmundsson, Sigfusson and Bjarnason, 1979; Arnesen, Arason and Jonsson, 1981; Lall, 1991).

Fish silage is a relatively stable product, but it is useful to have knowledge of some of the chemical changes taking place during storage. Since sufficient acid is present to halt bacterial growth, liquefaction is almost totally due to autolytic proteolysis. The biochemistry of the process has not been studied in detail but it is thought to be influenced in part by the relative amounts of muscle, viscera, skin and other parts of fish. During initial liquefaction and storage the protein is broken down into smaller units, i.e. peptides and free amino acids. When autolysis is allowed to continue, the percentage of nitrogen in free amino acids increases while the nitrogen bound in polypeptides decreases. Further degradation of free amino acids produces ammonia

Table 11.1 Common composition of certain fish silage

Silage	*Moisture*	*Protein*	*Oil*	*Ash*
Capelin (March) (*Mallotus villosus*)	78.7	13.1	5.0	1.6
Herring (summer) (*Clupea harengus*)	67.8	17.1	13.4	1.9
Herring (winter)	72.5	18.2	8.2	2.0
Mackerel (summer) (*Scomber scombrus*)	56.4	15.1	27.4	1.6
Mackerel (spring)	74.6	17.9	6.1	1.4
Sprat (*Sprattus sprattus*)	69.4	15.5	13.0	2.2
Sandeel (*Ammodytidae*)	77.2	15.4	3.4	2.4

Table 11.2 Common composition of certain fish by-products silage

Silage	*Moisture*	*Protein*	*Oil*	*Ash*
White fish viscera	60.0	11.5	26.5	1.5
White fish offal[a]	78.9	15.0	0.5	4.2
Herring offal[b]	75.4	13.5	8.7	2.6
Lumpsucker offal (*Cyclopterus lumpus*)	86.8	7.8	3.2	1.0

[a]Without viscera.
[b]With viscera.

and causes losses of essential amino acids like tryptophan, although other amino acids are relatively stable in fish silage.

A crude measure of the extent to which protein hydrolysis has occurred can be obtained from the ratio of protein solubles to total nitrogen. Independent of the type of silage, up to 70% of the nitrogen present will be soluble within 1 week if storage temperatures are within the right limits. The development of soluble nitrogen has been found to be temperature dependent (Tatterson and Windsor, 1974). In Figure 11.1 the particularly rapid increases at the higher temperature during the first few days of storage for a silage prepared from sprats with formic acid can be seen (Tatterson, 1982). In Figure 11.2 the rates of formation of non-protein nitrogen at 30°C are shown for the different parts of fish in silage made with formic acid. The highest rate of proteolysis was in viscera and head, but lowest in flesh (Backhoff, 1976).

In Figure 11.3 the formation of non-protein nitrogen (NPN) is shown in silage produced from capelin by adding formic acid and ethoxyquin.

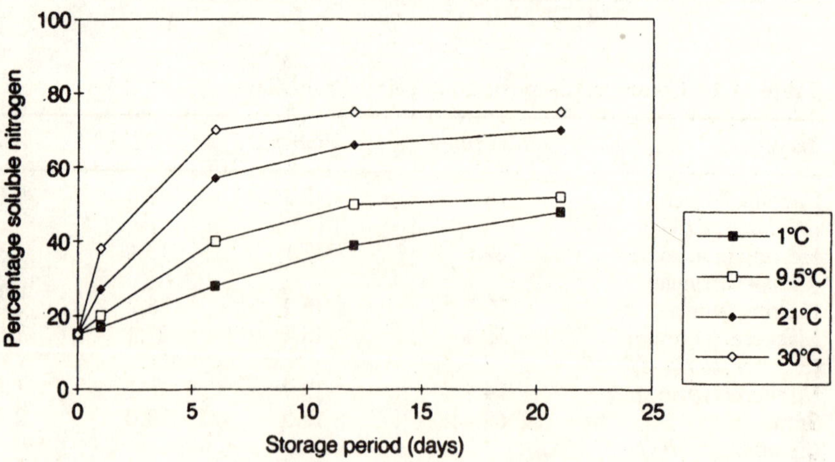

Figure 11.1 Soluble nitrogen as a percentage of total nitrogen during storage of sprat silage (Tatterson, 1982).

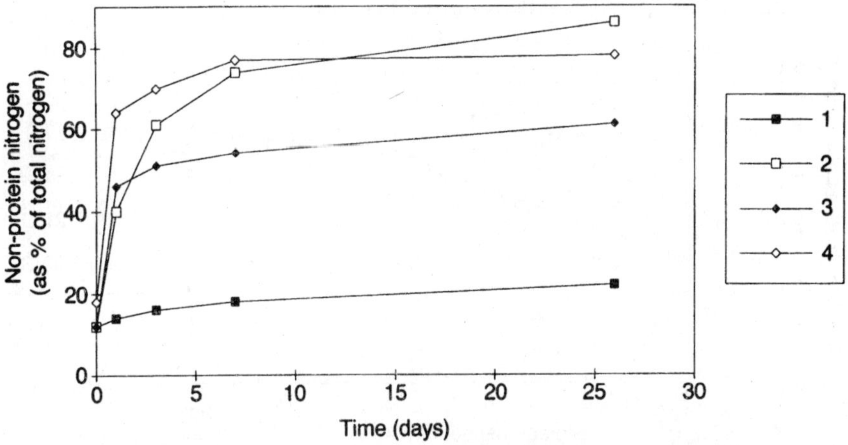

Figure 11.2 Formation of non-protein nitrogen in silages produced from different parts of cod at pH 3.9 and 30°C. (4) Silage from viscera; (3) silage from skin; (2) silage from head; (1) silage from flesh (Backhoff, 1976).

Figure 11.3 also shows the effect of heating the silage (Espe *et al.*, 1992). The increase in total volatile nitrogen (TVN) is mainly due to an increase in NH_3-N; amide-N decreased but trimethylamine (TMA) did not increase. Therefore it is possible to use TMA values as an indicator of the quality of the raw material before processing (Jangaard, 1987). TVN in silage from fish will reach a value of about 110 mg NH_3/100 g sample, mainly due to breakdown of the amino acids glutamine and asparagine (Arason 1986; Haarland and Njaa, 1988, 1989). Whether reduced contents of glutamine or asparagine in the final product have an influence

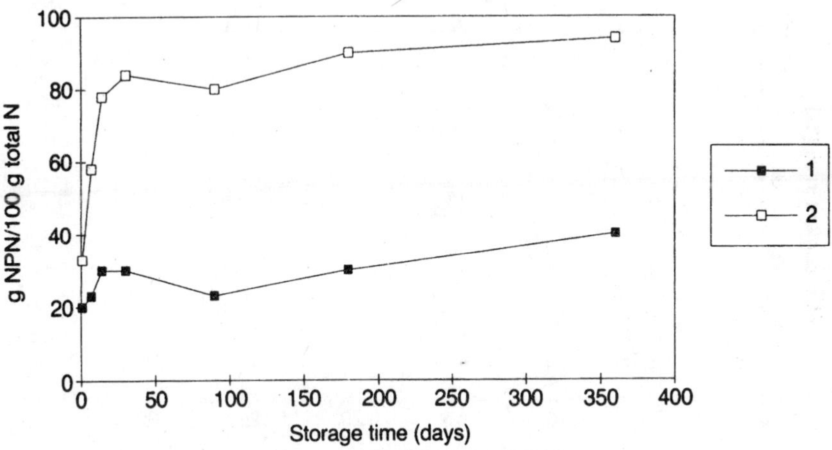

Figure 11.3 NPN as g soluble N/100 g total N for raw silage (2) and cooked silage (1) stored for 360 days at 12°C (Espe *et al.*, 1992).

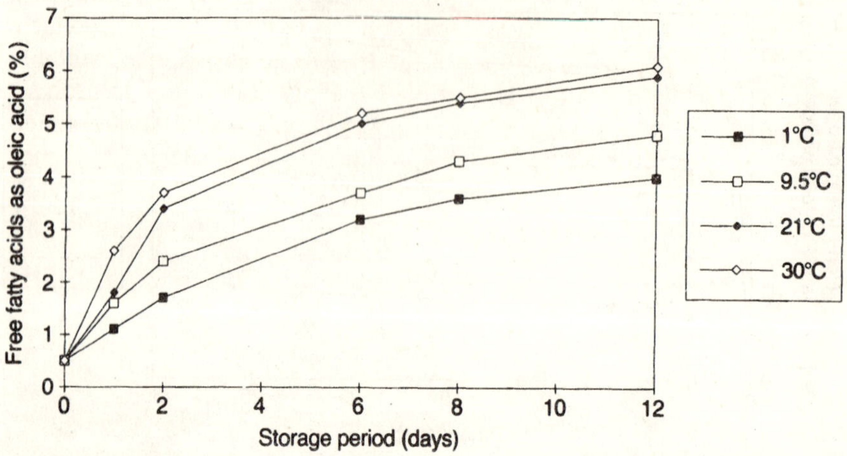

Figure 11.4 Free fatty acid content of extracted oil during storage of sprat silage (Tatterson, 1982).

on the biological value of the silage as a feed resource needs to be examined (Espe, Raa and Njaa, 1989; Espe, Haaland and Njaa, 1990; Espe *et al.*, 1992).

Some major chemical changes can occur in the oil phase during storage of silage. The high level of unsaturated fatty acids favors reaction with oxygen to form hydroperoxides and secondary reaction products. The addition of an antioxidant such as ethoxyquin (200 p.p.m.) to the silage will slow down the oxidation reactions.

Triglycerides will break down to free fatty acids as a result of lipolysis. The increase of free fatty acids in sprat silage stored at several temperatures has been measured, as shown in Figure 11.4. (Tatterson 1976,

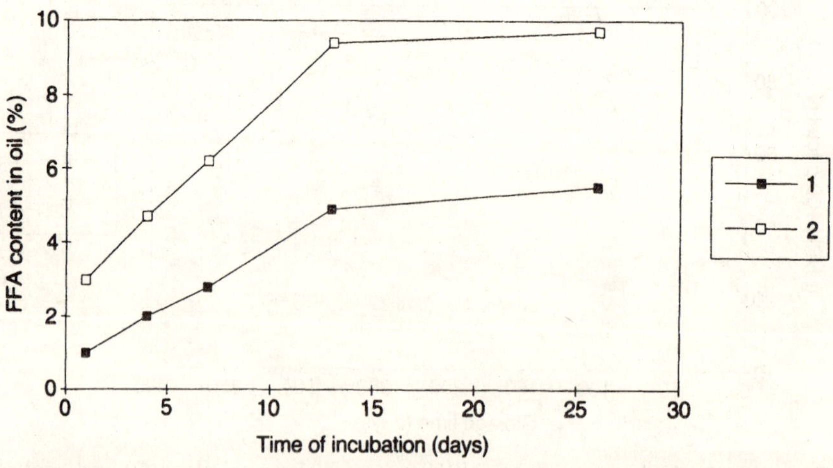

Figure 11.5 Increase of free fatty acid in mackerel (1) and sprat (2) silage oils (Reece, 1981).

1982). It can be seen that free fatty acid formation is most rapid during the first days of storage and is dependent on temperature. A free fatty acid content of about 3% or less is desirable in a good quality oil suitable for refining for edible purposes (Tatterson, 1982). Rates of oil hydrolysis in silage produced from different fish spices vary, e.g. in mackerel silage the rate of oil hydrolysis was much lower than in sprat silage, see Figure 11.5 (Reece, 1981). To protect the quality of the oil it has to be separated from the silage as soon as possible after production (Windsor and Barlow, 1981).

11.5 PRODUCTION METHODS

Silage production is a relatively simple process, as shown in Figure 11.6. The fish material is comminuted and mixed with preservatives, depending on the material and intended use (Windsor and Barlow, 1981; Arason and Hardarson, 1982a, b; Jonatansson, 1983; Arason and Gudmundsson, 1984; Arason, Asgeirsson and Hardarson, 1984; Arason, Gudmundsson and Runolfsson, 1984; Arason, Thoroddsson and Valdimarsson, 1990).

The enzymes already present in the fish will liquefy the fish mass to form a stable liquid with a malty odor and very good storage characteristics.

The quality of the silage depends on the freshness of the raw material, but toxins already formed are not destroyed during the process (Raa and Gildberg, 1982).

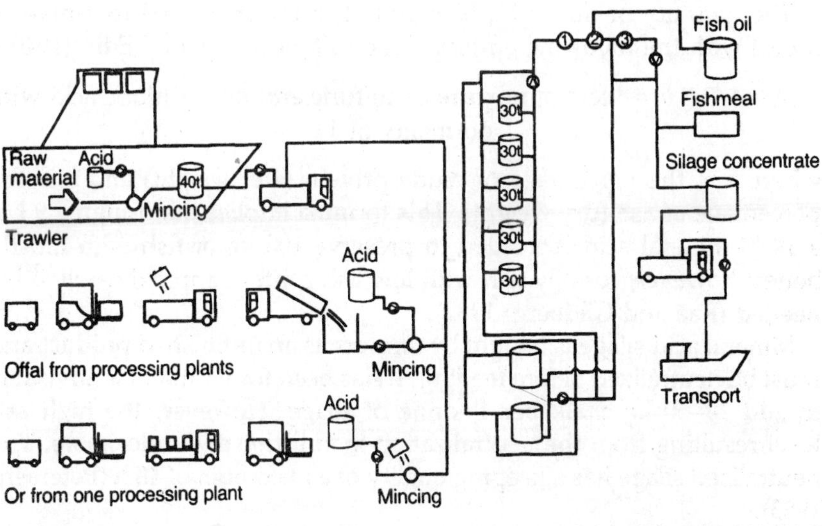

Figure 11.6 Production of fish silage and collection (Jonatansson, 1983).

11.5.1 Mincing

The types of equipment for silage production differ from one raw material to another. The raw material is usually first minced and this is best accomplished with a grinder able to produce as small particles as possible, but no greater than 3–4 mm in diameter. Mincing is necessary to distribute the enzymes throughout the mass of fish and also to assure thorough blending of the acid to avoid pockets of untreated fish where bacterial growth can continue (Windsor and Barlow, 1981; Arason, Throroddsson and Valdimarsson, 1990).

1. When producing silage from viscera, one can use high-speed, low-quantity mincers and keep the silage moving with a circulation pump or other inexpensive equipment. The pump can be any acid-proof, low-speed, positive-displacement pump.
2. When producing silage from filleting waste, the machinery must be more rigid and the mincer has to be a low speed type that can handle fish skin and bones, and will not break down even if foreign objects get into it. The pumps will also have to be very rigid. It has been found that piston pumps or force-fed, mono-type pumps are best suited for this operation and the mixing must be done by propellers.
3. When producing silage from industrial fish such as capelin the mincer can be highspeed and usually must have a high output to be able to handle large quantities of fish in a short time. The silage must be mixed by screws in the tanks.

11.5.2 Acid preservation

For acid preservation the choice of acids is between mineral acids, organic acids or a mixture of both.

The quantity of mineral acid required to lower the pH to 2 may be found with the following empirical formula developed by Edin (1940):

$0.14\,a + 0.9\,b$ = liters of mixture of sulfuric and hydrochloric acid with normality of 14

where a is the percentage of crude protein (wet weight) and b is the percentage of ash (dry weight). This formula implies that roughly 9 l of a 14 N mineral acid is needed to preserve 100 kg of fish with mostly bones; however, for oily fish with low ash content approximately 4 l is needed (Raa and Gildberg, 1982).

Mineral acid silage is bought by farmers as an unfinished product and must be neutralized before feeding. It has been found that it is advisable to add 20–50 kg chalk per 1 tonne of silage. However, the high salt level resulting from the neutralization is nutritionally undesirable. The neutralized silage has a keeping quality of a maximum of 48 h (Petersen, 1953).

Hydrogen ions (protons) cannot diffuse through cell membranes and therefore living cells are capable of keeping a neutral pH in the cytoplasm, in some acid silages. Protons may nevertheless be carried into the cell as a weak non-ionized organic acid, provided that the environment is acid. Weak organic acids have a relatively low ionization at low pH and can then freely pass through cell membranes. Inside the cell, where the pH is neutral, the same organic acid will ionize, gradually leading to a lower pH and accumulation of the anions. These factors contribute to the anti-microbial effects of weak organic acids (Arnesen, Arason and Jonsson, 1981; Raa and Gildberg, 1982; Lall, 1991).

The anti-microbial activity of organic acids increases markedly when the pH of silage falls below a value equal to the pK_a of the weak organic acid. For example, the pH of the silage must be below 3.75 and 4.86, respectively, for formic and propionic acids to have less than 50% of their concentration in the dissociated anti-microbial form ($pK_a = 4.86$ for propionic acid and $pK_a = 3.75$ for formic acid) (Arnesen, Arason and Jonsson, 1981; Raa and Gildberg, 1982; Lall, 1991).

The organic acids are more expensive than the common mineral acids but their usage gives stabilization at a higher pH (around 4.0) than stabilization with mineral acids. Silage produced with organic acids, can therefore be used in feed without neutralization (Tatterson and Windsor, 1973, 1974).

The acid dosage unit consists of a metering pump of adjustable quantity and a tank for the acid. It is very important when ensiling to use the right dosage of the preserving acid. For the acid to enter sufficiently into all the material, the dispersion has to be completely homogeneous. Immediately after the acid has been added, the pH reading will be lowered artificially due to the fact that the acid enters the cell structure within a short period. It will also have to be kept in mind that a high bone content will cause neutralization of the acid (Petersen, 1953; Gilberg and Raa, 1977; Arnesen, Arason and Jonsson, 1981; Viken, 1987).

11.5.3 Mixing

The fish mince tends to stiffen slightly on addition of the acid. The liquefaction will now proceed, and the rate at which the process takes place depends upon the nature of the raw material, digestive enzymes, composition (lipid, protein and moisture), temperature and the amount of acid (Windsor and Barlow, 1981; Arason, Asgeirsson and Hardarson, 1984).

Silage gradually liquefies due to the activity of tissue degrading enzymes which are naturally present in the fish. The digestive proteases have an optimal pH range of 2–4 and their activity decreases sharply above pH 4 (Raa and Gildberg, 1976).

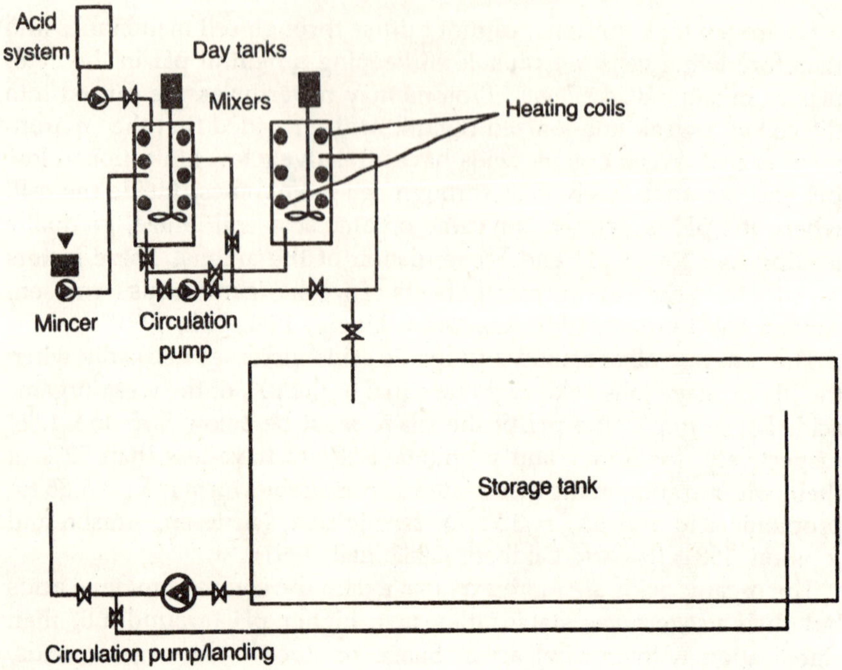

Figure 11.7 Fish silage process of fish by-products (Arason, Thoroddsson and Valdimarsson, 1990).

A digestion- or day-tank which is acid resistant is needed in the first steps. The mixture is stirred constantly at this time and the temperature is chosen for the right rate of autolysis. Digestive proteases from fish have maximal activity at 45–50°C (Raa and Gildberg, 1982). Approximately 80% of the protein in acid-preserved silage becomes solubilized after 1 week at temperatures between 23 and 30°C (Backhoff, 1976; Gildberg and Raa, 1977; Raa and Gildberg, 1982). As mentioned before, the rate of liquefaction is dependent on temperature. Silage made from fresh white fish offal, for example, might take about 2 days to liquefy at 25°C but about 5–10 days at 15°C and much longer at lower temperatures (Windsor and Barlow, 1981).

After the liquefaction step the silage can be stored in a tank and stirred occasionally. In the day tank a small sediment of insoluble tissue fragments, bone, sand and other heavy particles will accumulate in the bottom, see Figure 11.7 (Arason, Thoroddsson and Valdmarsson, 1990).

11.5.4 Storage

According to findings in former tests, silage can be stored for a long time due to the acid contents. If a factory already has much of the required equipment, silage production can be very economical. At certain times the quantity of material which arrives can exceed the capacity of tradi-

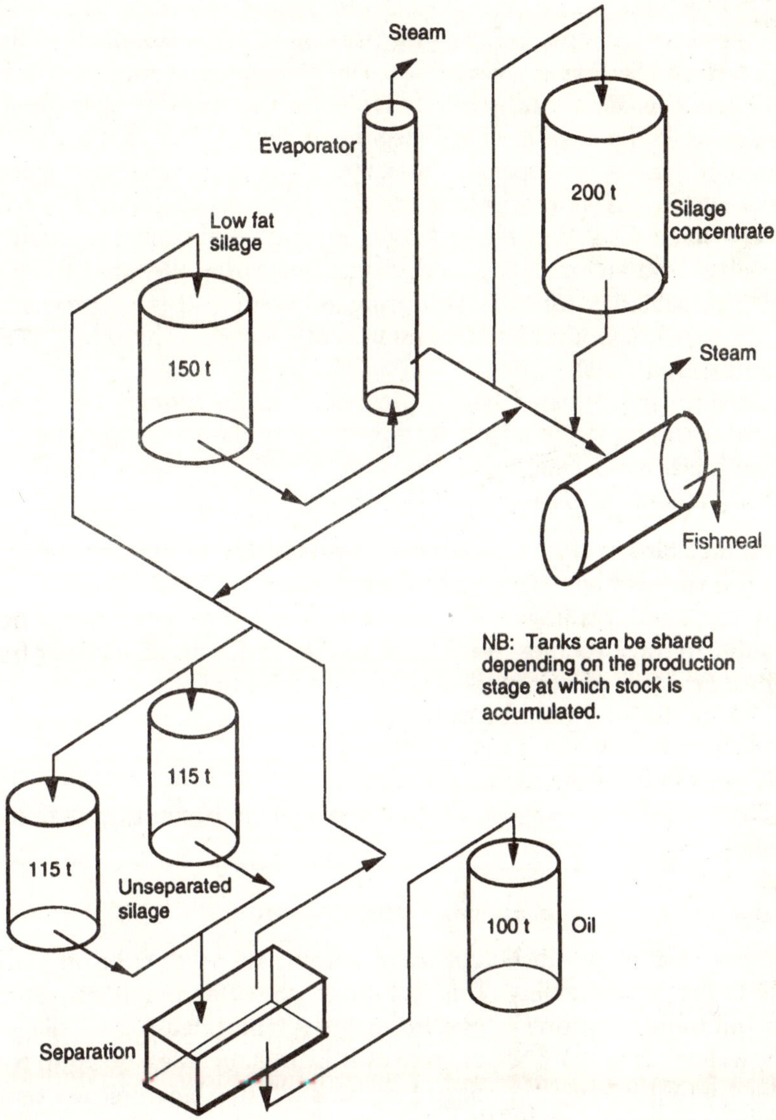

Figure 11.8 Fish silage processing plant (Arason, Asgeirsson and Hardarson, 1984).

tional methods used and then silage is an inexpensive means of processing fish waste in large quantities in a short time. It has been demonstrated that silage can be stored for at least 1.5–2 years, but it is necessary to add anti-oxidants to prevent oxidation of the fat (Raa and Gildberg, 1982).

Figure 11.8 shows the process for producing condensed silage. The space requirements are shown for a factory where the raw material for silage varies from 192 to 460 tons/month. In this case there are two tanks of 115 tons each which store 2 weeks' supply of raw material provided that the supply is regular during the peak month. The size of other tanks depends on discarding possibilities. The storage tanks for raw material can have a secondary function; as tanks for unprocessed silage and as storage tanks for low-fat silage (separated silage). The raw material is measured, the oil is separated and the silage is pumped into storage tanks where it is stored until it can be concentrated. Storage space is also required for concentrated silage. Separate tanks must be used for raw silage and low-fat (separated) silage. By sharing the tanks there are many possibilities for storage during different stages of production (Arason, Asgeirssonn and Hardarson, 1984; Arason, Thoroddsson and Valdimarsson, 1990).

Silage must be heated before processing in the storage tanks or in special warming tanks which are connected to the processing line.

The main requirements for the storage space for silage are shown below (Arason, Asgeirsson and Hardarson, 1984):

1. Storage tanks may be constructed from normal construction steel (st 37) if they contain silage at low temperatures ($T < 18°C$).
2. At temperatures above 30°C the tanks need to be constructed from anti-corrosive stainless steel (such as AMSE 318 or SIS 2343) or from heat resistant plastic (polyester).
3. Heating tanks should be insulated.
4. Must be located close to the factory.
5. It must be possible to mix the silage.
6. There must be no corners or dead spots where bacterial growth may start.

11.5.5 Transportation

It is clear that the production of silage concentrate needs to be on a large scale to reach profitability. This fact directs the attention to transportation and transportation systems for silage. As indicated before, silage is very watery, from 66% water content if liver is included or up to 80% water content in liver-free silage from white fish. Therefore, the transportation system must be efficient and inexpensive for silage products to be profitable. The transportation can be divided into several stages (Arason and Hardarson, 1982a, b; Arason, Asgeirsson and Hardarson, 1984):

1. Transportation from silage producers to silage processing plant:
   (a) from ship;
   (b) from producer ashore.
2. Transportation within the factory.
3. Transportation of silage products, such as fish oil and low-fat silage, silage concentrate and silage meal.

Each of these points will now be discussed.

1. Transportation from silage producers to silage processing plant. Two possibilities are discussed: transportation from trawler or from a silage station ashore. If the silage comes from a trawler, it is pumped ashore and into tanks which are separate and movable, or directly to the supply tanks of the silage factory. The separate tanks are emptied into supply tanks close to the factory. If the silage production takes place ashore, the transportation system depends on the conditions in each place. If the silage production takes place near the factory, the logical means is to pump it. When the distance from factory increases, pumping becomes less attractive and transportation in containers becomes more profitable. Finally, the point is reached when transportation of silage is unprofitable because of large distances.
2. Transportation within the factory. Transportation within the factory takes place only through pumping and the pipes must be made from anti-corrosive plastic or steel.
3. Transportation of silage products. Transportation of silage products, such as fish oil and low-fat silage or silage concentrate, is done with conventional methods for the transportation of liquids, i.e. in tanks. Recently, a new method of transportation was introduced which may be useful in transporting these products. This is a container with a balloon inside and the liquid is pumped into the balloon. Balloons with a capacity of 18 m^3 are now on the market. The product can be transported from the place of production to the user without the need to change the transportation tank/container.

11.6 PHYSICAL PROPERTIES OF SILAGE

Knowledge of the silage and raw material is important for those who wish to produce silage. Knowledge facilitates all the processes and thereby it is possible to predict the behavior of each substance.

The following work has been carried out to evaluate some important physical properties of fish silage, i.e. the viscosity, the heat convection coefficients and the termination of hydrolysis (Arason, Asgeirsson and Hardarson, 1984).

11.6.1 Viscosity

All liquids show resistance to change in shape and resist all changes in the substance. This characteristic is called viscosity and is the result of forces between molecules in the substance which hold them together. If portions of the substance are to be transported between places a certain force is needed to overcome the force that holds the substance together (Arason, Ageirsson and Hardarson, 1984).

Liquid flow in pipes is highly dependent on viscosity and speed, as well as the contact surface between the liquid and the pipes. In determining the transportation capacity of pipes, attention needs to be paid to forces of resistance and viscosity. The forces of resistance are in direct relation to dynamic pressure of the liquid.

The most common unit for measuring viscosity is centipoise (cP). One poise is equal to 1 g/m s. Viscosity of water at 20°C is about 1 cP and viscosity of cod liver oil is about 60 cP at 20°C.

Viscosity is very dependent on temperature and it is important to know the temperature for a given viscosity measurement. The viscosity of cod liver oil is 95 cP at 10°C and 20 cP at 50°C. Viscosity of water is 1.3 cP at 10°C but 0.55 at 50°C.

It is very important to know the viscosity of the liquid which is to flow through particular pipes. Thin liquid, i.e. with a low viscosity, flows much better than liquid with high viscosity.

Liquids are divided into two groups, Newtonian and non-Newtonian, and this classification depends on the measurements of viscosity. The viscosity of Newtonian liquids is dependent on temperature but independent of shear rate.

11.6.2 Viscosity of viscera and silage

The viscosity of liquids changes with the movements of the liquids such that the viscosity changes with the rate of flow in the pipes. Freshly minced viscera has the flow characteristics of pseudo-plastic liquids,

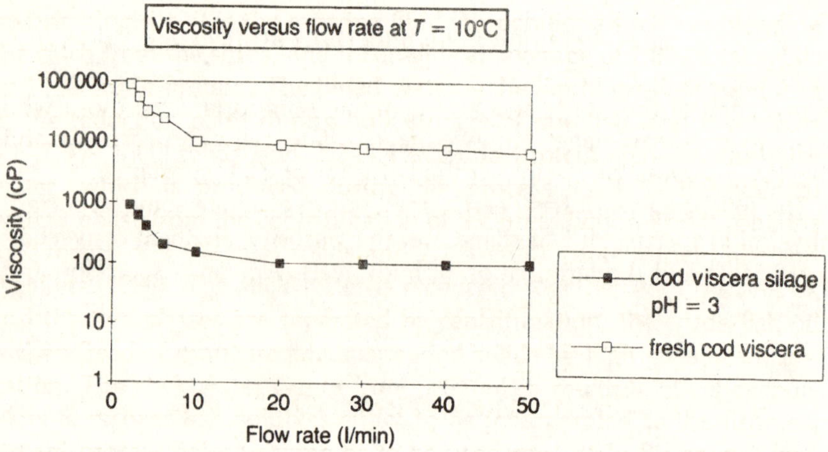

Figure 11.9 Viscosity of cod viscera silage and fresh cod viscera, measured with Brookfield synchro-lectric viscometer (Arason, Asgeirsson and Hardarson, 1984).

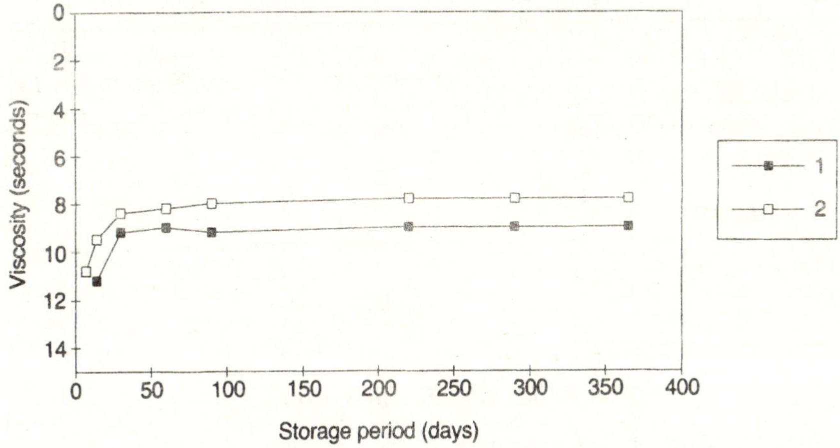

Figure 11.10 Viscosity of sprat silage during storage; +2°C (1); +23°C (2) (Tatterson and Windsor, 1974).

which means that the viscosity level is lower with a higher rate of flow. The result is that the resistance pressure is not increased as much at an increased flow rate as is the case, for instance, when water is pumped (Arason, Asgeirsson and Hardarson, 1984)

Viscosity is dependent on the composition of the transported liquid. The viscosity of fatty silage is much higher than for low-fat silage. The break down of substances in the liquid also affects the level of viscosity in the substances. The viscosity of silage is shown in Figure 11.9 (Arason, Asgeirsson and Hardarson, 1984).

Silages are normally sufficiently liquid to be handled within a few days and crude viscosity measurements are made. The method involves measuring the length of time taken for a fixed volume of silage to flow through a funnel with an orifice of diameter 13 mm. Figure 11.10 shows the results for sprat silage and it is seen that the decrease in viscosity is closely related to the increase in nitrogen solubility (Tatterson and Windsor, 1974)

11.6.3 Heat convection

Heat convection in and from silage is dependent on the composition, temperature and rate of flow of the silage. The heat convection coefficient for heat transfer from the heater in stock water evaporators to the silage was measured with the equipment of the experimental factory of the Icelandic Fisheries Laboratories. The measurements are shown in Table 11.3.

Table 11.3 Heat convection coefficient for heating silage in evaporator

	Evaporation pressure (bar)	*Heat convection coefficient* h(W/m^{2}K)
Low-fat silage from viscera[a]	0.12	4050
	0.5	7030
	0.8	7937
Fatty silage from viscera[b]	0.12	1808

[a]Water: 84%, fat: 0.9%.
[b]Water: 77%, fat: 8.5%.

11.6.4 Termination of hydrolysis

In silage production it is necessary to be able to forecast the termination of the hydrolysis, i.e. when the largest proportion of the substance is dissolved. A simple method to follow the solubility of substances is to use a glass separator and thereby produce sediment. Glass separators are available in many fishmeal factories and are convenient to use for these measurements.

The raw material which was used for these measurements was cod viscera. To the viscera were added 0.75% formic acid (85%) and 0.75% propionic acid, and the viscera was agitated occasionally during the whole processing period. The hydrolytic temperature was kept around 27°C in order to speed up the process (Gildberg and Raa, 1977; Arason, Asgeirsson and Hardarson, 1984).

The findings from these tests are shown in Figure 11.11.

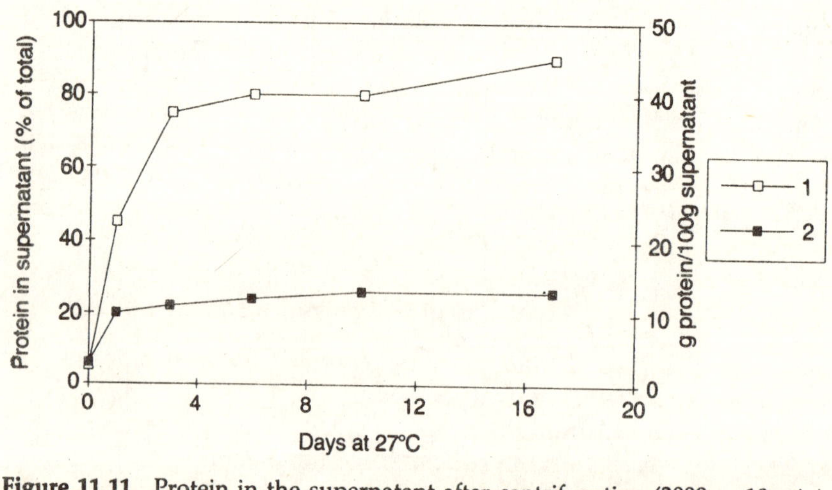

Figure 11.11 Protein in the supernatant after centrifugation (2000 g, 10 min), shown as per cent of total protein (1) and as g protein/100 g supernatant (2) (Gilberg and Raa 1977).

11.7 METAL CORROSION IN SILAGE PRODUCTION

Viscera silage without liver was produced with 3% of formic acid and the pH level was 3.5–3.8. The silage was placed into four glasses and magnetic propellers were used so that the viscera was moved constantly during the process. The temperature was 35 ± 2°C and the silage was kept that way for 6–7 days. In each glass two metallic pieces were inserted with known surfaces of four different substances, i.e. normal construction steel, high carbon steel, stainless steel and aluminum. The metallic pieces were kept in the glasses during the whole process (Arason, Asgeirsson and Hardarson, 1984).

According to the findings, a tank from normal commercial steel will corrode about 0.7 mm/year if the silage is kept at 35 ± 2°C. Such warming would not take place in silage production, nevertheless it is advisable to avoid tanks made from commercial steel for hot silage. A tank from stainless steel corroded only 0.004 mm/year, from which it can be concluded that such tanks have a very long lifetime.

The silage was later heated with constant stirring to 95–100°C. The silage was distilled at this temperature and it took about 8 h to condense

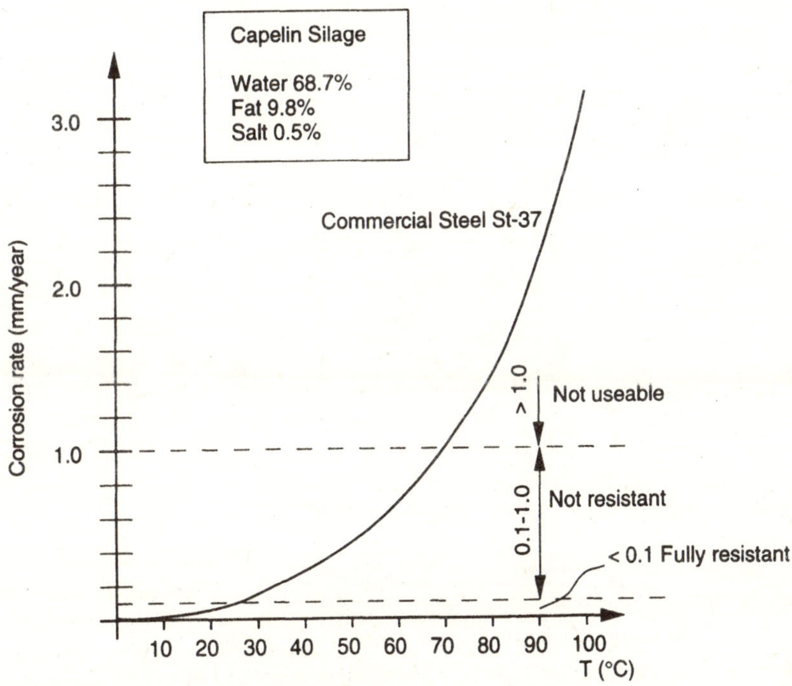

Figure 11.12 Metal corrosion in production of capelin silage (Arason, Asgeirsson and Hardarson, 1984).

the silage to the point that it could not be stirred by the magnetic propeller.

Walls of stock water evaporators made from commercial steel corroded about 10.6 mm annually if in use for 365 days a year. Stick water evaporators from stainless steel corrode about 0.005 mm/year. From this it can be seen that it is of no use to use evaporators from materials other than stainless steel for silage concentration.

Metal corrosion in silage production is shown in Figures 11.12 and 11.13.

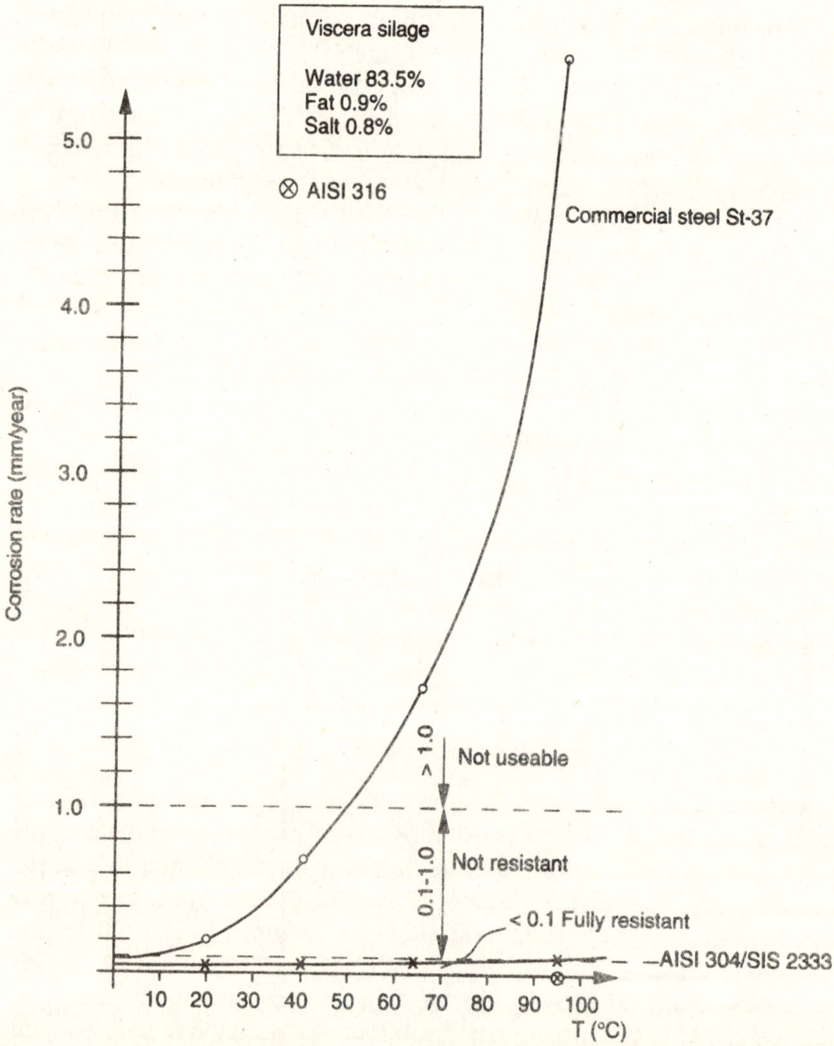

Figure 11.13 Metal corrosion in production of separated cod viscera silage (Arason, Asgeirsson and Hardarson, 1984).

11.8 PROCESSING OF SILAGE

In the storage tank silage can separate into three phases: a lipid–protein emulsion on top, an aqueous soluble phase in the middle, and a small sediment of heavy and insoluble fragments at the bottom. In a tank with good mixing the silage will not separate.

The oil can be separated from silage with a similar technique as used for the separation of oil during fishmeal manufacturing. The amount of oil recovered from fish silage by centrifuging increases with the time of incubation of the silage. For instance, for mackerel silage containing 12.5% oil, 6.5% was recovered without heating immediately after acidification, but 8.2% oil was recovered after 12 days of incubation. After heating the silage to 70°C a further 1.8% of oil could be recovered from the 12-day-old silage (Reece, 1981).

De-oiling of silage should take place as soon as possible after liquefaction. The first step is heating in a heat exchanger to 70–90°C. Coarse suspended solids are removed by decantation or by screening and the oil is then removed from the silage by centrifugation. In small-scale production it may be satisfactory to rely on self-sedimentation and manual decanting of the oil floating on top of the silage (Windsor and Barlow, 1981).

In the production of silage from low oil content raw material there is no problem to use the finished product to replace fishmeal in the diet of animals. White fish and by-products have about 0.5–3% oil content in the silage. Pelagic fish can have oil content between 2 and 30% depending on the species and season. Therefore much of the silage made from pelagic fish will be of little value in many instances for feeding unless the oil content is reduced (Potter, Tatterson and Wignall, 1980).

Good quality fish oil has a number of uses, e.g. in margarine which generally requires a free fatty acid content not exceeding 3%. De-oiled silage is of high commercial value for feed production and the use will run little risk of tainting. In general a level of 2% oil or less in the finished silage will be acceptable to the user.

In some de-oiling tests of viscera silage with an oil content of 9% and about 90% autolysis the oil content was easily lowered down to 0.2–0.5% of wet weight.

The action of enzymes on liquefaction of fish silage may be stopped at any degree of degradation by heating. Heating or pasteurization at approximately 85°C for 2 min will inactivate the enzymes and produce a silage with polypeptides of suitable length. To achieve this effect, the heating must occur early in the process and soon after the introduction of the acid. The next step is separating the oil from the silage. The pasteurization inhibits lipolysis of triglycerides during storage (Arason, 1986).

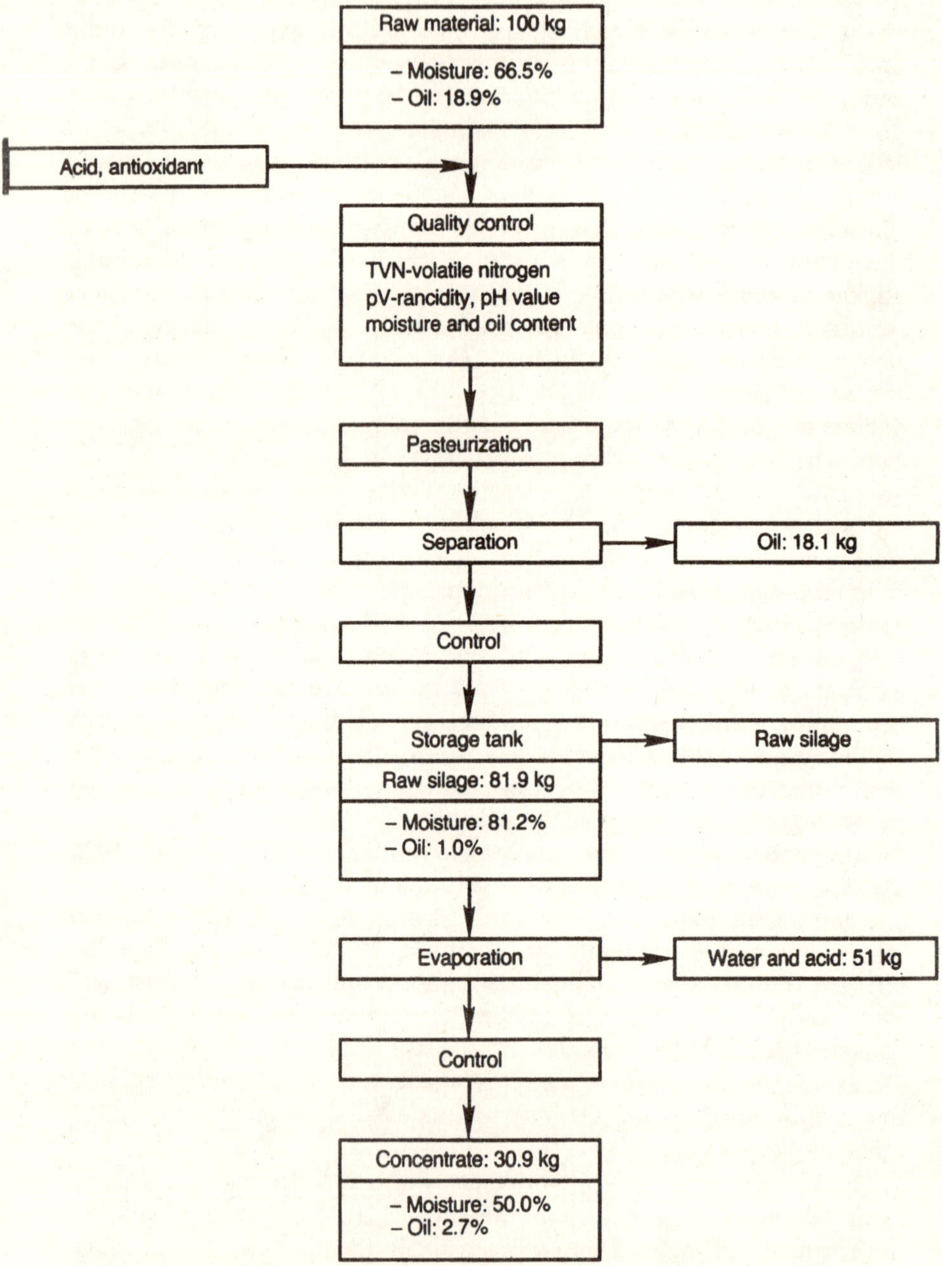

Figure 11.14 Flow diagram for processing fish silage from white fish by-products (Arason, Thorodosson and Valdimarsson, 1990).

Liquefied fish, in contrast, is made from slightly heated ground heads and viscera (60°C for 30 min). After liquefaction, the mixture is heated to 85°C for 15 min to destroy endogenous enzymatic activity and to pasteurize the liquid (Hardy, Shearer and Spinelli, 1984). After cooling the mixture is acidified to pH 4 to prevent spoilage. These liquefied proteins are more stable and better utilized than proteins in conventional silage.

To reduce the transport and storage cost of fish silage it is possible to produce concentrated silage. Silage can be concentrated to approximately 50% dry matter at such a low temperature to avoid denaturation of the proteins. Concentrated silage has several advantages, e.g. better nutritional value than ordinary silage. In the production of moist feed it is possible to reduce the amount of binder meal compared with ordinary silage, resulting in better economy. Fish silage or concentrate can be mixed with other raw material for fishmeal production (Arason, Thoroddsson and Valdimarsson, 1990).

A schematic diagram of such a process is shown in Figure 11.14. Various experiments have been performed to make direct use of the silage concentrate instead of processing it into fishmeal along with conventional raw material. The concentrate contains about 50–60% dry matter and has a syrupy consistency (Arason, Asgeirsson and Hardarson, 1984).

Silage concentrate has been used as an additive in grass pellets for feeding ruminants. The pellets contained 15–20% silage concentrate, which increases the protein content of the pellets and acts as a binding agent (Arason, Asgeirsson and Hardarson, 1984; Arason and Gudmundsson, 1984). Silage concentrate can also be dried by any of the conventional methods, e.g. spray drying, drum drying, dish drying etc.

Formic acid is mostly used for the production of fish silage. In the processing of silage all process equipment should be designed in stainless steel due to the use of formic acid.

The corrosive effects of silage were studied with respect to acid type, concentration, temperature and fat content. The results indicate that low-fat silage is more corrosive than silage with about 10% fat content. Silage can be stored in ordinary steel tanks if the temperature does not exceed 20°C. Above 20°C the silage must be stored in tanks made of stainless steel or plastic (Arason, Asgeirsson and Hardarson, 1984).

11.9 SILAGE QUALITY

Most fish and fish by-products are relatively unstable when fresh and under various storage conditions. Fresh raw fish produces the best quality silage.

One reason for the slow growth of the fish silage industry is inconsistent quality. Many thought that the silage process could be used to rejuvenate poor quality raw material. As a result some users were turned off by silage and must be convinced that a product of uniform quality and with guaranteed product specifications is available (Jangaard, 1987; Pedersen, 1987).

A product declaration should contain information of value to the user and it is recommended that users must get documentation of the silage content. After ensilaging a fish will become unidentifiable but the users should know the raw material which has been used.

Sampling is very important. Before sampling it is important that the silage is well mixed or the sample can just be out of one of three phases: oil–protein phase, aqueous phase and a bottom layer.

Several chemical tests for freshness of fish silage have been proposed to define the quality and value of the silage.

For standard information about fish silage content it is nessary to know the protein, fat, ash and dry matter. The value for ash will indicate if the silage was made from whole fish, mostly viscera or bony offal. The ash content of whole fish is usually in the 2–3% range (Jangaard, 1991).

The concept of the term quality is difficult to define. It is often limited in the case of fish silage to pH and the content of protein, fat, dry matter and ash, but it can also depend on the type of fish used.

Many different parameters have been evaluated and also been tested. Some of them are as follows:

Determination of amines, particularly TMA, TMA oxide (TMAO) and TVN. The TMAO concentration in fish varies widely (by species and age) and may range from 1 to 4% of total nitrogen. TMA is the decomposition product of TMAO. It has been found that TMA is mainly formed through bacterial decomposition of TMAO. When acid is added, the bacteria are inactivated or killed and the TMA value is stable and reflects the freshness of raw material used for silage production (Pedersen, 1987; Jangaard, 1991).

Measurements of TVN or NH_3 are used as an indication of freshness of the raw material for high-quality fish meal. As for fishmeal, raw material with a value higher than 50 mg TVN/100 g should not be accepted for high-quality fish silage. TVN is the sum of TMA, ammonia, dimethylamine and others. The TVN value will increase in unheated silage and then come from ammonia formed from the amide groups of glutamine and asparagine. The amide groups of free glutamine are very labile, as NH_3 is detected soon after dissolving it in water (Haaland and Njaa, 1988, 1989; Espe, Raa and Njaa, 1989).

Most fish, particularly those containing high levels of fat, are susceptible to oxidation. The rate of lipid oxidation depends on the fatty acid profile and oil containing a high proportion of unsaturated fatty acid is more susceptible to oxidation (Lall, 1991). Peroxide value is a measure of the early stage of oxidation of oils while the thiobarbituric acid (TBA) value will indicate one end-product of oxidation.

It is possible to use formaldehyde to inhibit protein hydrolysis in fish silage but it may prove toxic to some animals. Mixing 10 l of 37% formaldehyde in a tonne of fish silage will stop protein autolysis and lipid oxidation (Haard *et al.*, 1985). Studies with ruminants and trout have shown that amino nitrogen is used more readily from peptides of comparable length (two to six amino acids) than the corresponding free amino acids (Stone and Hardy, 1986).

11.10 NUTRITIONAL VALUE

Several workers have successfully utilized acid-preserved silage obtained from different raw materials in diets for different animal species. Some reports show that silage is a good source of protein and that its nutritional value is comparable with that of fishmeal (Stormo and Ström, 1978; Skrede, 1981; Ström and Eggum, 1981; Raa and Gildberg, 1982; Jackson, Kerr and Cowey, 1984; Krogdahl, 1985).

The amino acid compositions of raw minced saithe waste, silage made from minced saithe waste, raw minced herring and silage made from minced herring stored for 5 months are shown in Table 11.4 (Espe *et al.*, 1992). The amino acid composition of saithe waste silage is fairly similar to that of the protein of white fishmeal (Gildberg and Almås, 1986).

Ensiling and autolysis does not alter the amino acid composition significantly. There are indications, however, that tryptophan is unstable in acid conditions and it is clear that tryptophan is apparently the first limiting amino acid in the fish silage. In an animal diet the tryptophan level can be improved by addition of cereal protein (Johnsen and Skrede, 1981; Gildberg and Almås, 1986).

Feeding trials with rainbow trout on diets containing 60% silage preserved with hydrochloric acid observed growth equivalent to that on a diet of fresh fish; similar good results have been reported for Atlantic salmon (Lall, 1991).

Silage produced with propionic acid, either alone or in combination with formic, sulfuric or other acid, is not palatable to Atlantic salmon. However, rainbow trout fed diets containing the same silage preserved with propionic acid did not show aversion to feed nor growth depression (Lall, 1991).

The apparent digestibility of silage produced from whole herring, dogfish and cod indicates that all three types of fish silage are efficiently utilized by Atlantic salmon. The protein and lipid digestibility of all silage-based diets were significantly higher than for those based on raw fish (Lall, 1991).

Table 11.4 Amino acid composition (mg/100 g protein) in raw minced saithe waste, silage made from minced saithe waste, raw minced herring and silage made from minced herring stored for 5 months (Espe *et al.*, 1992)

	Saithe		*Herring*	
Amino acid	*not ensiled*	*ensilage*	*not ensiled*	*ensilage*
Aspartic acid	80.6	82.6	103.1	93.8
Glutamic acid	126.3	124.0	139.1	129.3
Serine	48.3	47.6	45.9	44.0
Glycine	87.1	78.6	58.6	64.0
Histidine	18.7	20.0	28.8	29.3
Arginine	68.6	67.0	58.4	58.4
Threonine	44.2	46.8	49.6	46.1
Alanine	62.1	58.2	62.2	61.8
Proline	54.8	51.1	ND	ND
Tyrosine	29.7	33.1	35.1	33.7
Valine	44.0	46.9	47.8	49.5
Methionine	28.8	29.1	31.8	33.3
Isoleucine	39.4	42.6	37.7	40.7
Leucine	72.4	72.7	74.6	75.0
Phenylalanine	40.9	41.8	39.7	38.7
Lysine	71.5	74.5	92.6	88.4

Fish silage can be used in the processing of moist and semimoist feeds. They contain variable amounts of fish silage, or fresh or frozen fish, together with other ingredients, binders, vitamins and mineral mixtures, and are extruded (Lall, 1991).

Fish silage or concentrate can be used in dry feed. Dry feed processing involves either an extrusion or steam pelleting process. A small percentage (15–20%) of fish silage could be incorporated into dry extruded feeds which contain 8–10% moisture. For example, a large food manufacturer in Norway adds 7% of fish silage concentrate to dry feed and some of the water added is then removed in the extruder. The concentrate used in dry feed is said to cut down on losses from fines. Feeding trials with this concentrate are under way but earlier trials indicate a digestibility similar to low temperature fishmeal (Jangaard, 1991; Lall, 1991).

The nutritional quality of liquefied fish has been found to be superior to 30-day-old conventional silage and nearly equal to that of fishmeal (Stone and Hardy, 1986).

11.11 CONCLUSIONS

Tests and experiments have been done with processing silage as well as research on the silage itself at the Icelandic Fisheries Laboratories for many years. At the end of these experiments it is up to the industry to make use of the findings and experiences which have been gained. The employees of the Laboratories are willing to contribute to a successful outcome.

The production of silage is an interesting possibility to make use of fish waste which comes from the fishing industry. The processes which are used in the production of silage from viscera can be utilized in processing other types of waste including shrimp, slaughterhouse waste and also for liver. It is possible to produce a better quality fish oil from liver silage than silage mixed with liver.

The mixture of silage and silage concentrate with grass meal was very successful. It is conceivable that grass pellet factories could make use of silage concentrate to increase the protein content of the pellets as well as for a binding agent.

The costs of condensing the silage are considerable and need to be calculated in each instance to decide whether it is profitable to use the silage directly or in concentrated form; however, this depends on the cost of transportation, storage and consumption possibilities of ruminants.

Feed mixtures produced from silage, containing correct acids in the proper ratio, can be stored for a long time. It has been found that these mixtures can be stored for up to 1 year. Mixtures from silage concentrate can also be preserved for a long time.

In this paper it has been attempted to respond to many questions about silage production and hopefully the fishing industry can make use of these findings.

REFERENCES

Anonymous (1976) Processed protein foods and lysinoalanine. *PAG Bull.*, **4**(2), 34–6.

Anonymous (1977) Liquid hydrolyzed fish protein from enzymatic process. *Food Eng.*, **2**, 26–29.

Arason, S. (1986) *Unpuplished Report*, Icelandic Fisheries Laboratories, Reykjavik.

Arason, S. and Gudmundsson, O. (1984) Silage and silage concentrate in grass pellets. I. Production and storage, report presented at the *1984 Meeting of Agricultural Consultants*, Iceland, pp. 95–102.

Arason, S., Asgeirsson, L. and Hardarson, T. (1984) Silage processing. *Technical Report 152*, Icelandic Fisheries Laboratories, pp. 1–43.

Arason, S., Gudmundsson, O. and Runolfsson, S. (1984) Silage and whey concentrate in grass meal. II. Feeding experiments of bulls, report presented at the *1984 Meeting of Agricultural Consultants*, Iceland, pp. 106–14.

Arason, S. and Hardarson, V. (1982a) Fish silage on board stern trawler. *Technical Report 137*, Icelandic Fisheries Laboratories, pp. 1–23.

Arason, S. and Hardarson, V. (1982b) Technical information on silage processing. *Technical Report 138*, Icelandic Fisheries Laboratories, pp. 1–33.

Arason, S., Thoroddsson, G. and Valdimarsson G. (1990) The production of silage from waste and industrial fish: the Icelandic experience. Marketing profit out of seafood wastes. *Proceedings of the International Conference on Fish*

By-products, (ed. S. Keller), University of Alaska Fairbanks, Alaska, pp. 79–85.

Arnesen, G., Arason, S. and Jonsson, S. (1981) Silage from fish offal, a review. *Technical Report 126*, Icelandic Fisheries Laboratories, pp. 1–19.

Backhoff, H.P. (1976) Some chemical changes in fish silage. *J. Food Technol.*, **11**, 353–63.

Dagbjartsson, B., Arnesen, G., Thorsteinsson, J. *et al.* (1976). New method for utilization of viscera and trash fish. *Technical Report 77*, Icelandic Fisheries Laboratories, pp. 1–54.

Edin, H., (1940) Undersökningar angåande importavstengningens äggviteproblem. *Nord. Jordbr. Forsk.*, **22**, 142–58.

Espe, M., Raa, J. and Njaa, L.R. (1989) Nutrional value of stored fish silage as a protein source for young rats. *J. Sci. Food Agric.*, **49**, 259–70.

Espe, M., Haaland, H. and Njaa, L.R. (1990) Digestibility and utilisation in young growing rats given saithe offal silage stored for different length of time as the sole protein source. *Fisk. Dir. Skr. Ser. Ernaering. III*, **1**, 37–42.

Espe, M., Haaland, H., Njaa, L.R. and Raa J. (1992) Growth of young rats on diets based on fish silage with different degrees of hydrolysis. *Food Chem.*, **44**, 195–200.

Freeman, H.C. and Hoogland, P.L. (1956) Processing of cod and haddock viscera: 1. Laboratory experiments. *J. Fish Res. Board Can.*, **13**, 869–77.

Gildberg, A. and Almås, K.A. (1986) Utilization of fish viscera, in *Food Engineering and Process Applications*, (eds M. Le Maguer and P. Jelen), Elsevier Applied Science, London, Vol. 2, pp. 383–93.

Gildberg, A. and Raa, J. (1977) Properties of a propionic acid/formic acid preserved silage of cod viscera. *J.Sci. Food Agric.*, **28**, 647–53.

Gudmundsson, O., Sigfusson, S.H. and Bjarnason, J. (1979) Fish and whale silage as a protein supplement for beef cattle. RALA *Report 54*, Agricultural Research Institute, Iceland, pp. 1–16.

Haaland, H. and Njaa, L.R. (1988) Ammonia (NH_3) and total volatile nitrogen (TVN) in preserved and unpreserved stored, whole fish. *J.Sci. Food Agric.*, **44**, 335–42.

Haaland, H. and Njaa, L.R. (1989) Effect of temperature on the autolysis of capelin silages stored for one year. *Fisk. Dir. Skr. Ser. Ernæring. II*, **7**, 219–26.

Haard, N.F., Kariel, N., Herzberg, G. *et al.* (1985) Stabilisation of protein and oil in fish silage for use as a ruminant feed supplement. *J. Sci. Food Agric.*, **36**, 229–41.

Hardy, R.W., Shearer, K.D. and Spinelli, J. (1984) The nutritional properties of co-dried fish silage in rainbow trout (*Salmo gairdneri*) dry diets. *Aquaculture*, **38**, 35–44.

Jackson. A.J., Kerr. A.K. and Cowey, C.B. (1984) Fish silage as a dietary ingredient for salmon. I: Nutritional and storage characteristics. *Aquaculture*, **38**, 211–20.

Jangaard, P.M. (1987) Fish silage: a review and some recent developments, paper presented at the *Fish Silage Workshop*, (ed. G. Comeau), Halifax, Canada, pp. 8–33.

Jangaard, P.M. (1991) Fish silage in Norway: current activities, paper presented at the *Fish Silage Workshop*, (ed. J. Delabbio), Nova Scotia, Canada, pp. 28–53.

Johnsen, F. and Skrede, A. (1981) Evaluation of fish viscera silage as a feed resource. *Acta Agric. Scand.*, **31**, 21–8.

Jonatansson, E. (1983) Silage production in fish meal processing plant of Einar Gudfinnsson hf. in Bolungarvik. *BSc Thesis*, University of Iceland.

Kompiang, I.P., Darwanto, A. and Arifuddin, R. (1979) Nutritional value of fish silage, in *Fish Silage Production and its Use (FAO Fisheries Report 230)*, FAO, Rome, pp. 44–7.

Krogdahl, Å. (1985) Fish viscera silage as a protein source for poultry. I. Experiments with layer-type chicks and hens. *Acta Agric. Scand.*, **35**, 3.

Lall, S.P. (1991) Nutritional value of fish silage in salmonid diets, paper presented at the *Fish Silage Workshop*, (ed. J. Delabbie), Nova Scotia, Canada, pp. 63–74.

Lindgren, S. and Clevström, G. (1978) Antibacterial activity of lactic acid bacteria. 1. Activity of fish silage, a cereal starter and isolated organisms. *Swed. J. Agric. Res.*, **8**, 61–6.

Lisac, H. (1961) The liquid ensilage of Mediterranean sardines for animal feeding. *Proc. Gen. Fish. Coun. Medit.*, **6**, 111–19.

Nilsson, R. and Rydin, C. (1968) Fermentation as means of preserving organic materials. *Acta Chem. Scand.*, **17**, 174–9.

Pedersen, O. (1987). Fish silage and quality, product specifications and quality standard. I and II. *Norsk Fiskeoppdrett*, **12**(2), 74–5 and **12**(3), 50–1.

Petersen, H. (1951) Ensilage of fish and fish waste. *Report 95*, Technological Laboratory, Ministry of Fisheries, Copenhagen, Denmark.

Petersen, H. (1953) Acid preservation of fish and fish offal. *FAO Fish Bull.*, **6**, 18–25.

Potter, D., Tatterson, I. and Wignall J. (1980) Preliminary studies of two techniques for the removal of oil from fish silage using commercial equipment, in *Advances in Fish Science and Technology*, (ed. J.J. Connell), Jubilee Conference of the Torry Research Station, Aberdeen, pp. 338–43.

Raa, J. and Gildberg, A. (1976) Autolysis and proteolytic activity of cod viscera. *J. Food Technol.*, **11**, 619–28.

Raa, J. and Gilberg, A. (1982). Fish silage. A review. *CRC Crit. Rev. Food Sci. Nutr.*, **16**, 383–419.

Ramanathan, G. and Moorjani, M.N. (1975) Soluble fish protein by use of ammonia, *Proceeding of the Sympsium of the Fish Processing Industry in India*, Central Food Technological Reserach Institute, Mysore, India.

Reece, P. (1981). Recovery of high quality oil from mackerel and sprat by the silage process. *J. Sci. Food Agric.*, **32**, 531–8.

Schrøder, K., Clausen, E., Sandberg, M.A. and Raa, J. (1980) Psychrotrophic *Lactobacillus plantarum* from fish and its ability to produce antibiotic substances, in *Advances in Fish Science and Technology*, (ed. J.J. Connell), Fishing News (Books) Ltd., Farnham, UK.

Skrede, A. (1981) Utilization of fish and animal byproducts in mink nutrition. *Acta Agric. Scand.*, **31**, 171.

Stone, F.E. and Hardy, R.W. (1986) Nutritional value of acid stabilised silage and liquefied fish protein. *J. Sci. Food Agric.*, **37**, 797–803.

Stormo, B. (1993) Personal communication. Riber Company, Tromø, Norway.

Stormo, B. and Ström, T. (1978). Ensilering av Fiskeslö. *Report*, Institute of Fishery Technology Research, Tromsø, Norway.

Ström, T. and Eggum, B.O. (1981) Nutritional values of fish viscera silage. *J. Sci. Food Agric.*, **32**, 115–20.

Ström, T., Gildberg, A., Stormo, B. and Raa, J. (1980) Fish silage, why not use propionic and formic acid?, in *Advances in Fish Science and Technology*, (ed. J.J. Connell), Fishing News (Books) Ltd, Farnham, UK.

Tatterson, I.N. (1976) The preparation and storage of fish silage, in *Proceeding of the Torry Research Station Symposium on Fish Silage*, Torry Research Station, Aberdeen.

Tatterson, I.N. (1982) Fish silage – preparation, properties and uses. *Animal Feed Sci. Technol.*, **7**, 153–9.

Tatterson, I.N. and Windsor, M.L. (1973) Torry tests practical value of fish silage. *Fish Mark. Process. Packag.*, **February**, 16.

Tatterson I.N. and Windsor M.L. (1974). Fish silage. *J. Sci. Food Agric.*, **25**, 369–379.

Viken, N.I. (1987). Silage processing plants, paper presented at the *Fish Silage Workshop*, (ed. G. Comeau), Halifax, Canada, pp. 8–33.

Windsor, M. and Barlow, S. (1981) *Introduction to Fishery By-products*, Fishing News (Books) Ltd, Farnham, UK

Wirahadikusumah, S. (1969). The effect of fish silage on the quality of hen eggs and meat of broilers. *Lantbr. högsk. Annlr.*, **35**, 823–35.

Table 12.1 Cost of H^+ at pH 4.0[a]

H^+ source	*Concentration (%)*	*Cost (\$/lb)*	*Dissociation at pH 4.0 (%)*	pK_a	*Cost of H^+ (\$/lb)*
H_2SO_4	100	0.0375	100	NA[b]	1.83
HCl	3.15	0.035	100	NA[b]	4.06
H_3PO_4	53	0.070	99	1.96	13.03
Formic acid	85	0.407	63	3.76	34.96
Acetic acid	100	0.33	13	4.73	151.80
Propionic acid	99.9	0.44	12	4.86	319.00
glucose	100	0.25	12	4.86	134.89[d]
Lactic acid	88	1.06	58	3.86	186.90
glucose	100	0.25	58	3.86	45.18[e]
fructose	100	0.39	58	3.86	70.47[e]
sucrose	100	0.34[c]	58	3.86	61.41[e]
lactose	100	0.21[c]	58	3.86	37.91[e]
corn syrup D.E. 63	NA[b]	0.19[c]	58	3.86	44.15[e]
molasses D.E. 50	NA[b]	0.203[c]	58	3.86	66.47[e]
corn starch	100	0.20[c]	58	3.86	36.47[e]

[a]Based in part on bulk costs in *Chemical Marketing Reporter*, December 14, 1992.
[b]Not applicable.
[c]Based on the addition of one water molecule (molecular weight = 18) for every glycosidic bond hydrolyzed.
[d]Based on 57% conversion efficiency of propionic acid fermentation.
[e]Based on 85% conversion efficiency of lactic acid fermentation.

CHAPTER 12

Lactic acid and propionic acid fermentations of fish hydrolyzates

R.E. Levin

12.1 INTRODUCTION

Fish waste or gurry comprises up to 60% of the original weight of fish after fillets are removed. The entrails of certain fish species such as haddock, cod and flounder are usually removed and discarded at sea so that the gurry comprises of heads, tails and skeletons. Species of flat fish, which are landed intact, often yield gurry composed, in part, of about 20% entrails. At present, processors in North America have more fish waste than can be utilized by traditional means such as conversion to fishmeal for animal feeds. Disposal at sea along US coastal zones is presently restricted (Clean Water Act, 1991) which has resulted in a disposal problem for the fish processing industry. In addition, fish farms that also perform processing have recently developed an interest in converting gurry to a product of nutritional value for incorporation into fish feeds to reduce feed costs and to eliminate the problem of waste disposal. One of the more expedient methods for the conversion of gurry to a microbially stable product consists first of grinding, followed by proteolytic hydrolysis and removal of the bone chips by screening. Preservation of the resulting hydrolysates is most readily achieved by acidification. If a mineral acid such as sulfuric or phosphoric is used alone the pH must be reduced to 2.0 to achieve long-term preservation (Wignall and Tatterson, 1976). Reduction of the pH to 2.0 with sulfuric acid involves some level of hazard and usually requires neutralization for subsequent use as a feed or fertilizer. The addition of sulfuric or

Fisheries Processing: Biotechnological applications.
Edited by A.M. Martin.
Published in 1994 by Chapman & Hall, London. ISBN 0 412 58460 3

phosphoric acid to reduce the pH to 4.0 will achieve long-term preservation only if a microbial inhibitor such as potassium sorbate or propionic acid is added at a level of 0.1–0.2% (Winter and Feltham, 1983; Levin *et al.*, 1989). If organic acids such as lactic and propionic are used for acidification there is no increase in ash, which may be of importance if the hydrolyzate is to be used as a feed adjunct. If fish hydrolyzates are to be used as fertilizers, organic acids such as lactic and propionic acids used for acidification and preservation will be utilized by soil microorganisms, resulting in the absence of acidification of the soil. Major advantages in the fermentation of added sugars to lactic acid and propionic acid for preservation of fish hydrolysates include reduced costs for acids (Table 12.1), especially if the processor is able to prepare the bacterial inoculum to be used in the fermentations, and the elimination of the hazard involved in handling concentrated acids.

12.2 ENZYMATIC HYDROLYSIS OF FISH TISSUE

Proteolytic hydrolysis of ground fish tissue releases the tissue from the bones and allows subsequent removal of the bones by screening. This results in considerably less acid being required to achieve the final desired acid pH and also results in pH stability. An additional advantage of proteolysis before fermentation is that proteins are hydrolyzed to amino acids and peptides. The lactic and propionic acid bacteria require a number of free amino acids and will yield a more rapid and effective fermentation if the tissue is first hydrolyzed. Papain has the advantage of low cost and high activity at 60–75°C at near neutral pH values when compared with other commercially available proteases. In addition, we have observed that the use of ground fish entrails as a source of proteases suffers from the disadvantage that the endogenous proteases are notably thermally labile and are rapidly denatured when the hydrolyzing tissue is in contact with the surface of a steam jacketed kettle. This problem is never encountered with papain which is notably heat resistant (Levin *et al.*, 1990). Hydrolysis with 0.05% papain (w/w) at 71°C for 15 min with continual agitation is usually sufficient to achieve liquefaction of ground fish tissue and separation of bones.

12.3 EFFECTIVENESS OF LACTIC ACID AND PROPIONIC ACID FERMENTATIONS

The use of lactic acid producing bacteria in association with dairy products and certain processed meats is long established as is the use of propionic acid producing bacteria in the dairy industry. Both groups of bacteria are capable of fermenting sugars to significant levels of the respective organic acids so as to result in a notable reduction in pH. Propionic acid in particular is a widely used preservative and is especially effective at acidic pH values. If adequate sugar concentrations are present, it is possible with hydrolyzed fish tissue to achieve a final pH value of 3.6 resulting from a lactic acid fermentation yielding 1.8% lactic acid. For long-term preservation of fermented hydrolysates the addition of 0.2% potassium sorbate or propionic acid is necessary. We have consistently observed that when potassium sorbate is used with lactic acid fermented hydrolysates stored in polyethylene containers for several months, molds such as *Penicillium* eventually develop on the surface, which we attribute to the uptake of volatile sorbic acid by the polyethylene from the head space. This problem does not occur if storage is in glass containers. If 0.2% propionic acid is added after a lactic acid fermentation of hydrolyzed fish tissue, no such problem is observed regardless of the composition of the storage container. A significant financial saving in the cost of propionic acid can be realized if it is produced in hydrolyzates by fermentation (Table 12.1). Propionic acid bacteria are somewhat less tolerant of low pH values than the more acid tolerant lactic acid bacteria and will usually reduce the pH in fish hydrolysates to no less than 4.5 with a resulting propionic acid concentration of about 1.0%. This higher pH and lower organic acid content is, however, quite adequate to establish long-term preservation of such fermented hydrolysates due to the potency of propionic acid as an acidic preservative. The pK_a value of propionic acid is 4.86. By definition, 50% of the propionic acid at a pH of 4.86 is undissociated. This is of considerable significance since it is only the undissociated form of propionic acid that is inhibitory to microorganisms. It has been reported that spontaneous mutants of propionic acid bacteria have been isolated that are capable of producing higher levels of propionic acid from sugars than the wild type strains (Glatz, 1992). Strøm *et al.* (1980) found that while 1.0% sulfuric acid at a pH of 2.5 failed to completely inhibit the growth of *Aspergillus flavus*, 1.0% formic acid at a pH of 3.5 and below as well as 1.0% propionic acid at a pH of 5.5 and below completely inhibited growth (Table 12.2). Russel (1992) has reviewed the various mechanisms proposed for the inhibition of microorganisms by organic acids having non-polar properties.

The fraction of an organic acid present in the undissociated state is determined by the pK_a value for that acid and the pH of the solution and is conveniently calculated from the Henderson–Hasselbach equation:

$$\text{pH} = \text{p}K_a + \log \frac{[\text{H}^+]}{[\text{AH}]}$$

Such information is invaluable for arriving at comparable cost calculations for the acidified preservation of fish hydrolysates (Table 12.1).

Table 12.2 Effect of pH on the growth of *A. flavus* in the presence of 1.0% sulfuric, formic and propionic acid[a]

Acid added	pH 2.5	3.0	3.5	4.0	4.5	5.0	5.5	6.0
Sulfuric	+	+	+	+	+	+	+	+
Formic	0	0	0	+	+	+	+	+
Propionic	0	0	0	0	0	0	0	+

[a]Growth (+) or no growth (0) after 7 days at 30°C.
From Strøm *et al.* (1980) with permission from Blackwell Scientific Publications.

12.4 CHARACTERISTICS OF THE LACTIC ACID AND PROPIONIC ACID BACTERIA

12.4.1 Lactobacilli

Members of the genus *Lactobacillus* are Gram-positive facultatively anaerobic non-spore-forming rods. Most species are observed to consist of long and slender rods. A few isolates have been found to consist of short rods. Most strains are non-motile. Occasional isolates have been observed to be motile with peritrichous flagella (Kandler and Weiss, 1986). The lactobacilli are presently allocated to three metabolic groups:

1. Obligately homofermentative, producing at least 85% lactic acid from sugars.
2. Facultatively heterofermentative, producing at least 85% lactic acid when abundant sugar is present.
3. Obligately heterofermentative, producing no more than 50% lactic acid from sugars plus ethanol, acetic acid and CO_2.

The first two groups contain species that are of significance in the fermentation of fish hydrolyzates. The selection of species is dependant on the temperature of fermentation, the carbohydrate(s) added to hydrolyzates, and the ability to grow in fish hydrolyzates with sufficient lactic acid production to significantly reduce the pH. Some variation between species and strains of a given species is found regarding the final pH that is achieved with fish hydrolyzates fermented with lactobacilli (Table 12.3).

The bacterial production of lactic acid can result in the formation of the D(−) isomer, the L(+) isomer, or in a racemic DL mixture (Jehanno, Thuault and Bourgeois, 1992). The precise isomeric form of the lactic acid formed is dependant on the species of lactic acid bacteria used. The D(−) isomer is known to cause acidosis in infants (Drosse and Strolley, 1964) and in adults with short bowel syndrome (Oh *et al.*, 1979; Saton *et al.*, 1982). This may be of significance in the use of animal or fish feeds prepared with lactic acid fermented fish hydrolyzates if a significant level of D(−) lactic acid is present. The precise identity of the species of lactic acid bacterium used and the isomeric form of lactic acid produced may therefore be of importance. Tables 12.4 and 12.5 list the various homofermentative and facultatively heterofermentative lactic acid bacteria capable of producing significant concentrations of lactic acid and the isomeric form(s) produced. Methodology has been developed for the colorimetric detection of L(+) lactic acid producing bacterial colonies involving the use of NAD, diaphorase, D(−) specific lactic dehydrogenase, and *p*-iodonitrotetrazolium violet as a colorimetric hydrogen acceptor (Jehanno, Thuault and Bourgeois, 1992). Colonies producing D(−) or DL lactic acid are surrounded by red halos while colonies

Table 12.3 Reduction in pH of gurry hydrolyzates by various *Lactobacillus* species[a]

Lactobacillus *species*	*pH after 96 h incubation*
Control (uninoculated)	6.2
L. helveticus NCK 15004	4.1
L. helveticus 12046	3.7
L. casei ATCC 393	3.8
L. casei PLY 3537	4.2
L. plantarum (Stabisil)	3.65

[a]One loop of cells from a 48 h MRS agar slant was inoculated into 20 ml of gurry hydrolyzate (prepared as described in section 12.6 in 15 × 150 mm screw capped tubes and incubated at 30°C for 96 h.

producing only L(+) lactic lack halos. It should be noted that two species of lactobacilli *L. amylovorus* and *L. amylophilus* (Table 12.4) produce extracellular amylases which offer an advantage in terms of the relatively lower cost of corn starch compared to most other carbohydrates (Table 12.1). The α-amylase from *L. amylovorus* has recently been purified and characterized (Burgess-Cassler and Imam, 1991). An alternative to the use of these species with starch as carbohydrate is the addition of a commercially available crude amylase to achieve starch hydrolysis followed by the addition of a non-amylitic lactic acid species. Ideally, such an organism should be capable of utilizing maltose and maltotriose as well as glucose. It should also be realized that the amylitic hydrolysis of starch usually results in the formation of limit dextrins (from unhydrolyzed α(1–6) linkages at branch points) constituting 5–10% of the original starch which are not fermented.

Table 12.4 Characteristics of the homofermentative lactobacilli[a]

Species	Acid from[b,c] G	F	Ga	L	S	M	St	Lactic acid isomer(s) product	Growth at 45°C[c]
L. delbrueckii ssp. *delbrueckii*	+	+	−	−	+	±		D	+
L. delbrueckii ssp. *lactis*	+	+	±	+	+	+		D	
L. delbruekii ssp. *bulgaricus*	+	+	−	+	−	−		D	+
L. sharpeae	+	+	+	+	−	+	−	L	−
L. yamanashiensis	+	+	±	−	+	−		L	−
L. ruminis	+	+	+	+	+	+		L	−
L. vitulinus	+	+	+	+	+			D	
L. amylophilus	+	+	+	−	−	+	+	L	−
L. farciminis	+	+	+	+	+	+	−	L(D)	−
L. animalis	+	+	+	+	+	+		L	+
L. salivarius	+	+	+	+	+	+		L	+
L. helveticus	+	±	+	+	−	±		DL	+
L. amylovorus	+	+	+	−	+	+	+	DL	
L. acidophilus	+	+	+	+	+	+	+	DL	
L. crispatus	+	+	+	+	+	+		DL	+
L. gasseri	+	+	+	±	+	±		DL	+
L. jensenii	+	+	+	−	+	±		D	+

[a]From Kandler and Weiss (1986) with permission from Williams and Wilkins.
[b]G, glucose; F, fructose, Ga, galactose; L, lactose; S, sucrose; M, maltose; St, starch.
[c]( ) data not indicated, (+) positive, (−) negative, (±) variability among strains.

Table 12.5 Characteristics of the facultatively heterofermentative lactobacilli[a]

Species	Acid from[b]							Lactic acid isomer(s) products	Growth at 45°C
	G	F	Ga	L	S	M	St		
L. agilis	+	+	+	+	+	+		L	+
L. alimentarius	+	+	+	−	+	+		L(D)	−
L. bavaricus	+	+	+	+	+	+		L	−
L. casei ssp. *casei*	+	+	+	±[c]	+	+		L	−
L. casei ssp. *pseudoplantarum*	+	+	+	+	+	+		DL	−
L. casei spp. *rhamnosus*	+	+	+	+	+	+		L	+
L. casei spp. *tolerance*	+	+	+	+	−	−		L	
L. coryneformis spp. *coryniformis*	+	+	+	±	+	+		D(L)	−
L. coryniformis ssp. *torquens*	+	+	+	+	+	+		D	
L. curvatus	+	+	+	±	−	+		DL	−
L. homohiochii	+	+	−	−	−	+		DL	−
L. plantarum	+	+	+	+	+	+		DL	−
L. sake	+	+	+	+	+	+		DL	−

[a]From Kandler and Weiss (1986).
[b]See legend to Table 12.4.
[c]Lactose fermentation is plasmid mediated (Shimizu-Kadota, 1987).

12.4.2 Streptococci

Members of the genus *Streptococcus* are normally spherical or ovoid non-spore-forming Gram-positive cells occurring in pairs or chains when grown in liquid media. They are facultatively anaerobic and usually non-motile (Hardie, 1986a). All species are homofermentative and have been placed into six groups:

1. Pyogenic hemolytic (Rotta, 1986).
2. Oral (Hardie, 1986b).
3. Enterococci (Mundt, 1986a).
4. Anaerobic (Hardie, 1986c).
5. Lactic acid streptococci (Mundt, 1986b).
6. Other streptococci (Hardie, 1986d).

Most species in the first four groups are associated with pathological processes and their use is to be avoided. Four species are used in various commercial fermentations (Table 12.6): *S. faecium* (group 3), *S. lactis*, *S. cremoris*, *S. diacetylactis* (group 5) and *S. thermophilus* (group 6). *S. cremoris* and *S. diacetylactis* are now combined with *S. lactis* (Mundt, 1986b). The fermentation of lactose and sucrose by strains of *S. lactis* are both plasmid mediated which should be noted in terms of long-term fermentative stability of such cultures with these sugars (Anderson and McKay, 1984; Gonsalez and Kunka, 1985).

12.4.3 Pediococci

Members of the genus *Pediococcus* are Gram-positive, non-motile, non-spore-forming spheres, occurring in pairs or tetrads. The various species (Table 12.7) are all homofermentative and are found associated with beer and vegetative material. *P. damnosus* and *P. parvulus* require the most anaerobic conditions and grow slowly (Garvie, 1986) and are therefore not as convenient to culture and use as the other species. *P. pentocaseus* and *P. acidilactici* grow rapidly and result in final pH values of 4.0 or below, and are therefore among the pediococci of choice. *P. acidilactici* is the most heat tolerant and is capable of growth at 50°C (Garvie, 1986).

12.4.4 Propionic acid bacteria

The so-called classical propionic acid bacteria (Table 12.8) allocated to the genus *Propionibacterium* are Gram-positive, non-spore-forming, non-motile rods. Cultures usually comprise pleomorphic rods in that they are 'mostly bent, swollen, and sometimes branched' (van Niel, 1928). Most isolates are catalase positive which readily distinguishes them from the lactic acid bacteria. The optimum growth temperature is in the range 30–32°C (Cummins and Johnson, 1986). Cultures are

Table 12.6 Characteristics of the non-pathological streptococci[a]

Species	Acid from[b]						Lactic acid isomer(s) products	Growth at 45°C	Final pH with glucose
	G	F	Ga	L	S	M			
S. faecium	+	+	+	+	+	+		+	4.0–4.4
S. lactis[c]	+		+	+[d]	±[e]	+		–	4.0–4.5
S. cremoris	+			+[d]	±	±			
S. raffinolactis	+	+	+	+	+	+		–	
S. thermophilus	+	+		+	+	–	L	+	4.0–4.5

[a]From Hardie (1989d) and Mundt (1986a,b) with permission from Williams and Wilkins.
[b]See legend to Table 12.4.
[c]*S. lactis* ssp. *diacetylactis* and *S. cremoris* are presently combined with *S. lactis* (Mundt 1986b).
[d]Fermentation of lactose is plasmid mediated (Anderson and McKay, 1984).
[e]Fermentation of sucrose is plasmid mediated (Gonzalez and Kunka, 1985).

Table 12.7 Characteristics of the pediococci[a,b]

Species	Acid from						Lactic acid isomer(s) products	Growth at 40°C	Final pH with glucose
	G	F	Ga	L	S	M			
P. damnosus	+	+	+	−	+	+	DL	−	<4.0
P. parvulus	+	+	+	−	−	+	DL	−	4.0
P. inopinatus	+	+	+	+	±	±	DL	−	4.0
P. dextranicus	+	+	+	±	±	+	L	+	4.4
P. pentocaceus	+	+	+	±	−	+	DL	+	<4.0
P. acidilactici	+	+	+	±	−	−	DL	+	<4.0
P. halophilus	+	+	+	+	−	+	D	−	5.0
P. urinaeequi	+	+	+	±	+	+	L	+	5.0

[a]From Garvie (1986) with permission from Williams and Wilkins.
[b]See legend to Table 12.4.

Table 12.8 Fermentation characteristics of classical dairy propionic acid bacteria[a]

Species	Acid from[b,c]						Final pH with glucose
	G	F	Ga	L	S	M	
P. freudenreichii	+	+	+	±[b]	−	−	4.5–4.9
P. jensenii	+	+	+	±	+	±	4.4–4.9
P. thoenii	+	+	+	±	±	±	4.7–4.9
P. acidipropionici	+	+	+	+	+	+	4.1–4.9

[a]From Cummins and Johnson (1986) with permission from Williams and Wilkins.
[b]*P. freudenreichii* ssp. *freudenreichii* (lactose −), *P. freudenreichi* ssp. *shermanii* (lactose +).
[c]See legend to Table 12.4.

anaerobic to aerotolerant and generally grow more luxuriantly under anaerobic conditions, often developing little or no growth aerobically and never producing large surface colonies. The fermentation of glucose is initiated via the Embden–Myerhoff pathway and results in the production of propionic acid according to the following equation (van Niel, 1928):

$$1.5\ \underset{\text{Glucose}}{C_6H_{12}O_6} \rightarrow 2.04\ \underset{\text{propionic acid}}{CH_3CH_2COOH} + 0.82\ \underset{\text{acetic acid}}{CH_3COOH} + 0.8\ CO_2$$

From lactic acid (Fitz, 1878):

$$3\ \underset{\text{Lactic acid}}{CH_3CHOHCOOH} \rightarrow 2\ \underset{\text{propionic acid}}{CH_3CH_2COOH} + \underset{\text{acetic acid}}{CH_3COOH} + CO_2 + H_2O$$

Lewis and Yang (1992) observed that with *P. acidipropionici* higher yields of propionic acid were obtained from 4.0% lactate than from 4.0% glucose or 4.0% lactose in part due to the formation of significant amounts of succinate from the sugars but not from lactate.

12.5 GROWTH MEDIA AND CULTURE MAINTENANCE

12.5.1 Media

Several commercially available culture media have been developed over the years for the cultivation of the lactic acid bacteria. Lactobacilli MRS (MRS) broth (de Man, Rogosa and Sharpe, 1960) supports the growth of all lactobacilli and most streptococci. Elliker Broth (Elliker, Anderson and Hannesson, 1956) is used for the cultivation of streptococci and lactobacilli associated with the dairy industry. The addition of 1.5% agar to both media can be used for the preparation of agar slants and plates.

MRS broth and agar can be used for growing all pediococci (Garvie, 1986).

A typical medium for the isolation and maintenance of propionibacteria is the yeast extract–sodium lactate (YEL) medium of Malik, Reinbold and Vedamuthu, 1968). Excellent growth of all propionibacteria can be obtained in a trypticase–yeast extract–glucose (TYEG) medium containing 0.05% Tween 80 (TYEG, Cummins and Johnson, 1981). The author has found that MRS broth or agar deeps support luxurious growth of *P. acidipropionici*. The addition of 0.05% L-cysteine to media for the growth of propionic acid bacteria is recommended to enhance anaerobiasis.

MRS broth

Proteose peptone no. 3	1.0%
Beef extract	1.0%
Yeast extract	0.5%
Dextrose	2.0%
Tween 80	0.1%
Ammonium citrate	0.2%
Sodium acetate	0.5%
$MgSO_4$	0.01%
$MnSO_4$	0.005%
Disodium phosphate	0.2%
pH 6.5	

Elliker broth

Tryptone	2.0%
Yeast extract	0.5%
Gelatin	0.25%
Dextrose	0.5%
Lactose	0.5%
Sucrose	0.5%
NaC1	0.4%
Sodium acetate	0.15%
Ascorbic acid	0.05%
pH 6.8	

YEL

Trypticase	1.0%
Yeast extract	1.0%
Sodium lactate	1.0%
KH_2PO_4	0.25%
$MnSO_4$	0.0005%
Agar	1.5%
pH 7.0	

From Malik, Reinbold and Vedamuthu (1968) with permission from the National Research Council of Canada.

TYGE

Trypticase	1.0%
Yeast extract	0.5%
Glucose	1.0%
$CaCl_2$	0.002%
$MgSO_4$	0.002%
NaCl	0.002%
KH_2PO_4	0.34%
K_2HPO_4	0.57%
Tween 80	0.05%
Sodium formaldehyde sulfoxalate	0.05%
$NaHCO_3$ (presterilized)	0.1%
pH 7.0	

From Cummins and Johnson (1981) with permission from Springer-Verlag.

12.5.2 Culture maintenance

Cultures of lactic acid and propionic acid bacteria can be conveniently maintained for many months at room temperature using chopped meat medium devoid of glucose if overlaid with mineral oil to avoid entry of oxygen. Permanent stocks can be maintained by lyophilizing or storage at −80°C in 10% glycerol or 10% dimethylsulfoxide (DMSO).

12.6 MATERIALS AND METHODS

12.6.1 Preparation and hydrolysis of gurry

Cod (*Gadus morhua*) gurry consisting of heads, tails and skeletal components were ground once through an 8.5 inch diameter steel plate with holes of 0.5 inch diameter. The antioxidant ethoxyquin (Monsanto Chemical Co., St Louis, MO) at 0.0125% and 0.05% papain (w/w 1000 milk clotting units; Biocon, Lexington, KY) were then added to 150 lb of the ground gurry. The papain was added as a dispersion in 100 ml of water. The gurry was then heated for 15 min at an internal temperature of 71°C with constant mechanical agitation in a steam jacketed kettle to achieve removal of the tissue from the bones and liquefaction as a result of proteolysis. The temperature was then raised to an internal temperature of 90°C for 10 min to inactivate the papain and achieve pasteurization and then rapidly reduced to 40–45° C. The bone chips were then removed by passing the hydrolyzate through a nylon screen (20 mesh) and the hydrolyzate was then frozen and held at −20°C until used.

12.6.2 Preparation of bacterial cultures

Stabisil (Medipharm, Des Moines, IA) is a commercial dry inoculum of 50% *P. acidilactici*, 30% *S. faecium* and 20% *L. plantarum* by cell number. Each of these organisms was isolated from hydrated Stabisil on MRS agar (MRS broth, Difco, Detroit, MI, plus 1.5% agar) and identified by Gram staining, microscopic morphology and fermentation tests. *L. helveticus* NCK 15009 was a gift from T. Klaenhammer. *L. helveticus* 12046 was a gift from C. Corre. *L. casei* ATCC 393 and *L. casei* PLY 3537 were gifts from M. Posno. *P. acidipropionici* ATCC 4875 was obtained from the American Type Culture Collection. Acid production from carbohydrates was determined in tubes containing 10 ml of MRS broth prepared from individual components (proteose peptone no. 3, 1.0%; beef extract, 1.0%; yeast extract 0.5%; Tween 80, 0.1%; ammonium citrate, 0.2%; sodium acetate, 0.5%; magnesium sulfate, 0.01%; manganese sulfate,

0.005%; and disodium phosphate, 0.2%) with specific single carbohydrates added at a concentration of 2.0%. Inoculated tubes and uninoculated controls were incubated at 37°C for 72 h prior to pH measurements. Pure cultures were maintained on slants of MRS agar at 4°C and in 10% DMSO at −80°C. Cultures were inoculated into 9 ml tubes of MRS broth and incubated 16 h at 37°C. Blackstrap molasses (6.5, 7.5 and 8.5%; w/w) and corn syrup (3.0, 3.5, 4.0 and 4.5%; w/w) were added to 200 g portions of hydrolyzate in 250 ml polycarbonate beakers which were then heated in a microwave oven to 90°C for 5 min to achieve pasteurization, cooled to 37°C and then inoculated. The inocula consisted of overnight MRS broth cultures diluted in MRS broth to 3.0×10^6 cells/ml. A 1 ml aliquot of the adjusted cell suspension was then added with mixing. The beakers were then sealed with screw cap lids and incubated at 37°C. The blackstrap molasses, analyzed by high performance liquid chromatography (HPLC), contained 23% sucrose, 15% glucose and 16% fructose (w/w). The corn syrup, similarly assayed, contained 24.6% glucose, 19.0% maltose, 13.9% maltotriose and 23.5% higher saccharides (w/w). Stabisil inoculum was prepared by reconstituting 3.0 g in 100 ml of water, and incubating for 2 h at 37°C before addition to hydrolyzates at a level of 0.5% (v/w). Cell numbers in broth cultures and hydrated Stabisil inocula were determined microscopically with the use of a Petroff–Hausser bacteria counting chamber and during gurry fermentation by spreading 0.1 ml of samples decimally diluted with MRS broth on to the surface of duplicate MRS agar plates incubated at 37°C for 48 h.

12.6.3 Quantitation of carbohydrates, lactic acid and propionic acid

An Aminex PHX-87H hydrogen column (Bio Rad, Melvill, NY) was used for HPLC analysis to quantitate carbohydrates and lactic acid. Portions (5.0 ml) of hydrolyzates were centrifuged at 10 000 *g* for 10 min and the top liquid clarified by passage through a 0.45 μm porosity nylon membrane filter. The HPLC column was maintained at 65°C and injected with 200 μl of sample which was then eluted at 1250 p.s.i with 0.012 M sulfuric acid at a rate of 0.8 ml/min. Detection of elution peaks was with the use of a refractive index monitor (LDC Refracto Monitor; Thermo Separation Products, Riviera Beach, FL) and an integrating recorder (HP 3394A Integrator; Hewlett Packard, Avondale, PA). Peak areas were then used for quantitation in conjunction with standard curves produced with glucose, fructose, galactose, sucrose, maltose, maltotriose and lithium lactate (Sigma, St Louis, MO). The utilization of carbohydrates and lactic acid production in hydrolyzate fermentations by Stabisil were determined only with 7.5% molasses and 3.5% corn syrup.

12.7 RESULTS

12.7.1 Utilization of individual sugars in MRS broth

The pattern of carbohydrate utilization by *S. faecium* (Table 12.9) was similar to that previously described for this organism (Mundt, 1986a). It is of particular interest that sucrose and fructose were readily utilized when present as individual carbohydrates in the culture broth. The pattern of carbohydrate utilization by *P. acidilactici* (Table 12.9) was similar to that previously described for this organism (Garvie, 1986) except for the utilization of maltose which reportedly should have been negative. *P. acidilactici* (Table 12.9) failed to utilize sucrose, which is not inconsistent with the reported variability of sucrose utilization among strains of this organism (Garvie, 1986), while fructose was readily utilized. *L. plantarum* (Table 12.9) adhered to the pattern of carbohydrate utilization previously reported for this organism (Kandler and Weiss, 1986). *S. faecium* and *L. plantarum* fermented galactose at significantly slower rates than the other sugars (Table 12.9) whereas *P. acidilactici* fermented lactose more slowly (Table 12.9).

12.7.2 Development of colony forming units (c.f.u.) and pH reduction in hydrolyzates containing molasses

We purposely used initial inoculum levels of 1.4×10^4 total cells/ml in hydrolyzates to ensure that differences in initial fermentation rates would be clearly discernable between the four inocula used. The use of Stabisil as inoculum also involved initial total c.f.u. of 1.4×10^4/ml of

Table 12.9 Acid production from sugars in MRS broth by *S. faecium*, *P. acidilactici* and *L. plantarum*

		pH					
		S. faecium		P. acidilactici		L. plantarum	
Sugar	*Blanks*[a]	*24 h*	*72 h*	*24 h*	*72 h*	*24 h*	*72 h*
Control[b]	–[c]	6.7	6.5	6.7	6.4	6.4	6.4
Glucose	6.3	4.6	4.3	4.2	3.7	4.0	3.8
Fructose	6.2	4.7	4.2	4.3	3.7	4.5	4.0
Galactose	6.1	5.5	4.6	4.3	3.8	5.0	4.0
Lactose	6.5	4.9	4.7	5.1	4.4	4.2	3.8
Sucrose	6.7	4.7	4.5	6.4	6.1	3.9	3.8
Maltose	6.5	4.8	4.5	4.3	3.9	4.2	3.8

[a]Blank tubes were uninoculated.
[b]Control tubes were prepared without a carbohydrate and inoculated.
[c]No tube.

hydrolyzates (composed of 50% *P. acidilactici*, 30% *S. faecium* and 20% *L. plantarum*). The use of 6.5% molasses (Figure 12.1a) resulted in the highest level of c.f.u. achieved at the cessation of growth by *S. faecium* (1.3×10^9/ml), *L. plantarum* was next (4.0×10^8/ml), while *P. acidilactici* yielded a significantly lower level of c.f.u. (1.6×10^8/ml). *L. plantarum* decreased the pH at a significantly greater rate than did *P. acidilactici*, *S. faecium* or Stabisil (Figure 12.1a and Table 12.10). A pH value of 4.5 was attained with *L. plantarum* in 11.5 h and a final pH of 4.0 only with *L. plantarum* (Figure 12.1a and Table 12.10). *S. faecium* resulted in the highest pH value (4.3) after 72 h of incubation (Figure 12.1a and Table 12.10) most probably reflecting less tolerance to low pH values than *L. plantarum* and *P. acidilactici*.

The use of 7.5% molasses (Figure 12.1b) resulted again in *P. acidilactici* yielding the lowest density of c.f.u. (4.0×10^8/ml) compared with *L. plantarum* (4.0×10^9/ml) and *S. faecium* (2.5×10^9/ml). A pH value of 4.5 (10.5 h) was achieved most rapidly by *L. plantarum* while *P. acidilactici* required 15 h to reduce the pH to 4.5 (Figure 12.1b and Table 12.10). *S. faecium* required the longest period of incubation to reduce the pH to 4.5 and was unable to reduce the pH to 4.0 (Figure 12.1b). The time for Stabisil to reduce the pH to 4.5 was 13 h (Figure 12.1b and Table 12.10). Final pH values of pH 4.0 or slightly below were attained by *L. plantarum*, *P. acidilactici* and Stabisil (Figure 12.1b and Table 12.10).

The use of 8.5% molasses (Figure 12.1c) resulted once more in *P. acidilactici* yielding the lowest density of c.f.u./ml (2.7×10^8/ml) com-

Table 12.10 Fermentation of cod gurry hydrolyzates containing 6.5, 7.5 and 8.5% molasses

Culture	*Hours to reach pH 4.5*	*Final pH*[a]
6.5% molasses		
L. plantarum	11.5	4.0
P. acidilactici	17.5	4.1
S. faecium	23.5	4.3
Stabisil	17.5	4.1
7.5% molasses		
L. plantarum	10.5	3.9
P. acidilactici	15.0	4.0
S. faecium	18.0	4.2
Stabisil	13.0	4.0
8.5 molasses		
L. plantarum	10.0	3.9
P. acidilactici	11.5	4.0
S. faecium	15.0	4.2
Stabisil	12.0	3.9

[a]Obtained after 72 h incubation.

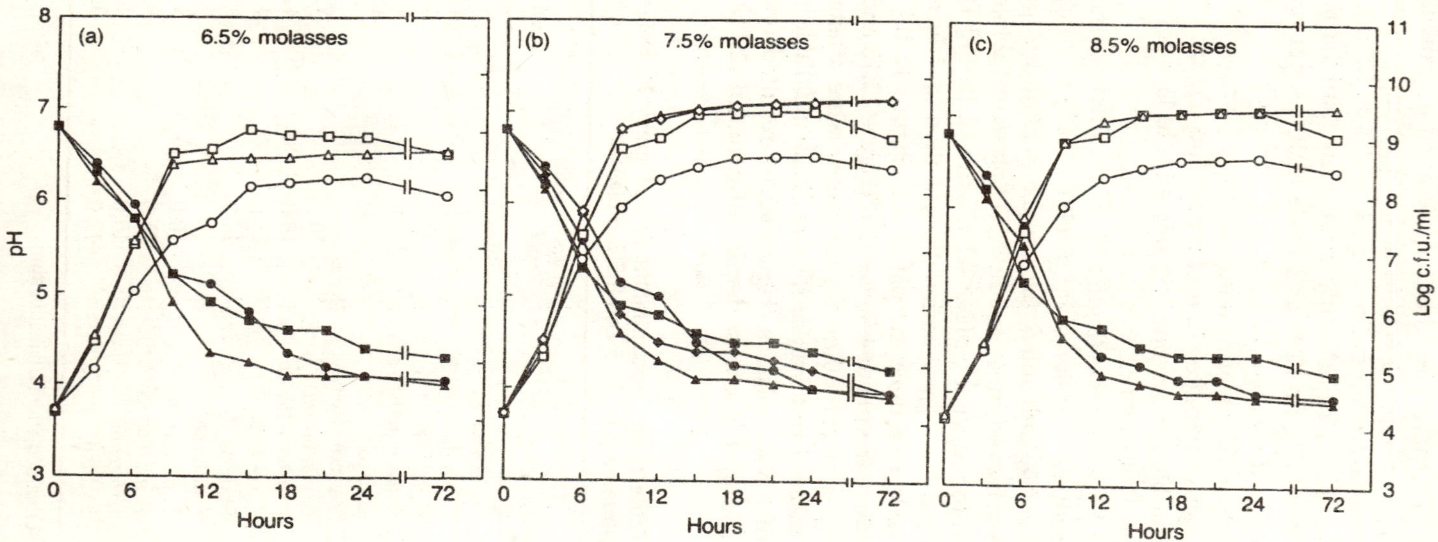

Figure 12.1 Rates of c.f.u. development and pH decrease by *S. faecium*, *P. acidilactici* and *L. plantarum* in cod hydrolysates containing 6.5, 7.5 and 8.5% blackstrap molasses. The c.f.u. values with Stabisil are the total c.f.u. derived from the mixed culture. pH: (■) *S. faecium*, (●) *P. acidilactici*, (▲) *L. plantarum*, (◆) Stabisil, c.f.u.: (□) *S. faecium*, (○) *P. acidilactici*, (△) *L. plantarum*, (◇) Stabisil.

pared to *S. faecium* (1.1×10^9/ml) and *L. plantarum* (3.5×10^9/ml). A pH value of 4.5 was again attained most rapidly by *L. plantarum* in 10 h (Figure 12.1c and Table 12.10). A final pH value of 3.9 resulted with *L. plantarum* and Stabisil while *P. acidilactici* yielded a final pH of 4.0 (Figure 12.1c and Table 12.10). It should be noted that *S. faecium* again was unable to reduce the pH to 4.0 and resulted in a final pH of 4.2 after 72 h of incubation (Figure 12.1c and Table 12.10).

12.7.3 Utilization of carbohydrates in hydrolyzates containing molasses

All three microorganisms and Stabisil utilized glucose in preference to fructose in the initial stages of the fermentations and failed throughout the fermentations to utilize sucrose significantly when molasses was the carbohydrate source (Figure. 12.2–12.5 and Table 12.11). This observation is of economic importance since sucrose constituted 42.6% of the total available carbohydrate present (sucrose, glucose and fructose) in the molasses.

S. faecium utilized fructose throughout the course of the fermentations but at significantly lower rates than glucose with all three concentrations of molasses (Figure 12.2). After 21 h, however, the rate of fructose

Table 12.11 Percent of initial carbohydrates in molasses remaining unfermented in gurry hydrolyzates after 72 h incubation

	Initial concentration of molasses		
Culture	*6.5%*	*7.5%*	*8.5%*
S. faecium			
glucose	16.8	10.8	4.0
fructose	61.4	65.6	34.7
sucrose	95.1	96.5	97.6
P. acidilactici			
glucose	12.4	4.8	3.5
fructose	72.9	81.7	24.3
sucrose	93.5	96.3	98.2
L. plantarum			
glucose	6.0	3.3	2.4
fructose	73.4	50.2	19.3
sucrose	100	93.6	93.3
Stabisil			
glucose	–[a]	3.9	–[a]
fructose	–[a]	52.6	–[a]
sucrose	–[a]	91.6	–[a]

[a]Not performed, no data.

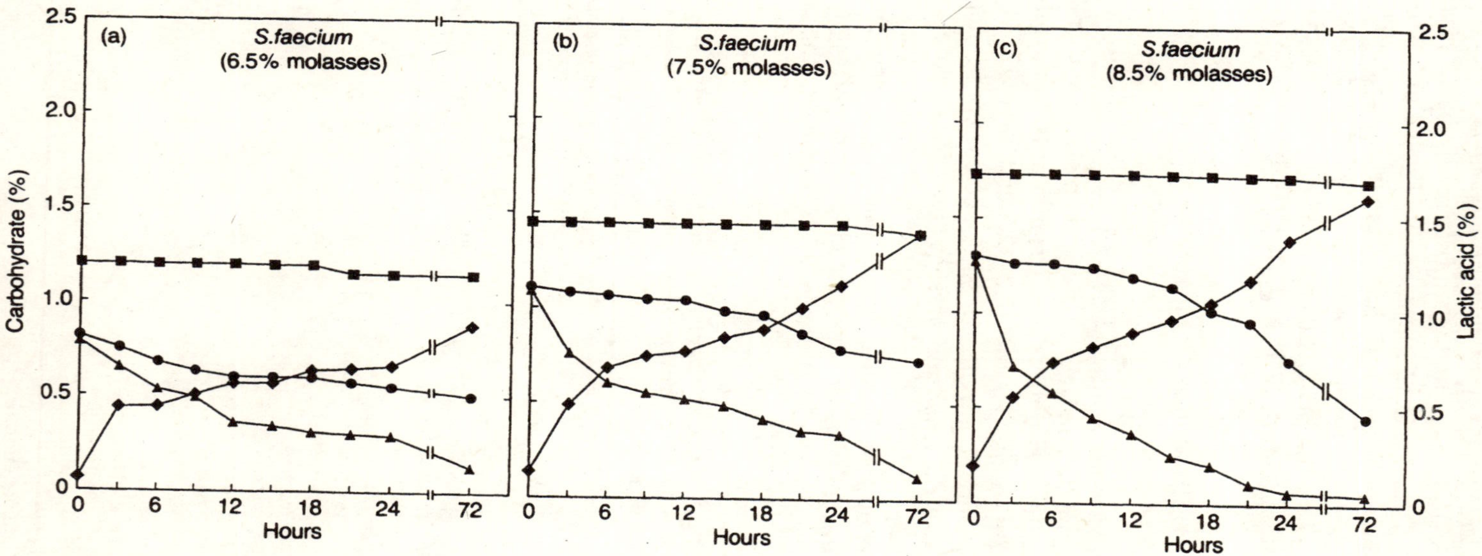

Figure 12.2 Rates of carbohydrate utilization and lactic acid production by *S. faecium* in cod hydrolysates containing 6.5, 7.5 and 8.5% blackstrap molasses. (■) sucrose, (●) fructose, (▲) glucose, (◆) lactic acid.

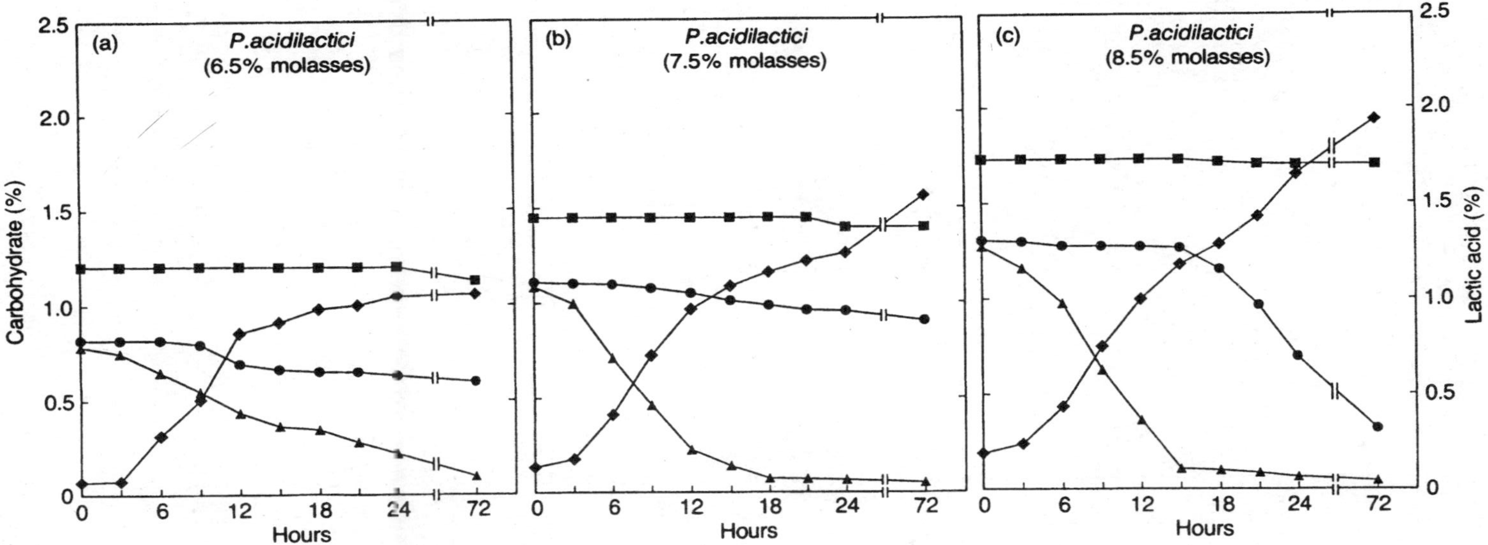

Figure 12.3 Rates of carbohydrate utilization and lactic acid production by *P. acidilactici* in cod hydrolyzates containing 6.5, 7.5 and 8.5% blackstrap molasses. See legend to Figure 12.2.

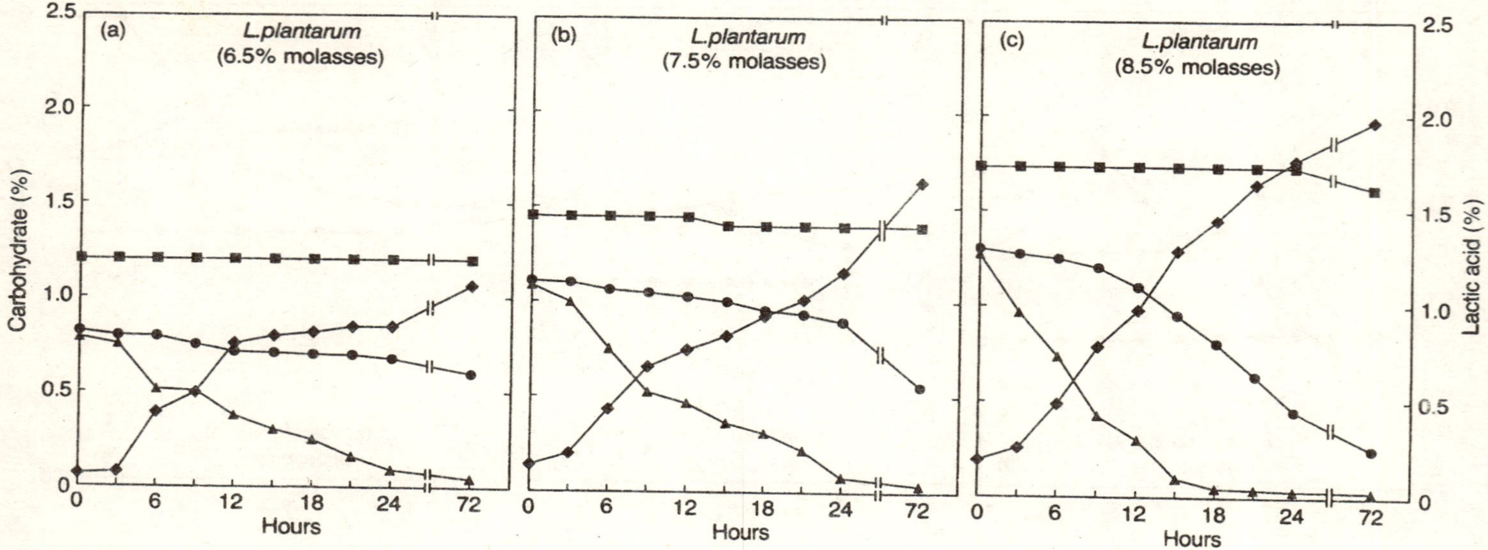

Figure 12.4 Rates of carbohydrate utilization and lactic acid production by *L. plantarum* in cod hydrolyzates containing 6.5, 7.5 and 8.5% blackstrap molasses. See legend to Figure 12.2.

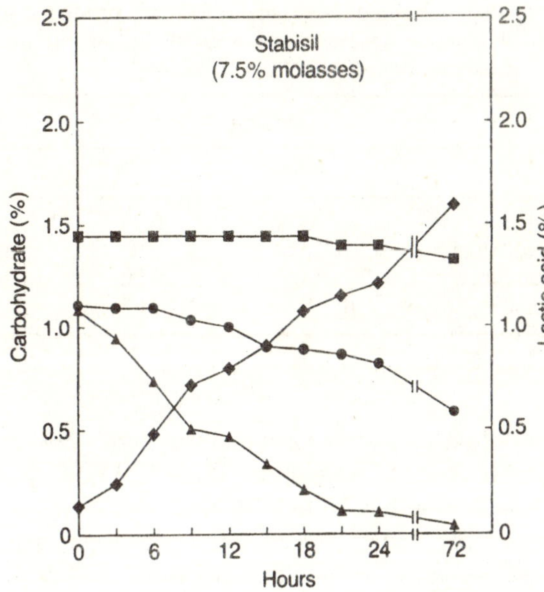

Figure 12.5 Rates of carbohydrate utilization and lactic acid production by Stabisil in cod hydrolyzates containing 7.5% blackstrap molasses. See legend to Figure 12.2.

utilization was significantly greater with 8.5% molasses than with either 6.5 or 7.5% molasses (Figure 12.2a–c). With all three concentrations of molasses, *S. faecium* utilized less glucose than did *P. acidilactici* or *L. plantarum* (Table 12.11).

P. acidilactici also utilized fructose at significantly lower rates than glucose with 6.5 and 7.5% molasses (Figure 12.3a and b). However, with 8.5% molasses (Figure 12.3c) the rate of fructose utilization after 18 h was similar to the maximum rate of glucose utilization. This result may be related to a significantly larger ratio of fructose to glucose at 18 h of incubation compared with the initial ratio.

L. plantrum utilized fructose at a notably lower rate than glucose throughout the course of the fermentations with 6.5 and 7.5% molasses (Figure 12.4a and b). After 9 h of incubation with 8.5% molasses, the rate of fructose utilization approached the maximum rate of glucose utilization (Figure 12.4c). *L. plantarum* exhibited the highest levels of glucose utilization (Table 12.11).

The fermentation profile with Stabisil and 7.5% molasses (Figure 12.5) resembled that of *L. plantarum* (Figure 12.4b) more closely than that of *S. faecium* (Figure 12.2b) or *P. acidilactici* (Figure 12.3b), particularly with respect to the rate of fructose utilization. This observation may reflect the dominance of *L. plantarum*, particularly towards the end (72 h) of the fermentation due to a greater tolerance of low pH values. The levels

Table 12.12 Ratio of lactic acid (g/100 ml) produced to the sum of glucose, fructose and sucrose (g/100 ml) utilized from molasses in cod gurry hydrolyzates[a]

	Molasses		
Culture	*6.5%*	*7.5%*	*8.5%*
S. faecium	0.786	0.95	0.67
p. acidilactici	1.00	0.90	0.79
L. plantarum	1.06	0.90	0.74
Stabisil		0.85	–

[a]See legend to Table 12.11.

of residual carbohydrates after fermentation with 7.5% molasses were nearly identical to that of *L.plantarum* (Table 12.11).

It is of interest that with 6.5 and 8.5% molasses, *S. faecium* yielded the lowest ratios for lactic acid produced to the sum of glucose, fructose and sucrose utilized, whereas with 7.5% molasses, *S. faecium* yielded the highest ratio (Table 12.12). Ratios of 1.0 and slightly higher may possibly be due to the presence in the hydrolyzates of alanine that has undergone reductive deamination to lactic acid.

12.7.4 Lactic acid production in hydrolyzates containing molasses

S. faecium (Figure 12.2) produced significantly lower levels of lactic acid than did *P. acidilactici* at the end of 72 h (Figure 12.3) and *L. plantarum* (Figure 12.4) with all three concentrations of molasses. The highest level of lactic acid produced by *S. faecium* (1.61%) was with 8.5% molasses (Figure 12.2). The final levels of lactic acid produced by *P. acidilactici* (Figure 12.3) with all three concentrations of molasses closely approached the final levels produced by *L. plantarum* with the respective concentrations of molasses (Figure 12.4). The highest level of lactic acid produced (1.97%) was achieved by *L. plantarum* with 8.5% molasses (Figure 12.4c). The profile of lactic acid production by Stabisil with 7.5% molasses (Figure 12.5) resembled that of *L. plantarum* (Figure 12.4b) more closely than that of *S. faecium* (Figure 12.2b) or *P. acidilactici* (Figure 12.3b). The final level of 1.59% lactic acid obtained with Stabisil and 7.5% molasses (Figure 12.5) was similar to the final level of 1.63% lactic acid obtained with *L. plantarum* and 7.5% molasses (Figure 12.4b). These results correlate closely with the final pH values attained after 72 h of incubation (Figure 12.1b and Table 12.10). The ratios of lactic acid produced to the sums of glucose, fructose and sucrose utilized indicate that both *P. acidlactici* and *L. plantarum* yielded maximum ratios with 6.5% molasses (Table 12.12). All three organisms yielded significantly reduced ratios with 8.5% molasses. Ratios of 1.0 and slightly higher may possibly be due to tissue components such as alanine reductively deaminated to lactic acid.

12.7.5 Development of c.f.u. and rates of pH reduction in hydrolyzates containing corn syrup

The use of 3.0% corn syrup failed to yield a pH of 4.0 with all four of the inocula (Table 12.13). *L. plantarum,* however, resulted in a final pH of 4.1 after 72 h of incubation (Table 12.13). The next lowest pH value after 72 h of incubation (pH 4.5), which we consider to be marginal, was achieved with Stabisil (Table 12.13). *S. faecium* and *P. acidilactici* resulted in final pH values above 4.5. We therefore considered the fermentation with 3.0% corn syrup borderline or unreliable with respect to the rate of pH reduction and final pH values achieved and did not investigate fermentations with 3.0% corn syrup further.

Corn syrup at a level of 3.5% resulted in the highest level of c.f.u. achieved at the cessation of growth by *S. faecium* (1.0×10^9/ml; Figure

Table 12.13 Fermentation of corn syrup in cod gurry hydrolyzates[a]

Culture	*Hours to reach pH 4.5*	*Final pH*[b]
Corn syrup (3.0%)		
L. plantarum	14	4.1
P. acidilactici	–[c]	4.7
S. faecium	–[c]	4.8
Stabisil	48	4.5
Corn syrup (3.5%)		
L. plantarum	11	3.9
P. acidilactici	19	4.1
S. faecium	21	4.3
Stabisil	15	4.0
Corn syrup (4.0%)		
L. plantarum	7	3.8
P. acidilactici	15	3.9
S. faecium	14	4.2
Stabisil	15	3.9
Corn syrup (4.5%)		
L. plantarum	7	3.6
P. acidilactici	15	3.8
S. faecium	12	4.2
Stabisil	9	3.9

[a]pH readings were taken ever 3 h for the first 24 h of incubation and at 24 h intervals thereafter.
[b]Obtained after 72 h incubation.
[c]pH 4.5 not attained.

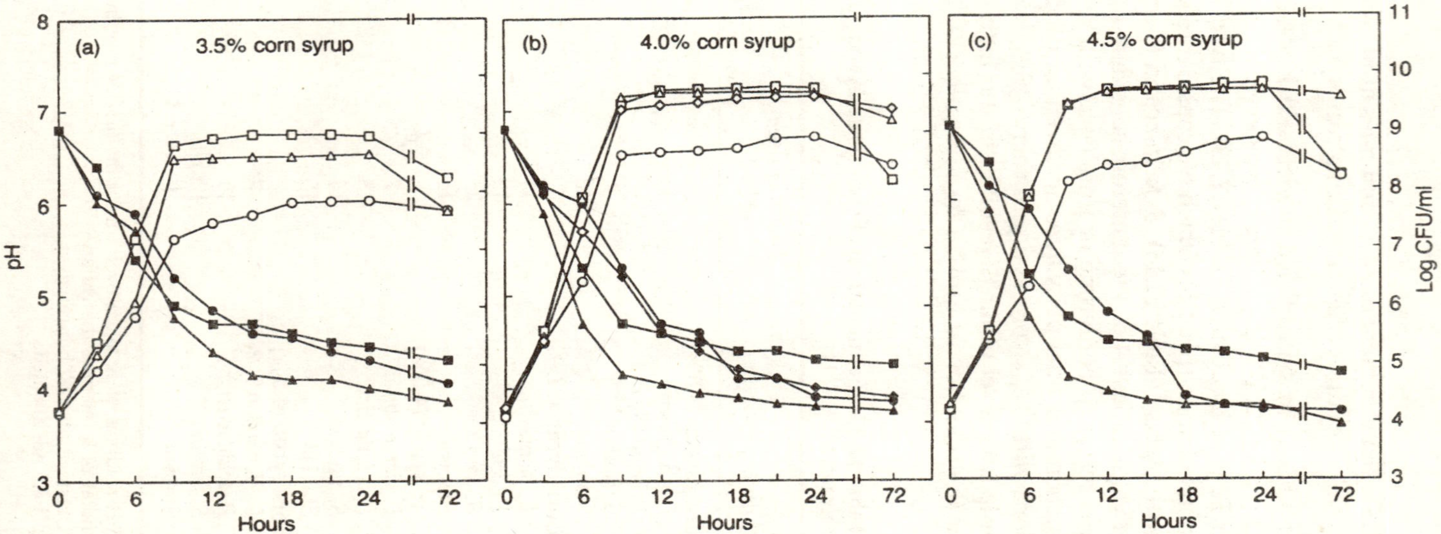

Figure 12.6 Rates of c.f.u. development and pH decrease by *S. faecium*, *P. acidilactici* and *L. plantarum* in cod hydrolysates containing 3.5, 4.0 and 4.5% corn syrup. The c.f.u. values with Stabisil are the total c.f.u. derived from the mixed culture. pH: (■) *S. faecium*, (●) *P. acidilactici*, (▲) *L. plantarum*, (◆) Stabisil; log c.f.u.: (□) *S. faecium*, (○) *P. acidilactici*, (△) *L. plantarum*, (◇) Stabisil.

12.6a). The use of *L. plantarum* resulted in a pH of 4.5 attained most rapidly (11 h) among the inocula used and a final pH of 3.9 after 72 h (Table 12.13). Both *P. acidilactici* and *S. faecium* failed to attain a pH value of 4.0. Stabisil resulted in the reduction of the pH to 4.5 in 15 h and a final pH of 4.0 (Figure 12.6a and Table 12.13).

The use of 4.0% corn syrup (Figure 12.6b) resulted again in *P. acidilactici* attaining the lowest density of c.f.u. compared with *L. plantarum* and *S. faecium*. A pH of 4.5 was achieved most rapidly with *L. plantarum* in 7 h whereas *P. acidilactici*, *S. faecium* and Stabisil required approximately twice the incubation time to reduce the pH to 4.5 (Figure 12.6b and Table 12.13). Final pH values of 3.9 to 4.2 were achieved with the remaining three inocula (Figure 12.6b and Table 12.13).

The use of 4.5% corn syrup offered little advantage over 4.0% corn syrup with respect to *L. plantarum* and *P. acidilactici* (Figure 12.6b and c and Table 12.13) although the final pH with *L. plantarum* was 3.6 (Figure 12.6c and Table 12.1). *S. faecium* again failed to reduce the pH to 4.0. Stabisil reduced the pH to 4.5 in 9 h and again resulted in a final pH of 3.9 (Table 12.13).

12.7.6 Utilization of carbohydrates in hydrolyzates containing corn syrup

S. faecium did not utilize maltose until after 6–9 h of incubation (Figure 12.7a–c) and failed to utilize maltotriose significantly throughout the fermentations. With 3.5, 4.0 and 4.5% corn syrup, significant levels of glucose, maltose and maltotriose remained unfermented after 72 h of incubation (Table 12.14).

P. acidilactici utilized maltotriose only slightly throughout the fermentations (Figure 12.8a–c). With 3.5% corn syrup maltose was not utilized until after the first 6 h of incubation, whereas with 4.0 and 4.5% corn syrup the initial rates of maltose utilization closely approximated those for glucose utilization (Figure 12.8a–c). With 3.5, 4.0 and 4.5% corn syrup, significant levels of glucose, maltose and maltotriose remained unfermented after 72 h of incubation (Table 12.14).

L. plantarum utilized only glucose to a notable extent during the first 24 h of incubation with 3.5% corn syrup and thereafter significantly utilized maltose while maltotriose was only slightly utilized (Figure 12.9a). At the end of 72 h of incubation, only 13% of the original glucose remained unfermented, while notably higher percentages of the initial maltose and maltotriose remained unfermented (Table 12.14). The use of 4.0% corn syrup resulted in *L. plantarum* utilizing maltose throughout the fermentation but at a notably lower rate than glucose, with only 10% of the original glucose remaining unfermented after 72 h of incubation, while notably higher percentages of the initial maltose and maltotriose remained unfermented (Table 12.14). The general profile of carbohy-

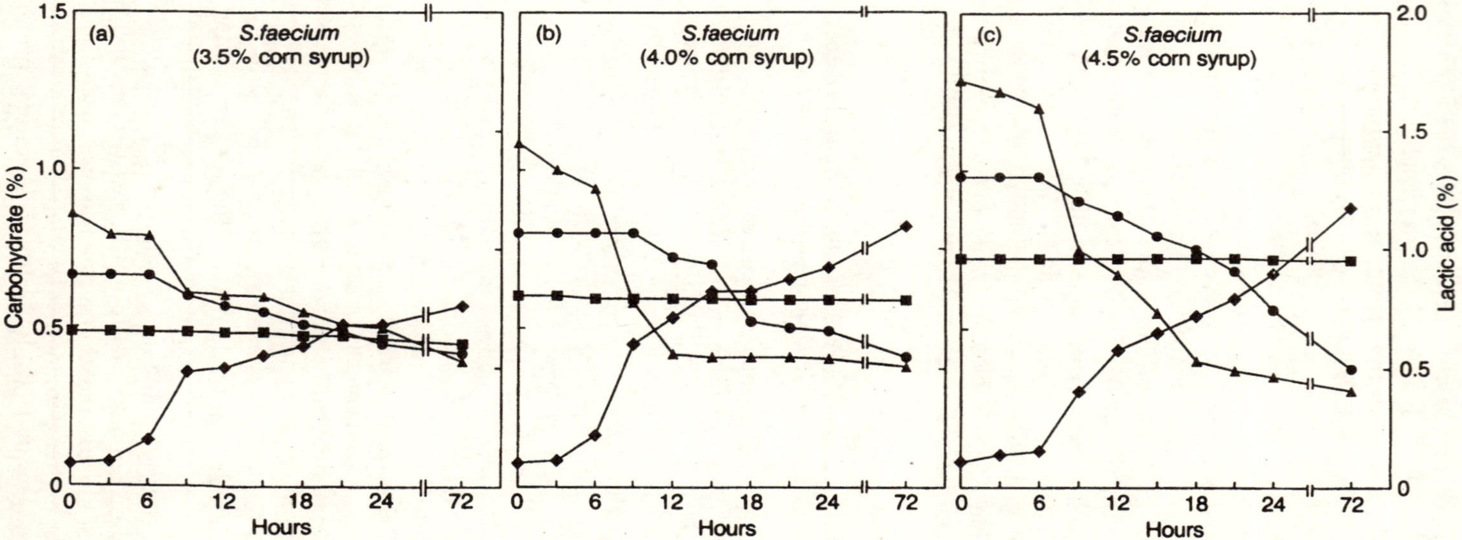

Figure 12.7 Rates of carbohydrate utilization and lactic acid production by *S. faecium* in cod hydrolyzates containing 3.5, 4.0 and 4.5% corn syrup. (■) maltotriose, (●) maltose, (▲) glucose, (◆) lactic acid.

Table 12.14 Percent of initial carbohydrates from corn syrup in cod gurry hydrolyzates remaining unfermented after 72 h incubation[a]

	Initial concentration of corn syrup		
Culture	*3.5%*	*4.0%*	*4.5%*
S. faecium			
glucose	46	35	24
maltose	63	51	38
maltotriose	93	97	97
P. acidilactici			
glucose	27	19	15
maltose	45	31	26
maltotriose	88	93	92
L. plantarum			
glucose	13	10	7.2
maltose	28	31	36
maltotriose	82	65	64
Stabisil			
glucose	–	19	–
maltose	–	33	–
maltotriose	–	68	–

[a]See legend to Table 12.11.

drate utilization with 4.5% corn syrup resembled that with 4.0% corn syrup with 7.2, 36 and 64%, respectively, of the initial glucose, maltose and maltotriose remaining unfermented after 72 h of incubation (Figure 12.9c and Table 12.14). It is of interest that a significantly higher percent of glucose was utilized by *L. planarum* with 3.5, 4.0 and 4.5% corn syrup than by either of the other three inocula (Table 12.14). The ratios for lactic acid produced to the sums of glucose, maltose and maltotriose utilized indicate that *L. plantarum* consistently yielded the highest conversion efficiencies and that *S. faecium* consistently yielded the lowest (Table 12.15).

Table 12.15 Ratio of lactic acid (g/100 ml) produced to the sum of glucose, maltose, and maltotriose (g/100 ml) utilized from corn syrup in hydrolyzed cod gurry[a]

	Corn syrup		
Culture	*3.5%*	*4.0%*	*4.5%*
S. faecium	0.62	0.69	0.58
P. acidilactici	0.88	0.80	0.66
L. plantarum	1.00	0.93	0.86
Stabisil	–	0.71	–

[a]See legend to Table 12.11.

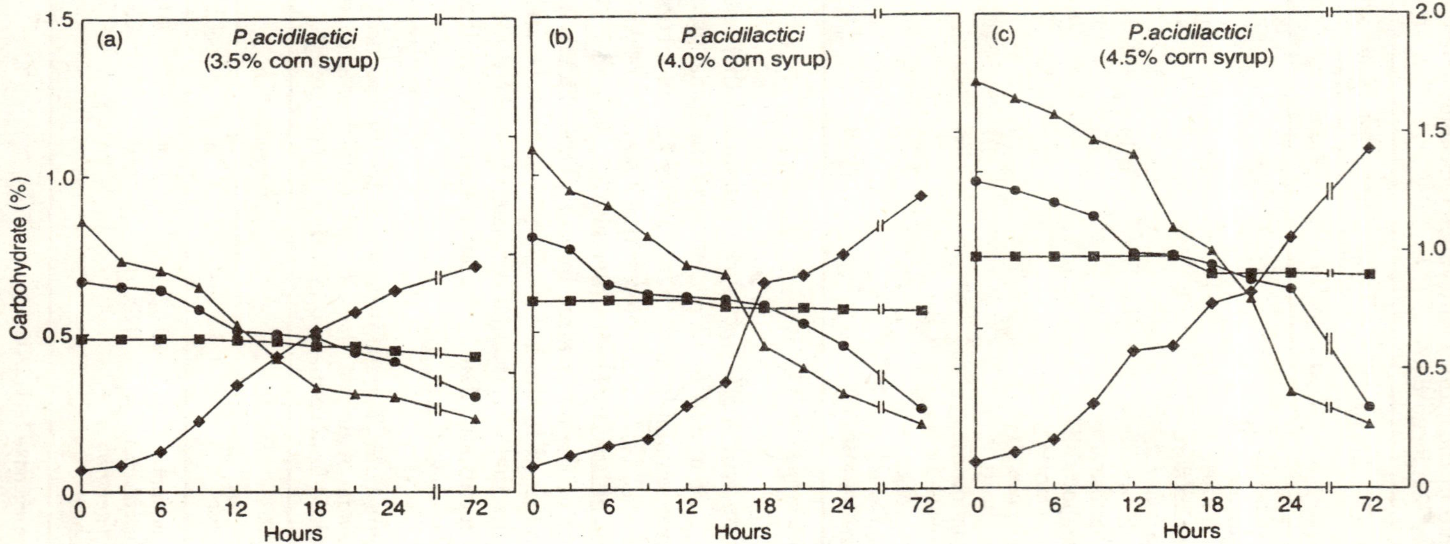

Figure 12.8 Rates of carbohydrate utilization and lactic acid production by *P. acidilactici* in cod hydrolyzates containing 3.5, 4.0 and 4.5% corn syrup. See legend to Figure 12.7.

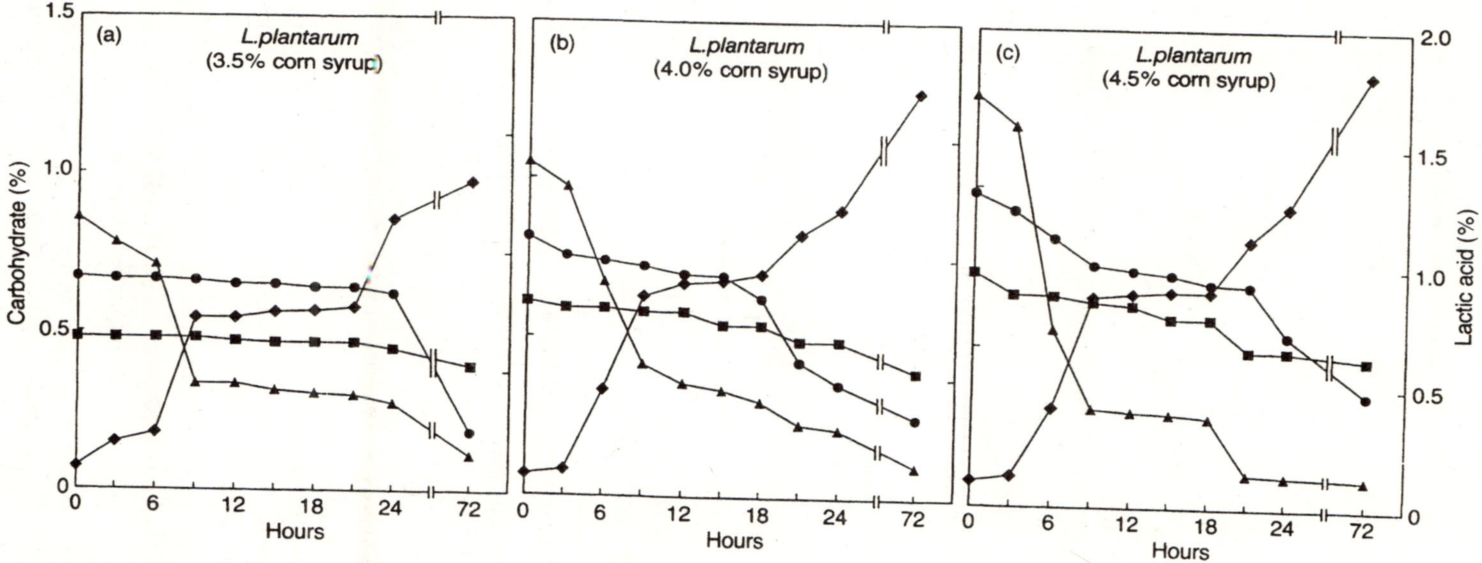

Figure 12.9 Rates of carbohydrate utilization and lactic acid production by *L. plantarum* in cod hydrolyzates containing 3.5, 4.0 and 4.5% corn syrup. See legend to Figure 12.7.

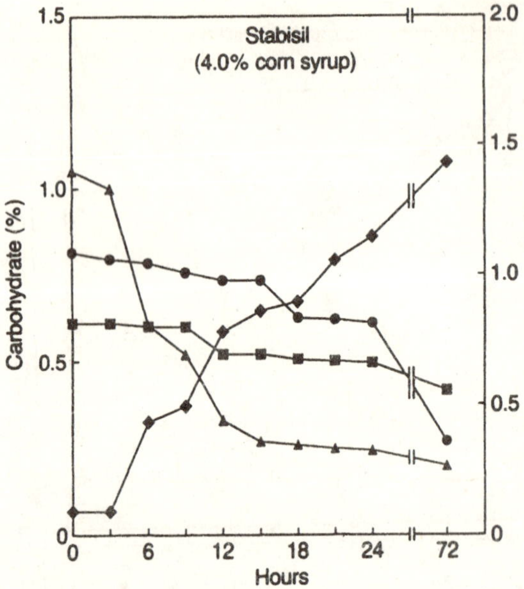

Figure 12.10 Rates of carbohydrate utilization and lactic acid production by Stabisil in cod hydrolyzates containing 4.0% corn syrup. See legend to Figure 12.7.

The profile of carbohydrate utilization for Stabisil with 4.0% corn syrup resembled that for *L. plantarum* with 4.0% corn syrup (Figures 12.10 and 12.9b) except for a significantly higher level (19%) of unfermented glucose after 72 h of incubation.

12.7.7 Lactic acid production in hydrolyzates containing corn syrup

S. faecium produced the lowest levels of lactic acid from 3.5, 4.0 and 4.5% corn syrup after 72 h of incubation (0.76, 1.1 and 1.2%, respectively; Figures 12.7–12.9) while *L. plantarum* produced the highest levels of lactic acid (1.3, 1.7 and 1.8%, respectively). *P. acidilactici* produced intermediate levels of lactic acid while Stabisil produced 1.4% lactic acid from 4.0% corn syrup (Figure 12.10) which was slightly higher than the 1.2% lactic acid produced by *P. acidilactici* with 4.0% corn syrup. These results closely correlate with the final pH values attained after 72 h of incubation (Table 12.13).

12.7.8 Propionic acid production and pH reduction in hydrolyzates containing glucose

An initial inoculum of 1×10^6 cells/ml of *P. acidilactici* was found to be optimum for the propionic acid fermentation of hydrolyzed cod gurry

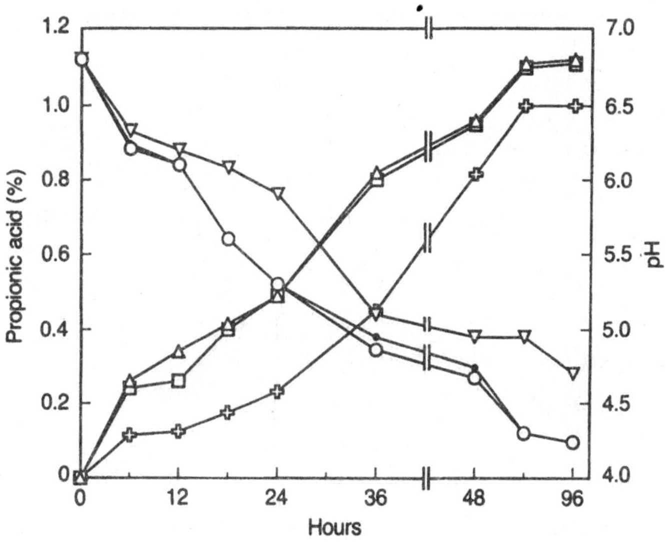

Figure 12.11 Rates of propionic acid production and pH reduction in cod hydrolyzates with varying initial levels of *P. acidipropionici*. Propionic acid: (✣) 1 × 10^5 c.f.u./ml, (□) 1 × 10^6 c.f.u./ml, (△) 1 × 10^7 c.f.u./ml; pH: (▽) 1 × 10^5 c.f.u./ml, (●) 1 × 10^6 c.f.u./ml, (○) 1 × 10^7 c.f.u./ml.

containing 3.5% added glucose and incubated at 30°C (Figure 12.11). When cod hydrolyzates containing 2.0, 3.0 and 4.0% added glucose were inoculated with *P. acidipropionici* the resulting fermentations and rates at which the pH declined were notably slower than with the lactic acid fermentations (Figure 12.12). With 2.0% glucose the fermentation was more rapid than with 3.0 and 4.0% glucose but the final yield of propionic acid was 0.94% compared with 1.1% with 3.0 and 4.0% glucose. In addition, 2.0% glucose resulted in a pH of 4.7 after 244 h and then rose to a pH of 5.5 after 488 h (data not shown) whereas with 3.0 and 4.0% glucose the final pH values were both 4.5 which remained stable. Bacterial surface plate counts on Tryptic soy Agar with 0.5% glucose indicated that with 3.5% glucose and an inoculum of 1 × 10^7 c.f.u./ml unidentified non-propionic acid bacteria had developed from 50/ml at zero time to 6.3 × 10^7/ml after 48 h of incubation and decreased thereafter (data not given). The number of c.f.u. of *P. acidipropionici* after 48 h of incubation was 7.2 × 10^9/ml. The development of these contaminants is presumably due to the characteristically slow rate of growth of *P. acidipropionici*, resulting in slow rates of propionic acid production and pH reduction. Although organoleptic spoilage was not encountered with any of the glucose concentrations used, the instability of the pH at the end of the fermentation with 2.0% glucose indicated that 3.0% glucose was optimum.

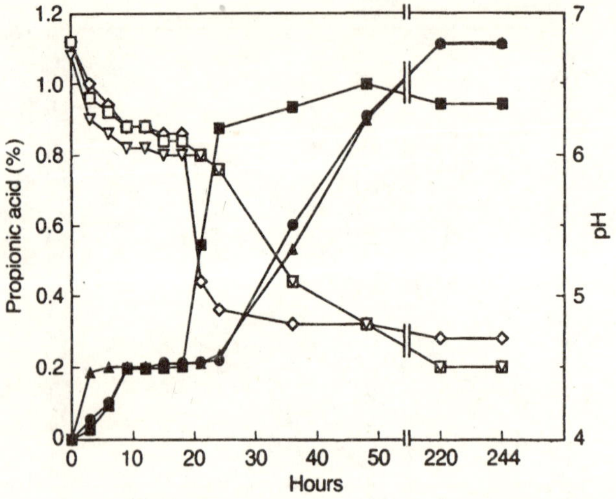

Figure 12.12 Rates of propionic acid production and pH decrease in cod hydrolyzates with varying gucose concentrations. Propionic acid: (■) 2.0% glucose, (●) 3.0% glucose, (▲) 4.0% glucose; pH: ◇) 2.0% glucose, (□) 3.0% glucose, (▽) 4.0% glucose.

12.8 DISCUSSION

The reduction of pH to 4.0 within 72 h was chosen as the criterion of a succcessful lactic aicd fermentation. A rapid lactic fermentation is particularly important in order to avoid malfermentations resulting from the ascendance of contaminating microorganisms in the non-sterile but pasteurized hydrolyzates. Slow lactic acid fermentations or fermentations that fail to result in a rapid reduction of the pH significantly below 5.0 can be expected to result in the development of spoilage organisms and putrefactive odors.

A molasses concentration of 7.5% was found to be optimum with the lactic acid fermentation. The use of *S. faecium* with molasses is questionable with respect to its inability to achieve a pH below 4.2 and more importantly because of its significantly lower rate of pH reduction compared to that achieved with *P. acidilactici* and *L. plantarum*. Among the three microorganisms studies, *L. plantarum* exhibited the most rapid rate of pH reduction with molasses and consistently reduced the pH to 4.0 with all three concentrations of molasses used. *L. plantarum* is therefore the organism of preference in the fermentation of fish hydrolyzates containing blackstrap molasses as a carbohydrate source.

All three of the microorganisms studied failed to utilize sucrose when molasses was added to cod hydrolyzates. *P. acidilactici* was unable to reduce the pH signficantly with sucrose as the sole carbohydrate in MRS broth while *S. faecium* and *L. plantarum* notably reduced the pH in 24 h.

These observations suggest that the small residual levels of glucose present after 24 h of incubation with cod hydrolyzates containing molasses were sufficient to repress the utilization of sucrose by *S. faecium* and *L. plantarum*. Glucose is recognized as an efficient repressor of the synthesis of numerous bacterial enzymes and may function to repress invertase synthesis by *S. faecium* and *L. plantarum*, even at notably low glucose concentrations. The use of such wild-type microorganisms therefore detracts significantly from the effective cost and efficiency of the fermentation since sucrose is usually the most abundant carbohydrate in blackstrap molasses. The employment of mutants of *L. plantarum* derepressed for glucose repression of invertase synthesis should theoretically allow a reduction in the molasses required from 7.5 to about 4.3% so as to still achieve a successful fermentation with a final pH of 4.0.

With corn syrup, *L. plantarum* consistently reduced the pH to 4.5 in notably less time than *P. acidilacticici* and *S. faecium*. *L. plantarum* also resulted in the lowest pH values at the end of 72 h, and yielded a consistently higher conversion of the lower carbohydrates in corn syrup to lactic acid. These results indicate that among the three microorganisms studied, *L. plantarum* is the organism of choice with corn syrup. The optimum corn syrup concentration with *L. plantarum* was 4.0%.

All three organisms failed to utilize maltose as rapidly as glucose in the initial stages of the fermentations with corn syrup, although all three organisms rapidly fermented maltose as a sole carbohydrate source (Table 12.9). This may be due to partial repression of the formation of an intracellular maltose phosphorylase (maltase) by high levels of glucose or to a greater affinity of glucose permease for glucose than for maltose, assuming that a common permease is involved in transporting both carbohydrates into the cell.

Maltotriose consituted 13.9% of the corn syrup (w/w) or 24.2% of the fermentable carbohydrate. The ability of *L. pantarum* to utilize maltotriose although at a notably slow rate in contrast to the failure of *S. faecium* and *P. acidilactici* to significantly utilize maltotriose, may have contributed to the consistently lower pH values achieved by *L. plantarum* with each of the corn syrup concentrations used. The use of strains of *L. plantarum* capable of utilizing maltotriose more efficiently in the presence of glucose should allow a rapid and successful fermentation resulting in a final pH of 4.0 with 3.0–3.5% corn syrup, having the same content of lower saccharides as the corn syrup used in this study.

The presence of several bacterial species in commercial starter cultures is made use of in anticipation that one or two of the strains may be attacked by bacteriophage under conditions of industrial use. The use of a mixture of several strains of *L. plantarum*, each having a different pattern of phage susceptibility or mutated to loss of phage receptors may be of considerable advantage over the presence of *S. faecium* and *P. acidilactici* in a mixed culture containing *L. plantarum* because of the more rapid fermentation achieved with *L. plantarum* with both molasses and corn syrup and its ability to utilize maltotriose.

The propionic acid fermentation by itself was found to allow the development of contaminating organisms in hydrolyzates. This was most probably due to the characteristically slow development of *P. acidilactici*. Co-cultures of *L. acidophilus* and *P. shermanii* were found to have a shorter fermentation time and higher propionic acid yields than the propionibacterium grown alone (Liu and Moon, 1982). An ideal co-fermentation of hydrolyzed fish tissue would result in a pH of 4.0 or below due to lactic acid with a resulting propionic acid content of 0.2% for the purpose of edible acceptance if the fermented hydrolysate is to be used as a fish or animal feed adjunct.

12.9 COMMERCIAL SOURCES OF LACTIC ACID BACTERIA, PROPIONIC ACID BACTERIA AND PAPAIN

Lactic acid and propionic acid bacteria

American Type Culture Collection
12301 Parklawn Drive
Suite 210
Rockville, MD 20852, USA

Chr. Hansen's Laboratory
9015 W. Maple Street
Milwauki,
WI 53124, USA

Vivolac Cultures Corporation
3862 E. Washington
Indianapolis, IN 46201, USA

Integrated Ingredients
1420 Barbor Bay Parkway
Suite 210
Alameda, CA 94501, USA

Burlington Biomedical Corp.
222 Sherwood Avenue
Farmingdale, NY 11735, USA

Medifarm USA
10215 Denis Drive
Des Moines, IA 50322, USA

Quest International
Flavors and Ingredients Co.,
Bioproducts
1833 57th Street
Sarasota, FL 34243, USA

Sanofi Bio-Industries, Inc.
620 Progress Avenue
PO Box 1609
Waukesha, WI 53187, USA

Papain

Amano International Enzyme Co.
PO Box 1000
Troy, VA 22974, USA

Biocon (US) Inc.
518 Codell Drive
Lexington, KY 40509, USA

Papain

Rohm Tech, Inc.
195 Canal Street
Malden, MA 02148, USA

Miles Laboratories, Inc.
PO Box 932
Elkhart IN 46515, USA

Novo Laboratories
59 Danbury Road
Wilton, CT 06897, USA

REFERENCES

Anderson, D.G. and McKay, L.I. (1984) *In vivo* cloning of *lac* genes in *Streptococcus lactis* ML3. *Appl. Environ. Microbiol.*, **47**, 245–49.

Burgess-Cassler, A. and Imam, S. (1991) Partial purification and comparative characterization of α-amylase secreted by *Lactobacillus amylovorus*. *Current Microbiol.*, **23**, 207–13.

Clean Water Act (1991) *40 Code of Federal Regulations, Part 220*, Office of the Federal Register, National Archives and Records Administration, Washington, DC, pp. 147–9.

Cummins, C. and Johnson, J.L. (1981) The genus *Propionibacterium*, in *The Procaryotes. A Handbook on Habitats, Isolation, and Identification of Bacteria*, (eds M.P. Starr, H. Stolp, H.G. Trüper, A. Balows and H.G. Schlegel), Springer-Verlag, New York, Vol. II, pp. 1894–902.

Cummins, C.S. and Johnson, J.L. (1986) Genus 1. Orla-Lensen 1909, 337. *Propionibacterium*, in *Bergey's Manual of Systematic Bacteriology*, Vol. 2. (eds P.H.A. Sneath, N.S. Mair, M.E. Sharpe and J.G. Holt), Williams and Wilkins, Baltimore, MD, pp. 1346–350.

de Man, J.C., Rogosa, M. and Sharpe, M.E. (1960) A medium for the cultivation of lactobacilli. *J. Appl. Bacteriol.*, **23**, 130–5.

Drosse, W. and Strolley, H. (1964) Funktionelle Gesichtpunkte bei der Beurteilung von Milch-mischungen für den jungen gesunden Sariling. *Med. Ernaehr.*, **5**, 169–173.

Elliker, P.R., Anderson, A.W. and Hannesson, G. (1956) An agar medium for lactic acid streptococci and lactobacilli. *J. Dairy Sci.*, **39**, 1611–12.

Fitz, A. (1878) Über Spaltpilzgährungen, *Ber. deut. chem. Ges.*, **11**, 1890–99.

Garvie, E.I. (1986) Genus Pediococcus Claussen 1903, 68, in *Bergey's Manual of Systematic Bacteriology*, (ed. P.H.A. Sneath, N.S. Mair, M.E. Sharpe and J.G. Holt), Williams and Wilkins, Baltimore, MD, Vol. 2, pp. 1075–9.

Glatz, B.A. (1992) The classical propionibacteria: their past, present, and future as industrial organisms. *ASM News*, **58**, 197–201.

Gonzalez, C.F. and Kunka, B. (1985) Transfer of sucrose-fermenting ability and nisin production phenotype among lactic streptococci. *Appl. Environ. Microbiol.*, **49**, 627–33.

Hardie, J.M. (1986a) Genus Streptococcus Rosenbach 1884, 22, in *Bergey's Manual of Systematic Bacteriology.*, (eds. P.H.A. Sneath, N.S. Mair, M.E. Sharpe, and J.G. Holt), Williams and Wilkins, Baltimore, MD, Vol. 2, pp. 1043–53.

Hardie, J.M. (1986b) Oral streptococci, in *Bergey's Manual of Systematic Bacteriology*, (eds. P.H.A. Sneath, N.S. Mair, M.E. Sharpe, and J.G. Holt), Williams and Wilkins, Baltimore, MD, Vol. 2, pp. 1054–63.

Hardie, J.M. (1986c) Anaerobic streptococci, in *Bergey's Manual of Systematic Bacteriology*, (eds. P.H.A. Sneath, N.S. Mair, M.E. Sharpe, and J.G. Holt), Williams and Wilkins, Baltimore, MD, Vol. 2, pp. 1066–8.

Hardie, J.M. (1986d) Other streptococci, in *Bergey's Manual of Systematic Bacteriology*, (eds P.H.A. Sneath, N.S. Mair, M.E. Sharpe and J.G. Holt), Williams and Wilkins, Baltimore, MD, Vol. 2, pp. 1068–71.

Jehanno, D., Thuault, D. and Bourgeois, C.M. (1992) Development of a method for detection of lactic acid bacteria producing exclusively the L-(+)-isomer of lactic acid. *Appl. Environ. Microbiol.*, **58**, 4064–7.

Kandler, O. and Weiss, N. (1986) Genus Lactobacillus Beijerinck 1901, 212, in *Bergey's Manual of Systematic Bacteriology*, (eds P.H. A. Sneath, N.S. Mair, M.E. Sharpe and J.G. Holt), Williams and Wilkins, Baltimore, MD, Vol. 2, pp. 1206–34.

Levin, R.E., Witkowski, R., Meirong, Y. and Goldhor, S. (1989) Research note: preparation of fish silage with phosphoric acid and potassium sorbate. *J. Food Biochem.*, **12**, 253–59.

Levin, R.E., Fagerson, I.S., Karp, D. *et al.* (1990) In *Advances in Fisheries Technology and Biotechnology for Increased Profitability*, (eds M.N. Voigt and J.R. Botta), Technomic, Lancaster, PA, pp. 251–61.

Lewis, V.P. and Yang, S. (1992) Propionic acid fermentation by *Propionibacterium acidipropionici*: effect of growth substrate. *Appl. Microbiol. Biotechnol.*, **37**, 437–42.

Liu, J.A.P. and Moon, N.J. (1982) Commensalistic interaction between *Lactobacillus acidophilus* and *Propionibacterium shermanii*. *Appl. Environ. Microbiol.*, **44**, 715–22.

Malik, A.C., Reinbold, G.W. and Vedamuthu, E.R. (1968) An evaluation of the taxonomy of *Propionibacterium*. *Can. J. Microbiol.*, **14**, 1185–91.

Mundt, J.O. (1986a) Enterococci in *Bergey's Manual of Systematic Bacteriology*, (eds P.H.A. Sneath, N. Mair, M.E. Sharpe, and J.G. Holt), Williams and Wilkins, Baltimore, MD, Vol. 2, pp. 1063–5.

Mundt, J.O. (1986b) Lactic acid streptococci in *Bergey's Manual of Systematic Bacteriology*, (eds P.H.A. Sneath, N. Mair, M.E. Sharpe, and J.G. Holt), Williams and Wilkins, Baltimore, MD, Vol. 2, pp. 1065–6.

Oh, M.S., Kenneth, R.P., Traube, M. *et al.* (1979). D-(−) lactic acidosis in a man with the short bowel syndrome. *N. Engl. J. Med.*, **301**, 249–52.

Rotta, J. (1986) Pyogenic hemolytic streptococci in *Bergey's Manual of Systematic Bacteriology*, (eds P.H.A. Sneath, N. Mair, M.E. Sharpe, and J.G. Holt), Williams and Wilkins, Baltimore, MD, Vol. 2, pp. 1047–54.

Russell, J.B. (1992) A review: another explanation for the toxicity of fermention acids at low pH: anion accumulation versus uncoupling. *J. Appl. Bacteriol.* **73**, 363–70.

Saton, T., Nariswa, K., Konno, T. *et al.* (1982) D-(−) lactic acidosis in two patients with short bowel syndrome: bacteriological analysis of the fecal flora. *Eur. J. Pediatr.* **138**, 324–6.

Shimizu-Kadota, M. (1987) Properties of lactose plasmid pLY101 in *Lactobacillus casei*. *Appl. Environ. Microbiol.* **53**, 2987–91.

Strøm, T., Gildberg, A., Stormo, B. and Ras, J. (1980) *Advances in Fish Science and Technology*, (ed. J.J. Connell), Fishing News (Books) Ltd, Farnham, UK.

van Niel, C.B. (1928) *The Propionic Acid Bacteria*, J.W. Boissevain, Haarlem.

Wignall, J. and Tatterson, I. (1976) Fish silage. *Process Biochem.* **11**, 17–19.

Winter, K.A. and Feltham, L.A.W. (1983) Fish silage: the protein solution. *Contribution 1983–6E*, Research Branch, Agriculture Canada.

CHAPTER 13

Mussel processing wastes as a fermentation substrate

M.A. Murado, Ma.P. González and L. Pastrana

13.1 MUSSEL AND MUSSEL PROCESSING WASTES IN THE GALICIAN RÍAS

The Galician Rías Baixas (NW Spain) are a series of estuarine systems affected by the Galician upwelling (Fraga, 1981, 1991), whose primary productivity, reaching 260 g carbon/m^2/year (Fraga, 1976), is among the highest in the world. The intense semiculture of mussels developed with regard to these conditions, now generates approximately 250 000 metric tons/year (almost half the world's production), two thirds of which is found in a single system (the Ría of Arousa). Only part of this production is directed to fresh consumption, more than 60% – with a tendency to increase – being frozen or canned, by means of various processes involving a common first step of steam treatment that releases 300–500 l of effluents per metric ton of raw treated mollusc. Due to a complete lack of insight, these mussel processing wastes (MPW), rich in organic matter, particularly glycogen, are dumped in the coastal waters, where they constitute an eutrophication factor that seriously affects the very medium that generates the resource.

Thirty years ago, when production was greatly inferior to the present rate, the recovery of the proteic fraction of the MPW by precipitation with sulfuric acid was posed for the first time (Fraga, 1963). However, neither this nor later attempts that did not even contemplate large-scale treatment (cosmetic applications, use of the solid residue as seasoning, separation of taurin or glycogen) were economically feasible.

The summary presented here includes the results obtained in the attemp to develop a depuration and exploitation system for the MPW, which, based on their use as a microbial culture medium, can be directed

Fisheries Processing: Biotechnological applications.
Edited by A.M. Martin.
Published in 1994 by Chapman & Hall, London. ISBN 0 412 58460 3

simultaneously or alternatively towards diverse objectives characterized by differing requirements in terms of complexity and time of process, as well as by the different added values of the corresponding bioproductions (see Figure 13.18).

13.2 DIRECT TREATMENT OF NON-CONCENTRATED MPW

Regarding the alternative of using the MPW as a microbial culture medium, the more obvious possibility undoubtedly consists of the production of single-cell protein (SCP). To this end, the raw effluent, with pH near neutrality, is acidified to pH 4.5 with HC1 (about 0.5 ml of 5 N HC1 per litre of MPW is, in general, sufficient). After a 5 h period, the spontaneous decantation of the proteic precipitate formed as a consequence of acidification enables the superior clarified fraction to be separated from the sediment with ease, e.g. by using a siphon. With

Table 13.1 Global average composition (with range of variation) and average amino acid content of clarified MPW or medium M (in g/l)

Glycogen	10.00	(7.00–12.50)
Reducing sugars	0.15	(0.10–0.80)
Proteins	3.50	(2.00–4.00)
Total nitrogen	1.60	(1.00–2.00)
Total phosphorus	0.09	(0.08–0.11)
NaCl	18.00	(17.00–22.00)
COD (O_2)	25.00	(19.00–26.00)
Aspartic acid	0.14	
Glutamic acid	0.17	
Serine	0.10	
Histidine	0.12	
Glycine	0.42	
Threonine	0.08	
Arginine	0.39	
Taurine	2.43	
Alanine	0.31	
Tyrosine	0.06	
Methionine	0.02	
Valine	0.04	
Phenylalanine	0.02	
Isoleucine	0.03	
Leucine	0.02	
Ornithine	0.03	
Lysine	0.13	

Total amino acids (with taurine). Total 'useful' amino acids: 2.08 g/l.

Table 13.2 Composition[a] of the sediment 2 (about 2 g/l of wastes, see Figure 13.18) obtained in the pretreatment of the MPW

Organic matter	91.49
Ash	8.51
Crude protein (N × 6.25)	77.82
Lowry protein	70.64
Total carbohydrates	2.91
In vitro digestibility (%)	82.67
Cysteic acid	1.35
Aspartic acid	13.05
Glutamic acid	19.16
Serine	4.97
Histidine	4.65
Glycine	8.64
Threonine	3.34
Arginine	7.09
Alanine	7.01
Tyrosine	2.85
Valine	5.05
Phenilalanine	4.24
Isoleucine	4.81
Leucine	5.87
Lysine	7.91

[a]Values of amino acids are given as % (w/w) with respect to the protein fraction. The other values are given as % (w/w) with respect to the whole sediment 2 or, in digestibilities, the total nitrogen. From Murado *et al.* (1992).

slight variations, the composition of the clarified effluent is that given in Table 13.1, basically characterized by its level of glycogen. From this point onwards, a clarified effluent with a glycogen content of 10 g/l will be denoted as medium M.

The sediment may be purified by resuspension in water and a new precipitation at acid pH, which leads to the production of a material (see composition in Table 13.2) whose suitability as a feed supplement is obvious. Although of small volume (about 2 g/l of treated effluent), it is easy to obtain this co-product, equivalent to that of Fraga's (1963) treatment, and it adds no important expenses to the main process.

13.2.1 Microfungal and yeast pure cultures

After testing medium M against a series of yeasts and microfungi of general acceptance in the production of SCP, the results (Figure 13.1) reflected what was easily foreseeable *a priori*: although because of its reducing sugars and amino acid content, the medium allowed the detectable growth of all the microorganisms, only those species of recognized amylolytic character (even with some exceptions: compare Table 13.3 and Figure 13.1) developed appreciably (Murado *et al.*, 1986, 1989). Among the latter, four microfungi were outstanding (*Fusarium semitectum* and three *Aspergillus* spp.), as well as the fungus-like yeast *Endomyces fibuliger*. Figure 13.2 represents the principal features of the evolution of the corresponding cultures, which, in all five cases, promote a reduction of over 85% in the COD of the medium.

Although this subject will receive further attention, it can already be pointed out that on comparing these five cases (Gonzalez *et al.*, 1992) it became obvious, firstly, that the evolution of the pH throughout the

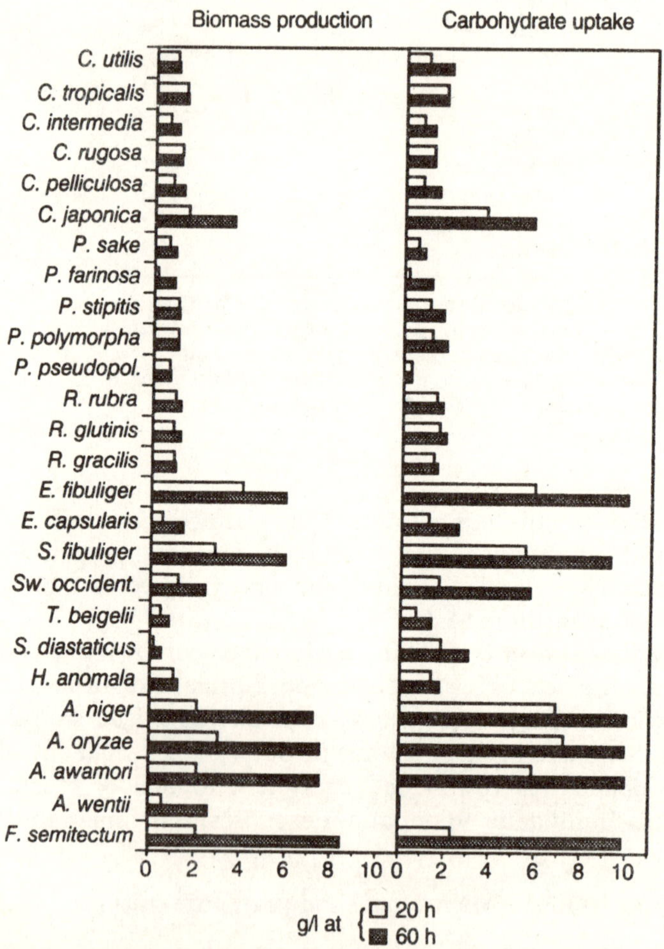

Figure 13.1 Growth on M medium of diverse species of yeasts and microfungi acceptable in the production of SCP.

Table 13.3 Microorganisms used

Species	*Key 1*	*Key 2*
Candida utilis	IFI 1256	C1
C. tropicalis	CBS 94	C2
C. intermedia	CBS 572	C3
C. rugosa	CBS 613	C4
C. pelliculosa	CBS 605	C5
*C. japonica**	CBS 1906	C6
Pichia sake	CBS 4710	P1
P. farinosa	CBS 185	P2
P. stipitis	CBS 5773	P3
P. polymorpha	CBS 186	P4
P. pseudopolymorpha	CBS 2008	P5
Rhodotorula rubra	IFI 1390	R1
R. glutinis	IFI 1416	R2
R. gracilis	CBS 14	R3
Hansenula anomala	CBS 5759	H1
Trichosporon beigelii	CBS 2466	Tr1
Saccharomyces cerevissiae	IFI 240	S1
Saccharomyces cerevissiae	CBS 1907	S2
Saccharomyces cerevissiae	IFI 87	S3
*Sa. diastaticus**	CBS 1782	S4
*Endomyces fibuliger**	CBS 2521	E1
*Endomycopsis capsularis**	CBS 2519	E2
*E. fibuliger**	CBS 329–83	E3
*Schwanniomyces occidentalis**	CBS 819	Sw1
*Aspergillus niger**	CBS 554–65	A1
*A. oryzae**	CBS 125–59	A2.01
*A. oryzae**	CBS 110–47	A2.02
*A. oryzae**	CBS 112–51	A2.03
*A. oryzae**	CBS 115–33	A2.04
*A. oryzae**	CBS 125–49	A2.05
*A. oryzae**	CBS 201–75	A2.06
*A. oryzae**	CBS 570–65	A2.07
*A. oryzae**	CBS 816–72	A2.08
*A. oryzae**	CBS 819–72	A2.09
*A. awamori**	CBS 139–52	A3
*A. wentii**	CBS 118–34	A4
*Fusarium semitectum**	CBS 479–83	F1
*Gibberella fujikuroi**	NRRL 2284	G1

Key 1: CBS: Centraalbureau voor Schimmelcultures (Baarn, The Netherlands). IFI: Instituto de Fermentaciones Industriales (Madrid, Spain). NRRL: North Research Center (Peoria, IL). Key 2: abbreviated notation used in this work. Amylolytic species are asterisked.

incubation period exerted a strong effect on the profile of the total amylolytic activity [TAA: the result of the joint action of α-amylase and glucoamylase; for determination of TAA and glucoamylase, conversion of units and comparison between activities obtained by different methods, see Miranda *et al.* (1987) and Gonzalez *et al.* (1992)], that fell when the pH dropped below 4.0.

This effect can be explained without any difficulty, at least in A1, A2 and E1, producers of the highest levels of TAA, according to the greater sensitivity of α-amylase to acid pH (Miranda *et al.*, 1987), as well as the different proportions of both enzymes in the different species (Fogarty and Kelly, 1980; Siso *et al.*, 1988; Franco *et al.*, 1989), with predominance of α-amylase in A2 and of glucoamylase in A1 and E1, as was repeatedly verified in this laboratory by differential thermal deactivation (Miranda *et al.*, 1987). The explanation was confirmed on repeating the

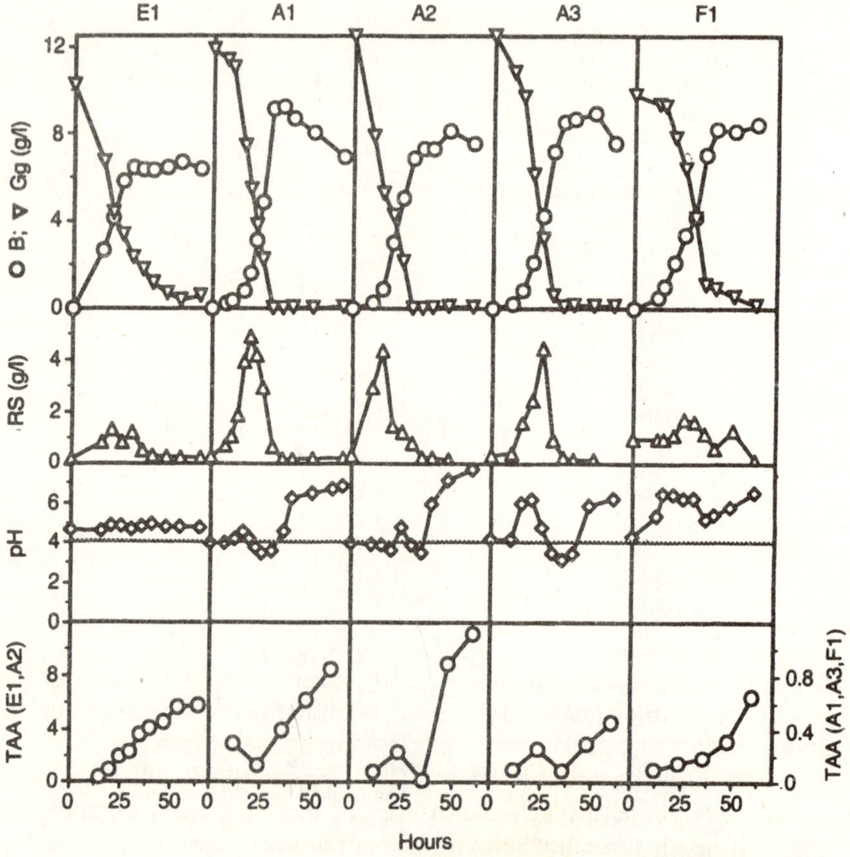

Figure 13.2 Principal features of the cultures of five amylolytic microorganisms on M medium. B: biomass; Gg: glycogen; RS: reducing sugars; TAA: total amylolytic activity. Key to the species in Table 13.3. From González *et al.* (1992).

incubations in medium M buffered with 0.05 M biphthalate/NaOH at an initial pH of 6.0 (Figure 13.3), or preventing the pH fall by alkali addition, conditions that noticeably raised the levels of TAA in A1 and A2, not having any effect on E1 (pH profile always above 4.0, even in non-buffered media).

Then again, the development of the five species in medium M was seen to take place according to two models, basically distinguishable by the production of a high transient reducing sugar surplus (A1, A2, A3)

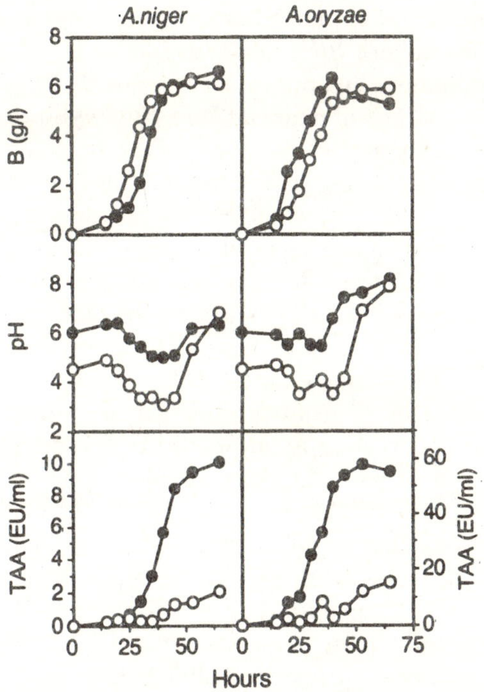

Figure 13.3 Cultures of *A. niger* and *A. oryzae* on unbuffered M medium (open symbols) and partially buffered with biphthalate/NaOH (closed symbols). Notation as in Figure 13.2.

or its absence (E1, F1), which suggests differences in the intensity of the regulation of amylolysis by the products of hydrolysis (inhibition of the amylolytic activity and/or repression of amylase synthesis).

Finally, it can be pointed out that the taurine in the medium, contrary to the other amino acids, was not used by any of the five microorganisms. The main difficulty in the recovery of this compound (of increasing interest in the food sector; Gaull, 1982) from the MPW, due to the presence of glycogen, is thus eliminated when starting from the post-incubated media.

13.2.2 Yeast and microfungus–yeast mixed cultures

In different situations mixed cultures have proved to be a good resort for improvement of the quality of the biomass, to raise the efficiency of the use of the nutrients or even to promote the growth of a microorganism incapable of doing so alone in a determined medium. Although this procedure can be carried out from different perspectives and with different combinations of species (e.g. Akaki, 1965; Peitersen, 1975; Achremowicz, Kosikowski and Masuyama, 1977; Harrison, 1978; De Angelis *et al.*, 1979; Abouzied and Reddy, 1987; Fellows and Worgan, 1987a,b), perhaps the most successful option consists of the so called Symba Yeast Process (Jarl, 1969; Skogman, 1976), in which the reducing sugar surplus generated by an amylolytic yeast of the genus *Endomyces* allows the development of another non-amylolytic species, of faster growth, higher quality and majority contribution to the final biomass.

The application of this expedient to medium M, with *E. fibuliger* as amylolytic species (Murado *et al.*, 1986), demonstrated, within a certain variety of responses (Figure 13.4), the special adequacy of the genus *Rhodotorula* species for the role of non-amylolytic 'partner', obtaining with *R. rubra* improvements of over 10% in the biomass production and 30% in the protein content, with regard to pure cultures of *E. fibuliger*.

However, given the characteristics of the amylolytic species indicated in Figure 13.2, it seems logical to think that their adequacy to mixed culture with non-amylolytic species should be related to the capacity to produce high transient reducing sugar surpluses, which points towards the microfungi A1, A2 and A3 (Siso *et al.*, 1988). Although a quantitative estimation of this capacity is given precisely by the maximum surplus (very similar in all three cases and in the order A1 > A3 > A2), the relation between this value and the corresponding biomass at the same incubation time seems a more adequate criterion, since it evaluates more specifically the carbohydrate level available to the non-amylolytic entity. This second criterion furnishes the order A2 > A1 > A3, which was precisely what was found in the mixed cultures represented in Figure 13.5, where it can again be observed that the species of *Rhodotorula* are among the best non-amylolytic 'partners'.

An approach to the interactions that govern this type of culture was obtained in the case of *A. oryzae–R. glutinis* (Miron *et al.*, 1988) by means of an orthogonal first order design (Akhnazarova and Kafarov, 1982; Box, Hunter and Hunter, 1989) involving the following five variables: temperature (T), pH, initial level of reducing sugars (R), yeast inoculum/fungus inoculum ratio (I), and temporal delay between the inoculum of the fungus and that of the yeast (d). Within the experimental domain used (which, along with the coding criteria for the variables, is indicated in Table 13.4), the production of yeast biomass (Y in g/l when coded values are used for the independent variables) was adjusted to the

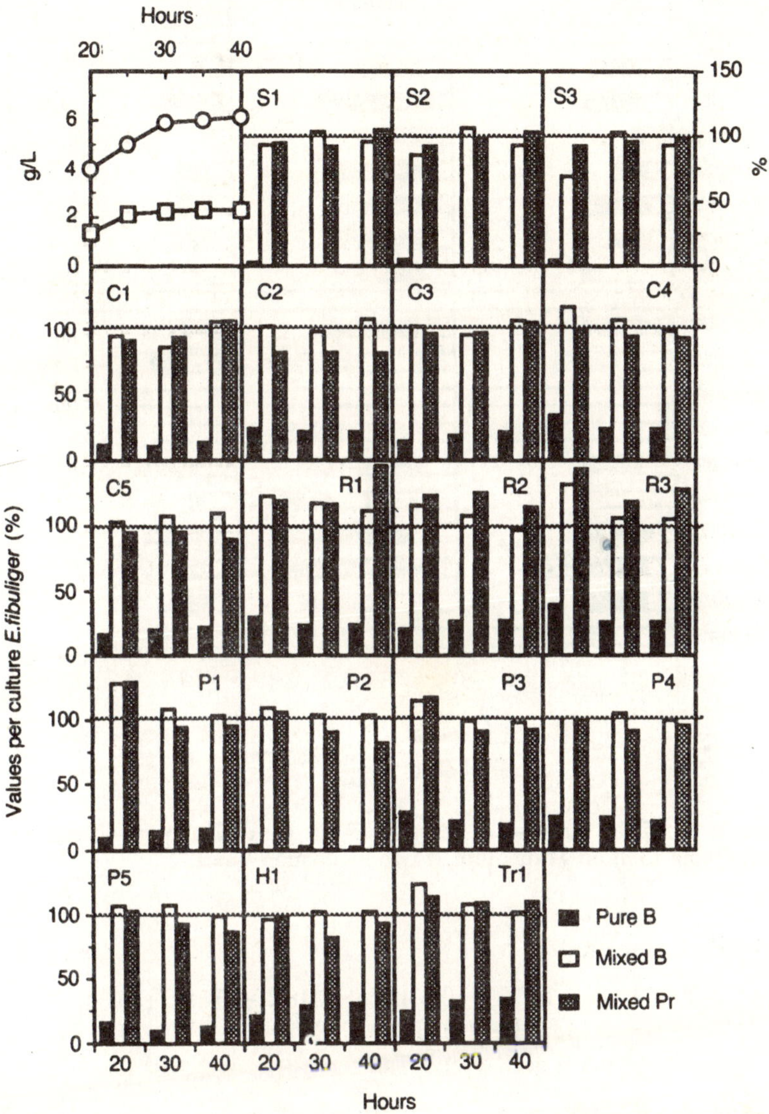

Figure 13.4 Mixed cultures of *E. fibuliger* with 18 non-amylolytic yeasts (key in Table 13.3) on M medium. The point diagram represents the average values (in g/l) of biomass (E) and proteins–Lowry (G) in pure cultures of *E. fibuliger*. The histograms show, at three incubation times, the production of biomass and proteins as percentages of the corresponding values in pure cultures of *E. fibuliger*.

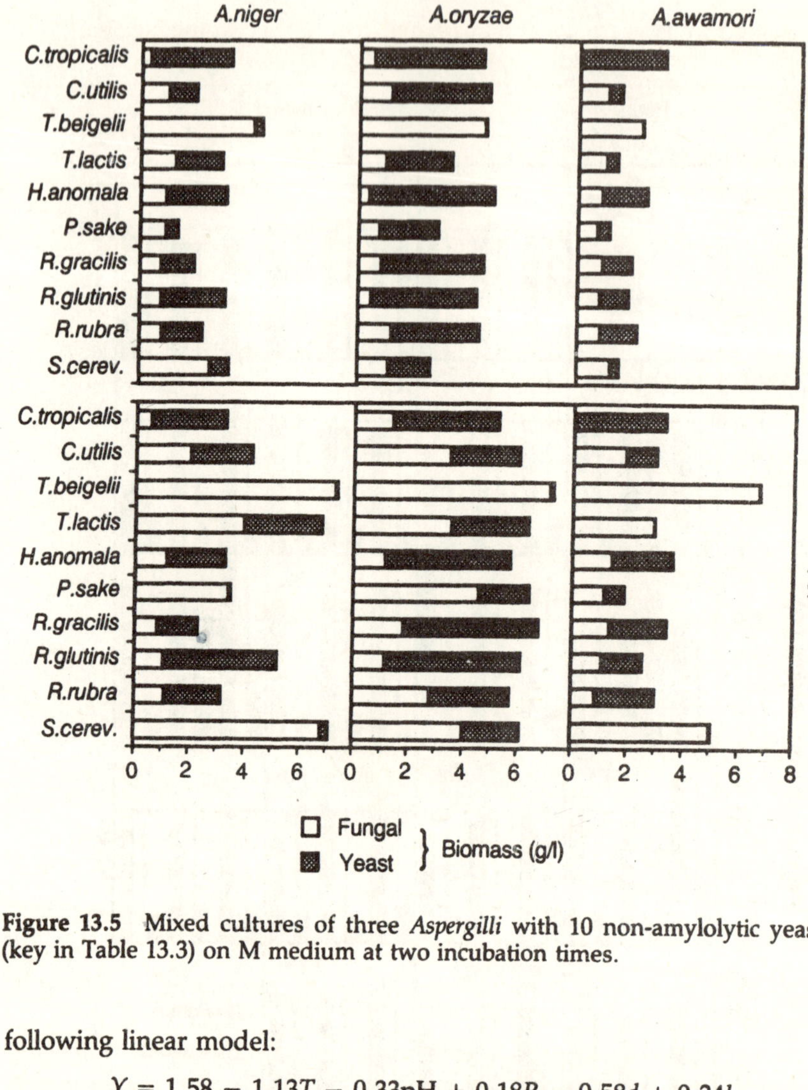

Figure 13.5 Mixed cultures of three *Aspergilli* with 10 non-amylolytic yeasts (key in Table 13.3) on M medium at two incubation times.

following linear model:

$$Y = 1.58 - 1.13T - 0.33\text{pH} + 0.18R - 0.58d + 0.34l + 0.14T\text{pH} - 0.10TR + 0.25Td - 0.11Tl - 0.11Rd - 0.13T\text{pH}R - 0.30Tdl \quad (1)$$

in which the coefficients (Student's t) as well as the model as a whole (Snedecor and Fisher's F) were significant with $\alpha < 0.05$.

Therefore, the primary characteristic of the system consisted of the improvement in the response associated with the decrease in pH and temperature (Figure 13.6). Although the microfungal biomasses and TAA corresponding to the same experimental domain were not adjustable to linear models – in both cases the introduction of second-order terms being necessary – it was possible to confirm these objectives were incompatible with the maximization of the yeast biomass, which seems

Table 13.4 Experimental domain and coding used in the factorial plan for the study of the mixed cultures *A. oryzae–R. glutinis*.

	Natural values				
Coded values	T(°C)	*pH*	R_0 *(g/l)*	d *(h)*	l
−1	25	4	0.2	0	1
0	30	5	0.8	5	5
1	35	6	1.4	10	9

R_0, initial concentration of reducing sugars; *d*, delay between inoculum of the microfungus and that of the yeast, *l*, inocula ratio (yeast/fungus); in all cases the fungal inoculum was 0.5×10^6 spores/ml.

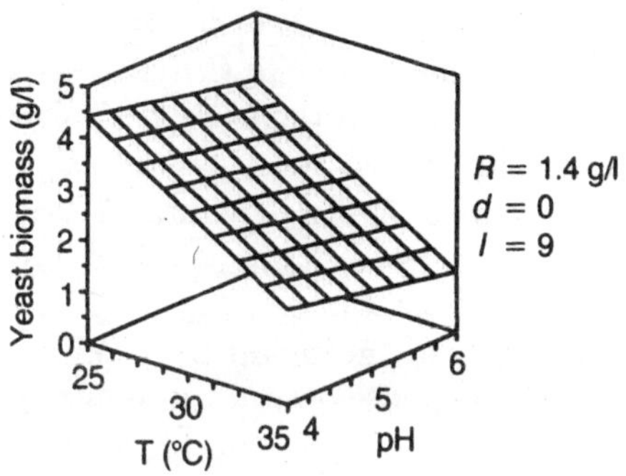

Figure 13.6 Effects of pH and temperature on the biomass of yeast produced in mixed cultures of *A. oryzae–R. glutinis* on M medium. See meanings of *R, d* and *I* in Table 13.4.

trivial in the case of the microfungal biomass, but not so much with regard to the enzymatic activity.

In view, therefore, of the conditions imposed by model (1) it may be concluded that the interest of the mixed fungus–yeast cultures in the case of MPW is restricted by the fact that the improvement of the response in terms of yeast biomass requires a pH incompatible with high levels of amylases in the post-incubate media (which, as will be seen, can constitute a priority goal over the production of SCP), as well as low temperatures that greatly extend the incubation periods.

13.2.3 Amylolytic capacity of the post-incubates

An additional possibility offered by the depuration of amylaceous effluents with the production of SCP is the initiation (e.g. Saha and Ueda, 1983) of other means of exploitation based on the amylolytic activity remaining in the post-incubated media. In the case of MPW, for example, such means could range from the recovery and purification of the corresponding enzymes to the direct use of cell-free media as amylolytic preparations for the saccharification (and consequent increase in their versatility as microbial substrates) of parallel effluent flows, eventually preconcentrated to levels adequate for the process being developed.

From this point of view the species of greater interest are, in principle, E1, A1 and A2. On studying the activities of their post-incubates it was found that, with similar dependencies with regard to pH and temperature, the highest stability corresponded to A2. Then again, on treating vacuum-concentrated MPW with different volumes of cell-free media from 48h cultures from the three species on biphthalate buffered M medium (Figure 13.7), it was noted that the exhaustive hydrolysis of glycogen was reached at lower enzyme/substrate ratios with A2 than with A1, and it was impossible with E1 in any of the situations tested. Furthermore, it was confirmed by HPLC that the glucose never represented less than 95% of the products of hydrolysis. Again the best option was A2, although, given the TAA levels present in the post-incubates (3, 10 and 50 EU/ml in E1, AA1 and A2, respectively), a more favorable α-amylase/glucoamylase ratio could be assumed in A1.

With regard to E1, the limitations found suggest the existence of an ezymatic system sensitive to inhibition by product, combined or not with condensing capacity (inversion of amylolysis), these being properties which could also explain the particular profile shown by the reducing sugars in the yeast cultures (Figure 13.2), characterized by a period of oscillating concentrations that descend two or three times after reaching values of 1.0–1.5 g/l, without a well defined maximum such as that produced in A1 and A2.

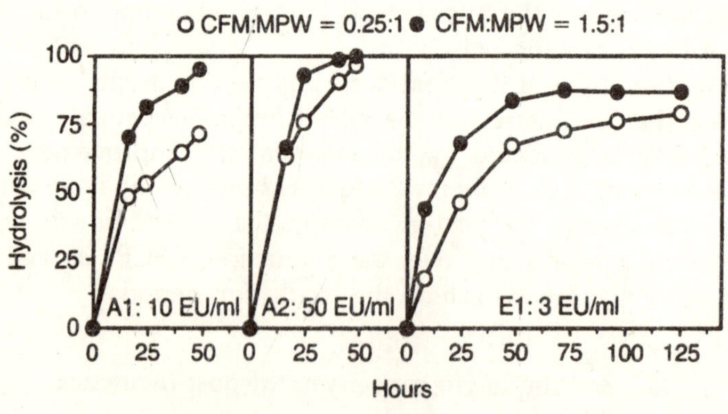

Figure 13.7 Amylolytic capacity of 48 h cell-free media (CFM) from *A. niger*, *A. oryzae* and *E. fibuliger* at 30°C for the treatment of concentrated MPW. In all cases, the initial concentration of glycogen in the incubation mixture was of 20 g/l. From González *et al.* (1992).

Both effects were in fact confirmed (González *et al.*, 1992). When the incubations with cell-free media from E1 were repeated in conditions analogous to those in Figure 13.7, but substituting glycogen with glucose, the appearance of low levels of maltose was detected, with a stable maximum of 0.36 g/l at about 80 h, without traces of oligosaccharides of higher polymerization degree. It was also confirmed that the glucose competitively inhibited (K_i = 1.51 mg/ml) the glucoamylase from E1, although not α-amylase, at least until a concentration of 20 g/l.

With cell-free media from A2 (the species differing most in behavior compared with E1) no condensing activity was observed, glucoamylase was inhibited by glucose (K_i = 1.10 mg/ml) and α-amylase was not inhibited by glucose (15 g/l) nor by maltose (30 g/l). Therefore, although the combination of the condensing and inhibitory effects could justify the differences between E1 and A2 indicated in Figure 13.7, they do not explain those observed in the profiles of the reducing sugars (Figure 13.2). Certainly, in A2 the levels of TAA are much higher and the predominant enzyme is α-amylase; however, in A1, with lower TAA and glucoamylase predominant, the reducing sugar profile resembles that of A2 and not E1. This suggests the possibility that a repressive effect of the amylase synthesis by glucose occurs in the yeast cultures.

This effect, whose possible confusion with inhibition has been widely discussed (e.g. Stepanov *et al.*, 1975; Sills and Stewart, 1982; Wilson, Khachatourians and Ingledew, 1982; Reddy and Abouzied, 1987; Pasari *et al.*, 1987), was also confirmed with regard to glucoamylase. Working with ultrafiltration retentates (30 kDa) to eliminate the inhibitory effect and comparing the evolution of glucoamylase activity during the period following the addition of different levels of glucose and cyclic AMP with cultures undergoing synthesis, it was demonstrated (González *et al.*, 1992) that the intensity of the repressive effect increases at least during 7 h following the addition (while the inhibitory effect decreases), that levels of 0.5 g/l glucose are already effective and that the effects of 2.0 g/l glucose are not reversible with doses of 0.066 mg/l cyclic AMP.

13.2.4 Biomasses and amylolytic preparations from microfungal cultures

The results described in the previous section suggest that, if in the treatment of MPW an attempt is made to make the securing of SCP compatible with other uses based on the remaining amylolytic activity in the post-incubates, *E. fibuliger* has less possibilities than the microfungi, particularly *A. oryzae*. Although, since the yeasts are the microorganisms traditionally preferred for SCP production, it is a pity that this alternative should be neglected, in fact the microfungi are also an appreciable material whose safety has been proven many times (e.g. Barnes, 1976; Barker, 1981; Anderson and Solomons, 1984; Udall *et al.*,

Table 13.5 Gross composition (% w/w) of the microbial biomasses

	A. niger	A. oryzae	A. awamori	F. semitectum
C:N:H	100:18:15	100:21:15	100:18:15	100:21:16
Ash	9.66	11.29	7.22	9.49
Reducing sugars	3.09	2.73	4.26	1.61
Total sugars	32.25	21.60	30.53	23.84
Total nitrogen	7.15	8.37	7.54	8.55
Crude protein (N × 6.25)	44.69	52.31	47.13	53.44
Protein (Lowry)	28.01	42.73	34.61	44.23
Non-protein nitrogen	2.67	1.53	2.00	1.47
Total fatty acids	0.90	1.29	1.56	2.24
DNA	0.65	0.62	0.72	0.99
RNA	2.43	2.67	3.50	4.59
In vitro digestibility (%)	75.17	79.60	82.98	80.84

From Murado *et al.* (1992).

Table 13.6 Amino acid composition (g amino acid/16 g nitrogen) of the microbial biomasses (classification according to FAO)

	A. niger	A. oryzae	A. awamori	F. semitectum	*FAO standard*
Essential					
phenylalanine	7.07	10.07	7.91	8.01	2.8
tyrosine	0.56	0.86	0.67	5.53	2.8
histidine	4.94	7.26	4.65	11.65	–
isoleucine	6.85	6.01	6.08	5.60	4.2
leucine	6.77	6.29	6.04	7.22	4.8
lysine	15.38	13.73	16.15	11.78	4.2
methionine	–	–	–	1.62	2.0
cysteic acid	0.79	0.90	0.80	0.30	2.2
threonine	5.63	2.98	3.62	2.45	2.8
valine	5.81	3.21	1.91	2.27	4.2
arginine	7.73	8.13	6.01	11.01	–
tryptophan	ND	ND	ND	ND	1.4
Total essential	61.53	59.44	53.84	67.44	31.4
Non-essential					
aspartic acid	6.28	5.65	5.66	7.60	
serine	6.40	7.67	5.97	5.25	
glutamic acid	5.22	8.63	6.91	9.39	
proline	ND	ND	ND	ND	
glycine	9.39	8.05	9.73	6.01	
alanine	8.79	4.53	6.67	4.63	

ND, not determined; –, not detected. From Murado *et al.* (1992).

Table 13.7 Fatty acid composition (% w/w; tr: traces) of the microbial biomasses

	A. niger	A. oryzae	A. awamori	F. semitectum
Palmitic [16:0]	0.042	0.191	0.208	0.307
Stearic [18:0]	–	tr	–	tr
Oleic [18:1]	0.348	0.130	0.139	0.268
Linoleic [18:2]	0.513	0.972	1.216	1.644
Arachidic [20:0]	tr	tr	tr	0.020

From Murado *et al.* (1992).

1984; Friedrich, Cimerman and Perdih, 1987) and they even offer advantages such as the ease with which the biomass can be separated from the medium (Pomeranz, 1976) as well as the textural properties of the mycelium (Worgan, 1976), which favor their acceptability.

The compositions of the biomasses corresponding to cultures (in general about 8 g/l in about 40 h) of the four microfungal species indicated in Figure 13.2 are given in Tables 13.5–13.7, where it can be seen that due to their protein content, digestibility, essential amino acid profiles, acceptable levels of nucleic acids and absence of unusual fatty acids, they all represent an interesting potential source of protein for animal feeding, particularly suitable as a ceral-based diet supplement, because of their high proportions of lysine.

With regard to the recovery of the remaining amylases in the post-incubated media of *A. oryzae*, a typical process starting with 8 l of a paper-filtered culture (Murado *et al.*, 1992) included the following steps:

1. Clarification of the filtrate in a continuous decanter centrifuge (about 4 min; negligible losses in volume and enzymatic activity).
2. Concentration to 2.5 by ultrafiltration with cut-off at 30 kDa. Subsequent diafiltration at constant volume, compensating the permeating flow by the continuous addition of tap water (5 l total). New phase of concentration until a final retentate of 0.5 l (about 10 min; recovery of 95.8% of the inital activity).
3. Lyophilization of the retentate to give 6.5 g of a dry powder (production = 0.81 g/l, a recovery of 89.5% of the inital activity in the whole process).

The powdered preparation thus obtained showed very similar characteristics (Table 13.8) to those of α-amylase Sigma A-0273. If the ultrafiltration at 30 kDa is applied to the permeate of a previous cut-off at 100 kDa, the specific activity equals that of the commercial product, since proteic materials coming from the MPW are excluded; however, the previous step also excludes the small quantities of glycogen remaining in the post-incubates, which, on concentrating in the preparation,

Table 13.8 Comparation between main characteristics of powdered enzymatic preparation (PAP) obtained from *A. oryzae* cultures on M medium and Sigma A-0273 α-amylase (S)

	PAP	*S*
Protein (Lowry) [%]	66.57	34.33
Total sugars [%]	19.72	64.69
Total amylolytic activity [EU/mg]	60.85	56.13
Glucoamylase activity [EU/mg]	0.43	0.27

From Murado *et al.* (1992).

favor its stability (Figure 13.8). When studying the joint influence of pH and temperature on TAA by a second-order rotatable design (Figure 13.9), a maximum is found at pH 5.7 and a temperature of 47.8°C, which represents a slightly lower thermostability than that suggested by the usual results with both variables studied separately.

Finally, it was observed that the increase in percentage oxygen saturation during the culture of *A. oryzae* not only increases the amylase production, but also the proportion of glucoamylase in the amylolytic system (Murado *et al.*, 1992), which enables the glucoamylase/TAA ratio to be controlled within an interesting range, as will be seen later on.

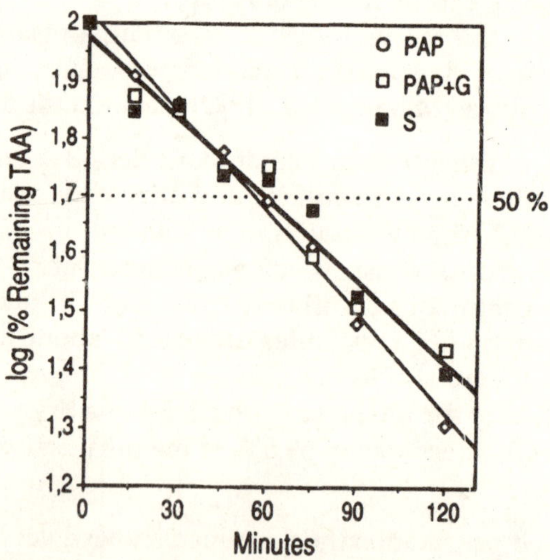

Figure 13.8 Comparison of the stabilities of diverse amylolytic preparations from *A. oryzae* in distilled water at 50°C. PAP: powdered preparation from cultures on M medium; S: α-amylase from Sigma; PAP+G: PAP supplemented with glycogen until a level of total sugars equivalent to that of the Sigma product. Initial TAA = 30 EU/ml in all cases. See also Table 13.8. From Murado *et al.* (1992).

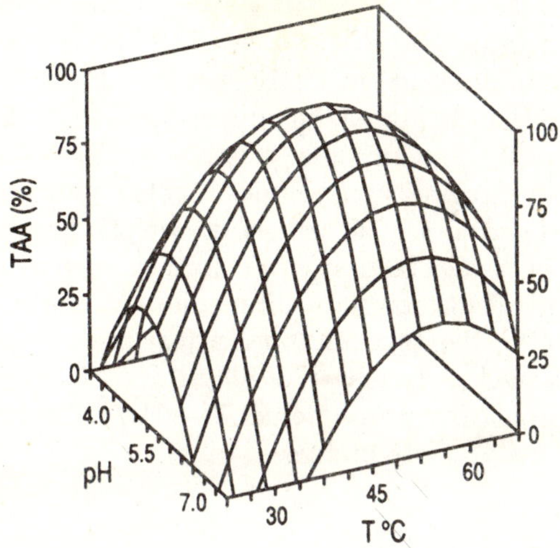

Figure 13.9 Joint effect of the pH and temperature on the total amylolytic activity of the powdered amylolytic preparations from *A. oryzae* cultures on M medium. From Murado *et al.* (1992).

13.3 TREATMENT OF CONCENTRATED MPW

Although medium M already constitutes an acceptable microbial subs-trate, it certainly does present two main disadvantages; the relatively low concentration of its principal carbon source and the fact that this is a polysaccharide. The first disadvantage limits its productive potential, the second limits its applicability to amylolytic microorganisms. Both can be overcome if the difficulties of the treatment required to eliminate them is justified by the value of the bioproduction that could thus be promoted.

13.3.1 Concentration of MPW by ultrafiltration

The high temperature of the emission (about 100°C) makes the concentration of MPW by evaporative methods easy; nevertheless, it is difficult to obtain volume reductions of more than half the quantity without additional energy. Moreover, these methods make it difficult to obtain the precipitation of the proteins during the clarification process due to the increased viscosity of the effluent, and they raise saline concentrations to unsuitable levels for the cultivation of numerous potentially interesting microorganisms.

In these circumstances it seems more appropriate to apply ultrafiltration techniques to the already clarified MPW (medium M), since these methods have already proved their feasability in the extensive treatment of other residual substrates (e.g. lactic whey; Lane, 1977; Maddox and Richert, 1977; Hossain, Brooks and Maddox, 1983) and offer certain advantages in this instance. In fact, given the absence of suspension materials in medium M and the high molecular weight (above 100 kDa) of the glycogen it contains, the process ocurrs both rapidly and efficiently. It also eliminates the salinity problem and permits securing media with a shortage of low molecular weight nutrients, which is important, as will be seen later on, for objectives such as the production of gibberellic acid. Thus, starting with medium M, ultrafiltration allows media *n*M (*n* being the concentration factor) to be obtained, characterized by a high level of glycogen (70–100 g/l) and a low nitrogen content (400–900 mg/l, proteic in a proportion of 75–90%), in an operation which includes the following three phases:

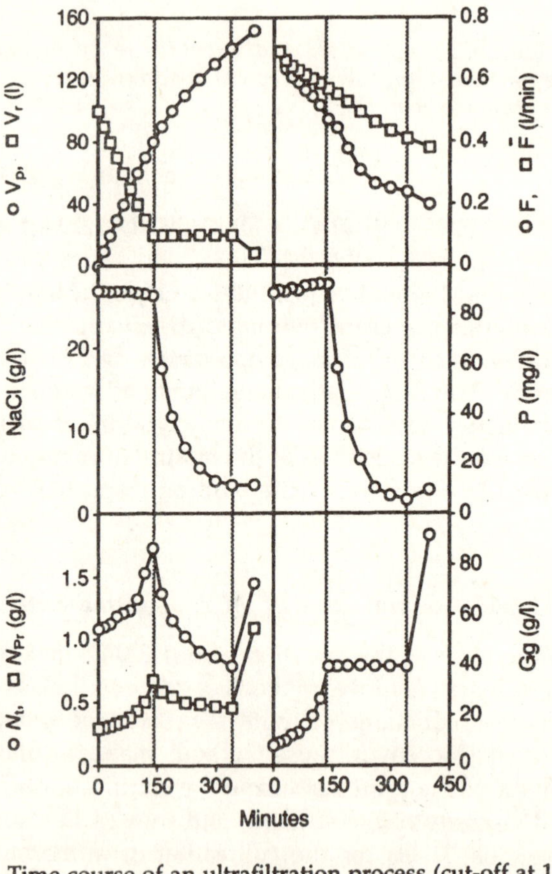

Figure 13.10 Time course of an ultrafiltration process (cut-off at 100 kDa) of M medium. V_p: permeate volume; V_r: retentate volume; *F*: flow (from two consecutive measurements); $\bar{F}$: flow (from zero time to actual measurement); *P*: total phosphorus; N_t: total nitrogen; N_{pr}: protein nitrogen; *Gg*: glycogen. Dotted lines indicate the starting of the different phases (see text). From Murado *et al.* (1993).

1. Concentration until a retentate volume equivalent to about 1/5 of initial is obtained. In this phase the levels of glycogen reach 40–60 g/l, the concentration of peptide materials slightly increases and that of the low molecular weight nutrients remains constant.
2. Diafiltration at constant volume, maintained by compensating the permeating flow with the continuous addition of tap water. In this way the glycogen levels are maintained and those of the low molecular weight components descend as desired.
3. Concentration until the final desired glycogen level is reached. The set of permeates, which is discarded, represents about 10% of the initial COD of the MPW.

Figure 13.10 shows the time course of a typical ultrafiltration process in these conditions, carried out in a single hollow fiber cartridge with cut-off at 100 kDa, without interruption or intermediate cleaning, and starting with a volume of 100 l. The use of two or three series of parallel cartridges allows them to be cleaned alternatively by inverse flushing and therefore improves the efficiency of the treatment.

13.3.2 Saccharification of concentrated MPW

If ultrafiltration transforms the MPW into a 'strong' medium, 'versatility' can be obtained by totally or partially saccharifying the polysaccharide in the concentrate (medium *n*M), in order to transform it into medium *n*M-*h*H, 10*n* being the concentration of total sugars (in g/l) and *h* the degree of hydrolysis (in %) (reducing sugars × 100/total sugars). In this respect, the activity of the amylolytic post-incubates studied in the preceding sections allows the requirements of the hydrolysis process to be covered within the context of the products generated from the effluent itself.

In fact, combining post-incubates from *A. niger* (low TAA, glucoamylase predominant), *A. oryzae* (high TAA, α-amylase predominant) and media 2M in different volumetric proportions, it was possible to obtain the total saccharification of glycogen (Murado *et al.*, 1993). It was also possible to make this hydrolysis process continuous in tubular reactors fed with a mixture of post-incubates and medium 2M, and they maintained a suitable piston flow due to the negative viscosity gradient generated by the progressive decrease in the molecular size of the polysaccharide.

As expected (see section 13.2.3 and Figure 13.7), this use of the post-incubates underlined the superiority of the cell-free media from *A. oryzae*, but it also proved that the contribution of the glucoamylase from *A. niger* improved the efficiency of the process in terms of the TAA/degree of hydrolysis ratio, particularly if hydrolysis was measured specifically as glucose. Nevertheless, it does not seem that the dilution of the cell-free media from *A. oryzae* (high TAA) with those of *A. niger* constitutes a good expedient, especially keeping in mind the relative control, already mentioned, to which the contribution of the glucoamylase to TAA can be submitted in cultures of *A. oryzae*.

The direct use of post-incubates is not recommendable either. Supposing that total hydrolysis with cell-free media from *A. oryzae* is guaranteed with TAA/glycogen proportions of about 0.44 EU/mg in the incubation mixture (the experimental data furnish a slightly inferior value), the initial glycogen/final glucose ratios would be of the kind indicated in Figure 13.11, which makes obtaining concentrated media for subsequent dilution with post-incubates rather meaningless.

The use of powdered enzymatic preparations from *A. oryzae* seems then the most logical option. Given the complexity of the expected interactions in these conditions, with the possibility of positive synergic effects (Fujii and Kawamura, 1985; Fujii, Homma and Taniguchi, 1988) and probably also negative ones (Miranda *et al.*, 1987), the process was studied by means of a complete factorial plan involving the following four sets of variables (experimental domain and coding in Table 13.9):

1. Initial concentration of glycogen in the incubation mixture (*S*). The levels of this variable, which represents a measure of the enzyme/substrate ratio, were obtained by adding the enzymatic solutions used to *n*M media (always in the proportion 0.025:1 v/v) with 50, 100 and 150 g/l of glycogen.
2. Glucoamylase/TAA ratio (*GA*), with values obtained by adding unitary increases of 11.3 EU/ml glucoamylase from *A. niger* to a base

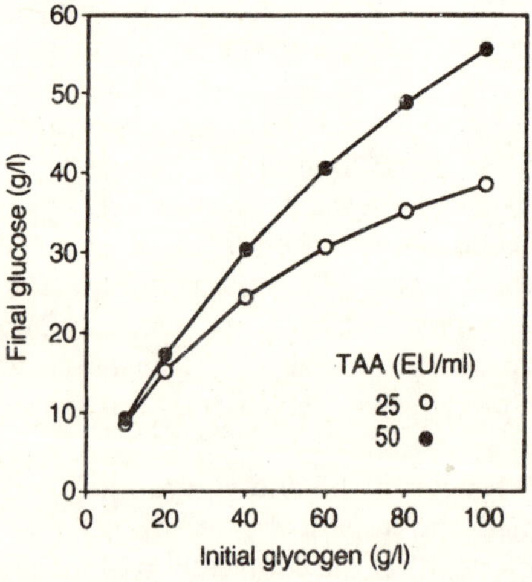

Figure 13.11 Relations between initial glycogen in *n*M media and final glucose in *n*M-*h*H media obtained by saccharification with cell-free media from *A. oryzae* at 50°C and two levels of total amylolytic activity. From Murado *et al.* (1993).

Table 13.9 Experimental domain and codification of the variables used in the factorial design for the study of saccharification

	Natural values			
Coded values	*substrate (S: g/l)*[a]	*glucomylase (GA: EU/ml)*[a]	*pH*	*time (t: h)*
−1	50	0.28	4.4	5
0	100	0.56	5.2	10
+1	150	0.84	6.0	25

[a]In incubation mixture. From Murado *et al.* (1993).

solution of a preparation of *A. oryzae* (TAA = 1600 EU/ml; glucoamylase = 11.3 EU/ml). Although the glucoamylase supplements represented a small contribution to the TAA, the TAA/glycogen ratios in the system were considered to be 40/50, 40/100 and 40/150.

3. Initial pH.
4. Time of hydrolysis (t), this last variable within a domain (5–25 h) in which a response deviating minimally from lineality was previously confirmed.

Although a slight convexity in the response hypersurface suggests a small second-order component, the process proved to be adjustable ($\alpha < 0.05$) to the next first-order model with interactions, in which Y is the percentage of hydrolysis when coded values are used for the independent variables:

$$Y = 69.43 + 5.78GA - 12.94S - 2.25\text{pH} + 14.30t + 2.34GASt - 2.30S\text{pH}t \qquad (2)$$

Figure 13.12 shows some of the response surfaces that best illustrate the essential characteristics of the system, including the fact that the experimental domain used involves certain conditions that allow the exhaustive hydrolysis of the substrate, even at concentrations of over 100 g/l. The strong effect of glucoamylase – despite its low level and especially at short periods and low pH – is particularly interesting, since it is not difficult to obtain enzymatic preparations from *A. oryzae* with levels of glucoamylase higher than the maximum used on this occasion, by reducing, as already mentioned, the oxygen limitation in the cultures during the growth phase at maximum rate. The increase in pH (which limits the activity of glucoamylase) obliges a reduction in the level of the substrate in order to maintain a high response at short periods only; at long periods and low levels of glucoamylase it seems more convenient to favor the α-amylase by raising the pH.

Finally, it can be pointed out that when securing media with hydrolysis degrees ranging from 25 to 100%, equation (2) furnishes a consistent criterion for the definition of operational alternatives in which economy

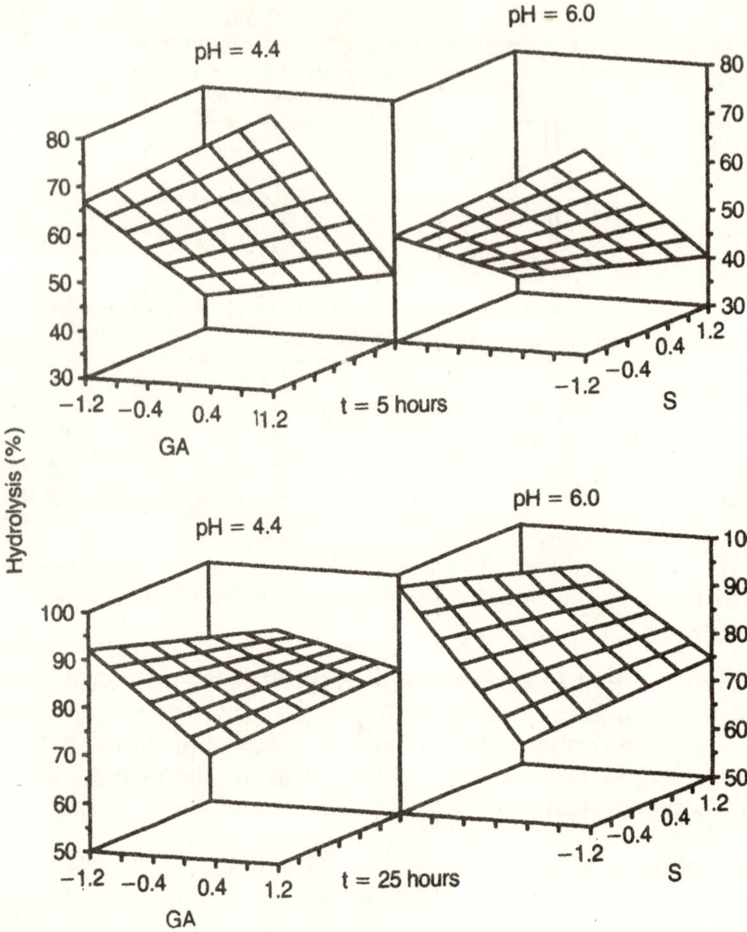

Figure 13.12 Joint effect of the substrate concentration (S) and proportion of glucoamylase (GA) on the degree of hydrolysis obtained in the saccharification of *n*M media. See also Table 13.9. From Murado *et al.* (1993).

of time or enzymatic preparation are a priority. Figure 13.13 shows the results of two of these alternatives and demonstrates a high degree of adjustment with the expected values, as well as the absence, in all cases, of dextrines with polymerization degree higher than maltotriose among the products of hydrolysis.

13.3.3 Utilities of saccharified media: production of gibberellic acid

The saccharified concentrates (media *n*M-*h*H) have already provided satisfactory preliminary results in various bioproductions such as gibberellic acid, ethanol, glucose oxidase and gluconic acid. Gibberellic acid (Murado *et al.*, 1992) is particularly interesting due not only to its

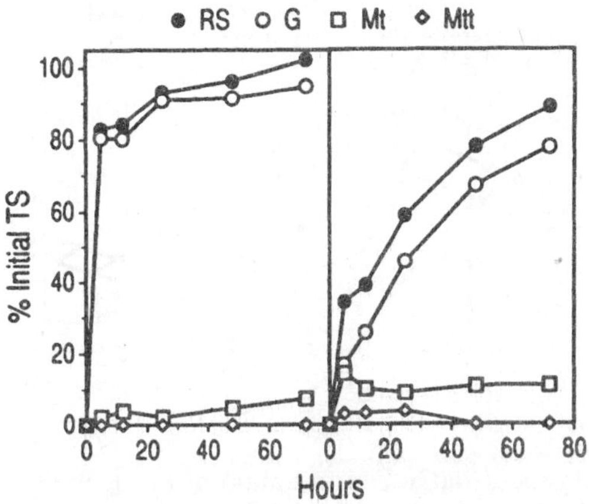

Figure 13.13 Saccharification of *n*M media according to two different operating modes. Left: GA = 1; S = −1; pH = −1. Right: GA = −1; S = 1; pH = 1 (coded values; see Table 13.9). TS: total sugars; RS: reducing sugars; G: glucose; Mt: maltose; Mtt: maltotriose. From Murado *et al.* (1993).

economic value but also because it illustrates the possibilities of media derived from MPW in situations that demand strict nutritional requirements.

As it is well known, the production of gibberellic acid (GA3) by *Giberella fujikuroi* depends to a large extent on the appearance of a limitation by nitrogen during the course of fermentation (e.g. Borrow *et al.*, 1955, 1961, 1964a,b), that determines the transition from an active growth phase of the microorganism, or trophophase, to another characterized by the synthesis of the secondary metabolite, or idiophase (Bu'lock *et al.*, 1974). Given this type of conditioning, the assay of different types of media derived from the MPW (Pastrana, 1991; Murado *et al.*, 1992) not only led to an appreciable production of GA3, but also, due to the peculiar composition of the wastes (glycogen and proteins as carbon and nitrogen sources respectively), allowed the characterization of certain metabolic particularities of the microorganism linked with the balance of essential nutrients.

(a) Use of totally saccharified media

First, it should be indicated that the *n*M media routinely obtained by ultrafiltration – where, by draining the nitrogen compounds of low molecular weight, the main fraction of nitrogen is proteic – are suitable, after saccharification (*n*M-100H), for the objective proposed with no more requirements than a slight phosphorous supplement. While some amylolytic microorganisms are not really capable of using the proteins from MPW (Murado *et al.*, 1989), *G. fujikuroi* uses them as the only

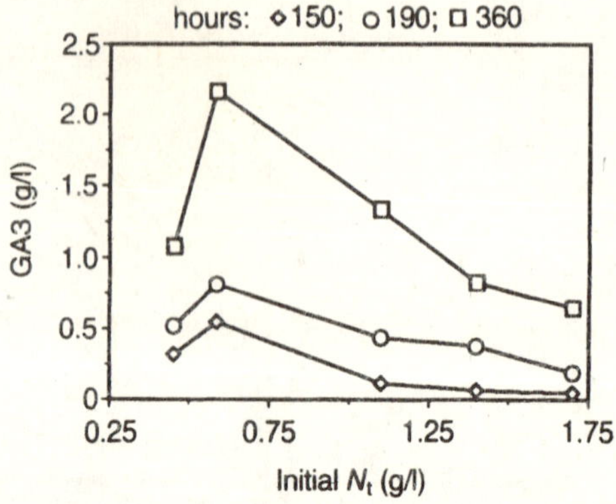

Figure 13.14 Effects of initial concentration of nitrogen on the production of gibberellic acid by *G. fujikuroi* on 7.5M-100H media at three incubation times. GA3: gibberellic acid; N_t: total nitrogen. From Pastrana *et al.* (1993).

nitrogen source, although in the presence of inorganic forms of the nutrient the latter are preferably consumed and at a greater rate.

The results obtained in media with different initial concentrations of nutrients but the same C/N ratio, indicated that going from 70 to 120 g/l glucose determined a decrease in productivity (GA3 production/ carbohydrate consumption) and the absolute production of the hormone, at the same time causing a decrease in the consumption of nitrogen per unit of carbon. Although the first effect could be attributed to inhibition by substrate similar to that indicated by Kumar and Lonsanne (1987) in solid state cultures, the second suggests, alternatively or additionally, that in the more diluted media the limitation of nitrogen necessary for beginning synthesis of GA3 is reached earlier.

The effect of initial levels of nitrogen on the synthesis of the hormone at different incubation times in media with 75 g/l of glucose (Figure 13.14) allowed the optimum concentration to be situated around 700 mg/ l (C/N ratio of 40–64), with greater decreases in production towards suboptimum rather than superoptimum values. These results coincide with those obtained by Bruckner, Blechsmidt and Schubert (1989) in synthetic media and suggest that optimum production requires a stricter nitrogen concentration interval than that usually accepted. Therefore, given that the production of GA3 requires a limitation of nitrogen with sufficient carbon available and that the increase in the level of nutrients on the basis of a constant C/N ratio does not furnish any real improvement, it seems clear that the increase in production demands maximization, at the same time and for each initial level of the carbon source of:

1. The deficit of nitrogen.
2. The concentration of carbon available at the time of the deficit.
3. The biomass at the same time.

(Note that (2) and (3) are contrasting objectives.)

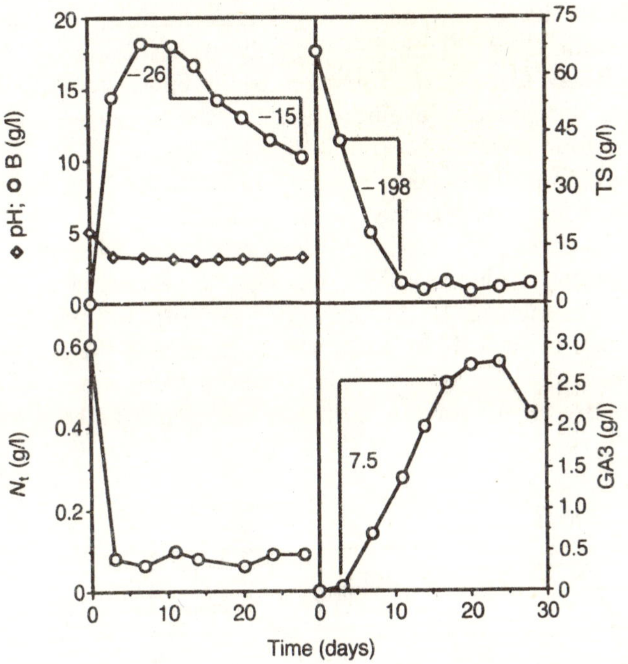

Figure 13.15 Growth of *G. fujikuroi* and production of GA3 on a 7M-100H medium with a ratio of C/N = 46. B: biomass; TS: total sugars; N_t: total nitrogen; GA3: gibberellic acid. The increments corresponding to each represented magnitude are given on the figure in g/l/h. From Pastrana *et al.* (1993).

The time course of a culture (Figure 13.15) in a 7M-100H medium with its composition situated in the optimum interval of nutrients (glucose = 70 g/l, C/N = 46, N 90% proteic) showed how the production of GA3 was maintained at an approximately constant rate of 7.5 mg/l/h from the beginning of the nitrogen deficit until 6 days after the exhaustion of carbon, reaching a level of 2.8 g/l after a 20 day incubation period.

(b) Use of partially saccharified media

The fact that the use of oligosaccharides or substrates based on flours or natural oils as the carbon source improves the production of GA3 (e.g. Darken, Jensen and Shu, 1959; Jefferys, 1970; Vass and Jefferys, 1979), suggested the possibility of obtaining analogous improvements using media with a degree of hydrolysis lower than 100%, this being an option that should offer the additional advantage of economy in securing the culture media either in terms of the consumption of the enzymatic preparation or time.

The hypothesis was confirmed by comparing the results obtained in media with the same initial concentrations of total carbohydrates (70 g/l) and nitrogen (600 mg/l) but with four different degrees of hydrolysis

(0, 50, 75 and 100%). As indicated in Figure 13.16, the reducing sugars were consumed in preference to the polysaccharide (in the media with 50% hydrolysis the effects of diauxia on the biomass can be detected), but the microorganism developed even in the media with glycogen as the only carbon source.

Although none of the cases in Figure 13.16 reached the moment when the hormone level declines (see Figure 13.15), the effects of the increased contribution of the polysaccharide are clearly noticeable in the tendential decrease in the production of the biomass, the accentuation of the lytic phase and the increase in carbohydrate concentration remaining in the final incubation period, this last aspect justifying the supposition of correlative extensions in the corresponding productive phases. Also, when the production of the hormone was represented as measured

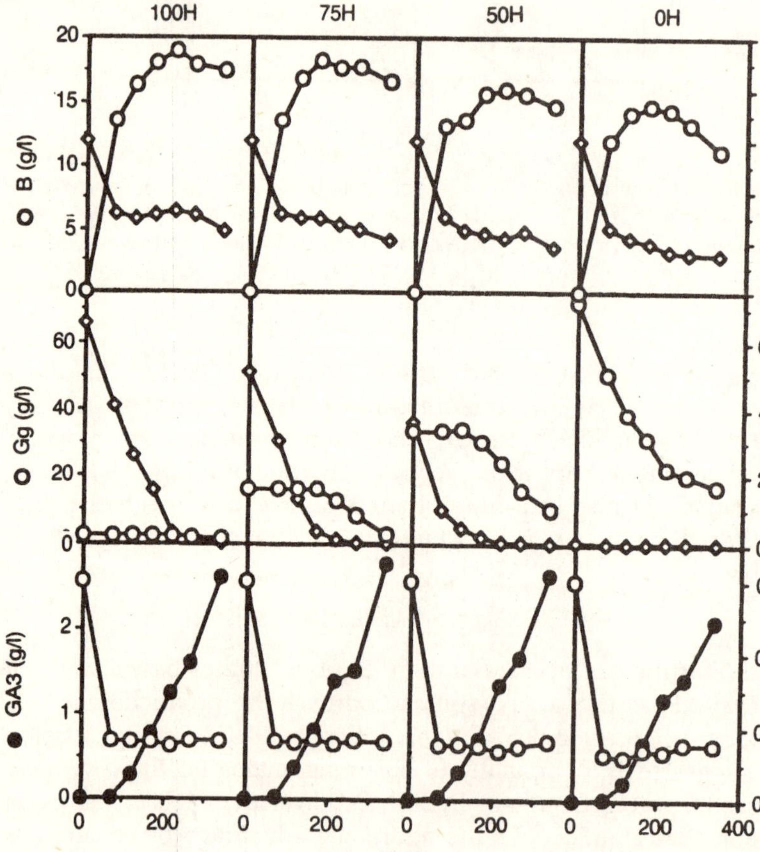

Figure 13.16 Growth of *G. fujikuroi* and production of GA3 on 7M media with different degrees of hydrolysis. B: biomass; Gg: glycogen; RS: reducing sugars; N_t: total nitrogen; GA3: gibberellic acid. From Pastrana *et al.* (1993).

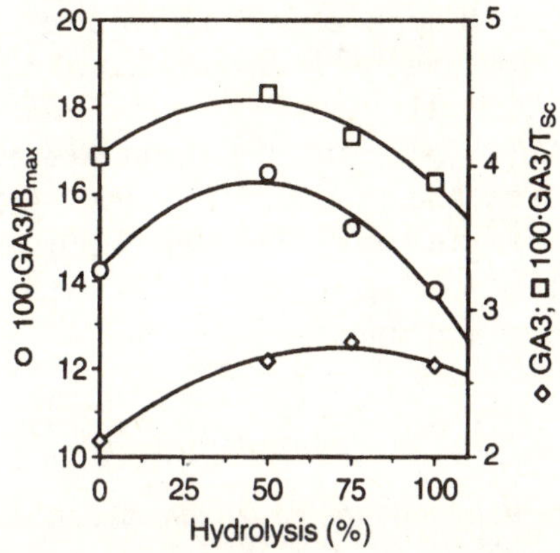

Figure 13.17 Effects of the hydrolysis degree of the substrate on the production of GA3 after a 330 h incubation period, measured according to different criteria. B_{max}: maximum biomass; T_{Sc}: Total sugars consumption. From Pastrana *et al.* (1993).

according to different criteria (*P*[GA3]), such as the function of the degree of hydrolysis (*H*), the results obtained (Figure 13.17) adjusted satisfactorily ($\alpha < 0.05$) to quadratic models such as:

$$P\,[\text{GA3}] = b_0 + b_1 H + b_2 H^2 \tag{3}$$

that provided the following degrees of hydrolysis at maximum ordinate points:

GA3 production (g/l; 330 h)	75.0%
GA3 production/total sugars consumption (330 h)	44.7%
GA3 production at 330 h/maximum biomass	46.9%

The capacity of *G. fujikuroi* to use glycogen also led to the supposition that the synthesis of extracellular amylases necessary in these conditions could modify the balance of carbon and nitrogen consumption during the trophophase; this hypothesis also proved to be correct. First, when comparing the development of the microorganism in synthetic media (Bu'lock *et al.*, 1974) and from MPW, using glucose and glycogen as exclusive carbon sources in each case (Pastrana, González and Murado, 1993), the inducible character of the amylases (which could not be detected in the glucose media) was demonstrated as well as the probable presence of components favoring their synthesis in the MPW, since in the synthetic media with glycogen, amylolytic avtivity could not be detected and hardly any growth took place.

Second, and on the basis of nitrogen consumption per unit of carbon used, it was seen that the glycogen represented a more 'expensive in nitrogen' source of carbon than glucose, probably due to the additional synthetic requirement (of extracellular amylases) imposed by the polysaccharide. Curiously though, everything seemed to indicate that, at a determined C/N ratio, the rate of consumption of nitrogen is a constant feature of *G. fujikuroi*, and if different sources of carbon impose different requirements of nitrogen, the rate of consumption of the carbon source and not the nitrogen source alters. Therefore, the use of glycogen determines the appearance of a deficiency of nitrogen in the presence of greater levels of remaining carbohydrates, extending the idiophase and increasing the production of the hormone.

(c) *Fed-batch culture*

This kind of culture, by means of which the effects of inhibition by substrate on the synthesis of GA3 can be forestalled (Sanchez-Marroquin, 1963; Jefferys, 1970), involved, in the case of the MPW, the danger of implicating an excessive contribution of nitrogen, with the consequent rupture of the deficit and interruption of the synthesis of the hormone.

Nevertheless, the assay with different feed rates (Murado *et al.*, 1992) provided consistently better results than those of discontinuous culture. No interruptions in the synthesis of GA3 were detected and with the contribution of 264 mg/l/h carbohydrates (50% degree of hydrolysis) and 1.2 mg/l/h of nitrogen added at 48 h intervals, it was found that increases of 83% were achieved in the net production of the hormone and 35% in its productivity with regard to the consumption of carbo-hydrates, with 40% prolongations to incubation times.

Apart from their practical interest, these results confirm the interpretation offered by Jones and Pharis (1987), who, working with immobilized cultures, suggested that the introduction of small levels of nitrogen once the synthesis of the hormone commences not only does not interrupt it, but, since it contributes to the maintainance of the basic cellular structures, it tends to extend the idiophase and improve production.

13.4 MPW AS RAW MATERIAL: OUTLOOK

As mentioned at the beginning of the chapter and indicated in Figure 13.8, the program here developed for the treatment of MPW includes two groups of different alternatives according to their requirements and potential interest. The first one, respresented by production that can be obtained directly from clarified MPW, is an essentially depurational option, and its main advantage is probably the possibility of substantially reducing COD in a short period of time. The second group, characterized by the necessity to obtain previously concentrated and saccharified – at least partially – effluents, tries to transform the MPW into a versatile raw material, which, in a similar manner to the process elaborated for sugar molasses, would be suitable to sustain diverse biotechnological processes.

It should nevertheless be pointed out that, apart from the objective difficulties of this kind of strategy, the food industry based on marine production is not characterized, at least in our country, by its openness to environmental problems or innovation. In contrast to its homologous terrestrial industry, which seems to have understood the utility of recycling its residues and already transforms various agricultural, livestock and forestal wastes, the numerous coastal installations whose

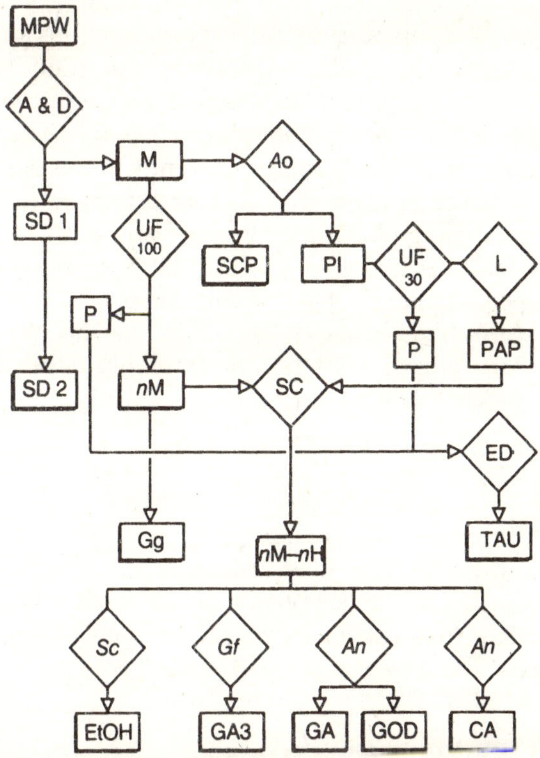

Figure 13.18 Diagram of the whole process. A&D: Acidification of the effluent and spontaneous decantation of the proteic precipitate formed. M: clarified wastes ('M' medium). *Ao*: culture of *A. oryzae*. SD: sediments (1: crude, 2: purified) obtained from clarification of MPW. SCP: microbial biomass. PI: cell-free medium (post-incubate) from *A. oryzae* culture. UF: ultrafiltration (cut-off 30 or 100 kDa). L: lyophilization. P: permeate. ED: electrodialysis. Tau: taurine. PAP: powdered amylolytic preparation. *n*M: concentrated 'M' medium. Gg: glycogen (precipitable from *n*M by ethanol treatment). SC: saccharification. *n*M-*h*H: concentrated and hydrolyzed 'M' medium. *Sc*: culture of *S. cerevissiae*. Production of ethanol (EtOH). *Gf*: culture of *G. fujikuroi*. Production of gibberellic acid (GA3). *An*: culture of *A. niger*. Production of citric acid (CA), gluconic acid (GA) and glucose oxidase (GOD). From Murado *et al.* (1993), modified.

effluents were studied in this paper do not seem to mind using the sea as a comfortable dumping ground. Even when the 'sea', as in the present case, is a semi-closed estuarine system that produces the very raw materials that they use.

Therefore, it is clear that if methodological simplicity is always well received, in this case that would be even more true. In this respect, solid state culture, which is technologically simpler and particularly adequate for the development of microfungi (Nigam, 1990; Pandey, 1990; Ramesh and Lonsane, 1990), could be an interesting alternative within an area of application based on concentrated MPW.

The production of GA3 (Pastrana, 1991) as well as that of amylases with *A. oryzae* and citric acid with *A. niger* could be carried out in systems obtained by imbibing different types of porous supports (corncob, synthetic foam) with media derived from the effluents. Apart from concrete advantages, such as a substantial increase in the contribution of the glucoamylase to the enzymatic production of *A. oryzae*, the results of the process, which this laboratory is currently developing, are leading to better performance in every case compared with those found in the corresponding submerged cultures. Preliminary results concerning the production of amylases and GA3 suggest that, by compensating for the increase in the liquid phase due to the addition of medium in a fed-batch regime with forced evaporation by continuous air flow, it will be possible to reduce substantially the concentrations to be reached in the ultrafiltration process.

ACKNOWLEDGMENTS

Ma. I.G. Siso, J. Mirón, Ma. I. Montemayor, J. Pintado and H. Farto have contributed to the development of the work summarized here to a greater extent than can be expressed in mere bibliographical acknowledgements. The financial support came from CICYT (projects PA 86–0432 and ALI 789–90).

REFERENCES

Abouzied, M.M. and Reddy, C.A. (1987) Fermentation of starch to ethanol by a complementary mixture of an amylolytic yeast and *Saccharomyces cerevissiae*. *Biotechnol. Lett.*, **9**(1), 59–62.

Achremowicz, B., Kosikowski, F.V. and Masuyama, K. (1977) Mixed cultures of different yeasts species and yeasts with filamentous fungi in the SCP production. *Acta Microbiol. Polonica*, **26**(3), 265–71.

Akaki, M. (1965) Studies on mixed yeast cultures in sulfite waste liquor medium *J. Ferment. Technol.*, **43**, 365–73.

Akhnazarova, S. and Kafarov, V., (1982) *Experiment Optimization in Chemistry and Chemical engineering*, MIR, Moscow.

Anderson, C. and Solomons, G.L. (1984) Primary metabolism and biomass production from *Fusarium*, in *The Applied Mycology of Fusarium*, (eds M.O. Moss and J.E. Smith), Cambridge University Press, Cambridge, pp. 231–50.

Barker, T.W., Drouliscos, N.J. and Worgan, J.T. (1981) Composition and nutritional evaluation of *Aspergillus oryzae* biomass grown on palm oil processing effluents. *J. Sci. Food Agric.*, **32**, 1014–20.

Borrow, A., Brian, P.W., Chester, V.E. *et al.* (1955). Gibberellic acid as metabolic product of the fungus *Gibberella fujikuroi*: some observations on its production and isolation. *J. Sci. Food Agric.*, **6**, 340–48.

Borrow, A., Jefferys, E.G., Kessell, R.H.J. *et al.* (1961) The metabolism of *Gibberella fujikuroi* in stirred culture. *Can. J. Microbiol.*, **7**, 227–76.

Borrow, A., Brown, S., Jefferys, E.G. *et al.* (1964a) The kinetics of metabolism of *Gibberella fujikuroi* in stirred culture. *Can. J. Microbiol.*, **10**, 407–44

Borrow, A., Brown, S., Jefferys, E.G. *et al.* (1964b) The effect of varied temperature on the kinetics of metabolism of *Gibberella fujikuroi* in stirred culture. *Can. J. Microbiol.*, **10**, 445–66.

Box, G.E.P., Hunter, W.G. and Hunter, J.S. (1989) In *Estadística para Investigadores*, Reverté, Barcelona.

Bruckner, B., Blechsmidt, D. and Schubert, B. (1989) *Fusarium moniliforme* Sheld. A fungus producing a broad spectrum of bioactive metabolites. *Zentralbl. Mikrobiol.*, **144**, 3–12.

Bu'Lock, J.D., Detroy, R.W. Hostalek, Z. and Munim-Al-Shakarchi, A. (1974) Regulation of secondary biosynthesis in *Gibberella fujikuroi*. *Trans. Br. Mycol. Soc.*, **2**, 377–89.

Darken, M.A., Jensen, A.L. and Shu, P. (1959) Production of gibberellic acid by fermentation. *Appl. Microbiol.*, **7**, 301–3

De Angelis, D.F., Kiyan, C. Corso, C.R. and Tauk, S.M. (1979) Emprego de leveduras em culturas puras e mistas objetivando o aproveitamento de vinhaça. *Brasil Açucareiro*, **6**, 401–6.

Fellows, P.J. and Worgan, J.T. (1987a). Growth of *Saccharomycopsis fibuliger* and *Candida utilis* in mixed culture on pectic materials. *Enzyme Microb. Technol.*, **9**(7), 430–33.

Fellows, P.J. and Worgan, J.T. (1987b). Growth of *Saccharomycopsis fibuliger* and *Candida utilis* in mixed culture on apple processing wastes. *Enzyme Microb. Technol.*, **9**(7), 434–37.

Fogarty, W.M. and Kelly, C.T. (1980). Amylases, amyloglucosidases and related glucanases, in *Microbial Enzymes and Bioconversions: Economic Microbiology*, (ed. A.H. Rose), Academic Press, London, pp. 115–70.

Fraga, F. (1963). Possible utilization of the cooking water mussel. *Proc. Gen. Coun. Medit.*, **7**, 325–27.

Fraga, F. (1976). Fotosíntesis en la Ría de Vigo. *Inv. Pesquera*, **40**, 151–57.

Fraga, F. (1981). Upwelling off the Galician coast, Northwest Spain, in *Coastal Upwelling*, (ed. F.A. Richards), American Geophysical Union, Washington, pp. 176–82.

Fraga, F. (1991). El afloramiento costero en la costa atlántica de la Península Ibérica. *Rev. Academia Galega de Ciencias*, **10**, 144–52.

Franco, J.M., Murado, M.A., Siso, M.I.G. *et al.* (1989) A HPLC method for specific determination of α-amylase and glucoamylase in complex enzymatic preparations. *Chromatographia*, **27** (7/8), 328–32.

Friedrich, J., Cimerman, A. and Perdih, A. (1987) Mixed culture of *Aspergillus awamori* and *Trichoderma reesei* for bioconversion of apple distillery waste. *Appl. Microbiol Biotechnol.*, **26**, 299–303

Fujii, M. and Kawamura, Y. (1985). Synergistic action of α-amylase and glucoamylase on hydrolysis of starch. *Biotechnol. Bioeng.*, **27**(3), 260–65.

Fujii, M., Homma, T. and Taniguchi, M. (1988). Synergim of α-amylase and glucoamylase on hydrolysis of native starch granules. *Biotechnol. Bioeng.*, 32, 9105.

Gaull, G.E. (1982). Taurine in the nutrition of the human infant. *Adv. Appl. Microbiol.*, **24**, 129–64.

González, Ma.P., Siso, Ma.I.G. Murado, M.A. *et al.* (1992). Depuration and valuation of mussel processing wastes. Characterization of amylolytic post-incubates from different species grown on an effluent. *Bioresource Technol.*, **42**, 133–40.

Harrison, D.E.F. (1978). Mixed cultures in industrial fermentation processes. *Adv. Appl. Microbiol.*, **24**, 129–64.

Hossain, M., Brooks, J.D. and Maddox, I.S. (1983). Production of citric acid from whey permeate by fermentation using *Aspergillus niger*. *N.Z. Dairy Sci. Technol.*, **18**, 161–8.

Jarl, K. (1969). Production of microbial food from low-cost starch materials and purification of industry's waste starch effluents through the Symba Yeast Process. *Food Technol.*, **23**, 23–6.

Jefferys, E.G. (1970). The gibberellin fermentation. *Adv. Appl. Microbiol.*, **13**, 283–323.

Jones, A. and Pharis, R.P. (1987). Production of gibberellins and bikaverin by cells of *Gibberella fujikuroi* immobilized in carragenaan. *J. Ferment. Technol.*, **65**, 717–22.

Kumar, P.K.R. and Lonsane, B.K. (1987). Potential of fed-batch culture in solid state fermentation for production of gibberellic acid. *Biotechnol. Lett.*, **9**, 179–82.

Lane, A.G. (1977). Production of food yeast from whey ultrafiltrate by dialysis culture. *J. Appl. Chem. Biotechnol.*, **27**, 165–9.

Maddox, I.S. and Richert, S.H. (1977). Production of gibberellic acid using a dairy waste as the basal medium. *Appl. Environ. Microbiol.*, **33**, (1), 201–2.

Miranda, M., Siso, Ma.I.G., González, Ma.P. *et al.* (1987). Amylolysis in systems with α-amylase and glucoamylase: a comparative study of six methods of evaluation. *Biotechnol. Tech.*, **1** (3), 195–200.

Miron, J., Siso, Ma.I.G., Murado, M.A. and Gonzalez, Ma.P. (1988). Microfungus-yeast mixed cultures in the degradation of an amylaceous waste. II: an experimental design for optimization of yeast production. *Biotechnol Lett.*, **2**(3), 171–76.

Murado, M.A., González, Ma.P., Montemayor, Ma.I. *et al.* (1986). Producción de proteína unicelular en efluentes del procesado de mejillón I: utilización de levaduras en cultivos axénicos y mixtos. *Invest. Pesquera*, **50**(4), 571–605.

Murado, M.A., Siso, Ma.I.G., González, Ma.P. *et al.* (1989). Producción de proteína unicelular a partir de efluentes de la elaboración del mejillón. II: utilización de microhongos. *Rev. Agroquím. Tecnol. Aliment.*, **29**(2), 209–20.

Murado, M.A., Pastrana, L., González, Ma.P. and Montemayor, Ma. I. (1992). Un procedimiento para la producción de ácido giberélico, en cultivo sumergido, a partir de efluentes del procesado industrial del mejillón. *Spanish Patent* 9 201 240.

Murado, M.A., Siso, Ma.I.G., González, Ma.P. *et al.* (1993) Characterization of microbial biomasses and amylolytic preparations obtained from mussel processing wastes treatment. *Bioresource Technol.*, **43**(2); 117–25.

Murado, M.A., González, Ma.P., Pastrana, L. *et al.* (1993). Enhancement of the bioproduction potential of an amylaceous effluent. *Bioresource Technol.*, **44**(2), 155–63.

Nigam, P. (1990). Investigation of some factors important for solid-state fermentation of sugar cane bagasse for animal feed production. *Enzyme Microb. Technol.*, **12**, 808–11.

Pandey, A. (1990). Improvement in solid-state fermentation for glucoamylase production. *Biological Wastes*, **34**, 11–19.

Pasari, A.B., Korus, R.A., Willians, R.A. and Heimsch, R.C. (1987). Catabolite repression of amylase synthesis in yeast. *Biotechnol. Bioeng.*, **30**, 363–7.

Pastrana, L.M. (1991). Producción de ácido giberélico a partir de efluentes de la industria mejillonera en cultivo sumergido y en estado sólido. *Thesis*, University of Santiago de Compostela, Spain.

Pastrana, L.M., González, MaP. and Murado, M.A. (1993). Production of gibberellic acid from mussel processing wastes in submerged batch culture. *Bioresource Technol.*, **45**(3), 213–21.

Peitersen, S.N. (1975). Cellulase and protein production from mixed cultures of *Trichoderma viride* and a yeast. *Biotechnol. Bioeng.*, **17**, 1291–99.

Pomeranz, Y. (1976). SCP from by-products of malting and brewing. *Brewers Digest*, **51**(1): 49–55.

Rameshm, V. and Lonsane, B.K. (1990). Critical importance of moisture content of the medium in α-amylase production by *Bacillus licheniformis* M27 in a solid-state fermentation system. *Appl. Microbiol. Biotechnol.*, **33**, 501–5.

Reddy, C.A. and Abouzied, M.M. (1986). Glucose feedback inhibition of amylase activity in *Aspergillus* sp. and release of this inhibition when cocultured with *Saccharomyces cerevisae*. *Enzyme Microb. Technol.*, **8**, 659–64

Saha, B.C. and Ueda, S. (1983). Alcoholic fermentation of raw sweet potato by a non conventional method using *Endomycopsis fibuligera* glucoamylase preparation. *Biotechnol. Bioeng.*, **25**, 1181–6.

Sanchez-Marroquin, A. (1963). Microbiological production of gibberellic acid in glucose media. *Appl. Microbiol.*, **11**, 523–28.

Skogman, H. (1976). Production of Symba-yeast from potato wastes, in *Food from Waste*, (eds G.G. Birch, K.J. Parker and J.T. Worgan), Applied Science, London, pp. 167–79.

Sills, A.M. and Stewart, G.G. (1982). Production of amylolytic enzymes by several yeast species. *J. Inst. Brew.*, **88**, 313–16.

Siso, Ma.I.G., Murado, M.A., Franco J.M., *et al.* (1988). Microfungus-yeast mixed cultures in the degradation of amylaceous wastes. I: Interactions affecting amylolytic activity. *Biotechnol. Lett.*, **10**(6), 431–36.

Stepanov, A.I., Afanasieva, V.P., Zaitseva, G.V. and Mednikova, A.P. (1975). Regulation of biosynthesis of amylolytic enzymes from *Endomycopsis fibuligera* STR. 20–9. *Prikl. Biokhim. Mikrobiol.*, **11**(5), 682–5.

Udall, J.N., Lo, C.W., Young, V.R. and Scrimshaw, N.S. (1984). The tolerance and nutritional value of two microfungal foods in human subjects. *Am. J. Clin. Nutr.*, **40**, 285–92.

Vass, R.C. and Jefferys, E.G. (1979). Gibberellic acid, in *Economic Microbiology: Secondary Products of Metabolism*, (ed. A.H. Rose), Academic Press, London, Vol. 3, pp. 421–34.

Wilson, J.J., Khachatourians, G.G. and Ingledew, W.M. (1982) *Schwanniomyces*: SCP and ethanol production from starch. *Biotechnol. Lett.*, **4**(5), 333–8.

Worgan, J.T. (1976) Wastes from crop plants as raw materials for conversion by fungi to food or livestock feed, in *Food from Waste*, (eds G.G. Birch, K.J. Parker and J.T. Worgan), Applied Science, London, pp. 23–41.

CHAPTER 14

Waste water treatment for fisheries operations

M.C. Veiga, R. Méndez and J.M. Lema

14.1 INTRODUCTION

The different plant operations of the seafood processing industry generate large quantities of effluents. The main plant processes are defrosting, cooking, canning and cleaning. The characteristics of the waste waters produced are defined and conditioned by the operations and production lines of the factory. Some factories work practically the whole year round and produce only one type of product (Veiga, 1989), such as the tuna processing industries, others possess different mixed seasonal production systems (Soto, 1990).

The pollution effect of these effluents is due to their high content in soluble organic matter and their different concentrations of suspended solids, depending on the raw material and the characteristics of the industrial process. A peculiar characteristic is concentration of salts (Na^+, Cl^-, SO_4^{2-}, K^+, Ca^{2+}, Mg^{2+}) which may lead to two types of effluents (Méndez *et al.*, 1988):

1. High salinity waste waters when the salts come from the products and/or the utilization of seawater in the process.
2. Low salinity waste waters when fresh water is used as a process water and no salinity is produced by the product.

Among these ions, sodium, chloride and sulfate in high concentrations can cause inhibition or toxicity effects during the biological treatment (Méndez, ten Brummeler and Hulshoff-Pol, 1989; Soto, Méndez and Lema, 1991). Some of the waste waters from the seafood processing industry which have a high protein content give a high organic nitrogen concentration, that will be converted into ammonium in the anaerobic

Fisheries Processing: Biotechnological applications.
Edited by A.M. Martin.
Published in 1994 by Chapman & Hall, London. ISBN 0 412 58460 3

process and can cause toxicity above a certain level (Anderson, Donnelly and McKeown, 1982; Veiga, 1989). Seafood processing waste waters are also characterized by strong fluctuations in flow rates depending mainly on the plant operations.

Since most of the fish processing industries are located along the coasts, they have historically discharged their waste waters directly into the sea or into the estuaries, with little or no treatment. Such an attitude threatens the seafood and in some cases the turistic richness of the implicated areas.

Variations in the composition of the raw material and the process water complicates the treatment process of the effluents. There are few data in the literature on the characterization and treatment of these effluents. Such information is, however, essential for elaborating optimal treatment strategies. The latter requires us to consider the need to reduce the volumes and load of the effluents, by modifying the process. The possibilities of recycling, the utilization of the residue as a source of raw material or energy (Ziminska, 1987; Martin and Patel, 1991), and the possibility to apply the methods of treatment in practical cases should also be taken into account. All these aspects should be considered globally, with the objective of minimizing the final costs.

Several alternatives have been reported for the treatment of these effluents, involving mainly processes such as solid–liquid separation, physico-chemical methods and biological processes (Aguiar and San-t'Anna, 1988; McComis and Litchfield, 1988). Aerobic treatments have been mainly developed so far, although anaerobic processes (Carozzi, 1988; Lema *et al.*, 1987; Veiga, Mendez and Lema, 1990; Guerrero, 1992) have emerged as a practical and economical alternative to aerobic processing, and few references of studies at pilot plant scale can be found in the literature (Balslev-Olesen, Lyngaard-Jensen and Nicklesen, 1990; Veiga *et al.*, 1991; Veiga, Méndez and Lema, 1992a, 1994; Méndez *et al.*, 1992). Ultrafiltration is also used to recover the proteins, reducing the pollution (McComis and Lichfield, 1988).

This chapter presents the work performed up to now on the characterization and possible treatment alternatives of the effluents generated by the seafood processing industry.

14.2 SOURCES AND CHARACTERISTICS OF SEAFOOD PROCESSING WASTE WATERS

The management and treatment strategies of the fish processing effluents, in centralized or in individual plants, require in each processing factory a previous study of characterization of the different effluents that constitute the general effluent. The seasonal fluctuations must be characterized as well. The data obtained during these studies allow one to define a strategy of operation, focusing on a reduction in contaminant organic matter and in the volumes of effluents.

The waste water compositions depend on the kind of fish and type of seafood obtained. The chemical oxygen demand (COD) reported for herring and salmon process effluents is higher than 2.5 g/l (Balslev-Olesen, Lyngaard-Jensen and Nickelsen, 1990; Battistoni, Fava and Gratto, 1992), for tuna it ranges from 10 to 50 g/l (Frankel and Phongsphetraratana, 1986; Veiga, 1989), while values of several orders lower have been found for clams, crabs and oysters (Pohland and Hudson, 1976; Brinsfield, Winn and Phillips, 1978; Wheaton *et al.*, 1984).

The waste waters can be classified into three classes according to their origin in the seafood processing factories: canning, filleting and fishmeal production waste waters. Each category is described below.

14.2.1 Fish canning

Fish processed for canning operations produce a considerable number of effluents of diverse compositions and concentrations, originating from processes such as defrosting, cooking, canning itself and cleaning (Frankel and Phongsphetraratana, 1986; Veiga, 1989). The pollution potential of these industries can reach up to 44.5 kg BOD/ton raw material (Aguiar and Sant'Anna, 1988). Frequently, factories operate on a seasonal basis and sometimes with several simultaneous production lines for mussel, tuna, sardine, octopus, etc. (Soto, Mendez and Lema, 1990).

The different sources of waste waters for a factory with different production lines of fish and cephalopods are shown in Figure 14.1 and are described below. The average values of stream characteristics from a tuna processing plant are presented in Table 14.1. Similar streams with different characteristics are generated in other fish processing factories.

Among canning processing industries it is possible to distinguish between fish and shellfish processing industries. Among the different fishes and shellfishes to be considered in the canning industry, the main types are tuna, sardine, mackerel and mussel. The amount of raw products converted into waste can reach up to 50% and often some of these wastes are transformed into fishmeal (Aguiar and Sant'Anna, 1988). The highest total COD content is usually characteristic of the streams coming from the cooking process with typical values of 12.4–50.0, 14.5 and 16.9 g/l, respectively, for the tuna, octopus and mussel processing industries.

(a) Tuna plant

During manufacturing, the frozen tuna undergoes several processes, such as defrosting, cooking and canning. A large number of effluents

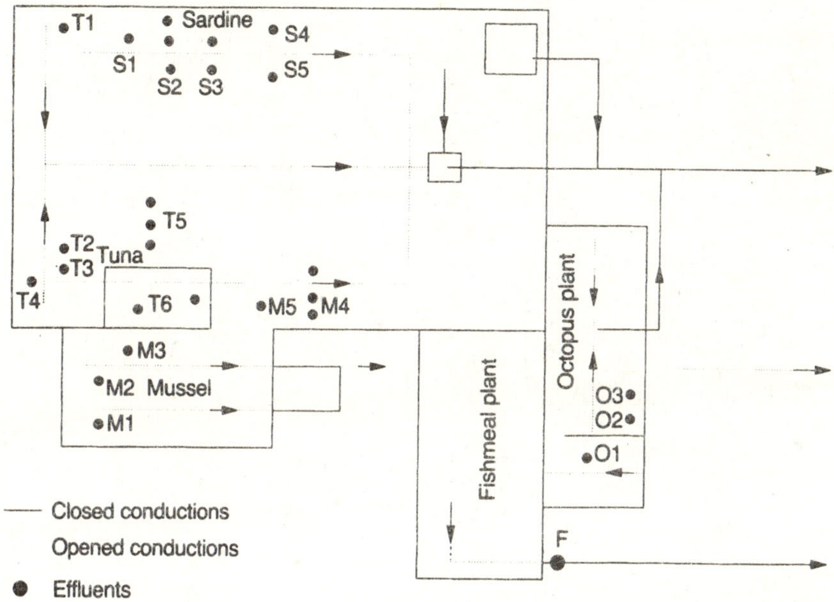

Figure 14.1 Representation of a seafood processing plant with several production lines.

are generated, as can be seen in the results obtained from a typical fish processing factory located in the northwestern part of Spain (Galicia) (Table 14.1). More than 10 tons of water are utilized per ton of fish processed, practically all of it going to the waste treatment system afterwards (Frankel and Phongsphetraratana, 1986). Depending on the water utilized in the factory, the effluents may have a relatively low salinity (5 g Cl^-/l) (Veiga, 1989) or, on the contrary, a high salinity ranging from about 16.8 (cooking) to 19.7 g Cl^-/l (defrosting) (Soto, 1990).

After being defrosted the tuna of lower size is peeled with sodium hydroxide. The effluent resulting from the peeling process has therefore a high pH value of 10.4 and also a high total suspended solids (TSS) content (5.0 g/l). Cooking is realized by injection of steam, using showers of water at the end of the process to obtain a quick cooling. The dumping of that effluent takes place in two stages, first before the process of cooling and later at the end of it.

The cooking effluent possesses the highest COD (12.4–50.0 g/l). Its flow may vary depending on the fish processing plant. Up to 77% of the COD in the tuna effluent is protein, with variable concentrations of fats, which can reach 12% of the COD (Soto, Mendez and Lema, 1990). Frankel and Phongsphetraratana (1986) found that almost 60% of the BOD of the effluent from a tuna canning industry in Thailand came from the precooking operations, having a COD content of 49.12 g/l. Similar

Table 14.1 Characteristics of a tuna stream in a fish processing factory using fresh water as process water

	T(°C)	*pH*	COD_t	COD_S	*TSS*	*VSS*	$N\text{-}NH_4^+$	$P\text{-}PO_4^{3-}$	*Prt*	Cl^-
Cooking	85	5.7	50.5	44.4	2.7	2.6	0.2	0.5	14.8	5.6
Grating	36	11.9	0.9	0.5	0.3	0.3	–	–	0.3	–
Debleeding	10	6.4	3.7	2.0	1.2	1.2	0.1	0.1	1.5	0.4
Defrosting	8	6.6	2.0	1.6	0.6	0.5	0.1	0.1	0.6	3.7
Washing	13	6.6	3.4	1.3	2.0	1.9	0.0	–	0.6	0.1

COD, chemical oxygen demand (t, total; s, soluble); TSS, total suspended solids; VSS, volatile suspended solids, Prt, proteins. Concentrations are means values, given in g/l.

results were obtained for a tuna processing factory in Galicia (Spain) (Veiga, 1989). Tuna plants operate usually the whole year without interruption.

(b) *Sardine plant*

The stream from cutting the sardine also contains refuse pieces which are later collected and sent to the fishmeal plant. Another stream with high organic load content (4.1 g COD/l) coming from the cookers is also produced. It contains mainly fats and is sent directly to the fishmeal plant to extract the oil. The periods of operation of this plant are very irregular throughout the year, depending mainly on the availability of the raw material.

(c) *Octopus and squid plant*

Three main well defined streams are generated from the octopus and squid processing plant, corresponding to the processes of defrosting, cooking and washing. The stream with the highest organic matter content, coming from the cookers, uses steam in a raft of salty water, but has a low and variable flow depending on the raw material used. The COD levels from factories canning octopus and squid in Galicia (Spain) range from 1.0 to 14.5 g/l. In addition the combined waste contains suspended solids up to 2.0 g/l (Soto, Méndez and Lema, 1990). Proteins (80%) and fats (20%) were the main components of the effluent.

(d) *Mussel plant*

In the estuarine systems known as the Rías Baixas in Galicia (Spain) approximately 250 000 tons of mussel semicultures are produced, 60% of which are processed in coastal plants, almost without seasonal interruption. This productivity is among the highest in the world (Fraga, 1976).

Five streams are produced in the process of mussel canning, each presenting very different characteristics. The highest flows correspond to the processes of trimming and washing the mussels, although they present quite low total COD contents. The first stream has a relatively high content of suspended solids, responsible for the turbidity of the waste water, the second one is almost clean. The highest organic load corresponds to the cooking process where the opening of the valves is realized simultaneously. About 77% of the COD are carbohydrates, 20% proteins and 3% fats. This stream is very rich in glycogen (10 g/l) and reaches volumes of 300–400 l/ton of raw mussel treated (González *et al.*, 1992).

The seasonal character of operation of this plant is a consequence of the productive cycle of this shellfish. Continuous operation from June to November is usual and occurs occasionally during the rest of the year.

14.2.2 Fish filleting

The fish filleting process generates three different streams resulting from the operations of cutting, cooking, and cleaning of floors and tables. The main characteristics of those streams and of the global effluent are presented in Table 14.2. The content of total suspended solids and volatile suspended solids can be considered as low, being the average percentage of volatiles (68%). The organic load of these effluents is mainly due to the presence of proteins and fats in high concentrations, while the carbohydrate fraction and the volatile fatty acids contribute to less than 5%.

14.2.3 Fishmeal

Depending on the raw material and the process used to produce fishmeal, different types of effluents are generated. One type of stream is produced when solid residues are used as raw material. In this process approximately 2 tons of waste water are generated per ton of fishmeal produced. On the one hand, meat can be produced from the residues and, on the other hand, oils can be extracted from it, mainly in the case of heads and tails of sardine and tuna, and thorns of tuna. Occasionally other residues are utilized. This material undergoes a process of cooking with steam, trituration and compression, generating a stream that, after extracting the oil, produces the residual effluent which is studied. The flow of this stream comes mainly from the material itself, which has a humidity of 65–70%.

The effluent is eliminated in only one stream having a high organic load (30–90 g COD/l), an important part of it being suspended solids (5–40 g VSS/l). The characteristics of the effluent also vary according to the residue used. The average characteristics of the fishmeal waste water are shown in Table 14.3. The contents of volatile suspended solids are 7, 13 and 30 g VSS/l for the residues of sardine, thorns and heads of tuna, respectively, while the soluble organic load and salinity are similar in the three cases. Proteins are the main components of the solids. The temperature of the effluent is 80°C. The operation period is determined by the production lines that supply the raw material and can practically be considered as continuous throughout the year, due to the important contribution of the tuna section in the entire plant (Guerrero, 1992).

Another different type of stream is produced when the whole fish is used to produce fishmeal. This process is widely developed in Chile. In

Table 14.2 Waste water characteristics of the different streams in the fish filleting process

	TSS	*VSS*	COD_t	COD_s	Prt_t	Prt_s	*Sugar*	*Fat*	Cl^-	SO_4^{2-}
Cutting	0.2	0.1	2.8	2.5	0.9	0.8	–	0	10.0	0.1
Cooking	1.1	0.6	13.7	12.4	5.9	5.3	0.6	8.3	52.7	0.5
Cleaning	4.9	4.7	19.1	4.8	3.5	0.6	–	6.4	2.9	0.2
Total	0.7	0.5	7.0	5.7	2.6	2.2	0.2	3.0	22.8	0.3

See Table 14.1.

Table 14.3 Waste water characteristics of the stream from the fishmeal process

	pH	*TSS*	*VSS*	COD_t	Prt_t	Prt_s	$Sugar_t$	Fat_t	Fat_s	*Volatile fatty acid*	Cl^-	SO_4^{2-}	$P\text{-}PO_4^{3-}$	$N\text{-}NH_4^+$
Tuna thorns	6.4	13.0	12.8	46.7	13.8	8.5	1.5	6.1	0.4	0.8	8.3	0.4	0.2	0.2
Tuna heads	6.3	29.9	29.5	81.3	27.3	8.6	2.5	8.8	3.6	1.2	8.7	0.7	0.2	0.2
Sardine heads	6.5	6.6	6.5	38.2	16.5	11.7	1.2	2.1	0.6	2.2	6.3	0.7	0.3	0.6
Average	6.4	16.5	16.5	55.4	19.2	9.6	1.7	1.5	1.5	1.4	7.8	0.6	0.2	0.3

See Table 14.1.

Table 14.4 Main characteristics of the waste waters from the discharge and the evaporator in the production of fishmeal from mackerel

	pH	*BOD_5*	*COD*	*TS*	*Prt*	T(°C)
Evaporator	8.3	0.1	0.8	43.7	0.3	29
Discharge	6.3	2.8	11.0	60.7	1.6	17

See Table 14.1.

1989, 93% of the fish captured in that country was destinated to the production of fishmeal. The mackerel is the most common fish used as a source of fishmeal in this kind of process; anchovies and hakes also being utilized.

The process involves the hydraulic transport of fishes, and the discharge, cooking and evaporation of the liquids present in the fish. The steam produced in the latest step is condensed into fish condensate. Thus the process generates three different types of streams which can be classified, according to their decreasing organic load, in the following order: discharge (86% of total COD), washing (5.5%), and process water (8.5%) that includes the condensate from the evaporators and water used to deodorize (Becerra, Aspe and Roeckel, 1990; Sandberg and Ahring, 1992). In the process, 1 ton of fish generates approximately 690 kg of waste water.

The main characteristics of the waste water from the discharge and the evaporator are presented in Table 14.4. The streams from this process may also vary when the fish processed is the same, due to the different conditions of discharge of the fish. Christoffersen and Petersen (1991) reported condensate waste water with an organic content of 2–13 g COD/l and nitrogen content of 0.2–1.6 g N/l.

The main cause of pollution in the transport of fish is the transformation of the insoluble proteins into soluble proteins. Therefore part of the solids from the fish are transformed in an aqueous solution that in a normal process is discharged, releasing the proteins and at the same time representing an important source of contamination for the sea.

14.3 TREATMENT TECHNOLOGIES

14.3.1 Waste water treatment alternatives

Until recently, the effluents were considered as subproducts with no value; however, in the actual circumstances, the effluent must be considered as an additional by-product to be included in the fabrication cost, and it is then important to reduce as much as possible its volume and organic matter.

Consequently, the need appears to diminish the consumption of processing and cleaning water, so that the effluents are less contaminating. The application of systems to recover proteins, fats and by-products allows one to reduce the contribution of contaminant organic matter. So, the substitution of mechanical pumps by pneumatic ones applied together with the recirculation of water and the use of centrifugation allow us to reduce the pollution notably.

In the fish canning factories, by modifying the production lines and segregating the effluents, it is possible to obtain an optimum result from the different treatment processes available that will minimize the treatment costs and the environmental impact of the fish processing industries.

Concentration systems are not a practical alternative due to the elevated energetic costs and to the high volumes to be treated. The physico-chemical and biological methods, aerobic and anaerobic digestion, are the most suitable.

The usual treatment systems, such as the aerobic biological oxidation and physico-chemical methods do not allow us to treat efficiently the medium and high organic loaded effluents, due to the high energetic and maintenance costs; in addition, they are sometimes not technically practicable. Therefore the adopted solution in most of the cases is their elimination by emissary way, using the diluting power of the marine currents. Another alternative method for avoiding this environmental contamination is anaerobic digestion which permits the treatment of the medium and high organic load effluents. The use of aerobic and physico-chemical processes would be reserved for the effluents of low load or as a post-treatment method of effluents from the anaerobic process. The most important aspects of each treatment process are described below.

14.3.2 Aerobic treatment processes

A major asset of the aerobic systems is their capability to handle all kinds of waste waters, especially those with extremely variable compositions. Traditionally, the aerobic system has been used for the treatment of high strength waste waters. In aerobic systems oxygen transfer to the liquid phase is generally the rate limiting stage. High strength waste waters require such long residence times or high rates of aeration that for increasing waste strength aerobic processes are not economically viable. Thus, aerobic processes are only suitable for the treatment of low strength effluents and also as a post-treatment method.

14.3.3 Anaerobic treatment processes

The possibility of applying treatment methods based on the anaerobic digestion process has been considered recently (Lema *et al.*, 1987;

Carozzi, 1988; Veiga, 1989, Balslev-Olesen, Lyngaard-Jensen and Nickelsen, 1990; Soto, 1990). The anaerobic processes are not suitable as a post-treatment method or to treat low strength waste waters, but are adequate for the treatment of high strength waste waters. For the treatment of high organic strength waste waters they present the advantages of a very low sludge production and production of energy that can be used to cover part of the energetic needs of the factory.

Taking into account the characteristics of the fish processing waste water (possible presence of high concentrations of salts and ammonium, and large volumes of effluents) high rate bioreactors should be used, that allow accumulation of high biomass concentrations and the use of relatively low hydraulic retention times (Hickey *et al.*, 1991). This method is suitable for waste waters with a COD concentration higher than 5 g/l.

14.3.4 Physical and chemical treatment processes

The physico-chemical methods are recommended for waste waters with a high suspended solid content. The suspended solids are efficiently eliminated by the coagulation/flocculation process followed by a sedimentation step. The chemical precipitation presents three disavantages respective to the biological methods:

1. An economic aspect, the cost of the reagents needed.
2. The fact that the presence in the effluent of the added reagents can increase the pollution.
3. The production of big volumes of sludges.

The physico-chemical methods can usefully be applied in combination with other methods, as a pre-treatment or a post-treatement method (Ramos, 1991).

14.3.5 Emissary

An alternative to the treatment methods is to take advantage of the diluting capacity of the sea to eliminate waste waters by emissary when the factories are located on the coast and provided effluents have a low organic load. Using this method, the industry should, however, make sure not to exceed the legally authorized upper contamination limits.

14.4 TECHNOLOGY SELECTION

The choice of individual or combined treatment methods is determined following evaluation of the waste water characteristics in terms of organic strength, soluble and insoluble solids fraction, and nutrient content. Thus effluents from the seafood processing industry can be divided into effluents with a high suspended solids content, with a high organic strength and with a low organic strength.

A case study of a fish processing plant operating with different production lines is described below, giving the treatment strategy. The treatment processes for the three types of waste waters indicated are also shown.

14.4.1 Treatment strategy

Fish processing usually takes place in small factories which makes it difficult to treat individually and specifically each waste stream. Nevertheless, because of the proximity of the various factories, one alternative is to imagine a global treatment of similar streams from different local factories.

A fish processing plant from Galicia (Spain), where several production lines operate in parallel, is described as an example of treatment strategy. It allows one to characterize separately the several effluents originating from the main process of the canning industry, elaborate and discuss the gestion of the effluents as well as an adequate treatment strategy for the plant, which can be generalized to the sector.

The factory is located on the coast and uses mainly seawater as process water with a flow of 26 000 m^3/week during the high production periods. Most of that water returns ultimately to the sea. The selected fish-canning factory processess a wide variety of raw materials and therefore produces several effluents with different characteristics. The following production lines have to be considered: fishmeal production plant (F), octopus and squid processing (O), mussel processing (M), tuna processing (T), and sardine processing (S).

There is also a group of effluents that are difficult to separate, due to the spatial situation of the several production sectors and to the fact that they often correspond to a common operation in all of them. The effluent designated by 'other waters' consists mainly of effluents from washing the machines and rooms, periodic operations of disinfection, effluents from refrigerators and other operations common to several lines and other marginal effluents.

The schematic process diagram is shown in Figure 14.1, where the situation of the different production lines can be observed as well as the waste water effluents generated in each one. The description of each process and their main characteristics have already been given above. Each stream contributes to a different extent to the final effluent characteristics, the organic load being more important in the streams produced during the cooking process.

The suspended solids content can be considered as low and is belo 1.5 g/l, except for the stream from the fishmeal process. Referring to t composition of the organic matter, carbohydrates, proteins and fats represent all together about 100% of the effluent COD, the proportion being variable from one stream to another. The salinity, resulting from the use of seawater or coming from the raw material itself, is high in most of the streams.

In Figure 14.2 one can observe the contribution of the different production lines to the global effluent of the factory, in terms of flow, organic load and suspended solids content. The situation where all the production lines are operating, which corresponds to the periods of operation of the mussel plant, is represented in this figure.

Two general alternative treatment strategies can be considered:

1. The global treatment of the effluent volume, using one or several methods in series.
2. The application of different methods to each effluent or group of effluents in parallel, depending on their characteristics.

In general, the viability or convenience of applying a method of treatment depends directly on the characteristics of the effluents and on their volume and strength. Strategy (1) is suitable for global effluents composed of individual effluents with similar characteristics, and also when the global volume is small, because a duplication of the installations, in this case, would be unfavorable. Strategy (2) should be considered whenever there is a considerable difference between the

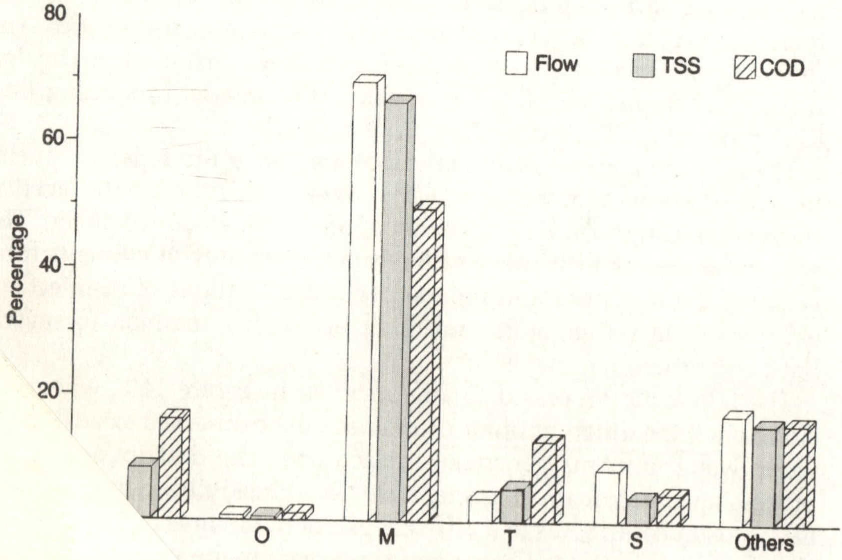

...bution of the flow, total suspended solids (TSS) content and ... each production line (F, fishmeal; O, octopus; M, mussel; ... the global effluent of a seafood processing factory.

contribution of each individual effluent discharged to the final effluent, respective to their flow or their organic load. This strategy is also recommended in the case of having different factories located close to each other which are interested in treating their effluents jointly at a local level.

According to the criteria mentioned for selecting the treatment method for each particular effluent, it is possible to separate the streams into several groups. However, it is not always feasible and would probably require important structural changes in the plant, with the consequent economical costs.

Therefore an index 'S' was defined that takes into account the percentages of organic load and flow that contribute to the global effluent. It gives information about the relative importance of each stream and permits one to minimize the separation of effluents (Soto, Méndez and Lema, 1990).

$$S = \%OLR - \%Q = \%Q \{(COD/\langle COD \rangle) - 1\}$$

where ⟨COD⟩ is the average value of the global effluent.

The order of the effluents established according to their S value is shown in Figure 14.3 and is divided into four groups. Group I, suitable for treatment by anaerobic digestion, is formed by effluents from fishmeal plant, mussel cooking and tuna cooking, and represents 3% of the total flow and contributes to 42% of the global organic load. Group II, is formed by effluents from mussel trimming and carries 51% of the total suspended solids and 64% of inorganic solids, requiring treatment by sedimentation before being dumped. Group III is generated from segregation of the rest of the groups and has an average COD of 1.4 g/l, which makes it suitable for aerobic treatment. Group IV, formed by effluents from mussel washing, represents 30% of the total flow with an average COD of 0.2 g/l and could be dumped straight away or partially recycled.

14.4.2 High suspended solids waste waters

Some of the effluents contain high levels of inorganic suspended solids (i.e. waste water from washing of mussels). Normally these components are easily removed by settling (Nagamine, 1986; Sheffield, 1992). When the waste water has a high content of organic suspended solids, physico-chemical treatment should be applied in order to separate and possibly recycle them into the process, thus improving its yield. Simple sedimentation can be effective for the removal of floatable fats and grease. The precipitation method can allow additional removal of oil and solids, increasing the removal efficiency. Aguiar and Sant'Anna (1988) investigated the treatment of waste waters from the fish processing industry with 4.7–13.0 g COD/l, 0.9–2.3 g VSS/l and 0.7–2.1 g/l of oil

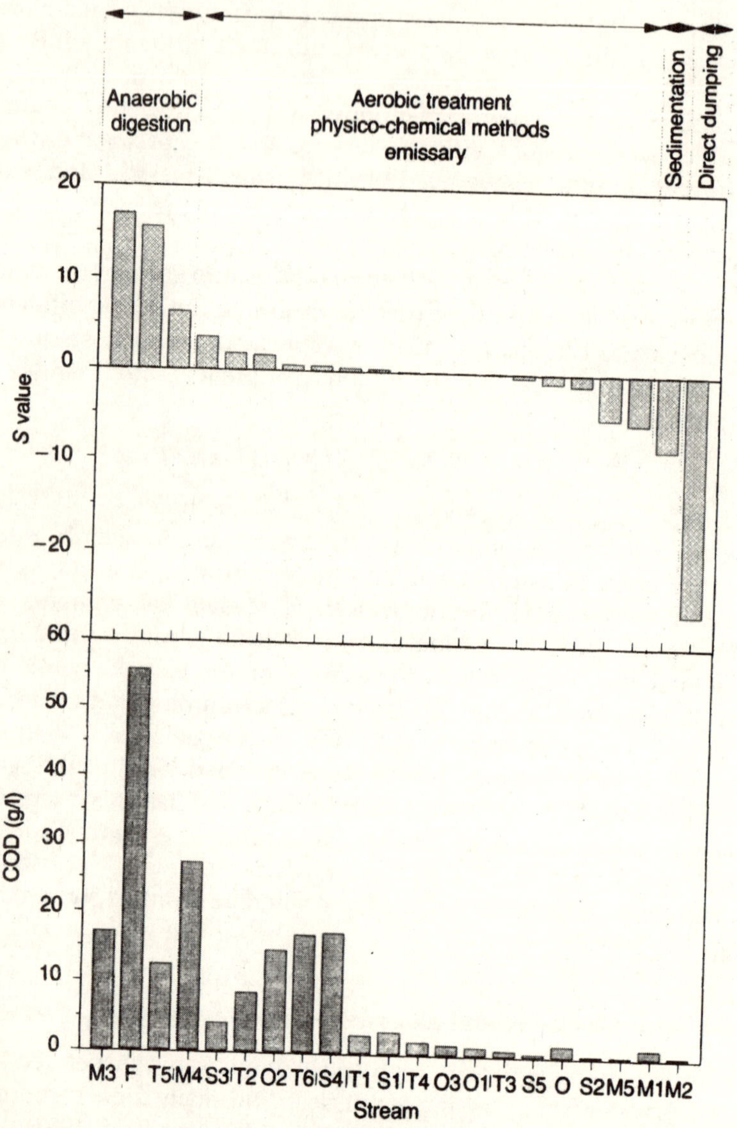

Figure 14.3 Classification of the streams from the different production lines according to their *S* value, comparison with the COD value and suggestion for the most adequate treatment. F, fishmeal; M, mussel, M1, trimming; M2, washing; M3, cooking; M4, dehydration; M5, transport; T, tuna, T1, defrosting, T2, peeling, T3, secondary peeling, T4, washing, T5, cooking, T6, grill washing; S, sardine, S1, scale removing, S2, transport, S3, cutting, S4, grill washing, S5, rinsing; O, octopus, O1, defrosting, O2, cooking, O3, washing; O, others.

and grease. They attained removals of 32% of the COD, 67% of oil and grease, and 50% of suspended solids by a simple sedimentation, and removals of 31% of the COD, 86% of oil and grease, and 70% of suspended solids by the coagulation/flocculation step using aluminum sulfate as coagulant in a concentration range of 0.07–0.30 g/l at pH 5.0–5.5.

The fishmeal processing waste water contains a high concentration of suspended solids, which can be separated by a physico-chemical method and recycled into the process. The effluent generated can be treated by a biological process. The coagulation–flocculation method is used to recover proteins. The nature of the proteins determines the coagulants to be selected and the conditions of operation (Becerra, Aspe and Roeckel, 1990). Different types of salts, such as ferric chloride, calcium chloride, aluminum sulfate and lignosulfonates, are used, the best results being obtained with ferric chloride. Other compounds, such as chitosan, have also been used to recover the proteins (Johnson and Gallanger, 1984).

Some authors have utilized the ultrafiltration to separate the solids, reutilizing the water for the process. Flotation and adsorption methods have also been applied (Iggleden and van Staa, 1984). Since the fishmeal processing industry utilizes centrifugation in one step of the process, a reduction in the suspended solids present in the effluent can be reached by optimizing the centrifugation step (Guerrero, 1992).

Guerrero (1992), applying a coagulation–flocculation process to the fishmeal processing waste waters, obtained a reduction of 62% of the suspended solids at pH 8. This reduction was due in part to the separation of the suspended solids and also to the hydrolysis of the insoluble compounds.

Another treatment alternative is anaerobic digestion in two phases. In the first phase the organic polymers, mainly insolubles, are hydrolyzed into soluble volatile fatty acids. Under thermophilic and mesophilic conditions, respectively, 55 and 45% solubilization of fishmeal waste waters at an hydraulic retention time of 24 h was obtained (Guerrero, 1992).

14.4.3 High organic strength waste waters

As was mentioned above, anaerobic digestion is the most suitable method for waste waters with a high content in soluble organic matter, because it allows one to produce energy that covers part of the energy demand of the factory. However, until recently this strategy had not been applied to wastes from seafood processing industries, probably due to the volumes and characteristics of the waste waters generated. Among seafood processing industries, the fish and shellfish processing industries generate high organic load effluents. They contain the required concentration of nutrients (nitrogen and phosphorus) for the anaerobic digestion process.

Some of the fish processing effluents present high concentrations of salts, sodium, chloride and sulfate which can be inhibitory or toxic for the microorganisms. Several batch and semicontinuous assays were carried out using three very different types of effluents: from mussel cooking with a high content of sugars, from octopus cooking with a high content of proteins and salts, and a stream from fishmeal processing with a high content of volatile suspended solids, proteins and fats. The anaerobic biodegradability was higher than 90% of the soluble fraction in all cases. In the case of mussel effluents, the carbohydrate fraction was almost completely removed, while removal of the proteinic fraction was dependent on the operational temperature. Only a fraction of the fishmeal effluents, rich in suspended solids, was solubilized during the anaerobic treatment.

The presence of sulfate, at ratios of COD/SO_4^{2-} close to 8, causes toxicity when it is reduced to sulfide during the anaerobic process. Destabilization of the process occurs when the concentration of H_2S reaches 5% in the gas phase below a pH of 7 in the liquid phase. However, at pH 7.25 or higher the system operates well, because the sulfide concentration is much lower in alkaline media. Methanogenic activities of 0.2 and 0.4 g COD/g VSS day were obtained under conditions where the concentrations of NH_4^+, H_2S, and Na^+ cause inhibition problems (Soto, Méndez and Lema, 1991). Adaptation of the microorganisms to sodium concentrations up to 9 g Na^+/l were possible. The toxicity of Na^+ could be reduced due to the antagonistic effect of different cations present in the waste water (Feijoo, 1991). The biodegradation of waste waters with a high protein content, such as tuna, octopus and fishmeal waste waters, with concentrations of ammonium up to 4 g/l is only possible after an adaption period (Veiga, 1989).

Studies carried out at bench scale showed better biodegradation of octopus, mussel and fishmeal effluents under thermophilic conditions than under mesophilic conditions, but worse removal of volatile fatty acids. The same comparison was studied at laboratory scale in two anaerobic filters (AF) treating mussel processing waste waters (Lema, Méndez and Blázquez, 1988; Soto, 1990). The waste waters treated were very saline (10–20 g chloride/l) and had a high content of organic matter (10–50 g COD/L). After a prolonged start-up period of about 9 months, to allow the microorganisms to adapt to the high salt concentrations, both reactors reached a steady-state with a similar performance for organic loading rates of 9 and 24 kg COD/m^3/day in the thermophilic and mesophilic reactor, respectively. They removed 73 and 63% of the COD, and reached 69 and 62% methanization, respectively. In both cases, practically all the sulfate was removed, leading to H_2S concentrations between 3 and 4% in the biogas (Table 14.5).

Table 14.5 Anaerobic reactor studies of seafood processing waste waters

Origin of waste water	*Reactor type*	*Volume (l)*	T (°C)	*COD (g/l)*	*HRT (d)*	*OLR (kg/m³ day)*	*COD removal (%)*	*Cl⁻ (g/l)*	*References*
Fish	FBR	373	–	5.1	2	2.6	53	–	(1)
Fish	AF	201	–	5.1	2	2.7	79	–	(1)
Tuna	DSFF	1	37	20–50	–	2	75	5.5	(2)
Tuna	DSFF	15000	37	20–50	–	2	75	5.5	(2)
Tuna	AF	1	37	12–25	2.2	5	90	3.0	(2)
Tuna	AF	1	37	12–25	1.3	12	80	3.0	(2)
Mussel	AF	0.9	37	11.4	2.7	4.2	77.0	14	(3)
Mussel	AF	0.9	37	25.6	3.1	8.6	73.4	14	(3)
Mussel	AF	0.9	37	17.5	1.0	17.1	65.1	14	(3)
Mussel	AF	0.9	37	18.2	0.7	24.7	62.9	14	(3)
Mussel	AF	0.9	55	14.5	5.7	25	72.0	14	(3)
Mussel	AF	0.9	55	14.7	2.3	6.4	77.7	14	(3)
Mussel	AF	0.9	55	25.6	2.8	9.2	72.7	14	(3)
Mussel	AF	0.9	55	20.6	1.9	10.7	63.5	14	(3)
Fish caning	AF	365	35	13.4	0.6	10	74	11.5	(4)
Fish caning	FBR	359	35	13.4	0.6	10	86	11.5	(4)
Tuna	CAD	15000	37	20–25	5.0	4.5	80	14	(5)
Tuna/mussel (1/1)	CAD	15000	37	20–22	7.5	3.2	90–95	14	(5)
Tuna/mussel (1/2)	CAD	15000	37	18–20	5.6	3.8	90–95	14	(5)
Tuna/mussel	CAD	15000	37	16–18	5.0	4.2	75–84	14	(5)
Fishmeal	UASB	4.6	32	2–13	–	30	95	–	(6)

References: 1, Carozzi, 1988; 2, Veiga, Méndez and Lema, 1992a,b; 3, Soto, 1990; 4, Balslev-Olesen, 1990; 5, Méndez *et al.*, 1992; 6, Christoffersen and Petersen, 1991.

Based on the laboratory scale results, showing the viability of the anaerobic process, a pilot plant was built near the factory. It consisted of a predigester of 7 m^3, a suspended sludge digester or CAD (central activity digester) of 15 m^3 and a clarifier of 3 m^3. The anaerobic treatment of a tuna cooking effluent and of a mixture of tuna and mussel effluents were carried out. The tuna effluent contained 14 g Cl^-/l, 34.5 g COD/l and 4.0 g TSS/l, being 77, 12 and 11% of the COD proteins, fats and carbohydrates, respectively. The mussel effluent contained 13 g Cl^-/l, 18.5 g COD/l and 1.4 g TSS/l, with 22, 4 and 74% of the COD being proteins, fats and carbohydrates, respectively. Good adaptation of the microorganisms to the salinity was obtained by feeding the pilot reactor with diluted tuna waste water and by progressively decreasing the dilution, which allowed the chloride concentration in the feed to increase until reaching the natural concentration of the effluent. In all cases, treating effluents from an individual line (mussel or tuna) or from mixtures of them in any proportion, the reactor reached an organic loading rate (OLR) of 4 kg/m^3/day with a COD removal of 75%, even when the feed composition was changed suddenly (Table 14.5). Higher OLR could be reached by improving the solids retention time in the reactor (Méndez *et al.*, 1992; Veiga, Méndez and Lema, 1992b).

Several laboratory and pilot studies were undertaken to determine the biodegradability of tuna processing waste waters (25–50 g COD/l) with about 80% of their total COD being proteinaceous material and with a relatively high salt content (9.2 g/l). The performances of an AF and a downflow stationary fixed film (DSFF) reactor were tested under mesophilic conditions (Veiga, 1989; Veiga *et al.*, 1990; Veiga, Méndez and Lema, 1992a, 1993). In the first step of anaerobic digestion, the degradation of proteins leads to the formation of high amounts of ammonium-nitrogen which can be inhibitory. A laboratory scale (1 l) and a pilot scale (15 m^3) DSFF reactor, using polystyrene as support material, both operated under stable conditions, achieved COD removal efficiencies of 75% at OLR of 2 kg/m^3 day (Table 14.5). Other supports that were tested were PVC and clay. No significant biomass adhesion was observed on any of the support materials; however, a wide variety of microorganisms was present in all the supports, the dominant methanogenic species being *Methanosarcina* and *Methanogenium*. *Methanosaeta* spp. were not able to grow in this kind of waste water, which could be due to the relatively high ammonium concentration (5 g/l). Laboratory studies carried out in an AF (1 l) showed much better performance than the DSFF reactor, allowing higher organic loadings and COD removal efficiencies, up to 12 kg COD/m^3/day and 80% respectively. It thus appears better to use an AF, which can entrap the biomass allowing a higher retention of the microorganisms, rather than the DSFF digester, where the only biomass present is that attached to the support media.

Frankel and Phongsphetraratana (1986) have also treated tuna precooking (3–6 g COD/l) waste by anaerobic digestion. The COD reduction obtained exceeded 65% and the gas produced reached 70% methane, corresponding to 18 m^3 biogas/m^3 precooking waste.

Carozzi (1988) evaluated the effectiveness of the anaerobic digestion process for the treatment of medium organic load fish processing waste waters in four different reactor configurations, one contact process, two fixed bed reactors with different support media and one fluidized bed reactor. All the reactors gave good COD removals. The fixed bed reactor was more stable at a short hydraulic retention time than the contact process and the fluidized bed reactor.

Condensate waste water, with 2–13 g COD/l, from a fishmeal production plant was treated by anaerobic digestion in an upflow anaerobic sludge blanket (UASB) reactor (4.65 l) in mesophilic conditions, achieving a COD conversion of 95% at organic loading rates up to 30 kg COD/m^3/day. The methane content in the produced gas was 80–90%. Kameoka, Sakimoto and Inno (1988) have also reported the anaerobic treatment of fishmeal plant effluents.

14.4.4 Low organic strength waste waters

The aerobic process can either be used to treat fish effluents with a low concentration of organic matter (Brinsfield, Winn and Phillips, 1978; Takumizawa, Inoue and Honda, 1986; Takumizawa, *et al.*, 1986a, 1986b) or as a post-treatment method for the effluents coming from the anaerobic process (Nair, 1990).

Dissolved organics from crab waste water could be reduced effectively by a biological treatment using extended aeration, provided that a high level of biomass was present in the reactor and settling of the biological mass occurred prior to effluent discharge. A retention time of 16–24 h removed approximately 90% of the dissolved organics (Geiger *et al.*, 1985).

Traditionally, most of the waste water from the fish processing industry in Japan is treated by the aerobic activated sludge process (Matsuyama and Izumi, 1988).

The treatability of raw shellfish processing waste waters by anaerobic methods is limited to the moderate to low organic strength waste waters, due to the relative nutrient concentrations which do not permit high biomass generation or biomass retention under normal hydraulic loadings. The aerobic treatment reduced the soluble organic constituents to levels acceptable for discharge (Pohland and Hudson, 1976; Hudson, Pohland and Pendergrass, 1978).

14.4.5 Nitrogen removal

Seafood and, consequently, seafood processing waste waters are rich in proteins. This results in the formation of ammonium during the anaerobic treatment of the effluents, which must therefore undergo a post-treatment step. Ammoniacal nitrogen is an environmental contaminant which may lead to problems of eutrophication. It may be toxic for animals at concentrations of 1 mg/l or even below and can lead to the appearence of nitrates, nitrites and nitrosamines, which are toxic and potential carcinogenic substances. Two kind of methods can be considered for nitrogen removal, i.e. physico-chemical methods and biological methods.

Physical and/or chemical processes can efficiently eliminate ammonium and nitrate (Dahba, 1987) although they are usually more expensive than biological processes and less interesting for industrial waste waters such as seafood processing waste waters since they are not suitable for the treatment of large volumes.

The physico-chemical methods include, among others, *reverse osmosis* which allowed removals of 80% ammonium and 50% nitrate from contaminated groundwater (Dahba, 1987), *membrane filtration*, *ion exchange*, which is based on the exchange of ions (for instance nitrate and nitrite) from a solution for ions contained in an anionic resin, and *stripping* with air.

The biological methods include nitrification–denitrification, algae production and immobilization of microorganisms (Guerrero, 1992). The first one is usually the process of choice for nitrogen removal from waste waters and will be discussed more in detail. Each method also allows the elimination of phosphorus.

In the nitrification–denitrification method, the nitrification step is carried out by two groups of aerobic chemolithotrophic bacteria (Focht and Verstraete, 1977). The first one, represented mainly by *Nitrosomonas* oxidizes ammonium to nitrite (1), the second group containing *Nitrobacter* as the dominant bacterium, oxidizes nitrite to nitrate (2):

$$NH_4^+ + 1.5\,O_2 \rightarrow NO_2^- + 2\,H^+ + H_2O \qquad (1)$$
$$NO_2^- + 0.5\,O_2 \rightarrow NO_3^- \qquad (2)$$

$$NH_4^+ + 2O_2 \rightarrow NO_3^- + 2\,H^+ + H_2O \qquad (3)$$

During the denitrification step nitrate plays the role of electron acceptor, oxygen being limited (Focht and Verstraete, 1977). Nitrate is degraded stepwise to molecular nitrogen (4):

$$NO_3^- \rightarrow NO_2^- \rightarrow NO \rightarrow N_2O \rightarrow N_2 \qquad (4)$$

The organic electron donor (carbon source) is simultaneously transformed to CO_2 mainly by facultative anaerobic heterotrophes (Knowles, 1982).

The nitrification–denitrification method was chosen as the nitrogen removing technique for the post-treatment of waste water from a fishmeal producing factory (Guerrero, 1992). The effluent was previously treated in an anaerobic filter. Since seafood waste waters contain high concentrations of proteins, high ammonium concentrations will be produced during the anaerobic treatment. Ammonium levels of 7 g/l were reached in the effluent leaving the anaerobic reactor.

Bacteria such as *Nitrobacter* carrying out reaction (2) are inhibited by high concentrations of ammonium (Anthonisen *et al.*, 1976). In order to lower that concentration a stripping system was tested as a possible preliminary step to be used before the nitrification–denitrification post-treatment. Stripping has been widely used in industrial waters treatment to eliminate dissolved gases such as H_2S, CO_2 or ammonium. This technique usually gives best results when adding an alkaline solution to reach pH values around 11, allowing better ammonium transfers from the liquid phase to the gas phase (air). For the fishmeal waste water, laboratory scale experiments were undertaken at pH 8.5, 10 and 12.5, reaching, respectively, 47, 69 and 96% ammonium removal after 5 h. Although such removal levels seem very atractive, one has to take into account the need to lower again the pH value close to neutrality to allow the subsequent nitrification process to take place. Even if the effluent is not to be treated further after the stripping stage, the pH should be lowered before dumping the effluent and a cost evaluation should first be undertaken. Another alternative would be to dilute the effluent as much as necessary to reach non-toxic ammonium-nitrogen concentrations.

During nitrification–denitrification experiments undertaken with the fishmeal waste water, the effluent was highly diluted (95–57%) during the first weeks of reactor operation, to use a mixture of 60% effluent/ 40% water at the beginning of the second month. Under such conditions, *Nitrobacter* is partly inhibited, but not *Nitrosomonas*. This results in relatively higher nitrite concentrations than nitrate concentrations in the medium to be denitrified, which may be favorable for denitrification (Suthersan and Ganczarczyk, 1986a,b).

At the beginning of the nitrification process, the effluent being very dilute, the nitrogen load was low and about 60% nitrification could be reached. However, as soon as the nitrogen loading rate reached 1 kg N/m^3/day or higher values, the nitrification percentages dropped below 20%. Removal of about 1.25 kg COD/kg VSS/day were reached under steady-state conditions. The presence of organic matter after the nitrification step is necessary for the denitrifying bacteria, consuming about 38 g COD for each gram of nitrate to be reduced.

14.5 CONCLUSIONS

Waste waters coming from fish processing factories can be classified into three categories: canning, filleting and fishmeal production waste waters.

Waste water characteristics from the canning factories are conditioned by the diversity of the raw material and by the fact that they are generated at many points of the factory and with very different properties. While some of them, especially those from cooking, present a high COD, others have a low COD but a high content in organic and/or inorganic solids. The presence of seawater that can strongly influence the treatment is also common.

Fish filleting uses a more homogeneous raw material, so that the diversity of problems is lower than in canning factories, the COD being mainly due to the presence of proteins and fats.

The use of either residual material or the whole fish to produce fishmeal strongly influences the characteristics of waste waters from fishmeal production because the technology used in each case is different. In any case, one of the most important problems is the presence of organic solids. They would be eliminated and recycled to the production process, in order both to improve the waste water treatment process and to diminish the costs.

The data obtained from the characterization of the processing plant effluents allow us to define the most suitable strategy to treat these effluents by anaerobic, aerobic or physico-chemical systems, depending on the production conditions and the location of the factories, in order to minimize contamination and reach levels established by legislation.

The anaerobic studies carried out on laboratory and pilot scales showed:

1. The viability of treating anaerobically effluents from the fish processing industry, with levels of chloride reaching 14 g/l, without modifying the pH or adding nutrients.
2. Depending on the waste water characteristics, reactor configuration and operation conditions it is possible to reach different organic loading rates. For high strength waste waters with a high content of proteins or carbohydrates, organic loading rates of 13–24 kg/m^3/day were obtained.

From the results obtained, it can be pointed out that to apply anaerobic treatment of these waters successfully several factors are very important:

1. It is recommended to remove the solids contained in the waste waters before feeding it into the digester. These solids can be recycled to produce fishmeal.
2. Design would include the possibility to operate with a certain degree of inhibition.
3. The start-up of the process should be relatively slow in order to adapt the sludges to this specific kind of waste water.
4. Special attention would be paid to control the high concentrations of ammonia in the effluent and, especially, to remove it by stripping and nitrification–denitrification processes.

ACKNOWLEDGMENTS

This work was financed by DGICYT (Spain) (Project number PB89–0555).

REFERENCES

Aguiar, A.L.C. and Sant'Anna G.L., Jr (1988) Liquid effluents of the fish canning industries of Rio de Janeiro state – treatment alternatives. *Environ Technol. Lett.*, **9**, 421.

Anderson, G. K., Donnelly, T. and McKeown, K.J. (1982) Identification and control of inhibition in the anaerobic treatment of industrial wastewaters. *Process Biochem.*, **17**, 28.

Anthonisen, A.C., Loehr, R.C., Prakasam, T.B.S. and Srinath, E.G. (1976) Inhibition of nitrification by ammonia and nitrous acid. *J. Water Pollution Centrol Fed.* **48**, 835.

Balslev-Olesen, P., Lyngaard-Jensen, A. and Nickelsen, C. (1990) Pilot-scale experiments on anaerobic treatment of wastewater from a fish processing plant. *Water Sci. Technol.*, **22**, 463.

Battistoni, P., Fava, G. and Gatto, A. (1992) Fish processing wastewater: emission factors and high load trickling filters evaluation. *Water Sci. Technol.*, **25**, 1.

Becerra, F., Aspe, E. and Roeckel, M. (1990) Characterization and preliminary essays of treatment of liquid effluents in the fish industry. *Alimentación Equipos y Tecnología*, 147–51 (in Spanish).

Brinsfield, R.B., Winn, P.N. and Phillips, D.G. (1978) Characterization, treatment and disposal of wastewater from Maryland seadfood plants. *J. WPCF*, **50**, 1943.

Carozzi, A. (1988). Anaerobic treatment of fish processing wastewater, in *Poster Papers of the Fifth International Symposium on Anaerobic Digestion*, (eds A. Tilche and A. Rozzi), Monduzzi Editore, Bologna, Italy, p. 477.

Christoffersen, D. and Petersen, G. (1991) Anaerobic treatment of polluted condensate from fishmeal production, in *Poster Papers of the Sixth International Symposium on Anaerobic Digestion*, Sao Paulo, Brazil, p. 83.

Dahba, M.F. (1987) Treatment alternatives for nitrate contaminated groundwater supplies. *J. Environ. Systems*, **17**, 65.

Feijoo, G. (1991) Sodium toxicity in anaerobic digestion processes: phenomena of antagonism and adaptation. *MSc dissertation*, University of Santiago de Compostela, Spain (in Galician).

Focht, D.D. and Verstraete, W. (1977) Biochemical ecology of nitrification and denitrification. *Adv. Microb. Ecol.*, **1**, 135.

Fraga, F. (1976) Photosynthesis in the Ria of Vigo. *Inv. Pesquera*, **430**, 151 (in Spanish).

Frankel, R.J. and Phongsphetraratana, A. (1986) Effects of water reuse, recycling and resource recovery on food processing waste treatment in Thailand. *Water Sci. Technol.*, **18**, 23.

Geiger, E.L., Wheaton. F.W., Brinsfield, R.B. and Alleman, J.E. (1985). Biological treatment of crab processing plant wastewater. *J. WPCF*, **57**, 1128.

González, Ma.P., Siso, Ma.I.G., Murado, M.A. *et al.* (1992) Depuration and valuation of mussel-processing wastes. Characterization of amylolytic postincubates from different species grown on an effluent. *Bioresour. Technol.*, **42**, 133.

Guerrero, L. (1992) Integral treatment of wastewaters from the fishmeal production. *PhD Thesis*, University of Santiago de Compostela, Spain (in Spanish).

Hickey, R.F., Wu, W.M., Veiga, M.C. and Jones, R. (1991) Start-up, operation, monitoring and control of high-rate anaerobic treatment systems. *Water Sci. Technol.*, **24**, 207.

Hudson, J.W., Pohland, F.G. and Pendergrass, R.P. (1978). Anaerobic packed column treatment of shellfish processing wastewaters, in *Proceedings of the 33rd Purdue Industrial Waste Conference*, Lewis Publishers, Chelsea, MI, p. 560.

Iggleden, G.J. and van Staa, R. (1984) The use of chemical dosing and dissolved air flotation in the recovery of protein from wastewater. *Inst. Chem. Eng. Symp. Ser.*, **84**, 381.

Johnson, R.A. and Gallanger, S.M. (1984) Use of coagulants to treat seafood processing wastewater. *J. WPCF*, **56**, 970.

Kameoka, T., Sakimoto, M. and Inno, Y. (1988) Methane gas production from fish meal factory wastewater. *Chem. Abst.*, **108**, 137247d.

Knowles, R. (1982) Denitrification. *Microb. Rev.*, **46**, 47.

Lema, J.M., Soto, M., Veiga, M.C. and Méndez, R. (1987) Mesophilic and thermophilic anaerobic treatment of wastewaters from industrial processing of mussel, in *Biomass for Energy and Industry*, (eds G. Grassi, B. Delmon, J.F. Molle and H. Zibetta), Elsevier, London, p. 872.

Lema, J.M., Mendez, R. and Blazquez, R. (1988) Comparison of mesophilic and thermophilic anaerobic filters treating very high saline wastewaters, in *Proceedings of the fifth International Symposium on Anaerobic Digestion*, (eds A. Tilche and A. Rozzi), Monduzzi Editore, Bologna, Italy, p. 547.

Martin, A.M. and Patel, T.R. (1991) Bioconversion of wastes from marine organisms, in *Bioconversion of waste materials to industrial products*, (ed. A.M. Martin), Elsevier, New York, p. 417.

Matsuyama, H. and Izumi, K. (1988) Psychrophilic methane fermentation of excess sludge by enrichment culture. *J. Ferment. Technol.*, **66**, 229.

McComis, W.T. and Litchfield, J.H. (1988) Meat, fish and poultry processing wastes. *J. WPCF*, **60**, 868.

Méndez, R., Soto, M., Veiga, M.C. *et al.* (1988) Management and treatment of effluents from seafood canning-factories, in *Proceedings of the Second National Congress of Biotechnology, BIOTEC-88*, Barcelona, Spain.

Méndez, R., ten Brummeler, E. and Hulshoff-Pol, L.W. (1989) Start up of UASB reactors treating sucrose containing substrates with a low COD/sulfate ratio. *Environ. Technol. Lett.*, **10**, 83.

Méndez, R., Omil, E., Soto, M. and Lema, J.M. (1992) Pilot plant studies on the anaerobic treatment of different wastewaters from a fish-canning factory. *Water Sci. Technol.*, **25**, 37.

Nagamine, F. (1986) Sedimentation treatment test of blood-containing wastewater from sardine (processing plant). *Chem. Abstr.*, **105**, 213682x.

Nair, C. (1990) Pollution control through water conservation and wastewater reuse in the fish processing industry. *Water Sci. Technol.*, **22**, 113.

Pohland, F.G. and Hudson, J.W. (1976) Aerobic and anaerobic microbial treatment alternatives for shellfish processing wastewaters in continuous culture. *Biotechnol. Bioeng.*, **18**, 1219.

Ramos, M.B. (1991) Utilization of quitosan and other conventional coagulants in the post-treatment of effluents from anaerobic digesters. *MSc Dissertation*, University of Santiago de Compostela, Spain (in Galician).

Sandberg, M. and Ahring, B.K. (1992) Anaerobic treatment of fish meal process waste-water in a UASB reactor at high pH. *Appl. Microbiol. Biotechnol.*, **36**, 800.

Sheffield, C.W. (1992). Treatment of scallop waste, in *Proceedings of the 46th Purdue Industrial Waste Conference*, (eds C.S. Dalton and R.F. Wukasch), Lewis Publishers, Chelsea, MI, p. 429.

Soto, M. (1990) Depuration of wastewaters from a seafood processing industry with an anaerobic filter. *PhD Thesis*, University of Santiago de Compostela, Spain (in Galician).

Soto, M., Méndez, R. and Lema, J.M. (1990) Wastewaters from the seafood processing industry. *Ingeniería Química*, **22**, 203 (in Spanish).

Soto, M., Méndez, R. and Lema, J.M. (1991) Biodegradability and toxicity in the anaerobic treatment of fish canning wastewaters. *Environ. Technol.*, **12**, 669.

Suthersan, S. and Ganczarczyk, J.J. (1986a) Utilization of nitrite oxidation inhibition to improve the nitrogen elimination process, in *Proceedings of the Conference on Innovative Biological Treatment of Toxic Wastewaters*, (ed. R.J. Scholze), Consortium for Biological Waste Treatment Research and Technology, p. 423.

Suthersan, S. and Ganczarczyk, J.J. (1986b) Inhibition of nitrite oxidation during nitrification; some observations. *Water Pollut. Res. J. Canada*, **21**, 257.

Takumizawa, K., Inoue, Z. and Honda, A. (1986) Wastewater treatment of Kamaboko processing plant (2). Characteristics of wastewater biodegradation. *Chem. Abstr.*, **104**, 212541t.

Takumizawa, K., Yamamoto, O., Inoue, Z. and Honda, A. (1986) Treatment of wastewater from a fish paste processing plant. (3). Biological treatment characteristics of the wastewater. *Chem. Abstr.*, **105**, 11404x.

Veiga, M.C. (1989). Treatment by anaerobic digestion of effluents from tuna processing industries. *PhD Thesis*, University of Santiago de Compostela, Spain (in Spanish).

Veiga, M.C., Méndez, R. and Lema, J.M. (1990) Anaerobic treatment of wastewater from a tuna processing industry in an anaerobic filter, in *Proceedings of IAWPR Special Seminar on Industrial Wastewater Treatment and Disposal*, Poster, Nicosia, Cyprus.

Veiga, M.C., Blázquez, R., Méndez, R. and Lema, J.M. (1991) Pilot scale experiments on anaerobic treatment of tuna processing wastewaters, in *Proceedings of the fifth Galician–Portuguese Meeting of Chemistry*, Official Spanish College of Chemists and Portuguese Society of Chemists, La Coruña, Spain, p. 70.

Veiga, M.C., Méndez, R. and Lema, J.M. (1992a) Treatment of tuna processing wastewaters in laboratory and pilot-scale DSFF anaerobic reactors, in *Proceedings of the 46th Purdue Industrial Waste Conference*, (eds C.S. Dalton and R.F. Wukasch), Lewis Publishers, Chelsea, MI, p. 447.

Veiga, M.C., Méndez, R. and Lema, J.M. (1992b) Development and stability of biofilms in bioreactors, in *Biofilms–Science and Technology*, (eds L.F. Melo, T.R. Bott, M. Fletcher and B. Capdeville), Kluwer, Dordrecht, The Netherlands, p. 421.

Veiga, M.C., Méndez, R. and Lema, J.M. (1993) Anaerobic filter and DSFF reactors in anaerobic treatment of tuna processing wastewater, in press.

Wheaton, F.W., Brinsfield, R.B., Lomax, K.M. and Geiger, E.L. (1984) Blue crab processing plant effluent treatment. *J. WPCF*, **56**, 482.

Ziminska, H. (1987) Fish processing plant effluents – characteristics, treatment and utilization. *Chem Abstr.*, **107**, 64143j.

PART THREE

Aquaculture Applications

CHAPTER 15

Microbial biomass as a source of protein in the feeding of cultivated fish

A.M. Martin

15.1 INTRODUCTION

Fish culture or aquaculture is developing worldwide as a promising way to produce additional amounts of good-quality food and to compensate for the depletion of the fishing resources traditionally harvested. In countries such as Canada, aquaculture is also a means for an enhanced and controlled production of high-priced marine species such as shellfish and salmonids, among other products. Intensive aquacultural production systems require the availability of abundant and inexpensive sources of food for the marine animals being cultivated. Feeds for aquaculture processes account for some of the main costs associated with this economic activity. This fact is particularly true in northern areas, which need to import most of the feed components from distant places.

15.1.1 Methods to enhance the productivity of aquaculture processes

The addition of fertilizer, including organic wastes and manure, to ponds and fish farming operations is one way to provide either a direct source of feed for the fish or nutrients for the growth of plants and organisms upon which the fish can feed. Both organic and inorganic fertilizers can be employed in the fertilization of water basins. Both methods have limitations. Organic fertilizers (seaweed, manure, etc.) must be broken down to release their nutrients, and oxygen is required

Fisheries Processing: Biotechnological applications.
Edited by A.M. Martin.
Published in 1994 by Chapman & Hall, London. ISBN 0 412 58460 3

for this process. This could produce an oxygen depletion of the water. Such organic fertilizers as manure could also conduct diseases to which man is susceptible. On the other hand, inorganic fertilizers could cause algal blooms, which may deplete the oxygen supply at night or when the algae die off (Wheaton, 1977).

It is, therefore, problems related to the oxygen demand of organic wastes that impose limitations on the amount of nutrients that can be added by those methods. However, it has been claimed that if composted wastes are added instead of the original raw waste materials, greater amounts of nutrients can be provided without adversely affecting the oxygen level of the system. Banerjee and Srinivasan (1983) and Edwards *et al.* (1983) studied the use of compost in aquaculture pools and ponds. The former reported a 10-fold increase in the concentration of phytoplankton over that caused by conventional manure. The latter obtained yields of body weight gained by fish per unit of food consumed comparable to those obtained with the use of cereals.

It has been reported (Dalzell *et al.*, 1987) that compost feeding has been successful with fish of the genus *Tilapia* and of the carp family (Cyprinidae). However, it has also been noted that, even with fertilization, the total income from cultured fish crops is too small to make the system worthwhile (Hickling, 1971).

15.2 SUPPLEMENTARY FOODS IN FISH CULTURE

It appears that the most effective way of substantially increasing fish crops is by supplementary feeding. It has been stated that the capital costs associated with aquaculture systems are so high, that in order to economically cultivate fish, high fish yields must be obtained by supplementary feeding with what are called 'artificial foods' (Hickling, 1971).

15.2.1 Artificial feeding of fish

Artificial feeding is one of the principal methods of increasing production in fish cultivation. In intensive salmonid culture, artificial food may be the only basic food of the fish (Huet, 1986). Artificial feeding is essential for intensive methods of fish cultivation.

Many animal and vegetable sources are utilized in the artificial feeding of fish. The convenience of pelleted foods for fish has prompted much work on them (Huet, 1986). However, according to Brown (1993), expenses associated with using these feeds has sometimes adversely affected the affordability of the fish produced, and further advances in the cultivation of these species may be limited by the supply and price of necessary feed ingredients.

The majority of fish species farmed in intensive aquaculture systems are carnivorous or omnivorous, having a high requirement for protein in their diets. Generally, fishmeal is employed as the major protein supplier in feeds formulated for cold water fish rations (Tacon and Jackson, 1985). High-quality fishmeals supply the major portion of protein (30–40% by weight) in commercial diets for salmonids. It has been noted that the increasing emphasis on utilizing nutritionally valuable fisheries resources as human foods will mean a decrease in fishmeal supplies. With the development of aquaculture, there has been an increased demand for fish feeds. The use of fishmeal and other fisheries-related products as feed for cultivated fish most likely will not satisfy the demand that is expected, given the worldwide development of aquaculture as the fishery of the future. Therefore, new sources of protein for aquaculture processes should be found and evaluated (Spinelli, Mahnken and Steinberg, 1979).

15.3 MICROBIAL BIOMASS PROTEIN

The term 'single-cell protein' (SCP) has been used to identify the protein content of microbial biomass, its origin being attributed to workers at the Massachusetts Institute of Technology in 1968 (Vasey and Powell, 1984). In this chapter, the more general term 'microbial biomass protein' (MBP) will be employed (Moo-Young and Gregory, 1986) instead of SCP. Much research has been conducted on MBP production from a variety of substrate sources and many reviews have been published. Those studies have included nutritional and toxicological studies (Tannebaum and Wang, 1975). A comparison of the technical and economic aspects of the production of MBP for feed and food has been presented by Litchfield (1977). The author indicated that, for feed, the major competitors of SCP are soybean meal and fishmeal, having 44 and 60–65% protein content (on a dry weight basis), respectively.

The acceptance of MBP for food applications has been discussed by Scrimshaw (1985) and Udall and Scrimshaw (1986), who referred to the limitations imposed by the levels of nucleic acids in microbial protein. The acceptance of MBP for animal feeds was analyzed by Stringer (1985). Among the potential nutritional problems to be considered when feeding microbial protein to animals, from both nutritional and toxicological points of view, the most important cited are the high concentrations of nucleic acids in the final product and mineral balance. Some evidence has been given that dietary nucleotides inhibit gut motility in fish (Fänge and Grove, 1979). Nevertheless, Stringer (1985) concluded that MBP may be used as nutrient sources for feed in the same way as soybean meal, fishmeal or skimmed milk power, as long as the diets are adequately formulated to avoid nutritional deficiencies.

15.3.1 *Candida utilis* yeast

The choice of microorganisms to be employed in the production of MBP is critical and dependent on many factors. For simple carbohydrate substrates, the yeasts *Saccharomyces cerevisiae* and *Candida utilis* are attractive choices because they have been used as food ingredients for a long time (Vasey and Powell, 1984).

The genus *Candida* includes a heterogeneous group of non-spore-producing yeast species which do not qualify for classification in any other of the more homogeneous genera of the imperfect yeast. *C. utilis* yeast (for synonyms and taxonomic characteristics, see van Uden and Buckley, 1970; Meyer, Ahearn and Yarrow, 1984) has been regarded as one of the best sources of microbial protein due to, among other things, its favorable biomass composition when grown in appropriate substrates and its ability to metabolize a wider spectrum of sugars than such other commercial yeasts as *S. cerevisiae* (Shay and Wegner, 1985). *C. utilis* is able to metabolize pentoses in addition to hexoses. Pentoses are produced when lignocellulosic products are hydrolyzed, e.g. the production of xylose from hemicellulose. Lawford, Kligerman and Williams (1979) described the production of high-quality edible protein by cultivation of *C. utilis* in continuous culture, with a 50–55% protein content on a dry weight basis. Quierzy, Therien and LeDuy (1979) conducted experiments cultivating this yeast in peat extracts.

It is well known that this yeast is a very resistant, fast-growing microorganism and, because of these characteristics, it has been one of the most preferred microbial species for employment in the production of MBP (Shay and Wegner, 1985).

15.4 MICROBIAL BIOMASS AS FOOD FOR FISH CULTURE

One of the most notable examples of the usefulness of MBP in animal feeding is precisely the role of algal species in the feeding of fish. Natural seaweed vegetation (macroalgae) and cultivated microalgae, such as *Chlorella*, *Scenedesmus* and *Spirulina* spp., have all been used in fish feed. Regarding cultivated microalgal products, their existing industrial production is still relatively small, and some technological and toxicological problems remain to be solved before they attain a larger role in fish feeding (Soeder, 1976). The reported use of hydrolyzed yeast as a vitamin supplier in a cultivated fish diet provides another example of the potential of MBPs in fish feeding (Huet, 1986). It has been indicated that microbial biomass is a good source of vitamins, which make it superior to protein concentrates from plant origin (Moo-Young *et al.*, 1979).

Microbial biomass possesses the nutritive qualities required in fish diets, including color pigments synthesized by specific microorganisms. Microbial biomass productivity is higher than that of other food sources. Its production could be profitable if a stable market, such as an aquaculture industry, is found and if abundant renewable resources, such as organic waste biomass, is utilized as fermentation substrate for the microbial growth. One of the main factors in the production of microbial biomass is the requirement for a substrate which is inexpensive and abundant.

The production of microbial biomass to replace part of the fish protein in fish feed is considered to be a very promising and innovative solution to the above-mentioned problem of fishmeal scarcity. This is because microbial biomass, besides being a source of bulk protein, can also supply the feed with vitamins, minerals and components which will stimulate the disease resistance of fish (Raa, 1990).

15.4.1 Microbial biomass as a protein source

In view of the increasing cost of fishmeal and instability in long-term supply, a range of alternative sources of protein for cultivated fish have been evaluated. Diets containing yeasts, partially substituting for fishmeal, have been tested most extensively in the salmonids, particularly trout (Iida *et al.*, 1970; Andruetto, Vigliani and Ghittino, 1973). The use of yeast grown in petroleum was reported by Shimma and Nakada (1974a,b). These authors indicated that reports concerning hardness shown by yeast biomass produced in petroleum caused them to add three kinds of oils (including fish oil) to the feed preparation for rainbow trout fingerlings, before pelleting. Gropp *et al.* (1976) conducted experiments in both fresh and brackish water to determine the appropriate extent of replacement of fishmeal in diets for rainbow trout (*Salmo gairdneri*, now identified as *Oncorhynchus mykiss*). The authors found that half of the dietary fishmeal protein could be replaced by methionine-supplemented alkane-grown yeast. Shimma *et al.* (1976) conducted long-term feeding tests of the same species with a diet of 30% petroleum-grown yeast and a fishmeal control feed. In general, little difference was observed in fish growth, feed efficiency, number of matured fish and spawning conditions, between the two groups. Matty and Smith (1978) evaluated *Candida lypolytica*, *Pseudomonas* spp. and *Spirulina maxima* as protein sources for rainbow trout. The yeast protein was the best utilized for growth and gave higher food conversion efficiency (FCE) values. An inclusion rate of 35% of protein resulted in maximum growth and FCE. The maximum protein efficiency ratio (PER) reported, also obtained with the feed containing the yeast protein, was 2.12. The effect of replacement of dietary protein in the diet of rainbow trout by various levels of alkane-grown yeast was investigated by Beck *et al.* (1979). The results indicated that the yeast had a good nutritional value when supplemented with the limiting amino acids DL-methionine and L-arginine. Davies and Wareham (1988) presented a preliminary evaluation of the use of microbial biomass protein in practical diets for tilapia (*Oreochromis mossambicus*). Mahnken, Spinelli and Waknitz (1980) replaced up to 40% of the fishmeal portion in diets fed to rainbow trout with alkane-grown yeast (*Candida* spp.) biomass (22% of dietary protein). Also, the same authors replaced up to 100% of the fishmeal with the yeast supplemented with DL-methionine (54% dietary protein) in diets fed to coho salmon in salt water. The results were more favorable in the trout, where the yeast was an acceptable substitute for fishmeal at all levels tested. However, Higuera *et al.* (1981) found that food intake, weight increments and PERs decreased in trout fed on *Hansenula anomala* biomass as the only protein source, with white fishmeal as a control diet. A mixed microbial protein biomass from *H. anomala*, *Candida kruzei* and *Geotrichum candidum* was incorporated at increasing levels of substitution for casein in diets of 40% protein for rainbow trout (Murray and Marchant, 1986). The authors stated that the microbial biomass utilized required improved methods of processing to increase the total nitrogen and possibly the whole cell digestibility. Rumsey and Hughes (1990) fed lake trout *Salvelinus namaycush* with preparations of *S. cerevisiae* (baker's yeast). Fish fed with preparations of yeast with disrupted cell walls grew at a rate equal to that of the control group and faster than fish fed intact yeast cells. The authors indicated that it may be possible to replace up to 50% of the nitrogen in the diet of cultivated salmonids with disrupted yeast cells, without deleterious effects on growth and feed conversion.

Diets containing SCPs have also been fed experimentally to other fish species, including Atlantic salmon, *Salmo salar*. Bergström (1979) found that the addition of lysine and methionine improved fish growth in some diets containing microbial protein. This author suggested that although microbial protein-based diets were similar to the control diets in inducement of growth, a too-high proportion of such protein could cause metabolic disturbances. Various levels of protein were tested in trials with hybrid tilapia (*Oreochromis* spp.) by Viola and Zohar (1984). Microbial protein replaced 10% of fishmeal in a 30% protein control diet without resulting in differences in growth rate and body composition. Atack, Jauncey and Matty (1979) concluded that carp (*Cyprinus carpio*) require a diet containing high-quality protein, and that a major proportion of this could be provided by bacterial and yeast proteins. Spinelli, Mahnken and Steinberg (1979) reported that protein of yeast and bacterial origin successfully replaced fishmeal in the diets of smolted coho salmon reared in salt water without detrimental effects on growth and feed conversion. Arai *et al.* (1975) reported that the addition of calcium to a diet in which hydrocarbon-cultured yeast was the sole protein source was beneficial to rainbow trout. In conclusion, adding yeasts to salmonid diets at levels between 15 and 30% of the total diet, replacing between 25 and 50% of the fishmeal component, has produced satisfactory results (Gropp *et al.*, 1976; Beck *et al.*, 1979; Spinelli, Mahnken and Steinberg, 1979; Tiews *et al.*, 1979; Mahnken, Spinelli and Waknitz, 1980; Rumsey and Hughes, 1990). However, the use of yeasts as the sole protein source in salmonid diets has resulted in reductions in both feed utilization and growth performance (Matty and Smith, 1978; Beck *et al.*, 1979; Murray and Marchant, 1986). Poor performance has been attributed to deficiencies in some amino acids (Nose, 1974a,b; Schultz and Oslage, 1976). Supplementation with cystine and arginine (Nose, 1974a,b), methionine and arginine (Beck *et al.*, 1979) and methionine (Murray and Marchant, 1986) enhanced growth in rainbow trout.

Studies which have tested *Candida* spp. in diets for rainbow trout have indicated the following maximum inclusion levels for fishmeal replacement: 40% (Mahnken, Spinelli and Waknitz, 1980), 30% (Andruetto, Vigliani and Ghittino, 1973) and 35% (Beck *et al.*, 1979).

15.4.2 Microbial biomass as a pigment source

Products using colorants constitute an annual market of several billion dollars, and many of those products would not exist without colors (Noonan, 1985). In the food industry, the market for food colors is expected to grow in the near future, with new applications being investigated to adapt to changes in consumer trends (Dziezak, 1987). It has been forecast that there will be a 'continuing and sharp decline in the use of synthetic colors' in the food industry and that the use of natural materials will rise (Anonymous, 1987).

Carotenoids constitute an important group of color compounds, being employed in many food products such as margarine, orange drinks, beverage mixes, candies, salad dressings and ice cream, among other applications. A review of color compounds production in biological systems and their applications to the food industry has been presented by Hari, Patel and Martin (1994).

(a) Fish culture and flesh color

Colorless crustaceans can be produced by feeding these animals a carotenoid-free diet (Simpson, Klein-MacPhee and Beck, 1983). Also, trout and salmon raised in pens often lack their typical coloration, which would affect consumer acceptance. Their color is produced by the carotenoid pigment astaxanthin, which is commonly found in the zooplankton. Astaxanthin from ingested crustaceans is the main carotenoid in wild salmonids. However, intensive fish and crustacean cultivation requires feeds containing added color compounds (Trivedi, 1986), and feeds used in commercial fish farming should be supplemented with the carotenoids astaxanthin or canthaxanthin to achieve the desired coloration (Torrissen, 1986). Synthetic canthaxanthin and astaxanthin are available, but their inclusion in fish feed is not permitted by some regulatory agencies (Haard, 1988).

Studies conducted by some researchers indicate that cultured salmonid flesh can be pigmented by incorporating crustacea and crustacean processing wastes into their diets (Peterson, Jager and Savage, 1966; Saito and Regier, 1971; Spinelli, Lehman and Wieg, 1974). However, problems may be encountered in feeding moist diets that are nutritionally imbalanced due to the addition of large amounts of such materials, such as induction of mineral imbalances in the fish (Spinelli and Mahnken, 1978). Attempts to concentrate these materials to reduce their bulk density result in the loss of significant amounts of carotenoids while drying (Spinelli, Lehman and Wieg, 1974). Torrissen *et al.* (1981) presented a method of processing shrimp by-products to improve the uptake of astaxanthin in fish. Torrissen (1986) reported that the economic importance of colored flesh in salmonids makes it imperative to find new sources of pigment suitable for incorporation into salmonid feed. A review of carotenoid deposition and metabolism in salmonids has been presented by Torrissen, Hardy and Shearer (1989). One alternative method of food colorant production is the production of microbial pigments by fermentation processes.

(b) Microbial production of color compounds

It is known that some microorganisms, such as actinomycetes, yeasts and other fungal species, are able to produce colored pigments under various growth conditions, although many of those pigments are not suitable for consumption by either animals or humans (Trivedi, 1986).

One of the most promising carotenoid-producing microbial species is the yeast *Phaffia rhodozyma* (Miller, Yoneyama and Soneda *et al.*, 1976). This red yeast has been found to contain the carotenoid astaxanthin (Andrewes, Phaff and Starr, 1976), which imparts the red coloration to salmonid flesh, shellfish (Johnson, Conklin and Lewis, 1977) and poultry (Johnson, Lewis and Grau, 1980), thereby increasing their commercial value. The cultivation of *P. rhodozyma* on a large scale is therefore desirable, not only because it will provide natural astaxanthin, but also because it can serve as a valuable protein source for the fish farming industry (Beck *et al.*, 1979). It has been stated that astaxanthin is better utilized than canthaxanthin by Atlantic salmon (Storebakken *et al.*, 1987) and rainbow trout (Foss *et al.*, 1984; Torrissen, 1986). Studies on the absorption and deposition of astaxanthin and canthaxanthin in salmonid flesh have also been presented by Foss *et al.* (1987) and Tidemann *et al.* (1984). The latter reported that the coefficient of pigment accumulation of astaxanthin from shrimp waste (and its silage) appeared to be higher than that of canthaxanthin.

P. rhodozyma can accumulate a pigment concentration higher than that found in most crustacean meals or products, which are other important available alternatives for obtaining astaxanthin from natural sources (Johnson, Conklin and Lewis, 1977). The use of this yeast as a microbial colorant producer has been reported to '. . . seem ready for exploration and commercialization' (Trivedi, 1986). Other studies published on this yeast include those of Okagbue and Lewis (1983, 1984a–c, 1985), Johnson *et al.* (1978), Johnson and Lewis (1979), Johnson, Lewis and Grau (1980); Gentles and Haard (1989), An, Schuman and Johnson (1989), Sedmak, Weerasinghe and Jolly (1990) and Martin *et al.* (1993). This subject is presented in more detail elsewhere in this book. Recently, accounts of studies on other microorganisms with potential for the production of astaxantin have been presented (Kobayashi, Toshihide and Nagai, 1991; Hari, Patel and Martin, 1993b). Sommer, Potts and Morrissy (1991) reported significant deposition of astaxanthin and enhancement of the flesh coloration in trout when fed with spores from the algal species *Haematococcus pluvialis*.

15.5 A CASE STUDY: UTILIZATION OF *C. UTILIS* BIOMASS AS A FEED FOR AQUACULTURE

The following will review some experimental results related to the use of MBP from *C. utilis* in the feeding of rainbow trout, *O. mykiss*. The studies conducted on the production of yeast for this experiment are described in Martin *et al.* (1991) and Martin, Goddard and Bemister (1993).

Some of the main objectives of this work were:

1. To analyze chemically the *C. utilis* biomass produced to determine its composition and suitability as feed.
2. To produce trout feeds that incorporate varying amounts of yeast biomass as a source of protein, and to test them for digestibility and weight increase production in rainbow trout, *O. mykiss*.

15.5.1 Use of feeds containing *C. utilis* biomass protein in the feeding of trout, *O. mykiss*

Table 15.1 presents the composition of the *C. utilis* microbial biomass employed and compares it with other sources of microbial protein. The profile of amino acids for the yeast biomass produced is presented in Table 15.2, which also presents the amino acid contents of other micro-

Table 15.1 Composition of *C. utilis* biomass and of other microbial protein sources

Component	*Fodder yeast*[a] (%)	*SCP from Waterloo Process*[a] (%)	C. utilis[b] (%)	C. utilis[c] (%)
Ash	5	5	10.1	18.17 ± 2.09
Lipids	6	10	0.7	1.31 ± 0.23
Protein	45	45	48.1	50.12 ± 2.46

[a]Moo-Young *et al.* (1979).
[b]Le Duy (1981).
[c]Martin, Goddard and Bemister (1993). Mean values ± SD.

bial protein sources. In this study, the essential amino acid requirements of the fish, presented by Ogino (1980), were met by the combination of yeast biomass with fishmeal. Some properties of the yeast biomass product are indicated in Table 15.3, and a picture of the final dry yeast product is shown in Figure 15.1.

Table 15.2 Amino acids profile (g/100 g protein) of *C. utilis* biomass grown in compost extracts, and of other protein products

Amino acid	*Fodder yeast*[a]	*Waterloo Process–SCP*[a]	C. utilis[b]	C. utilis[c]
Alanine			10.7	5.44 ± 2.34
Arginine			6.6	5.06 ± 2.27
Aspartic acid			12.2	6.47 ± 2.73
Cystine				1.02 ± 0.82
Glutamic acid			15.2	8.87 ± 3.98
Glycine			4.7	3.87 ± 1.76
Histidine			2.6	2.38 ± 1.41
Isoleucine	5.3	4.7	3.6	4.54 ± 2.14
Leucine	7.0	7.5	7.0	5.25 ± 2.45
Lysine	6.7	6.8	12.1	6.68 ± 3.07
Methionine	1.9[d]	2.6[d]	0.9	1.14 ± 0.88
Phenylalanine	4.3	3.8	2.1	3.76 ± 1.38
Proline			6.3	4.43 ± 2.03
Serine			6.0	3.28 ± 1.15
Threonine	5.5	6.1	5.6	4.55 ± 2.20
Tryptophan	1.4	–[e]	–[e]	–[e]
Tyrosine	3.3	3.3	3.1	3.01 ± 1.91
Valine	6.3	5.8	4.6	5.81 ± 2.50

[a]Moo-Young *et al.* (1979).
[b]LeDuy (1981).
[c]Martin, Goddard and Bemister (1993). Mean values ± SD.
[d]Methionine and cystine.
[e]Not determined.

Table 15.3 Properties of dried, ground *C. utilis* yeast biomass[a]

Color	light brown
Moisture[b]	8.29 ± 0.02%, wet weight basis
Odor	sweet
Solubility[b]	40.60 ± 1.69
Texture	powdery

[a]Reproduced by permission of the Society of Chemical Industry (SCI) from Martin, Goddard and Bemister (1993).
[b]Mean Values ± SD.

(a) Feeding trials

As indicated by Martin, Goddard and Bemister (1993), diets containing approximately 40% crude protein (N × 6.25) were formulated, in which the yeast biomass replaced fishmeal at various percentages of the total dietary nitrogen. The compositions of the diet ingredients are given in Figure 15.2. An example of the formulation of a diet that included 20%

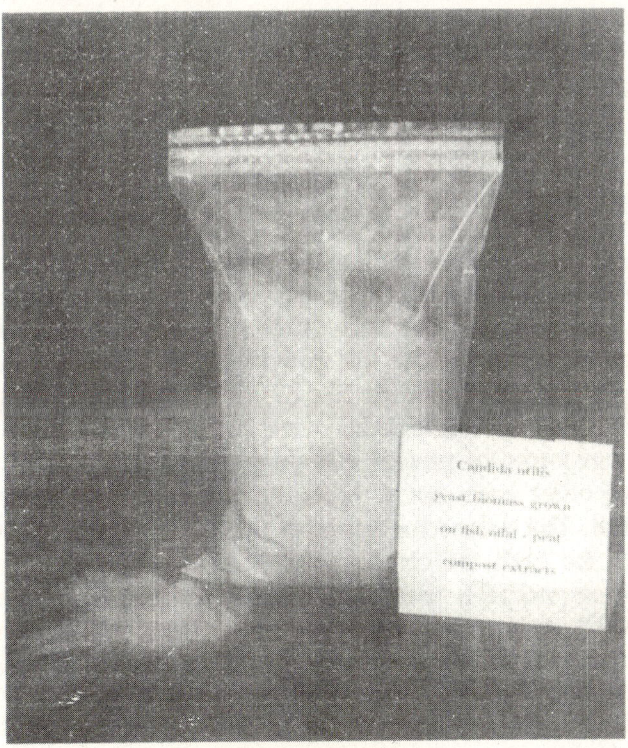

Figure 15.1 Final *C. utilis* product.

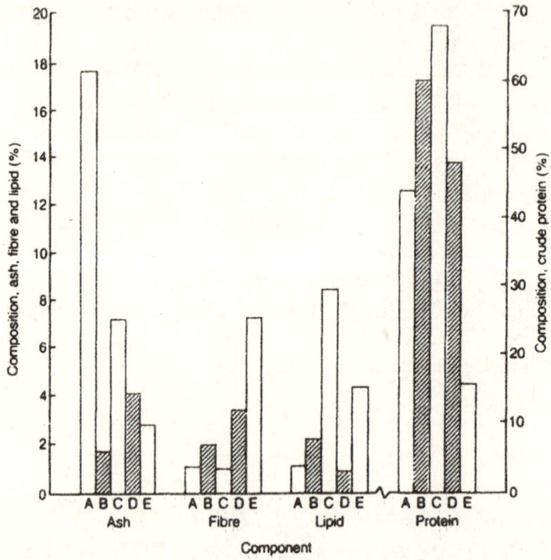

Figure 15.2 Proximate composition (%) of experimental feed ingredients. A, *C. utilis*; B, corn gluten; C, herring meal; D, soybean meal; E, wheat middlings. From Martin *et al*. (1991).

by weight of *C. utilis* biomass, which was the highest concentration of yeast tested, is presented in Table 15.4, together with the composition of the control diet employed (the diet without yeast biomass). The proximate compositions of those diets are presented in Table 15.5.

15.5.2 Growth study on trout, *O. mykiss*

There were no significant differences between the mean final weights of the control groups and those fed diets containing yeast biomass (Martin, Goddard and Bemister, 1993). Figure 15.3 compares some parameters representing the fish growth performances for the control diet and for the diet containing 20 g of yeast biomass per 100 g of feed. The food conversion efficiencies were higher for the control group; however, in terms of protein efficiency ratios, fish groups fed diets containing yeast were not significantly different from those fed the control diet. The values for apparent protein digestibility coefficients were high for both the control group and for the group with yeast biomass incorporated in their diets. The highest apparent protein digestibility (APD), estimated using the method of Nose (1960), was recorded for the group fed the diet in which the largest portion of the fishmeal component was replaced by yeast biomass (Martin *et al*., 1991).

Regarding the carcass composition of the fish (Figure 15.4), the results were similar between the fish in the control group and the fish fed the diet containing yeast biomass (Martin, Goddard and Bemister, 1993).

Table 15.4 An example of a diet formulation for rainbow trout (*O. mykiss*) utilizing yeast biomass and of a control diet, presented as g/100 g diet[a]

Ingredient	*Diet*	
	control	*with* C. utilis *biomass added*
C. utilis	0.0	20.0
Corn gluten	13.0	13.0
Fish oil	12.0	13.25
Guar gum	1	1
Herring meal	30	19.5
Mineral mix[b]	1	1
Soybean meal	17.0	17.0
Vitamin mix[c]	1	2
Wheat middlings	24.0	13.25
Total	100.0	100.0

[a]From Martin *et al* (1991). © Martin, A.M. and Goddard, S., 1991.
[b]Mineral premix supplying per kg dry diet: Fe, 600 mg; Cu, 17.5 mg; Zn, 750 mg; Mn, 600 mg; iodine, 50 mg; Co, 3 mg.
[c]Vitamin premix supplying per kg dry diet: vitamin A, 50 000 IU; vitamin D_3, 32 000 IU; vitamin E, 3200 IU; vitamin K, 400 mg; vitamin B_{12}, 1.6 mg; biotin, 6.4 mg; folic acid, 160 mg; pantothenic acid, 320 mg; pyridoxine, 320 mg; niacin, 800 mg; riboflavin, 200 mg; thiamine, 160 mg; ascorbic acid, 1100 mg.

Table 15.5 Proximate composition of experimental diets (g/100 g diet)[a]

Components	*Diets*	
	control	*with* C. utilis *biomass added*
Ash	5.2	7.6
Carbohydrates[b]	21.8	21.2
Fibre	2.5	2.2
Lipid	16.9	17.1
Moisture	12.9	11.3
Protein	40.7	40.5
Total	100.0	100.0
Gross energy (kcal/g)	4.791	4.775

[a]From Martin *et al*. (1991). © Martin, A.M. and Goddard, S., 1991.
[b]Carbohydrate determined by difference.

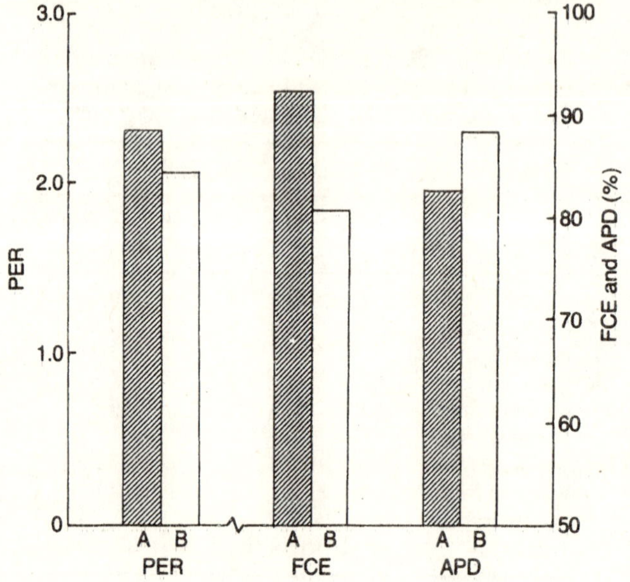

Figure 15.3 PER, FCE and APD for two diets fed to rainbow trout over a 50-day feeding trial. A, Control diet; B, diet containing 20% *C. utilis* by weight. Adapted with permission of the Society of Chemical Industry (SCI), from Martin, Goddard and Bemister (1993).

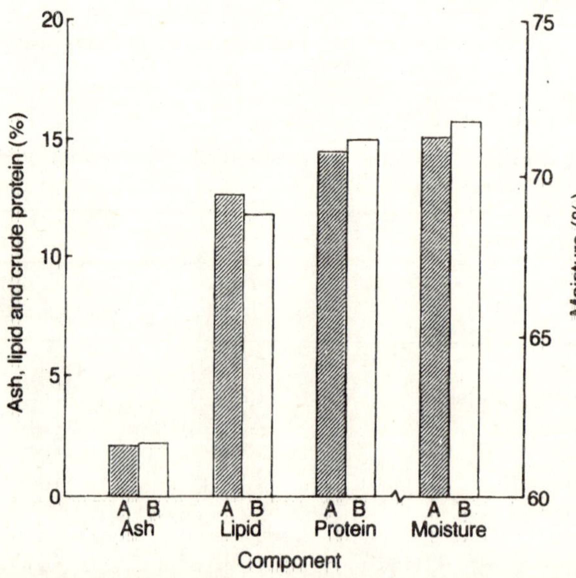

Figure 15.4 Carcass composition (wet weight basis) of rainbow trout fed diets containing *Candida utilis* biomass and a control diet. A, Control diet; B, diet containing 20% *C. utilis* by weight. From Martin *et al.* (1991) and Martin, Goddard and Bemister (1993).

15.6 CONCLUSIONS

From the case study presented (Martin, Goddard and Bemister, 1993), the following conclusions can be drawn:

1. The *C. utilis* biomass produced had high protein, mineral and essential amino acid contents, and a low lipid content.
2. The microbial biomass dried well and possessed a good degree of solubility and good organoleptic characteristics.
3. The fish accepted the diets tested in which microbial biomass was incorporated and their behavior seemed to indicate that these diets were palatable.
4. No nutritional problems were observed resulting from the use of yeast biomass in the diets of cultivated rainbow trout, *O. mykiss*.
5. The study gives an indication of the potential for using yeast biomass as a partial replacement for fishmeal in diets for rainbow trout.

ACKNOWLEDGMENTS

The experimental work for the case study reviewed in this paper on the use of feeds containing *C. utilis* biomass protein in the feeding of rainbow trout, *O. mykiss,* was funded by the Center for Fisheries Innovation, St John's, Newfoundland, Canada. The other investigator in this project was Dr S. Goddard, and technical assistance was provided by P. Bemister and G. DeVeau.

REFERENCES

An, G.-H., Schuman, D.B. and Johnson, E.A. (1989) Isolation of *Phaffia rhodozyma* mutants with increased astaxanthin content. *Appl. Environ. Microbiol.*, **55** (1), 116–24.

Andrewes, A.G., Phaff, H.J. and Starr, M.P. (1976) Carotenoids of *Phaffia rhodozyma,* a red pigmented fermenting yeast. *Phytochemistry,* **15**, 1003–7.

Andruetto, S., Vigliani, E. and Ghittino, P. (1973) Possibile uso nei pellets per trota di proteine da leiviti coltivati su idrocarburi (Proteine BP). *Riv. It. Piscic. Ittiop.*, **9** (4), 97–100 (English summary).

Anonymous (1987) 'The impact of food science and technology on the EEC food industry in the next 1–15 years', cited in *PDR Notes* **36**, December 4,Ministry of Agriculture and Food, Province of Ontario.

Arai, S., Müller, R., Shimma, Y. and Nose, T. (1975) Effects of calcium supplement to yeast grown on hydrocarbons as a feedstuff for rainbow trout. *Bull. Freshwater Fish. Res. Lab., Tokyo,* **25**(1), 33–40.

Atack, T.H., Jauncey, K. and Matty, A.J. (1979) The utilization of some single cell proteins by fingerling mirror carp (*Cyprinus carpio*). *Aquaculture,* **18**, 337–48.

Banerjee, R.K. and Srinivasan, K.V. (1983) Composted urban refuse and primary sewage sludge as a fish-pond manure. *Agric. Wastes,* **7**, 209–19.

Beck, H., Gropp, J., Koops, H. and Tiews, K. (1979) Single cell proteins in trout diets, in *Proceedings of the World Sympsium on Finfish Nutrition and Fishfeed Technology*, (eds J.E. Halver and K. Tiews), H. Heenemann GmbH & Co., Berlin, Vol. 2, pp. 269–80.

Bergström, E. (1979) Experiments on the use of single cell proteins in Atlantic salmon diets, in *Proceedings of the World Symposium on Finfish Nutrition and Fishfeed Technology*, (eds J.E. Halver and K. Tiews), H. Heenemann GmbH & Co., Berlin, Vol. 2, pp. 105–116.

Brown, E.E. (1983) *World Fish Farming: Cultivation and Economics*, 2nd edn, AVI Publishing, Westport, CT, pp. 2.

Dalzell, H.W., Biddlestone, A.J., Gray, K.R. and Thurairajan, K. (1987) Soil management: compost production and use in tropical and subtropical environments. *FAO Soils Bulletin* **56**, United Nations Food and Agriculture Organization, Rome.

Davies, S.J. and Wareham, M. (1988) A preliminary evaluation of an industrial single cell protein in practical diets for tilapia (*Oreochromis mossambicus* Peters). *Aquaculture*, **73**, 189–99.

Dziezak, J.D. (1987) Application of food colorants. *Food Technol.*, **14** (4), 78–88.

Edwards, P., Polprasert, C., Pacharaprakiti, C. *et al.* (1983) Compost as fish feed, a practical application of detritivory for the cultivation of tilapia. *Aquaculture*, **32**, 409–13.

Fänge, R. and Grove, D. (1979) Digestion, in *Fish Physiology, Vol. 8 Bioenergetics and Growth*, (eds W.S. Hoar, D.J. Randall and J.R. Brett), Academic Press, New York, pp. 162–260.

Foss, P., Storebakken, T., Scheidt, K. *et al.* (1984) Carotenoids in diets for salmonids I. Pigmentation of rainbow trout with individual optical isomers of astaxanthin in comparison with canthaxanthin. *Aquaculture*, **41**, 213–26.

Foss, P., Storebakken, T., Austreng, E. and Liaeen-Jensen, S. (1987) Carotenoids in diets for salmonids V. Pigmentation of rainbow trout with astaxanthin and astaxanthin dipalmitate in comparison with canthaxanthin. *Aquaculture*, **65**, 293–305.

Gentles, A. and Haard, N. (1989) Bioavailability of astaxanthin in *Phaffia rhodozyma*, in *Advances in Fisheries Technology and Biotechnology for Increasing Profitability*, (eds M.N. Voigt and J.R. Botta), Technomic, Lancaster, PA, pp. 373–86.

Gropp, J., Koops, H., Tiews, K. and Beck, H. (1976) Replacement of fishmeal in trout feeds by other feedstuffs, in *Advances in Aquaculture*, (eds T.V.R. Pillay and W.A. Dill), Fishing News (Books) Ltd, Farnham, Surrey, pp. 596–601.

Haard, N.F. (1988) Astaxanthin formation by the yeast *Phaffia rhodozyma* on molasses. *Biotechnol. Lett.*, **10**(9), 609–14.

Hari, R.P.K., Patel, T.R. and Martin, A.M. (1992) A new strain of *Rhodotorula rubra* isolated from yogurt. *J. Ind. Microbiol.*, **11**, 43–51.

Hari, R.K., Patel, T.R. and Martin, A.M. (1994) An overview of pigment production in biological systems: functions, biosynthesis and applications in food industry. *Food Rev. Int.*, **10**(1), 49–70.

Hickling, C.F. (1971) Supplementary feeding, in *Fish Culture*, 2nd edn, Faber and Faber, London, pp. 125–50.

Higuera, M. de la, Sánchez-Muñiz, F.J., Mataix, F.J. and Varela, G. (1981) Nitrogen utilization by rainbow trout (*Salmo gairdneri*) fed on the yeast *Hansenula anomala*. *Comp. Biochem. Physiol.*, **69A**, 583–6.

Huet, M. (1986) Artificial feeding of fish, in *Textbook of Fish Culture. Breeding and Cultivation of Fish*, 2nd edn, Fishing News (Books) Ltd, Farnham, Surrey, UK, pp. 334–55.

Iida, A., Tayama, T., Kumai, T. *et al.* (1970) Studies on the nutritive value of petroleum yeast in carp and rainbow trout feeding. *Suisan Zoshoku*, **18**, 35–43.

Johnson, E.A., Conklin, D.E. and Lewis, M.J. (1977) The yeast *Phaffia rhodozyma* as a dietary pigment source for salmonids and crustaceans. *J. Fish. Res. Board Can.*, **34**, 2417–21.

Johnson, E.A., Villa, T.G., Lewis, M.L. and Phaff, H.J. (1978). Simple method for the isolation of astaxanthin from the basidiomycetous yeast *Phaffia rhodozyma*. *Appl. Environ. Microbiol.*, **35**(6), 1155–9.

Johnson, E.A. and Lewis, M.J. (1979) Astaxanthin formation by the yeast *Phaffia rhodozyma*. *J. Gen. Microbiol.*, **115**, 173–83.

Johnson, E.A., Lewis, M.J. and Grau, C.R. (1980) Pigmentation of egg yolks with astaxanthin from the yeast *Phaffia rhodozyma*. *Poultry Sci.*, **59**(8), 1777–82.

Johnson, E.A., Villa, T.G. and Lewis, M.J. (1980) *Phaffia rhodozyma* as an astaxanthin source in salmonoid diets. *Aquaculture*, **20**, 123–34.

Kobayashi, M., Toshihide, K. and Nagai, S. (1991) Astaxanthin production by a green alga, *Haematococcus pluvialis* accompanied with morphological changes in acetate media. *J. Ferm. Bioeng.*, **71**(5), 335–9.

Lawford, G.R., Kligerman, A. and Williams, T. (1979) Production of high-quality edible protein from *Candida* yeast grown in continuous culture. *Biotechnol. Bioeng.*, **21**, 1163–74.

LeDuy, A. (1981) Microbial bioconversion of peat: a perspective, in *Proceedings of the Peat as an Energy Alternative Symposium*, The Institute of Gas Technology, pp. 479–493. Quoted in LeDuy, A. (1981) SCP from peat extracts – Chemical composition of substrates and products, in *Proceedings of the International Symposium on Peat Utilization*, (eds C.H. Fuchsman and S.A. Spigarelli), Bemidji State University, Bemidji, MN, pp. 89–116.

Litchfield, J.H. (1977) Comparative technical and economic aspects of single-cell protein process. *Adv. Appl. Microbiol.* **22**, (ed D. Perlman), Academic Press, New York, pp. 267–305.

Mahnken, C.V.W., Spinelli, J. and Waknitz, F.W. (1980) Evaluation of an alkane yeast (*Candida* sp.) as a substitute for fish meal in Oregon Moist Pellet: feeding trials with coho salmon (*Oncorhynchus kisutch*) and rainbow trout (*Salmo gairdneri*). *Aquaculture*, **20**, 41–56.

Martin, A.M., Goddard, S., Bemister, P. and DeVeau, G. (1991) *Internal Project Report*, Memorial University of Newfoundland. St. John's, Newfoundland, Canada.

Martin, A.M. Acheampong, E., Patel, T.R. and Chornet, E. (1993) Study of growth parameters for *Phaffia rhodozyma* cultivated in peat hydrolysates. *Appl. Biochem. Biotechnol.*, **37**, 235–41.

Martin, A.M., Goddard, S. and Bemister, P. (1993) Production of *Candida utilis* biomass as aquaculture feed. *J. Sci. Food Agric.*, **61**, 363–70.

Matty, A.J. and Smith, P. (1978) Evaluation of a yeast, a bacterium and an alga as a protein source for rainbow trout: I. Effect of protein level on growth, gross conversion efficiency and protein conversion efficiency. *Aquaculture*, **14**, 235–46.

Meyer, S.A., Ahearn, D.G. and Yarrow, D. (1984) Genus 4. *Candida* Berkhout, in *The Yeasts – A Taxonomic Study*, (ed. N.J.W. Kreger-van Rij), Elsevier, Amsterdam, pp. 585–844.

Miller, M.W., Yoneyama, M. and Soneda, M. (1976) *Phaffia* a new yeast genus in the *Deteromycotina* (*Blastomycetes*). *Int. J. Syst. Bacteriol.*, **26** (2), 286–91.

Moo-Young, M., Daugulis, A.J., Chahal, D.S. and Macdonald, D.G. (1979) The

Waterloo Process for SCP production from waste biomass. *Process Biochem.*, **14** (10), 38–40.

Moo-Young, M. and Gregory, K.F. (1986) In *Microbial Biomass Proteins*, (eds M. Moo-Young and K.F. Gregory), Elsevier Applied Science, London, pp. v–vi.

Murray, A.P. and Marchant, R. (1986) Nitrogen utilization in rainbow trout fingerlings (*Salmo gairdneri*) fed mixed microbial biomass. *Aquaculture*, **54**, 263–75.

Noonan, J.E. (1985) An analysis of the factors affecting the future of FD and C colors. *Cereal Foods World*, **30** (4), 265–67.

Nose, T. (1960) On the digestion of food protein by gold-fish (*Carassius auratus* L.) and rainbow trout (*Salmo irideus* G.). *Bull. Freshwater Fish. Res. Lab.*, **10**(1), 11–22.

Nose, T. (1974a) Effects of amino acids supplemented to petroleum yeast on growth of rainbow trout fingerlings – I. A preliminary experiment. *Bull. Freshwater Fish. Res. Lab.*, **24** (1), 57–63.

Nose, T. (1974b) Effects of amino acids supplemented to petroleum yeast on growth of rainbow trout fingerlings – II. Methionine and cystine. *Bull. Freshwater Fish. Res. Lab.*, **24** (2), 101–9.

Okagbue, R.N. and Lewis, M.J. (1983) Mixed culture of *Bacillus circulans* WL-12 and *Phaffia rhodozyma* on different carbon sources: yeast-wall lytic enzyme production and extractability of astaxanthin. *Biotechnol. Lett.*, **5**(11), 731–736.

Okagbue, R.N. and Lewis, M.J. (1984a) Use of alfalfa residual juice as a substrate for propagation of the red yeast *Phaffia rhodozyma*. *Appl. Microbiol. Biotechnol.* **20** (1), 33–9.

Okagbue, R.N. and Lewis, M.J. (1984b) Inhibition of the red yeast *Phaffia rhodozyma* by saponin. *Appl. Microbiol. Biotechnol.*, **20**(4), 278–80.

Okagbue, R.N. and Lewis, M.J. (1984c) Autolysis of the red yeast *Phaffia rhodozyma*: a potential tool to facilitate extraction of astaxanthin. *Biotechnol. Lett.*, **6**(4), 247–50.

Okagbue, R.N. and Lewis, M.J. (1985) Influence of mixed culture conditions on yeast-wall hydrolytic activity of *Bacillus circulans* WL-12 and on extractability of astaxanthin from the yeast *Phaffia rhodozyma*. *J. Appl. Bacteriol.* **59**(3), 243–55.

Ogino, C. (1980) Requirements of carp and rainbow trout for essential amino acids. *Bull. Jpn. Soc. Sci. Fish.*, **46**(2), 171–4.

Peterson, D.H., Jager, H.K. and Savage, G.M. (1966) Natural coloration of trout using xanthophylls. *Trans. Am. Fish. Soc.*, **95**, 408–14.

Quierzy, P., Therien, N. and LeDuy, A. (1979). Production of *Candida utilis* protein from peat extracts. *Biotechnol. Bioeng.*, **22**, 1175–90.

Raa, J. (1990) Biotechnology in aquaculture and the fish processing industry: a success story in Norway, in *Advances in Fisheries Technology and Biotechnology for Increased Profitability*, (ed. M.N. Voigt and J.R. Botta), Technomic, Lancaster, PA, pp. 509–24.

Rumsey, G.L. and Hughes, S.G. (1990) Use of dietary yeast, *Saccharomyces cerevisiae*, nitrogen by lake trout. *J. World Aquacult. Soc.*, **21**(3), 205–209.

Saito, A. and Regier, L.W. (1971) Pigmentation of brook trout (*Salvelinus fontinalis*) by feeding dried crustacean wastes. *J. Fish. Res. Board Can.*, **28**(4), 509–12.

Schultz, E. and Oslage, H.J. (1976) Composition and nutritive value of single cell protein (SCP). *Anim. Feed Sci. Technol.*, **1**, 9–24.

Scrimshaw, N.S. (1985) Acceptance of single-cell protein for human food applications, in *Comprehensive Biotechnology*, (ed. M. Moo-Young), Pergamon Press, Oxford, Vol. 4, pp. 673–84.

Sedmak, J.J., Weerasinghe, D.K. and Jolly, S.O. (1990) Extraction and quantitation of astaxanthin from *Phaffia rhodozyma*. *Biotechnol. Tech.*, **4**(2), 107–12.

Shay, L.K. and Wegner, G.H. (1985) Improved fermentation process for producing *Torula* yeast. *Food Technol.*, **39**(10), 61–66, 70.

Shimma, Y. and Nakada, M. (1974a) Utilization of petroleum yeast in fish feed – I. Effect of supplemental oil. *Bull. Freshwater Fish. Res. Lab., Tokyo*, **24**(1), 47–56.

Shimma, Y. and Nakada, M. (1974b) Utilization of petroleum yeast in fish feed – II. Effect of growth and body lipids of rainbow trout fingerlings raised in cages. *Bull. Freshwater Fish. Res. Lab., Tokyo*, **24**(2), 111–19.

Shimma Y., Kato, T., Fukuda, Y. *et al.* (1976) Utilisation of petroleum yeast in fish feed – III. Long-term feeding tests on rainbow trout at 9.5°C and 17°C. *Bull. Freshwater Fish. Res. Lab., Tokyo*, **26**(1), 45–57.

Simpson, K.L., Klein-MacPhee, G. and Beck, A.D. (1983) Zooplankton as a food source, in *Proceedings of the Second International Conference on Aquaculture Nutrition: Biochemical and Physiological Approaches to Shellfish Nutrition*, (eds G.D. Pruder, C. Langdon and D. Conklin), Louisiana State University, Baton Rouge, LA, pp. 180–201.

Soeder, C.J. (1976) The technical production of microalgae and its prospects in marine aquaculture, in *Harvesting of Polluted Waters*, (ed. O. Devik), Plenum Press, New York, pp. 11–26.

Sommer, T.R., Potts, W.T. and Morrissy, N.M. (1991) Utilization of microalga astaxanthin by rainbow trout (*Oncorhynchus mykiss*). *Aquaculture*, **94**, 79–88.

Spinelli, J. and Mahnken, C. (1978) Carotenoid deposition in pen-reared salmonids fed diets containing oil extracts of red crab (*Pleuroncodes planipes*). *Aquaculture*, **13**, 213–23.

Spinelli, J., Lehman, L. and Wieg, D.H. (1974) Composition, processing, utilization of red crab (*Pleuroncodes planipes*) as an aquacultural feed ingredient. *J. Fish. Res. Board Can.*, **31**, 1025–1029.

Spinelli, J., Mahnken, C. and Steinberg, M. (1979) Alternative sources of proteins for fish meal in salmonid diets, in *Proceedings the World Symposium on Finfish Nutrition and Fishfeed Technology*, (eds J.E. Halver and K. Tiews). H. Heenemann GmbH & Co., Berlin, Vol. 2, pp. 131–142.

Storebakken, T. Foss, P., Schiedt, K. *et al.* (1987) Carotenoids in diets for salmonids. IV. Pigmentation of Atlantic salmon with astaxanthin, astaxanthin dipalmitate and canthaxanthin. *Aquaculture*, **65**, 279–92.

Stringer, D.A. (1985) Acceptance of single-cell proteins for animal feeds, in *Comprehensive Biotechnology*, (ed. M. Moo-Young), Pergamon Press, Oxford, Vol. 4, pp. 685–94.

Tacon, A.G.J. and Jackson, A.J. (1985) Utilisation of conventional and unconventional protein sources in fish feeds, in *Nutrition and Feeding in Fish*, (eds C.B. Cowey, A.M. Mackie and J.G. Bell), Academic Press, London, pp. 119 –45.

Tannebaum, S.R. and Wang, D.I.C. (eds) (1975) *Single Cell Protein II*, The Massachusetts Institute of Technology, Cambridge, MA.

Tidemann, E., Raa, J., Stormo, B. and Torrissen, O. (1984) Processing and utilization of shrimp waste, in *Engineering and Food*, (ed. B.M. McKenna), Elsevier Applied Science Publishers, London, Vol. 2, pp. 583–94.

Tiews, K., Gropp, J., Beck, M. and Koops, K. (1979) Compilation of fish meal free diets obtained in rainbow trout (*Salmo gairdneri*) feeding experiments at Hamburg, in *Finfish Nutrition and Fishfeed Technology*, (eds J. Halver and K. Tiews), H. Heenemann GmbH & Co., Berlin, Vol. 2, pp. 219–28.

Torrissen, O.J. (1986) Pigmentation of salmonids – a comparison of astaxanthin and canthaxanthin as pigment sources for rainbow trout. *Aquaculture*, **53**, 271–8.

Torrissen, O.J., Tidemann, E., Hansen, F. and Raa, J. (1981) Ensiling in acid. A method to stabilize astaxanthin in shrimp processing by-products to improve uptake of this pigment by rainbow trout (*Salmo gairdneri*). *Aquaculture*, **26**, 77–83.

Torrissen, O.J., Hardy, R.W. and Shearer, K.D. (1989) Pigmentation of salmonids – Carotenoid deposition and metabolism. *Aquatic Sci.*, **1**(2), 209–25.

Trivedi, N. (1986) Use of microorganisms in the production of unique ingredients, in *Biotechnology in Food Processing*, (eds S.K. Harlander and T.P. Labuza), Noyes Publications, Park Ridge, NJ, pp. 115–32.

Udall, J.N. and Scrimshaw, N.S. (1986) Evaluating the safety and nutritional value of microbial biomass, in *Microbial Biomass Proteins*, (ed. M. Moo-Young and K.F. Gregory), Elsevier Applied Science, London, pp. 97–102.

Uden, N. van and Buckley, H. (1970) Genus 2. *Candida* Berkhout, in *The Yeasts – A Taxonomic Study*, (ed. J. Lodder), North-Holland, Amsterdam, pp. 893–1087.

Vasey, R.B. and Powell, K.A. (1984) Single-cell protein. *Biotechnol. Genetic Eng. Rev.*, **2**, 285–311.

Viola, S. and Zohar, G. (1984) Nutrition studies with market size hybrids of tilapia (*Oreochromis*) in intensive culture. *Bamidgeh* **36**(1), 3–15.

Wheaton, F.W. (1977) Modified systems, in *Aquacultural Engineering*, John Wiley, New York, pp. 213–28.

CHAPTER 16

Coloring of salmonids in aquaculture: the yeast *Phaffia rhodozyma* as a source of astaxanthin

A. Tangerås and E. Slinde

16.1 INTRODUCTION

The salmonid aquaculture industry has grown fast from its start in the early 1970s, Norway being the main producer, accounting for around 70% of the world production of Atlantic salmon (*Salmo salar*) in 1989 (Figure 16.1). One of the outstanding features of the salmonid fishes is the salmon-pink color of their flesh. This color is, in the consumer's mind, closely connected with the quality and esteem of the fish, and a correct coloring of the flesh is of great importance to the aquaculture industry. Farmed salmon that do not have 'a color that is natural for the species' will achieve a low classification and a low price in the market. Color has therefore become an important factor in the market competition, and we estimate that in salmon farming worldwide at least US$ 36 million/year is used to attain the desired color.

The color of the flesh of salmonids is due to the presence of astaxanthin (3,3′-dihydroxy-β,β-carotene-4,4′-dione) (Kanemitsu and Aoe, 1958a,b; Khare *et al.*, 1973; Schiedt, Leuenberger and Vecchi, 1981) which belongs to a large group of compounds named carotenoids (Figure 16.2). These are often colorful and widely distributed in nature. Salmonids are not able to synthesize astaxanthin and depend on an adequate supply through their feed to obtain the color. Wild salmonids obtain astaxanthin from small crustaceans like krill and shrimps

Fisheries Processing: Biotechnological applications.
Edited by A.M. Martin.
Published in 1994 by Chapman & Hall, London. ISBN 0 412 58460 3

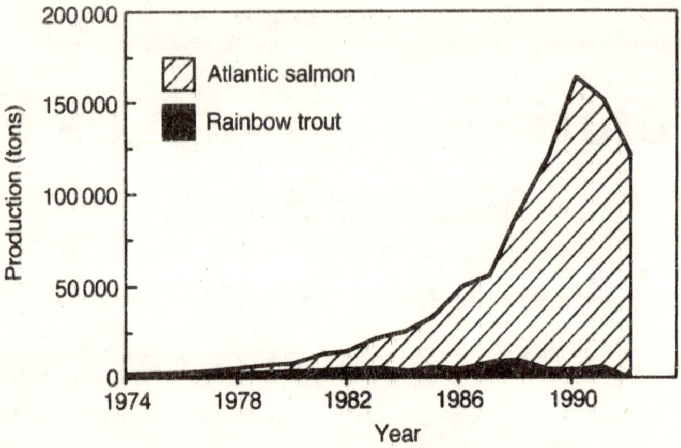

Figure 16.1 Yearly production of farmed Atlantic salmon (*Salmo salar*) and rainbow trout (*Oncorhynchus mykiss*) in Norway.

(Schiedt, Leuenberger and Vecchi, 1981; Foss, Renstrøm and Liaaen-Jensen, 1987), but in the oceans only microorganisms seem to be able to synthesize carotenoids *de novo* (Liaaen-Jensen, 1990). When aquaculture of salmonids was in its infancy, waste products from the shrimp industry containing astaxanthin were used as feed additives (Torrissen *et al.*, 1982; Choubert and Luquet, 1983). This was not without problems since the product was rather low in astaxanthin that deteriorated when stored (Torrissen *et al.*, 1982).

When the chemical company Hoffmann-La Roche (Switzerland) succeeded in producing synthetic canthaxanthin (Carophyll red), this was used since it was convenient and the amount of crustacean waste was limited. A few years later the same company introduced synthetic astaxanthin (Carophyll pink) to the market (Bernhard, 1990) and this is now becoming the dominating source of color in the salmonid aquaculture industry. Even though the synthetic astaxanthin is identical to astaxanthin found in wild salmonids, except for stereoisomer distribution (Figure 16.2), biologically produced astaxanthin would be preferred both by the aquaculture industry and by the end consumer. Synthetic astaxanthin is rather expensive and can account for as much as 10–15% of the total feed cost. A biological product would therefore be competitive in the market, provided the right organism and production process can be found.

Two types of microorganisms have been studied with the aim to develop an industrial production of astaxanthin. These are microalgae, mainly from the genus *Haematoccus* and the yeast *Phaffia rhodozyma* (Johnson and An, 1991; Nelis and De Leenheer, 1991). The former have the advantage of containing large amounts of astaxanthin when grown

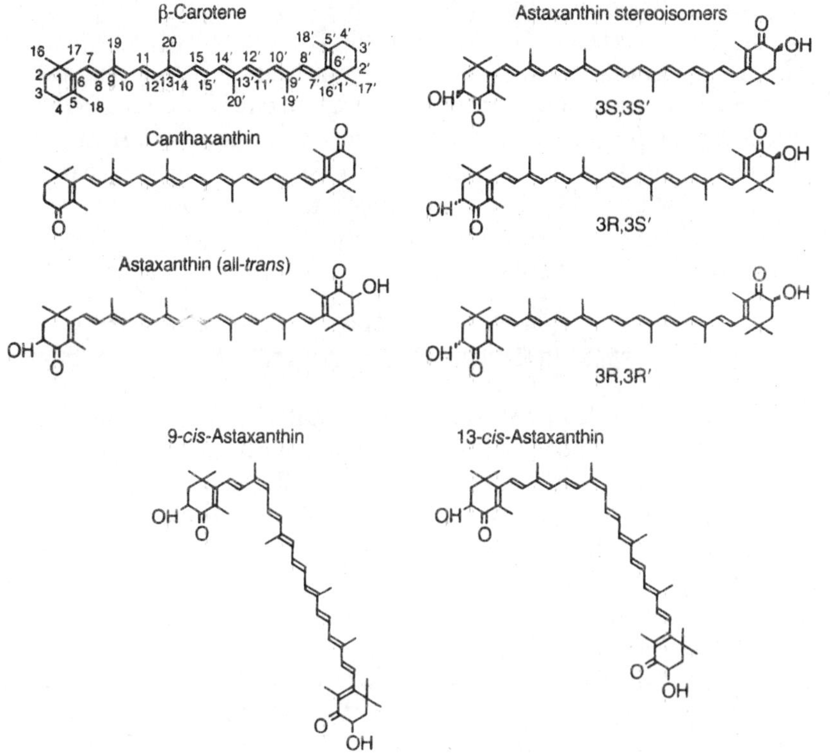

Figure 16.2 Chemical structures of astaxanthin and canthaxanthin which are used for coloring of salmonids, and of β-carotene which can be a precursor for these carotenoids both in biological and chemical synthesis. The three molecules are shown in their all-*trans* configuration. Two *cis* isomers of astaxanthin and the three possible stereoisomers of astaxanthin are also shown. The fourth theoretical stereoisomer, 3*S*,3*R*′, can be superimposed on the 3*R*,3*S*′ when the molecule is rotated and therefore does not represent a separate entity. They are named the *meso* form.

under the right conditions. Such conditions are difficult to obtain in a large-scale process. The yeast can be grown in large fermenters in a process similar to that used for commercial production of baker's yeast (*Saccharomyces cerevisiae*). The main problem is that *P. rhodozyma* as found in nature contains rather low amounts of astaxanthin when grown under such conditions. Efforts are today devoted to develop new strains that will both grow well and produce large amounts of astaxanthin within an acceptable time. So far no commercial supply of *P. rhodozyma* has been introduced as a regular supply to the market, but with the efforts several companies put into projects on *P. rhodozyma* it could well be in use within a few years.

This chapter review sources and quantitation of astaxanthin, and we focus on the yeast *P. rhodozyma* as an astaxanthin source for salmonids.

The growth properties and astaxanthin production of the yeast are described. We also show how astaxanthin from *P. rhodozyma* is utilized by both salmon and trout. We propose a process for production of *P. rhodozyma* by fermentation, pressure homogenization and spray drying, and discuss the problems involved. At present the economical limitation of the process is the limited production of astaxanthin by the yeast. We therefore show how mutation can improve the yield. Finally the economics and market for an astaxanthin production process by *P. rhodozyma* and its future potential are discussed.

16.2 SOURCES OF ASTAXANTHIN

Astaxanthin is found in bacteria, fungi, plantae and animaliae (Goodwin, 1980, 1984). In aquaculture different biological sources can be used, in addition to synthetic astaxanthin, all having advantages and disadvantages (Table 16.1). The astaxanthin level is generally low in an organism found in nature (50–400 p.p.m.). When grown artificially high amounts have been found in starved algae, and at stress conditions the content of astaxanthin may rise to percentage values (Bubrick, 1991; Johnson and An, 1991).

Crustacean waste that contains 50–200 mg astaxanthin/kg must be added to the feed in rather high amounts to give acceptable coloration of salmonids (Saito and Regier, 1971; Torrissen *et al.*, 1982; Chen and Meyers, 1983). Crustacean oil extracts have been produced to increase the astaxanthin concentration (Arai *et al.*, 1987; Inoue *et al.*, 1988).

Crustaceans contain mostly astaxanthin esters (Meyers and Bligh, 1981; Choubert and Luquet, 1983; Arai *et al.*, 1987) and salmonids utilize free astaxanthin better (Foss *et al.*, 1987; Storebakken *et al.*, 1987). Such products are seldom used, mainly due to limited availability and high

Table 16.1 Sources of astaxanthin and their main advantages and disadvantages

Source	*Astaxanthin (mg/kg dry weight)*	*Advantage*	*Disadvantage*
Synthetic	50 000–100 000	High in astaxanthin	Synthetic
Crustacean	50–200	Biological	Low in astaxanthin
Crustacean oil extract	1000–1500	Biological	Moderate in astaxanthin
Microalgae	<10 000	Biological, high in astaxanthin	Difficult to obtain high biomass with high astaxanthin
Yeast (wild strain)	<500	Biological, easy to grow	Low in astaxanthin

price. Krill has been looked upon as a possible feed and astaxanthin source (Ibrahim, Shimizu and Kono, 1984; Storebakken, 1988; Mori *et al.*, 1990). A group of these marine pelagic crustaceans exist, and they are found in large amounts both in Arctic and Antarctic seawater. However, so far this source has not found any use in aquaculture.

Growth of the green alga *Haematococcus pluvialis* has been scaled up in ponds of standard raceway design with paddlewheels, giving a commercial product with 1% astaxanthin (Bubrick, 1991). The problem with large-scale production is contamination with foreign algae and protozoan predators that may eliminate 90% of the biomass within 72 h. Bubrick (1991) states that production of *Haematococcus* and *Dunaniella*, the β-carotene producer (Aasen, Eimhjellen and Liaaen-Jensen 1969), is similar, and for astaxanthin production costs do not exceed US$ 2000/kg astaxanthin. The algae contain mainly astaxanthin esters (Grung *et al.*, 1992) deposited in spores that have to be disrupted in order to make astaxanthin available to salmonids (Sommer, Potts and Morrissy, 1990, 1991).

Bacteria such as *Mycobacterium lactiola* and a *Brevibacterium* spp. as well as animalia such as sea anemones, sea stars, insects, reptiles and birds such as flamingos contain astaxanthin (Bernhard, 1990; Johnson and An, 1991) but none of these species are at present of commercial interest.

Among the fungi, it is only the yeast *P. rhodozyma* that has been selected as a candidate for industrial production and mutants giving high contents of astaxanthin but with a low biomass have been observed (Johnson and An, 1991).

The synthetic products used for coloring salmonids are canthaxanthin and astaxanthin. Synthetic canthaxanthin was used in the Norwegian aquaculture industry until synthetic astaxanthin became available in 1984. In a review article (Bernhard, 1990) the total chemical synthesis of astaxanthin from β-ionone has been described, as well as its synthesis from β-carotene or canthaxanthin. Synthetic astaxanthin (Carophyll pink) is supplied as a 5–10% powder to the salmon feed manufacturers. Since this is the natural compound, as opposed to canthaxanthin, it is at present the one preferred by the salmonid industry. We believe that astaxanthin produced by a biological source would be preferred by the aquaculture industry and the consumers when such a product becomes available.

16.3 QUANTITATION OF ASTAXANTHIN

The bright color of astaxanthin is due to its strong absorbance of light in the visible region with an optimum around 475 nm and this allows it to be quantitated rather easily by visible absorption spectroscopy. The quantitation is not without problems for several reasons. The sensitivity of astaxanthin towards oxygen, light and low pH must be considered during preparation of samples prior to analysis (Liaaen-Jensen and Andrewes, 1985). Different extinction coefficients for astaxanthin ranging from $E_{1\%}$ = 1600 (Johnson, Conklin and Lewis, 1977) to $E_{1\%}$ = 2200 (Saito, 1969) have been used and the former gives a 27% higher astaxanthin concentration. Using crystalline astaxanthin in acetone we found an $E_{1\%}$ = 2000 at 478 nm, which is similar to that reported by Buchwald and Jencks (1968). In biological material astaxanthin is found together with other carotenoids that absorb light in the same region, occasionally giving high overestimations of astaxanthin when measured by spectrophotometry. This problem can be avoided using high performance liquid chromatography (HPLC) where astaxanthin is separated from the other carotenoids. The lack of commercially pure astaxanthin as a primary standard represents a problem when using HPLC. We have routinely used a mixture of all-*trans*- and *cis*-astaxanthin (Figure 16.2), extracted with diethylether:methanol (2:1) from Carophyll pink dissolved in 30% NaCl in water as a standard. The total amount of astaxanthin in the standard is estimated from the absorption at 478 nm and the proportion of all-*trans*-astaxanthin is estimated from the ratio between the two isomers as determined from the area of the peaks in the HPLC chromatogram, assuming that all-*trans*- and *cis*-astaxanthin have identical extinction coefficients at 478 nm. We find no difference when using this standard compared with crystalline all-*trans*-astaxanthin.

Quantitation of astaxanthin in *P. rhodozyma* requires disruption of the cell wall to make the carotenoids extractable. This can be achieved by mechanical methods such as grinding with sand, shaking with glass beads or by French pressing (Andrewes, Phaff and Starr, 1976; Haard, 1988), or chemically by the use of dimethylsulfoxide (Sedmak, Weerasinghe and Jolly, 1990). We have used incubation with Novozym™234, which is both convenient and gives complete extraction of carotenoids from *P. rhodozyma* grown under most conditions.

16.4 THE YEAST *P. RHODOZYMA*

In 1967–8 Herman J. Phaff with his college Miller at the University of California and Yoneyama and Soneda collected more than 800 isolates of yeasts from tree exudates in Japan and the West Coast of North America (Phaff, 1986). Nine of the isolates from Japan and one from Alaska were identified as a new yeast since it both synthesized the carotenoid astaxanthin and was able to grow on selected carbohydrates without the use of oxygen, i.e. it was fermentative. Initially it was called *Rhodozyma montanae* (Phaff *et al.*, 1972) since it was found mostly in high areas, but when a description according to the rules of the International Code of Botanical Nomenclature was given a few years later, it was renamed to P. *rhodozyma* (Miller, Yoneyama and Soneda, 1976). The yeast was thought to be unique with regard to the kind of stereoisomer it synthesized which is the 3*R*,3′*R*-isomer (Figure 16.2), since astaxanthin from all natural sources examined until then had been found to contain mainly the 3*S*,3′*S*-isomer (Andrewes and Starr, 1976). It is now known that *P. rhodozyma* is not the only organism that contains mainly 3*R*,3′*R*-astaxanthin since this is also true, for instance, for *Euphausia superba* (Maoka *et al.*, 1985; Foss, Renstrøm and Liaaen-Jensen, 1987).

P. rhodozyma abundantly develop in the spring in bleeding sap flows of birch (*Betula verrucosa*, Ehrh.) and other broadleafed trees (Phaff *et al.*, 1972; Miller, Yoneyama and Soneda, 1976; Golubev, Bab′eva and Norik, 1977). It can grow to high numbers at which the sap gains the red–orange color of astaxanthin. The yeast multiply by budding and no spore formation or sexual cycle has so far been identified. The characteristics of the genus *Phaffia* have been compared with other yeasts and fungi imperfecti by De Mot, Demeersman and Verachtert (1984), Walker (1984) and Yamada and Kawasaki (1989). The evolutionary affinities of *P. rhodozyma* to the heterobacidomycetes have also been estimated from 18S and 25S ribosomal RNA sequence divergence by Gueho, Kurtzman and Peterson (1989) and Yamada *et al.* (1990), and they conclude that *Phaffia* should be retained as a separate genus, which is in contrast to that suggested by Weijman, Rodrigues de Miranda and van der Walt (1988).

16.4.1 Growth of *P. rhodozyma*

The yeast is able to grow on a range of different carbohydrate sources (Miller, Yoneyama and Soneda, 1976; Johnson and Lewis, 1979; Barnett, Payne and Yarrow, 1983). From an industrial point of view the fact that sucrose gives a good growth (Johnson and Lewis, 1979) is of significance. Sucrose is the main carbohydrate of molasses, which is one of the most inexpensive carbon sources available and is used for baker's yeast production (Barford, 1987). The doubling time of *P. rhodozyma* in the exponential growth phase is around 5 h (μ = 0.14/h) with molasses as a carbon source at 22°C (Figure 16.3), which is the optimum growth temperature (Johnson and Lewis, 1979). Others have reported generation times as low as 3.6 h using different growth conditions (Johnson and Lewis, 1979). For comparison, generation times for baker's yeast can be as low as 90 min (Barford, 1987).

When the yeast is grown on molasses the disaccharide sucrose is split to the monosaccharides glucose and fructose by the enzyme sucrase which is very active in *P. rhodozyma*. The amount of sucrose in the medium therefore decreases rapidly and disappears early in the growth phase, and at the same time the glucose and fructose concentrations increase to approximately equimolar concentrations (Figure 16.4). Glu-

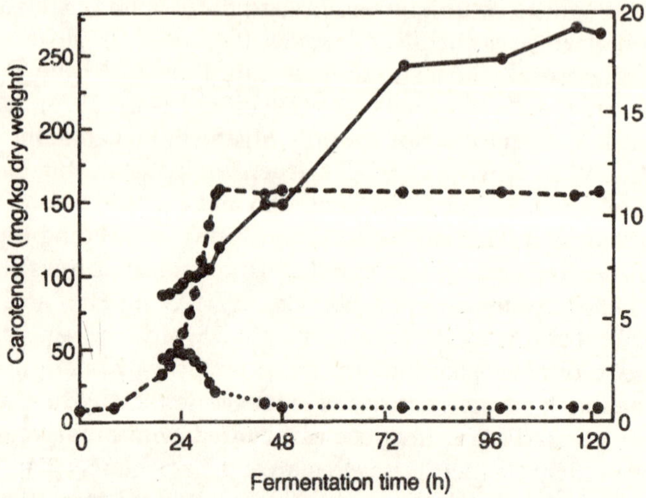

Figure 16.3 Formation of astaxanthin (—), β-carotene (...) and biomass (---) during batch fermentation of *P. rhodozyma* (ATCC 24202). The yeast was grown at 22°C in a 10 1 fermenter with a medium containing molasses (40 g/l), yeast extract (2 g/l), $(NH_4)_2SO_4$ (2 g/l), KH_2PO_4 (1 g/l), $MgCl_2{\cdot}6H_2O$ (0.5 g/l) and $CaCl_2{\cdot}2H_2O$ (0.1 g/l). During fermentation pH was kept constant at 4.5 by automatic addition of NH_4OH and H_2SO_4.

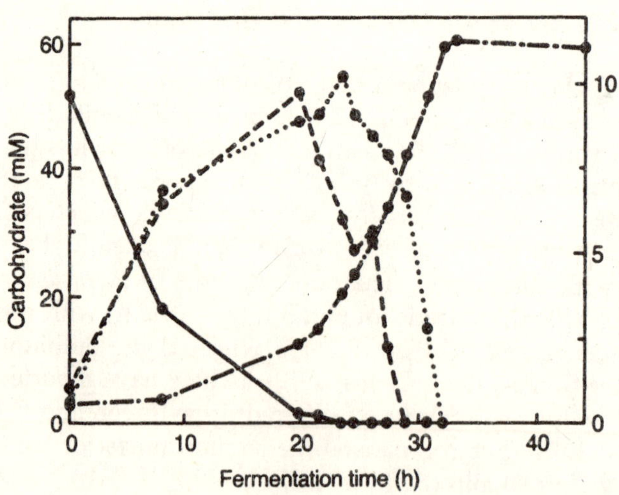

Figure 16.4 Concentration of sucrose (——), glucose (- - -), fructose (. . . .) and biomass (-·-·-) during growth of *P. rhodozyma*. The yeast was cultivated in a 10 l fermenter according to the conditions given in legend to Figure 16.3 and the carbohydrates quantitated using HPLC.

cose then starts to disappear and it is obviously preferred over fructose. When all the fructose is used, growth ceases (Figure 16.4).

16.4.2 Astaxanthin formation by *P. rhodozyma*

A recent review discusses the present knowledge of the metabolic pathways for astaxanthin synthesis in *P. rhodozyma* (Johnson and An, 1991). Several factors influence the rate and extent of astaxanthin synthesis in the yeast. Astaxanthin synthesis is highest at pH 5 but there is no sharp optimum (Johnson and Lewis, 1979). With regard to temperature there is an optimum in astaxanthin production at 22°C, where the highest growth rate is also observed (Johnson and Lewis, 1979). At temperatures above 27–28°C both yeast growth and astaxanthin synthesis are generally absent, but Polulyakh *et al.* (1991) reported that at 30°C torulene and torularhodin rather than astaxanthin were synthesized.

Astaxanthin synthesis varies with the carbon source used for *P. rhodozyma* growth and D-cellobiose was found to give the highest astaxanthin content (Johnson and Lewis, 1979). Sucrose also gave a high astaxanthin content (Johnson and Lewis, 1979) and Haard (1988) reported even higher values using molasses, suggesting it to contain additional factor(s) that stimulate astaxanthin synthesis. The usefulness of molasses for astaxanthin production was, however, not confirmed by Lewis (1990). Haard (1988) also found that with molasses the astaxanthin (or rather total carotenoids) in the yeast increased with the concentration of molasses used. This is in contrast to what others have found

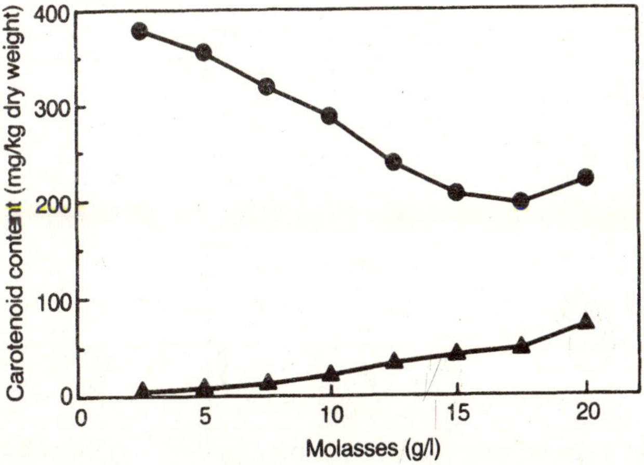

Figure 16.5 Amount of astaxanthin (●) and β-carotene (▲) in *P. rhodozyma* after 7 days' cultivation in shake flasks in a medium containing various concentrations of molasses as the carbon source. Other ingredients in the medium were as given in legend to Figure 16.3.

for glucose where a significant decrease with increasing sugar concentration was reported (Johnson and Lewis, 1979). Using moderate concentrations of molasses and shake flask cultures of *P. rhodozyma* we found a decrease in astaxanthin and an increase in β-carotene with increasing molasses concentrations (Figure 16.5).

Another factor that influences astaxanthin synthesis is the supply of oxygen (Johnson and Lewis, 1979). This is not easily controlled in shake flask cultures, especially at high yeast concentration where the oxygen demand is high. In a fermenter where oxygen supply can be both controlled and measured, we found the supply to have a strong influence on the astaxanthin synthesis when *P. rhodozyma* was grown on molasses (Figure 16.6). When oxygen supply was adequate, maximum astaxanthin level was reached shortly after yeast growth ceased.

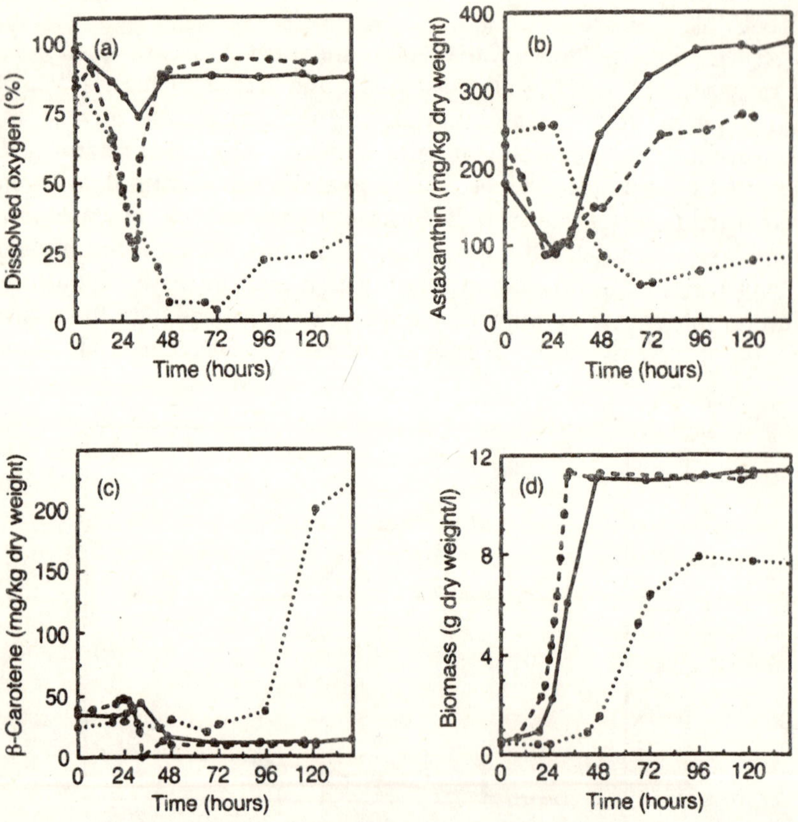

Figure 16.6 The influence of oxygen supply on the formation of astaxanthin in *P. rhodozyma*. The yeast was grown in a 10 l fermenter as specified in legend to Figure 16.3 with high (—), medium (---) and low (···) aeration rates obtained by varying the impeller rotation speed and by restricting the air supply. (a) Dissolved oxygen. (b) Amount of astaxanthin in the yeast. (c) Amount of β-carotene in the yeast. (d) Amount of yeast cells.

When oxygen was restricted slightly, there was a delay in astaxanthin synthesis and when strong restriction was applied, very slow astaxanthin formation was observed. Also, the final levels of astaxanthin were reduced when oxygen supply was restricted and with the strongest restriction β-carotene rather than astaxanthin accumulated (Figure 16.6). To optimize astaxanthin production it is important to have a high oxygen supply.

16.5 *P. RHODOZYMA* AS AN ASTAXANTHIN SOURCE FOR SALMONIDS

Shortly after its discovery the possibility that *P. rhodozyma* may be fed to salmonids, crustaceans and hens to invoke pigmentation was investigated (Johnson, Conklin and Lewis, 1977; Johnson, Lewis and Grau, 1980; Johnson, Villa and Lewis, 1980). In rainbow trout (*Oncorhynchus mykiss*) the amount of astaxanthin in the flesh increased when fish were fed a diet containing *P. rhodozyma* (Johnson, Conklin and Lewis, 1977; Johnson, Villa and Lewis, 1980) and, when compared with a commercial fish feed containing synthetic canthaxanthin, Johnson, Villa and Lewis (1980) found higher amounts of carotenoid in the *P. rhodozyma* fed fish. Astaxanthin is in general a better carotenoid source for salmonids than canthaxanthin (Foss *et al.*, 1984; Torrissen, 1986; Choubert and Storebakken, 1989; Torrissen, Hardy and Shearer, 1989; Bjerkeng, Storebakken and Liaaen-Jensen, 1990; Guillou, Choubert and de la Noue, 1992; Storebakken and No, 1992) and synthetic astaxanthin rather than synthetic canthaxanthin is the predominant color source in fish farming today. A comparison between astaxanthin from *P. rhodozyma* and synthetic astaxanthin is therefore of more interest. When three diets containing either *P. rhodozyma*, synthetic astaxanthin or no added astaxanthin were given to rainbow trout, the feed with synthetic astaxanthin, which also contained most astaxanthin (Figure 16.7a), gave the highest level in the flesh (Figure 16.7c). When we adjust for the differences in astaxanthin levels in the feeds by calculating the amount of astaxanthin accumulated in the flesh per weight unit of astaxanthin in the feed, *P. rhodozyma* seems to be a better astaxanthin source than the synthetic compound (Figure 16.7d). A prerequisite for this conclusion is that the amount of astaxanthin deposited in the flesh is linearly correlated to the amount of astaxanthin in the feed, which seems to be the case at such low levels (Choubert and Storebakken, 1989; Torrissen, Hardy and Shearer, 1989; Bjerkeng, Storebakken and Liaaen-Jensen, 1990; Storebakken and No, 1992). Others reported no significant differences in the amounts of astaxanthin in the flesh of rainbow trout when identical levels of astaxanthin (40 mg/kg) from *P. rhodozyma* and Carophyll pink were fed during a period of 73 days (Flenø, Christensen and Larsen,

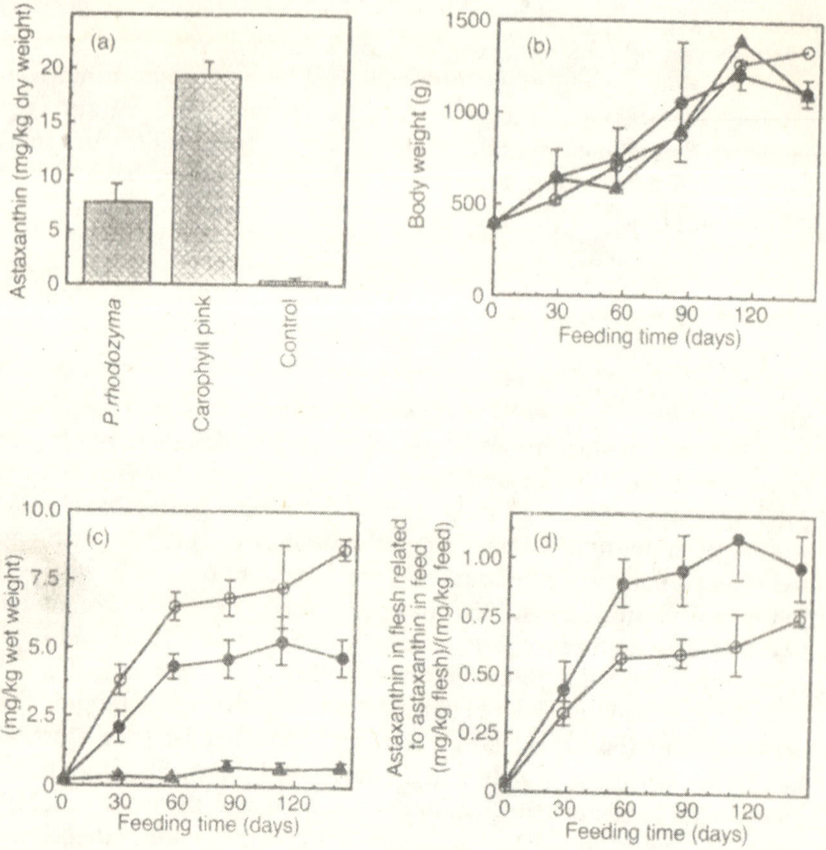

Figure 16.7 Astaxanthin deposition in rainbow trout (*O. mykiss*) fed a diet containing *P. rhodozyma* (●), Carophyll pink (○) or no astaxanthin added (▲). (a) Amount of astaxanthin in the various feeds. (b) Growth of fish fed the three feeds. (c) Amount of astaxanthin in the flesh of fish fed the various diets. (d) The ratio between amount of astaxanthin in flesh and feed for fish fed *P. rhodozyma* and Carophyll pink.

1988). More extensive feeding experiments are required in order to decide if *P. rhodozyma* is a better astaxanthin source than synthetic or other sources of astaxanthin.

Johnson, Villa and Lewis, (1980) found that only astaxanthin from disrupted *P. rhodozyma* cells was available to rainbow trout. They suggested that the thick cell wall known to be present in *P. rhodozyma* (Miller, Yoneyama and Soneda, 1976) restricted pigment availability, indicating that no enzyme able to degrade the components of the cell wall is present in the digestive tract of this fish. More recently, Gentles and Haard (1991) found that rainbow trout were able to utilize astaxanthin also from intact *P. rhodozyma* cells to a significant degree, even though a slight increase in uptake was found when the cells were broken by milling or enzyme treatment. Whether or not rainbow trout is able to accumulate astaxanthin from intact *P. rhodozyma* cells is thus controversial. It should be mentioned that rainbow trout seems not to be able to degrade the cell wall of brewer's yeast (*Saccharomyces cerevisiae*) completely since the digestibility was significantly increased by breaking its cell wall (Rumsey *et al.*, 1991). For Atlantic salmon we found that breaking the *P. rhodozyma* cells by high pressure homogenization very much increased the availability of astaxanthin (Figure 16.8). Johnson and An (1991), on the other hand, state that feeding trials with Atlantic salmon have indicated that it is able to utilize astaxanthin from non-disrupted *P. rhodozyma*. Since no data were given, there may be differences in fish size and rearing conditions used and further studies are certainly needed to clarify if salmonids can take up astaxanthin from unbroken *P. rhodozyma* cells under certain conditions. The data in Figure 16.8 show that this is not the case for Atlantic salmon reared under conditions used for commercial production in Norway.

Astaxanthin predominates in *P. rhodozyma* grown under most conditions, but it also contains other carotenoids that probably represent intermediates in the biosynthesis of astaxanthin (Andrewes, Phaff and Starr, 1976; Johnson and An, 1991). Fish that are given *P. rhodozyma* are therefore also exposed to other carotenoids that may influence the uptake of astaxanthin, for instance by competing for the same hypothesized binding sites in the gut. Great differences in the efficiency of

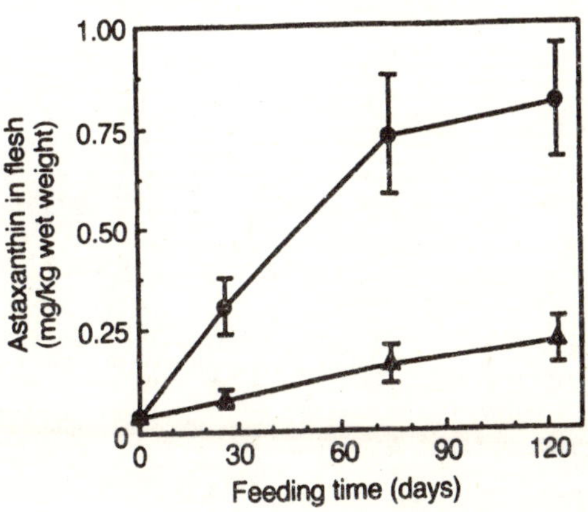

Figure 16.8 Deposition of astaxanthin in the flesh of Atlantic salmon (*S. salar*) fed a diet containing intact (▲) or broken (●) *P. rhodozyma* cells. Fish weighing about 1.3 kg at the start and 1.7 kg at the end of the experiment were given a diet containing 10% *P. rhodozyma* that was either disrupted by high pressure homogenization or spray dried directly after fermentation and washing. For feed with unhomogenized cells 10% of the astaxanthin could be extracted with diethylether/methanol, whereas for the homogenized cells 49% was extractable. Total astaxanthin in both feeds were 8 mg/kg dry weight when determined after treating the feeds with Novozym™ 234 (Novo Biolabs, Denmark).

uptake of various carotenoids exist in salmonids (Schiedt *et al.*, 1985; Torrissen, Hardy and Shearer, 1989), and even for rather similar carotenoids like astaxanthin and canthaxanthin there is a difference, since the former is deposited approximately 1.3 times more readily (Torrissen, 1986; Choubert and Storebakken, 1989; Bjerkeng, Storebakken and Liaaen-Jensen, 1990). From Figure 16.9 it is evident that astaxanthin from *P. rhodozyma* is the only carotenoid that is accumulated to any extent in the fish. Whereas astaxanthin represents only around 50% of the peak area in the HPLC chromatogram of carotenoids extracted from the *P. rhodozyma* feed, it represents more than 90% in the flesh of fish (Figure 16.9a and c). For comparison similar HPLC chromatograms for feed and flesh from an experiment using Carophyll pink are shown (Figure 16.9b and d), where astaxanthin is predominant both in feed and flesh.

In the feeds with *P. rhodozyma* and Carophyll pink there are two carotenoids with retention times longer than that of the main peak (Figure 16.9c and d). From the absorption spectra of these two peaks we conclude that they represent *cis* isomers of astaxanthin, probably 9-*cis*- and 13-*cis*-astaxanthin. The main peak in the chromatograms (Figure 16.9) is shown to be all-*trans*-astaxanthin (Figure 16.2). The two *cis* isomers represent 15–20% of total astaxanthin in both the *P. rhodozyma* and the Carophyll pink feed, the rest of astaxanthin being all-*trans*. For *P. rhodozyma* most of the *cis* isomerization takes place during spray

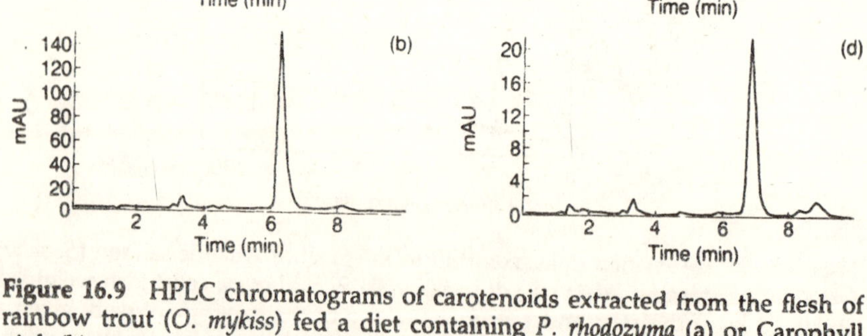

Figure 16.9 HPLC chromatograms of carotenoids extracted from the flesh of rainbow trout (*O. mykiss*) fed a diet containing *P. rhodozyma* (a) or Carophyll pink (b) as an astaxanthin source. In (c) and (d) the chromatograms of carotenoids extracted from the feeds with *P. rhodozyma* and Carophyll pink, respectively, are shown. The carotenoids were separated on a Spherisorb CN column using a system slightly modified from that described by Schiedt *et al.* (1981, their system C) with a mobile phase of hexane:2-propanol:*N*-ethyldiisopropylamine = 89:10:1. The absorption at 478 nm was monitored.

drying of the yeast, since prior to this step *cis* isomers represent only around 5% of the total astaxanthin. The *cis* isomers are obviously less readily taken up by the fish compared with all-*trans*-astaxanthin, since in the flesh of both *P. rhodozyma* and Carophyll pink fed fish only small amounts of *cis* isomers were found (Figure 16.9a and b). 11-*cis*- and 13-*cis*-astaxanthin have also been found in wild Atlantic salmon, representing as much as 20–30% of astaxanthin (Schiedt, Leuenberger and Vecchi, 1981; Schiedt *et al.*, 1989).

Astaxanthin can exist in three different stereoisomeric forms (Figure 16.2) and in *P. rhodozyma* the predominant form is of the *3R,3′R* type (Andrewes and Starr, 1976). Stereoisomeric molecules differ as little as a right and a left glove, but even so biological systems may treat them very differently. When discussing possible astaxanthin sources for salmonid farming it is therefore important to know whether the fish can differentiate between the three stereoisomers. In wild salmon caught at different places in Europe and Canada around 80% of the astaxanthin had the *3S,3S′*-configuration, around 15% was *3R,3R′* and only around 5% was *meso*-astaxanthin (Schiedt, Leuenberger and Vecchi, 1981; Schiedt *et al.*, 1989). This was very similar for chum salmon (*Oncorhynchus keta*) caught in Japan, except that the amount of *meso*-astaxanthin was even lower (Ando and Hatano, 1987). The salmonids perform no interconversion between the different astaxanthin stereoisomers (Foss *et al.*, 1984; Storebakken *et al.*, 1985; Katsuyama, Komori and Matsuno, 1987), and the three stereoisomers seem to be equally well deposited both by rainbow trout and Atlantic salmon (Foss *et al.*, 1984; Storebakken *et al.*, 1985). The astaxanthin stereoisomer distribution in the flesh of salmonids therefore simply reflects the distribution found in their diet.

Currently, farmed salmon are given synthetic astaxanthin (Carophyll pink) which is racemic and therefore contains the isomers in the ratio (*3S,3′S*:*3R,3′R*:*meso*) = (1:1:2). This has allowed Lura and Sægrov (1991a,b) to develop a method where eggs and alevins from farmed and wild Atlantic salmon can be differentiated based on their astaxanthin stereoisomer distribution. Differentiation between farmed and wild salmonids is of importance in the maintenance of a natural population of wild Atlantic salmon which is desired, at least in Norway, and this differentiation is even easier if the farmed salmons are given *P. rhodozyma* as an astaxanthin source.

16.6 PRODUCTION OF *P. RHODOZYMA*

Based on our experience with both pilot-scale and large-scale experiments with *P. rhodozyma* we foresee a scheme for industrial production of the yeast as shown in Figure 16.10. Fermentation should be per-

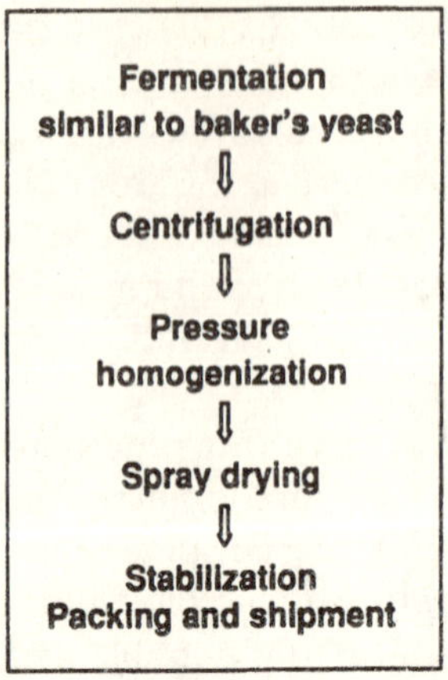

Figure 16.10 Schematic representation of how *P. rhodozyma* could be produced in a commercial process.

formed so that a high cell mass with optimum astaxanthin concentration is produced as quickly as possible. In this regard the carbon source used and the supply of oxygen are of particular importance. As discussed previously molasses is a good and inexpensive carbon source. Others have studied the potential of the by-product alfalfa residual juice for *P. rhodozyma* production, but found that even though it supported growth the astaxanthin synthesis was inhibited due to the presence of saponins (Okagbue and Lewis, 1984b,c). Deproteinized whey permeate has also been studied as a substrate for *P. rhodozyma* production (Lewis, 1990). Growth and astaxanthin synthesis were acceptable, but the enzyme β-galactosidase had to be added to split the disaccharide lactose into glucose and galactose, and the latter remained in the medium since *P. rhodozyma* cannot utilize this sugar. These factors will certainly increase the cost of production. At present none of the inexpensive carbohydrate sources tested can compete with molasses.

Three methods of aerobic fermentation can be considered for production of *P. rhodozyma*. These are batch fermentation, where all the substrate is present at the start of fermentation, fed-batch fermentation, where substrate is added according to demands during fermentation, and continuous fermentation, where some of the spent medium with produced yeast is continuously replaced with new medium. The latter technique is most likely not useable for production of *P. rhodozyma* since a continuous growth of yeast cells appears and astaxanthin production continues after growth has ceased. Furthermore, the process is more expensive and due to longer fermentation times requires stricter sterile conditions than the other two techniques. With batch fermentation the growth during the exponential phase is maximal and at the end of this phase when cell mass is high, the oxygen demand becomes very large

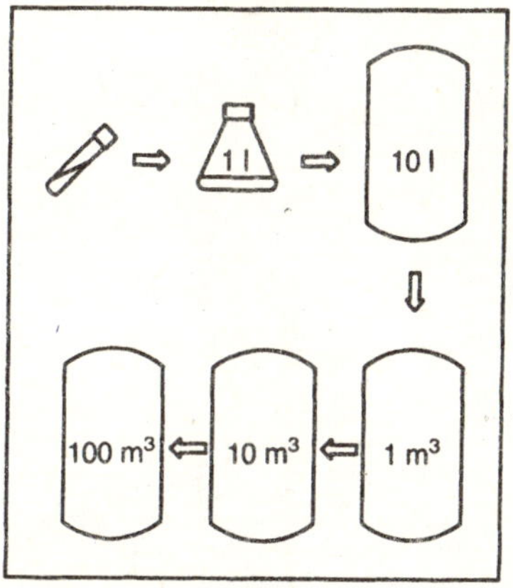

Figure 16.11 Schematic representation of propagation of *P. rhodozyma* in a commercial production plant with an end fermenter of 100 m^3.

and is difficult to satisfy technically. With a fed-batch technique growth is controlled by the rate of substrate addition and an adequate oxygen supply can be achieved (Barford, 1987). This technique gives optimal use of substrate for biomass production since ethanol production is negligible and astaxanthin production optimal due to high oxygen supply. The aerobic fed-batch technique, which has been optimized for years for baker's yeast production (Barford, 1987), can be used with a modified profile of molasses addition for large-scale production of *P. rhodozyma*. Successful production using a yeast propagation scheme as shown in Figure 16.11, where a 100 m^3 fermenter is used in the final step, has been performed. Following fermentation the cells can be washed and concentrated by continuous centrifugation at conditions identical to that used for baker's yeast production. The resulting yeast suspension typically contains 200–230 g dry matter/l.

In order to make the astaxanthin available to salmonids, at least for Atlantic salmon reared under commercial conditions, the cell wall of the yeast must be modified. This can be achieved by chemical, mechanical or enzymatic means. The process of autolysis of *P. rhodozyma* has been studied by Okagbue and Lewis (1984a). With this technique the yeast is left standing in suspension for a certain period of time during which the yeast's own enzymes partially degrade its cell wall, and after 26 h incubation at 37°C in distilled water 12–100% of the astaxanthin was extractable, depending on the strain of *P. rhodozyma* used. Cell wall degradation can be increased by adding hydrolytic enzymes to the yeast. Johnson *et al.* (1978) grew *Bacillus circulans* WL-12 on heat-inactivated *P. rhodozyma* cells and the induced hydrolytic enzymes were used to open *P. rhodozyma* cells. The *B. circulans* strain could also be grown in a mixed culture with *P. rhodozyma* and this system was optimized by Okagbue and Lewis (1983, 1985).

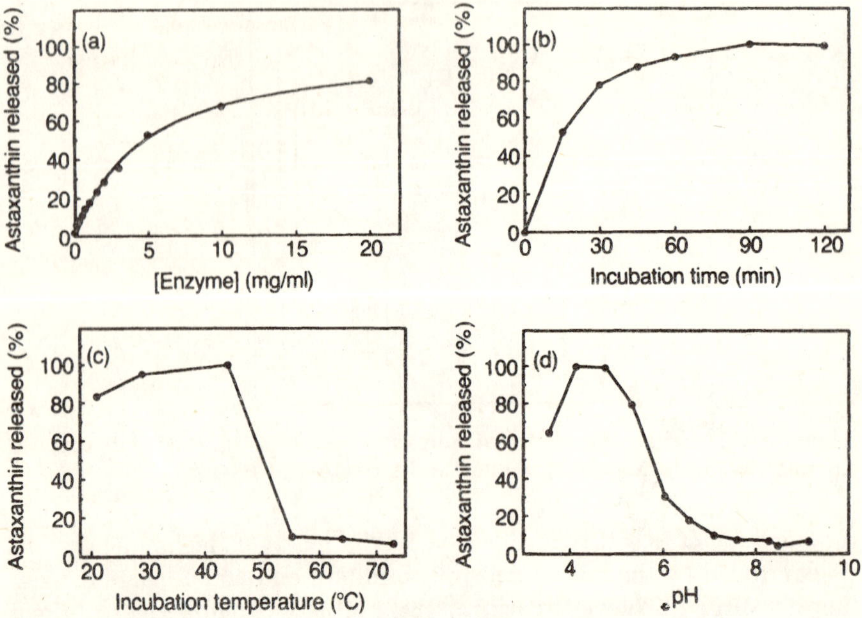

Figure 16.12 Release of astaxanthin from *P. rhodozyma* after treatment of the cells with Novozym™234. Effects of enzyme concentration (a), incubation time (b), incubation temperature (c) and pH (d) on the extractability of astaxanthin from the yeast are shown. *P. rhodozyma* was grown for 7 days in shake flasks at the conditions given in legend to Figure 16.3, collected by centrifugation and washed once with water. In the standard conditions the yeast (17 mg dry weight/ml) was suspended in 0.2 M sodium acetate buffer (pH 4.5) and the reaction started by adding Novozym™234 at a concentration of 5 mg/ml. After incubation for 60 min at 25°C astaxanthin was extracted and analyzed by HPLC. In (a) the incubation time was 15 min. In (d) different buffer systems with overlapping pH regions were used to obtain the given pH values.

Novozym™234 is a commercial multicomponent enzyme product isolated from *Trichoderma harzianum* that degrades the cell wall of yeast and fungi. Novozym™234 can also modify the cell wall of *P. rhodozyma* so that the astaxanthin becomes extractable (Figure 16.12). The major component responsible for the cell wall degradation is β(1–3)-glucanase (Sørum and Robertsen, 1991).

Opening of *P. rhodozyma* cells can be achieved using high pressure homogenization. With this technique the cell suspension is forced through a homogenization valve at high pressure (Figure. 16.13). The mechanism of cell breakage by homogenization is not known in detail, but disruption is mainly due to the abrupt pressure decrease and cavitation forces generated when the yeast passes the vent (Moore, Hoare and Dunnill, 1990). One important factor for the efficiency of the process is the pressure, and for *P. rhodozyma* we found a proportional increase in astaxanthin extractability with homogenization pressure

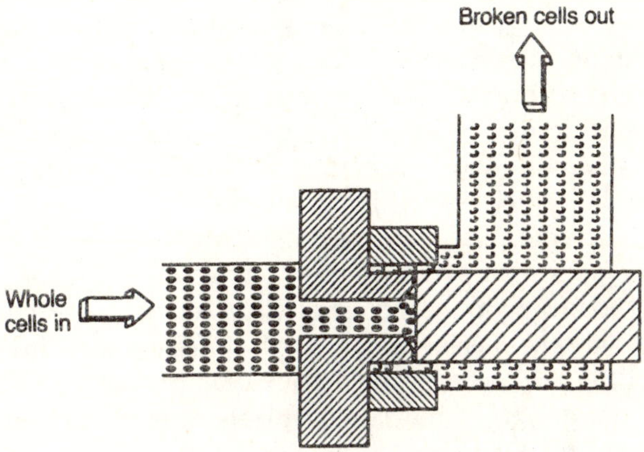

Figure 16.13 Schematic drawing of a homogenization valve, which is where the actual cell opening occurs in a high pressure homogenization machine.

(Figure 16.14a). Both linear, exponential and sigmoid shapes of such curves have been reported (Moore, Hoare and Dunnill, 1990; Schütte and Kula, 1990). Another factor generally found to be important is the shape of the homogenization valve (Moore, Hoare and Dunnill, 1990; Schütte and Kula, 1990). We tested three different laboratory-scale homogenizers equipped with different homogenization valves and

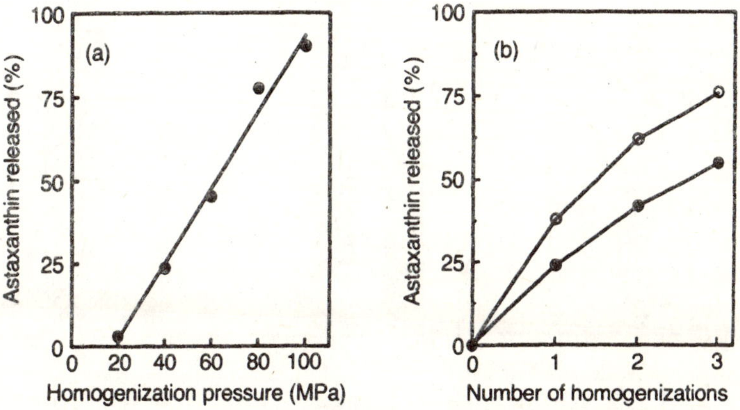

Figure 16.14 Release of astaxanthin from *P. rhodozyma* by high pressure homogenization. (a) Effect of homogenization pressure. A yeast suspension (170 g dry matter/l) was passed once through the homogenizer (machine III, valve E in Figure 16.15) at the pressures indicated. (b) The effect of running the yeast suspension through a homogenizer several times. Laboratory and industrial scale homogenizers with a capacity of 0.2 (o) and 3.5 (●) m^3 were used. The laboratory-scale machine was the same as that used in (a), but with valve A (Figure 16.15). The industrial-scale machine was from APV Rannie A/S, Denmark. Both machines were run at 100 MPa. The degree of cell opening was determined from the ratio between the amount of astaxanthin extracted with organic solvent before and after complete cell opening with Novozy™234.

found great variability in efficiency (Figure 16.15). There were no significant differences between the three homogenizers when the least efficient homogenization valve for each machine was used. With the most efficient homogenization valve for each machine differences appeared. When an industrial homogenizer with a capacity of 3.5 m^3/h was tested, the efficiency decreased to around 50%, compared with around 90% for the pilot-scale machine. One explanation for the drop in efficiency on scale-up is that the homogenization valve cannot be constructed according to the increase in flow capacity, since this requires forces exceeding the mechanical strength of certain machine parts. The efficiency of the production scale machine can probably be increased by optimizing the design of the homogenization valve. Cell opening can also be increased by running the suspension through the homogenizer several times (Figure 16.14b), which is in accordance with present theory (Moore, Hoare and Dunnill, 1990; Schütte and Kula, 1990) and is similar to that found for the release of protein from baker's yeast (Hetherington *et al.*, 1971; Moore, Hoare and Dunnill, 1990).

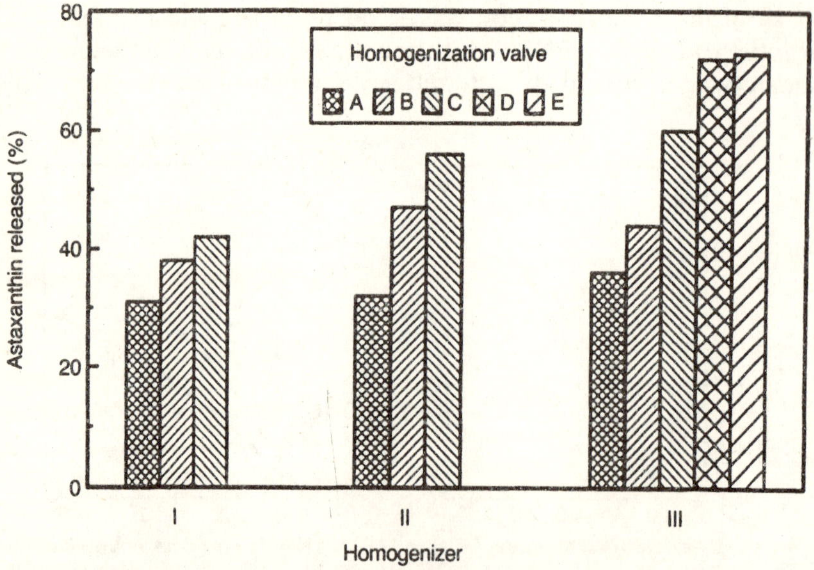

Figure 16.15 The efficiency of three different high pressure homogenizers tested for the release of astaxanthin from cells of *P. rhodozyma*. The following homogenizers were used: I: Bran & Luebbe (Germany), SHL 15 (capacity, 400 l/h). II: APV Gaulin (Netherlands), MC4 (capacity, 300 l/h). III: APV Rannie A/S (Denmark), Lab.12.51 (capacity 200 l/h). The machines were run at the maximal pressures recommended by the manufactures which were 80, 70 and 100 MPa for machines I, II and III, respectively. Each homogenizer was tested with from three to five different homogenization valves supplied with the machines. A yeast suspension of 170 g/l (dry weight) was passed once through the homogenizers. Relative amount of astaxanthin released was measured as given in legend to Figure 16.14.

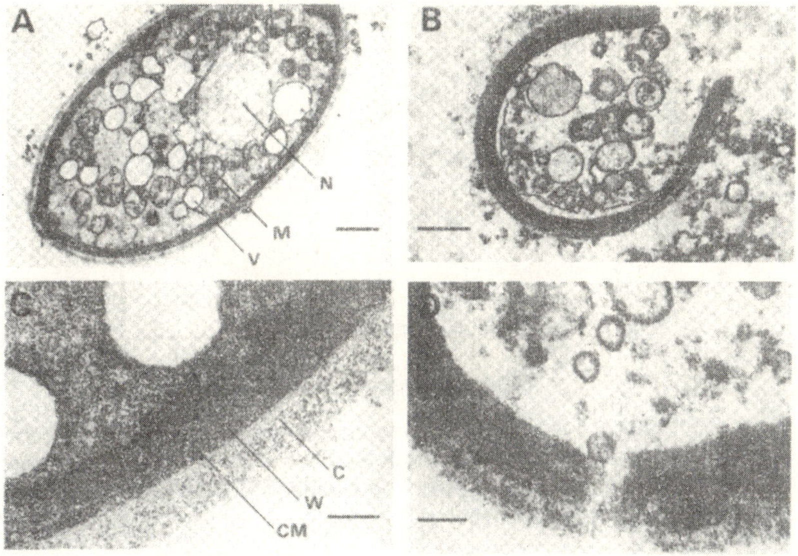

Figure 16.16 Transmission electron micrographs of cells of *P. rhodozyma*.(A) Intact yeast cell produced in a 10 m^3 fermenter (bar, 1 μm). (B) Broken cell obtained by high pressure homogenization at 100 MPa (bar, 1 μm). Panels (C) and (D) show a higher magnification of the cell wall of an intact and a broken yeast cell respectively (bar, 0.2 μm). In (A) nucleus (N), mitochondria (M) and vacuoles (V) are clearly visible. In (C) the cell membrane (CM), cell wall (W) and capsular material (C) are seen. Cells were fixed with potassium permanganate and stained with uranyl acetate and lead citrate.

Figure 16.16 shows electron micrographs of cells of *P. rhodozyma* before and after high pressure homogenization. After disruption of the cell wall, large holes and release of cell components can be seen in some cells (Figure 16.16B), whereas in other cells only small cracks were observed (Figure 16.16D). Probably also in the latter case the astaxanthin is available for extraction and uptake in the fish.

The dry feed used in salmon farming today is extruded and the high pressure used in this process could be sufficient to release astaxanthin from intact *P. rhodozyma* cells. When a suspension of *P. rhodozyma* was mixed with feed components and extruded, only about 40% of the astaxanthin was released (Figure 16.17). Also, the extrusion conditions that gave the highest astaxanthin release are not acceptable for commercial fish feed production. A separate cell opening step is therefore required.

In a *P. rhodozyma* suspension the astaxanthin stability is low, and a suspension is also impractical and expensive to transport. A drying step is therefore required. We have tested spray drying for this purpose since it is a rather gentle method where the material is exposed to high temperatures only for a short time. The loss of all-*trans*-astaxanthin

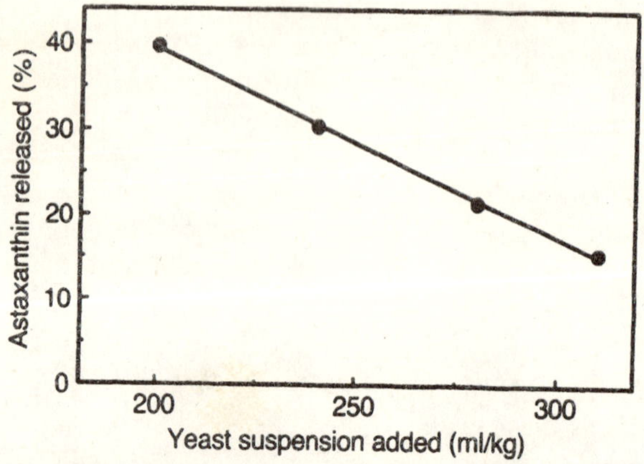

Figure 16.17 Effect of fish feed extrusion on the release of astaxanthin from *P. rhodozyma*. The dry ingredients of a fish feed were mixed and added various amounts of the yeast suspension prior to extrusion. The final fish feed produced contained about 90% dry matter.

during such a drying can be as low as a few per cent, depending on drying conditions, whereas a small increase in the *cis* forms of astaxanthin was found (Figure 16.18).

Astaxanthin, like other oxygenated carotenoids, is easily degraded during storage (Torrissen *et al.*, 1982; Liaaen-Jensen and Andrewes, 1985), and this is also the case with astaxanthin in homogenized and spray dried *P. rhodozyma* when stored at room temperature (Figure

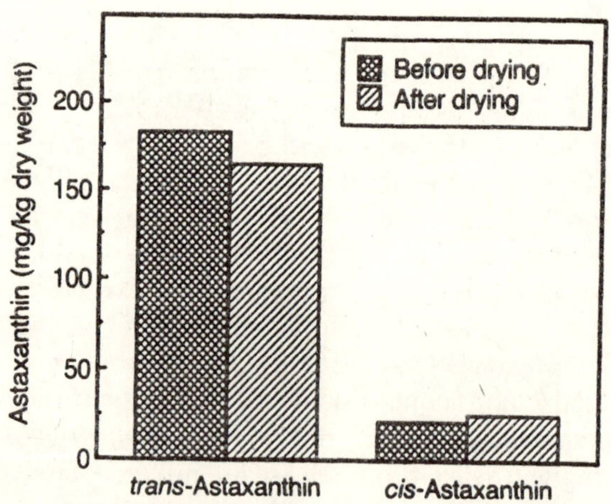

Figure 16.18 The influence of spray drying of *P. rhodozyma* on the content of all-*trans*-astaxanthin and *cis*-astaxanthin (9-*cis* plus 13-*cis*) in the yeast. A suspension of *P. rhodozyma* (170 g dry weight/l) that had been subjected to high pressure homogenization was spray dried in a pilot-scale dryer using an air inlet temperature of 190°C and an outlet temperature of 95°C. The dried yeast contained about 94% dry matter.

16.19a). The degradation was reduced at a temperature of 4°C and was almost eliminated at temperatures of −20°C or below. When the yeast was mixed with components normally found in a semimoist salmon feed containing about 60% dry matter, the stability of astaxanthin was even less (Figure 16.19b). Also in this case the stability was increased by lowering the temperature, but only at −80°C was the stability acceptable. Normally fish feed is stored at outdoor temperatures that on a sunny summer day can be rather high. The astaxanthin therefore needs

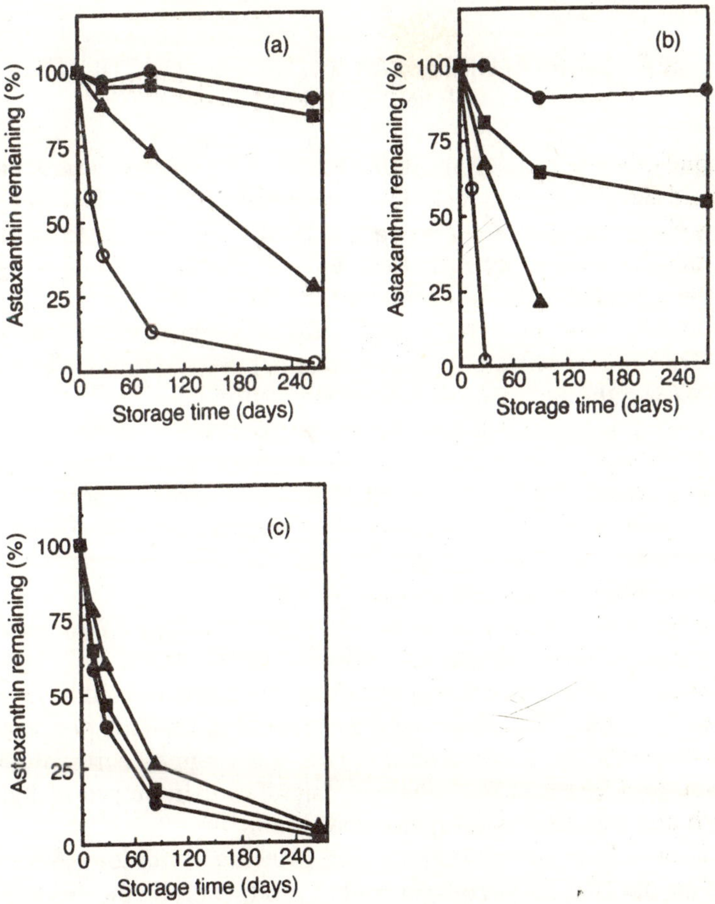

Figure 16.19 Degradation with time of astaxanthin in spray dried *P. rhodozyma* (a) and after mixing into a semimoist fish feed (b). The yeast was produced in a 10 m^3 fermenter, washed with water, homogenized at 100 MPa in a high pressure homogenizer and spray dried using the conditions given in legend to Figure 16.18. The dried yeast was then stored in the dark at 20°C (o), +4°C (▲), −20°C (■) or at −80°C (●). Panel (c) shows the influence of the stability of astaxanthin at 20°C in dried *P. rhodozyma* when spray drying was done in the absence of antioxidant (●) and in the presence of the antioxidants Ethoxyquin 100 (■) and Rexoform 33 (▲) (both from Grace Rexolin, Sweden).

to be stabilized. We have tested the commercial antioxidants, Etoxyquin 100 and Rexoform 33 for this purpose, and found only minor effects on astaxanthin stability (Figure 16.19c). The antioxidants were added to the yeast suspension prior to spray drying and their concentration during the storage period was not measured. It is possible that most of the antioxidants were oxidized at the start of the storage experiment and a more profound effect on astaxanthin stability may occur if the antioxidants are added to the dried yeast.

16.7 INCREASE IN ASTAXANTHIN PRODUCTION BY *P. RHODOZYMA*

We conclude that industrial production of *P. rhodozyma* is possible and when added to salmonid feed, the astaxanthin colors the fish well. Each step in the production process needs to be studied in detail for optimization, but the main problem is that the final product obtained contains too little astaxanthin to be competitive in the market. As far as we can judge all the companies that are involved in projects with *P. rhodozyma* have faced this obstacle and much effort has been put into the process of increasing the astaxanthin production by the yeast.

New strains of *P. rhodozyma* have been isolated from nature, but to our knowledge no strain with a sufficiently high astaxanthin production has been found. We have isolated strains of *P. rhodozyma* in Norway, but no great differences in astaxanthin content were found between these strains and the best strains obtained from culture collections produced more astaxanthin (Figure 16.20).

Different effects of light on astaxanthin synthesis by *P. rhodozyma* have been found. Evans, Adams and Wisdom (1990) found an increase in astaxanthin when a mutant was grown in the presence of light, whereas An and Johnson (1990) observed a decrease. Experiments performed in our laboratory are in accordance with the original observations by Johnson and Lewis (1979) where no significant influence of light on growth and astaxanthin formation was found.

Different chemical compounds and growing conditions have been tested for the stimulation of astaxanthin production by the yeast, but so far none have been reported to have a dramatic effect. The most fruitful approach has been to make new mutants with the capacity to synthesize more astaxanthin. Mutations in *P. rhodozyma* can be induced either by physical means, like UV light and X-rays, or by chemical agents. The cells can then be seeded on agar plates after proper dilution so that single colonies develop after a suitable time of incubation. The colonies may then be visually inspected for any change in color. When UV radiation was used, mutants from *P. rhodozyma* were generated (An,

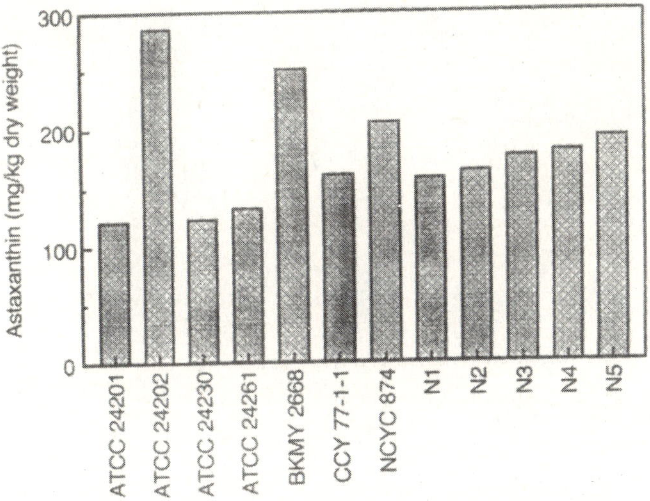

Figure 16.20 Amount of astaxanthin in various wild strains of *P. rhodozyma* after 7 days' cultivation in shake flasks with the medium given in legend to Figure 16.3, except that 20 g/l molasses was used. The seven first strains were obtained from various culture collections, and the last five strains (N1–N5) were isolates collected in Norway.

Schuman and Johnson, 1989; Johnson *et al.*, 1990; Prevatt, Dickson and Harris, 1991), but they were generally paler and had a lower astaxanthin content than the parent strain (An, Schuman and Johnson, 1989; Johnson *et al.*, 1990). The chemicals ethyl methanesulfonate and *N*-methyl-*N'*-nitro-*N*-nitrosoguanidine (NTG) generate mutations in *P. rhodozyma*, and the latter was more effective for increasing astaxanthin production (Flenø, Christensen and Larsen, 1988; An, Schuman and Johnson, 1989; Johnson *et al.*, 1990; Prevatt, Dickson and Harris, 1991). We have also found that NTG generates mutants in *P. rhodozyma* that vary significantly in color and when grown in shake flasks in a standard medium they contained increased amounts of astaxanthin (Figure 16.21). The best mutants obtained by the first round of NTG mutation can be mutated again with NTG, and cells with additional mutations and higher astaxanthin content can be isolated (Johnson and An, 1991; Prevatt, Dickson and Harris, 1991).

Mutants of *P. rhodozyma* with an increased astaxanthin synthesis have also been isolated on agar plates containing antimycin A (An, Schuman and Johnson, 1989). This chemical inhibits electron transport in the mitochondrial respiratory chain and is not generally recognized as a mutagen. Peculiar colonies with papillae were reported to develop on plates with antimycin A after several weeks of incubation and from one such papilla a mutant with a 2–3 fold increase in astaxanthin content was isolated.

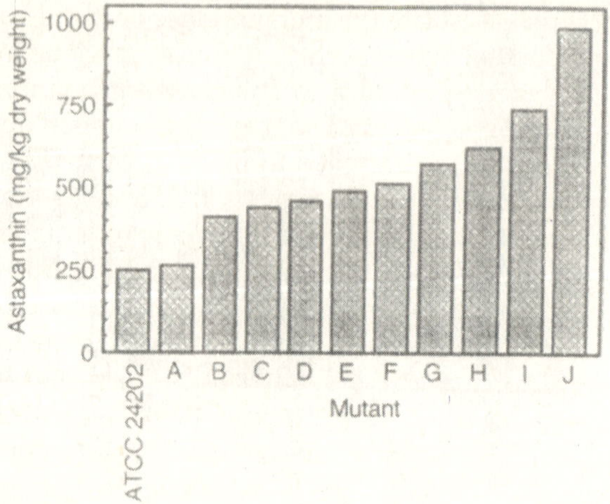

Figure 16.21 Astaxanthin content of various mutants of *P. rhodozyma* isolated after treatment of a wild strain with the mutagen NTG. The mutants were grown for 7 days in shake flasks as stated in legend to Figure 16.20.

Mutants which contain large amounts of astaxanthin when grown in liquid media may not do so when grown on solid media (Lewis *et al.*, 1990; Prevatt, Dickson and Harris, 1991), and on agar plates they can therefore not be differentiated from their progenitor. When β-ionone, which inhibits astaxanthin synthesis, was included in the plates, the amount of astaxanthin in the colonies was decreased. Since the parent strain was more sensitive to the inhibitor than the mutants, differences in astaxanthin content and thereby color of the colonies now appeared, allowing high astaxanthin producers to be isolated (Lewis *et al.*, 1990).

The problem with limited astaxanthin production on agar plates by favorable mutants has also been circumvented by growing the yeast in a liquid media and selecting mutants with a higher astaxanthin content using a flow cytometer (An *et al.*, 1991). With this technique the fluorescence of individual cells can be measured and when the flow cytometer is equipped with a cell sorter, it becomes possible to collect different populations of cells based on fluorescence. Unfortunately, astaxanthin has a very low fluorescence and flow cytometry is therefore not very sensitive to differences in astaxanthin content of the cells. An *et al.* (1991) were able to enrich a mixture of wild-type and mutant cells with regard to the latter, but only moderate increases in astaxanthin production were found when the sorted cells were grown under standard conditions. The authors (An *et al.*, 1991) expect that recent improvements in flow cytometer design will make the technique more useful for isolating astaxanthin overproducing mutants of *P. rhodozyma* in the future.

In mutation programs one generally looks for conditions that will let the desired mutant but not the progenitor grow. With such conditions it becomes very easy to isolate the mutants since only these will develop colonies on solid media. When such conditions cannot be found, conditions where the desired mutant grows faster then the unmutated cells can often be found. This is utilized so that after mutagen treatment cells with desired properties are enriched in liquid culture. The subsequent screening on solid media then becomes much less laborious. With *P. rhodozyma* this approach is not very easy (An, Schumann and Johnson, 1989) since we do not know the function of astaxanthin in the cell. Indeed, we have isolated mutants that do not synthesize colored carotenoids and in the laboratory they grow just as well as the parent strain. Work with another carotenoid synthesizing yeast and a bacterium (Tuveson, Larson and Kagan, 1988; Moore, Breedveld and Autor, 1989) suggested that conditions where the growth of astaxanthin-containing *P. rhodozyma* is favored exist. The bacterium *Escherichia coli* does not synthesize carotenoids but has been made to do so after cloning in genes from the bacterium *Erwinia herbicola* (Perry *et al.*, 1986). The mutant that produces the carotenoids β-cryptoxanthin glucoside, zeaxanthin monoglucoside and zeaxanthin diglucoside (Hundle *et al.*, 1991) was highly protected from light in the presence of photosensitizers compared with the uncloned *E. coli* cells (Tuveson, Larson and Kagan, 1988). The yeast *Rhodotorula mucilaginosa*, which synthesizes torularhodin, was also protected from radicals as long as carotenoid synthesis was not inhibited (Moore, Breedveld and Autor, 1989). Carotenoids in general are known to quench radicals (Krinsky, 1989) and, of the carotenoids, astaxanthin is in several systems shown to be one of the strongest quenchers (Terao, 1989; Kurashige *et al.*, 1990; Lim *et al.*, 1992; Palozza and Krinsky, 1992). When *P. rhodozyma* with a normal astaxanthin synthesis was mixed with an equal number of cells of an albino mutant, the culture was gradually enriched with the astaxanthin-synthesizing cells when exposed to light in the presence of riboflavin (Figure 16.22). This system is known to generate radicals as long as the light is turned on and, in the absence of light, no enrichment of the red *P. rhodozyma* cells occurred (Figure 16.22). Colorless mutants of *P. rhodozyma* may synthesize other carotenoids that are more saturated than astaxanthin, such as phytoene (Chun *et al.*, 1992), and these may also function as radical scavengers, although to a lesser extent than astaxanthin. Radical generating systems thus represent a selection pressure for astaxanthin-containing cells and the more astaxanthin, the stronger is the quenching capacity of the cell. Growth of *P. rhodozyma* with a high astaxanthin content is thereby favored and these mutants will be enriched in the culture.

A general problem in mutation programs is that not all mutants generated are stable. This is also the case with *P. rhodozyma* (An, Schuman and Johnson, 1989). Back mutation is due to repair mechanisms the cells possess in order to correct errors that may arise in its

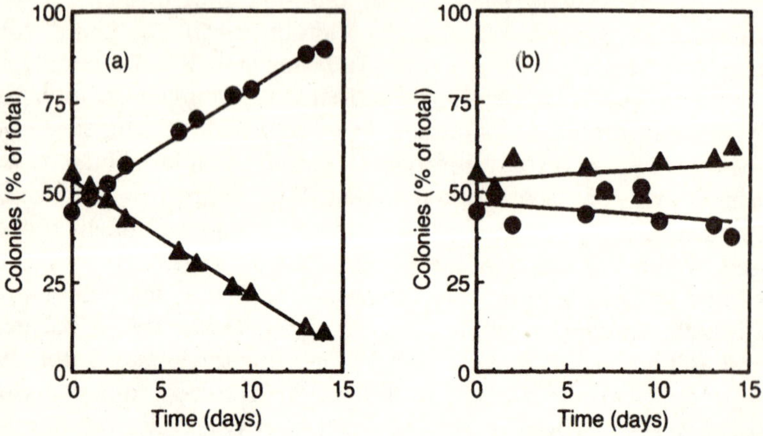

Figure 16.22 Protection of *P. rhodozyma* by astaxanthin against radicals generated by riboflavin in the presence of light. Equal numbers of cells of a wild strain of *P. rhodozyma* and an albino mutant unable to produce astaxanthin were mixed and cultivated in shake flasks with the conditions given in legend to Figure 16.20. At regular intervals a sample was taken and seeded on agar plates, and half of the yeast suspension in the shake flask was replaced with fresh medium. (a) After incubation of the agar plates red (●) and colorless (▲) colonies were counted. In (b) the results obtained when the shake flasks were protected from light during incubation are shown.

genome. In non-haploid cells, where there is more than one copy of each chromosome, a mutation in one chromosome can be eliminated by using the information stored in the equivalent part of another set of that chromosome. This particular type of repair mechanism, which is one out of many, is not present in haploid cells, and stable mutants are therefore more difficult to obtain with cells that are diploid or polyploid. The ploidy of *P. rhodozyma* is not quite settled and may vary with strain and growth conditions, but polyploid cells have been identified (Evans, Adams and Wisdom, 1990; Chun *et al.*, 1992). Haploid and stable strains that produced increasing amounts of astaxanthin have also been isolated and were suggested to be of use in mutation programs (Evans, Adams and Wisdom, 1990).

A sexual cycle has not yet been identified in *P. rhodozyma* and this precludes ordinary mating with the aim to transfer genetic information from one cell to another. The genetic barrier can be circumvented by protoplast fusion (Chun *et al.*, 1992). With this technique the cell wall of the yeast is removed by enzyme treatment and the naked cells can then be fused and will form colonies. *P. rhodozyma* strains obtained by this technique were reported to contain increased amounts of astaxanthin (Chun *et al.*, 1992).

Mutants of *P. rhodozyma* with the capacity to synthesize increasing amounts of astaxanthin can be isolated easily (Figure 16.21), but in

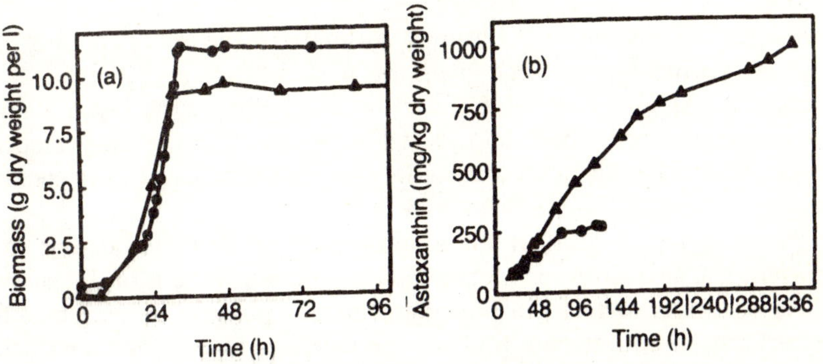

Figure 16.23 Growth (a) and astaxanthin formation (b) of a NTG-generated mutant of *P. rhodozyma* (▲) as compared with that of the wild strain (●). The yeasts were grown in 10 l fermenters with the conditions given in the legend to Figure 16.3.

general they need long incubation times to fully express their potential (Figure 16.23b). Even though some of the mutants may have a growth rate similar to the wild strain (Figure 16.23a), many of the mutants that are reported to have a high content of astaxanthin grow slowly and give a low biomass. The total amount of astaxanthin produced by such mutants may be even lower than the astaxanthin production of the parent strain and are consequently of no use in a commercial process. It is possible that protoplast fusion of a good grower and a mutant high in astaxanthin could result in a strain with both these characteristics (Prevatt, 1992).

By use of the various techniques discussed above there has been a gradual increase in the astaxanthin production by *P. rhodozyma* strains. We believe that within a short time a strain suitable for use in an industrial process will be obtained.

16.8 ECONOMICAL CONSIDERATIONS

16.8.1 The Atlantic salmon market

No sale statistics exist for astaxanthin. The market potential for Atlantic salmon farming depends on the following factors:

1. Feeding period.
2. Amount of astaxanthin in feed.
3. Feed factor.
4. Price of astaxanthin.
5. Production volume of fish.

With regard to feeding period it has been recommended that astaxanthin should be given to the salmon throughout the life span since it is such a feeding regime that gives the best coloration of the fillets. If astaxanthin is given to the fish during, for instance, the last 6 weeks, the fish appear red when slaughtered but upon cooking a large amount of the color is lost through oxidization and solubilization in the cooking water.

The amount of astaxanthin in the feed, feeding period and how efficient the astaxanthin is deposited determine the color of the salmon fillets. Color is one of the important quality parameters and that the correct amount of astaxanthin has been supplied can be evaluated using a color card developed for raw flesh of astaxanthin-fed salmon (Skrede *et al.*, 1990a). During the years the amount of astaxanthin in the feed has varied and today 50 p.p.m. is recommended in Norway.

The feed factor, amount of dry feed (in kg) used to produce 1 kg of fish, determines the amount of feed that is used. Depending on feed quality, which varies considerably throughout the world, and the feeding regime, the factor may vary from 0.9 to 3.0 and in general is around 1.5. For the astaxanthin producer knowledge of feed quality and feeding regimes are of importance.

The price of astaxanthin is at present around US$ 2000/kg and is determined according to the market and importance of the salmon color by Hoffmann-La Roche who is the only supplier of synthetic astaxanthin. Since the available amount of astaxanthin from biological sources like crustaceans, algae and yeast at present is small, this has limited influence on the astaxanthin price today.

The worldwide production volume of Atlantic salmon is approximately 240 000 tons (Tilseth, Hansen and Møller, 1991). Using a feed factor of 1.5, 50 p.p.m. of astaxanthin in the feed and a price for astaxanthin of US$ 2000/kg, this gives a market of US$ 36 million. This certainly represent a minimum value.

As shown in Figure 16.19, astaxanthin degrades rather fast in feed. Upon feed formulation this has to be taken into account (Bernhard, 1990). Today the actual amount of astaxanthin in the feed when given to the salmon on the farm can be very low due to prolonged and bad storage conditions. This may explain why pale salmon often results and why a rather high amount of astaxanthin in the feed is recommended. In general, the redder the product, the more attractive is the salmon and even with an increased stabilization it is difficult to foresee a reduction of astaxanthin in different feeds.

Atlantic salmon aquaculture is a growth market. Feed containing astaxanthin is also used for the production of 20 000 tons of Pacific salmon, 180 000 tons of rainbow trout and 600 000 tons of crustaceans each year. Production of these species is increasing through intensive aquaculture, and the demand for astaxanthin will therefore increase in the future.

16.8.2 Production costs of astaxanthin using *P. rhodozyma*

Figure 16.10 shows how *P. rhodozyma* could be produced commercially. The first step in the process is fermentation (Figure 16.11) which is similar to that used for production of baker's yeast. This yeast and other yeasts used as feed, processed into yeast extracts or taste fortifiers and other compounds, have a market price of US$ 1.5–2/kg dry weight. If *P. rhodozyma* contained 1000 p.p.m. of astaxanthin, this would be equivalent to a price of US$ 2/kg dry weight, assuming an astaxanthin price of US$ 2000/kg. Fermentation of *P. rhodozyma* with an optimal astaxanthin content will be more expensive than fermentation of baker's yeast since the growth of *P. rhodozyma* is slower. Furthermore, the production of astaxanthin is not limited to the exponential growth phase (Figure 16.3), increasing the fermentation time and production cost, especially for the end fermenter (Figure 16.11). The prerequisite that the temperature has to be below 25°C requires extra cooling, and an area with cold water and cheap electricity would be favorable. The longer production time increases the possibility of infection and the scale-up process has to be sterile up to 1 m^3, preferably 10 m^3, in order to make the final fermentation in 100 m^3 safe. The labor force used in the production would be very similar to that of baker's yeast production. This yeast has for years been grown on molasses, which is inexpensive and at present is also the most promising energy source for growth of *P. rhodozyma*.

After centrifugation, cell wall modification is necessary for making astaxanthin available to the fish (Figure 16.8). At present homogenization is the method of choice due to high enzyme costs and limited knowledge of co-cultivation with microorganisms producing lytic enzymes. Also, homogenized cells are reported to function somewhat better as an astaxanthin source for rainbow trout compared with enzyme-treated cells (Gentles and Haard, 1991).

The final step is drying of the yeast and stabilization of astaxanthin. We found evaporative drying to degrade too much of the astaxanthin, whereas spray drying was sufficiently gentle (Figure 16.18). This is also less expensive than even more gentle methods, like freeze drying. Storage at low temperatures stabilizes the astaxanthin (Figure 16.19), but distribution temperatures demand alternative stabilization methods.

Even though it is difficult to estimate the exact cost of the total process, we conclude that given the above prerequisites the amount of astaxanthin in the final product has to be in the range of 2000–2500 p.p.m. to justify a start-up of production. A moderate volume of, for instance, 1000 tons/year could be produced in an existing baker's yeast factory and such a volume would easily be used by one of the feed suppliers in the market today. This would introduce *P. rhodozyma* to the market at moderate cost and risk.

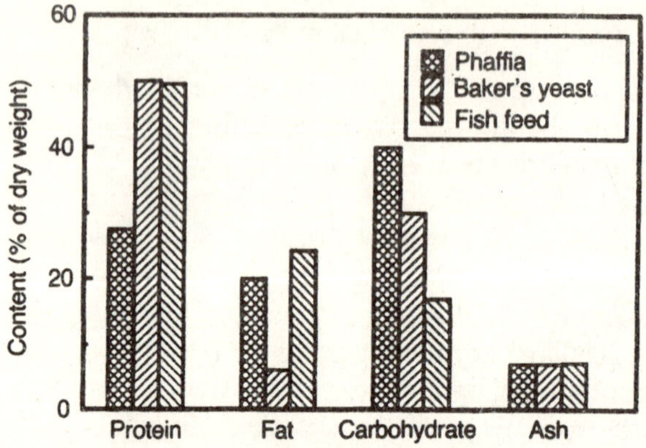

Figure 16.24 Gross composition of *P. rhodozyma* as compared with baker's yeast (*S. cerevisiae*) and a fish feed used in salmonid farming.

The yeast *P. rhodozyma* also contains minerals, vitamins, protein and fat (Johnson, Villa and Lewis, 1980; Lewis, 1990) which gives the product an added value as a nutritional source (Figure 16.24). The yeast cell wall contain β(1,3)-glucans that stimulate the immune apparatus of fish and is used to improve fish vaccines (Robertsen *et al.*, 1991; Aasjord and Slinde, 1994). In our opinion the economics of the process should only depend on the astaxanthin of the yeast and added values should be disregarded in the developing phase.

16.9 FUTURE USE OF *P. RHODOZYMA* AND ASTAXANTHIN IN SALMONID FARMING

Aquaculture of salmonids requires a coloring compound. The coloring has the objective to make the product look good but may also be of importance to the success of reproduction (Torrissen, 1984). Astaxanthin and canthaxanthin are today used to color trout and salmon, and a difference in the color of the fish can be measured but is difficult to detect visually unless a salmon fed canthaxanthin is placed opposite to one fed astaxanthin (Skrede, Storebakken and Næs, 1990). Canthaxanthin used to be the dominant color additive due to the fact that the synthetic compound was available commercially before synthetic astaxanthin and that it has a lower price. In Norway, and later also in other countries, synthetic astaxanthin has now gradually replaced canthaxanthin, which is sensible since astaxanthin is the compound found in wild salmonids. Indeed, the question of whether or not salmonids fed canthaxanthin are at all marketable has been raised (Nieper, 1990; Gundisch, 1991). Megadoses of canthaxanthin, gained by eating so-called 'artificial suntan' pills, are reported occasionally to be toxic, causing reduced vision (retinopathy) and anemia (Ros, Leyon and Wennersten, 1985;

Bluhm *et al.*, 1990; Herbert, 1991; Mathews-Roth, 1991). In a WHO committee report (Anonymous, 1988) the daily intake of canthaxanthin was recommended not to exceed 0.05 mg/kg body weight, which would represent about 350 g of salmonid flesh colored with canthaxanthin per day. There is a large, at least 10 times, safety factor incorporated in this recommendation. We would like to stress that the toxicity of megadoses of canthaxanthin is not very well documented, as discussed by Herbert (1991) and Mathews-Roth (1991), and the US Food and Drug Administration has classified canthaxanthin as a 'generally recognized as safe' (GRAS) substance. A similar toxicity of megadoses of astaxanthin has never been shown, but cannot be excluded. Even so, to our knowledge astaxanthin has not a GRAS classification in the US, since it has not been applied for. The situation may be different in other countries, and Universal Food who have announced the introduction of a *P. rhodozyma* product to the fish feed market, said they would start test marketing outside the US, partly because of this (Anonymous, 1992).

In our opinion the question is not whether astaxanthin or canthaxanthin should be used as a feed additive, but rather what kind of astaxanthin source to use. Astaxanthin produced biologically has to compete with synthetic astaxanthin, and in the future there may be a market for both biological and synthetic astaxanthin. As a feed ingredient the price could be somewhat equal, but due to the present green wave a somewhat higher price for biologically produced astaxanthin is probably justified for the food market. For β-carotene produced by algae (Borowitzka and Borowitzka, 1990) this seem to be the situation in Japan, since margarine colored with this product has a higher price than that colored with synthetic β-carotene.

Natural sources of astaxanthin from crustaceans are available in large quantities in the oceans (Storebakken, 1988). The use of astaxanthin-containing products based on crustacean waste is at present limited, partly because of price and availability which shows seasonal variations. Capture of crustaceans, such as krill, with the sole purpose to supply the aquaculture industry is so far not profitable.

The two most likely candidates for the biological production of astaxanthin are the alga *H. pluvialis* and the yeast *P. rhodozyma*. At present neither of them is used in salmonid farming, except for test purposes. This may soon change since there is at present active research with both. The advantage of the alga is its high content of astaxanthin. Its growth generally requires light (Kobayashi *et al.*, 1992a) which restricts industrial production to sunny areas, unless artificial light can be supplied at a very low cost. Recently it was shown that *H. pluvialis* can also synthesize astaxanthin in the absence of light when grown on acetate (Kobayashi *et al.*, 1992b) and if this can be performed under industrial conditions to give a high astaxanthin production, it would certainly bring the process closer to a commercial realization.

P. rhodozyma has growth conditions similar to that of *S. cereviciae* and can be produced in a baker's yeast factory, provided sterility can be achieved during the first scaling steps and that efficient cooling exists so that the temperature does not rise above 22–25°C. However, the astaxanthin yield has to be increased compared with the wild strains of *P. rhodozyma* without a loss of biomass or an increase in production time. From the results of strain improvements obtained so far it is reasonable to believe that strains with the necessary properties will soon be available. In the long run we believe that the competition between biologically and synthetically produced astaxanthin is in favor of the biological product, also since this process has the highest potential for cost reduction through continuous strain improvement made possible by the increased biological knowledge about the yeast.

There is a general demand for red coloring components in the food industry and astaxanthin may be one such candidate. The argument for the use of biological products in food is increasing and *P. rhodozyma* could in the future also supply astaxanthin to this market. A more pure form of additives is desired in food compared with feed and we have found that astaxanthin can be extracted from *P. rhodozyma* using liquid CO_2 in a process similar to that used by others for β-carotene extraction from an alga (Cygnarowicz and Seider, 1990). Since it has been shown that CO_2 at the pressures used disrupt the cell wall of other yeasts (Lin *et al.*, 1991), the homogenization step is probably in this case redundant. The use of astaxanthin in food may well increase in popularity in the future since recent data suggest that it prevents cancer and stimulates the immune system (Jyonouchi *et al.*, 1991).

P. rhodozyma has all the biological properties required for an astaxanthin source in aquaculture. It can be produced in large quantities using existing techniques and functions well as an astaxanthin source for salmonids. The only obstacle at present is the content of astaxanthin in the product. Research has increased this gradually during the last 5 years and at present different companies are launching test products. An economical process will, within a short time, be feasible due to improved strains, and this improvement will continue and increase the economics of the process even further. When *P. rhodozyma* has been accepted as a salmonid feed additive, further development can make the yeast usable as a red coloring source for the food market.

ACKNOWLEDGMENTS

We would like to thank Georg Backer, Nora A/S and Kjell Sandberg, Idun Fabrikker A/S, Oslo, Norway, for inspiring collaboration during this work, and Dr Geir Totland, Zoological Laboratory, University of Bergen, Norway, for kind help with the electron microscopic work.

Finally, the fine collaboration and skilful technical assistance of Mrs Sissel Wahlstrøm and Tove Marøy are highly appreciated. Crystalline astaxanthin was kindly provided by F. Hoffmann-La Roche & Co., Basle, Switzerland.

REFERENCES

Aasen, A.J., Eimhjellen, K.E. and Liaaen-Jensen, S. (1969) An extreme source of β-carotene. *Acta Chem. Scand.*, **23**, 2544–5.

Aasjord, P.M. and Slinde, E. (1994) Fish vaccines: Development, production and use of bacterial vaccines, with special reference to salmon, in *Fisheries Processing: Biotechnological Applications*, (ed. A.M. Martin), Chapman & Hall, London, pp. 430–64.

An, G.-H. and Johnson, E.A. (1990) Influence of light on growth and pigmentation of the yeast *Phaffia rhodozyma. Antonie van Leeuwenhoek*, **57**, 191–204.

An, G.-H., Schuman, D.B. and Johnson, E.A. (1989) Isolation of *Phaffia rhodozyma* mutants with increased astaxanthin content. *Appl. Environ. Microbiol.*, **55**, 116–24.

An, G.-H., Bielich, J., Auerbach, R. and Johnson, E.A. (1991) Isolation and characterization of carotenoid hyperproducing mutants of yeast by flow cytometry and cell sorting. *Bio/Technology*, **9**, 70–3.

Ando, S. and Hatano, M. (1987) Metabolic pathways of carotenoids in chum salmon *Oncorhynchus keta* during spawning migration. *Comp. Biochem. Physiol. B: Comp. Biochem.*, **87B**, 411–16.

Andrewes, A.G. and Starr, M.P. (1976) (3*R*,3′*R*)-Astaxanthin from the yeast *Phaffia rhodozyma. Phytochemistry*, **15**, 1009–11.

Andrewes, A.G., Phaff, H.J. and Starr, M.P. (1976) Carotenoids of *Phaffia rhodozyma*, a red-pigmented fermenting yeast. *Phytochemistry*, **15**, 1003–7.

Anonymous (1988). Canthaxanthin, in *Toxicological Evaluation of Certain Food Additives. WHO Food Additives Series* 22, Cambridge University Press, Cambridge, pp. 65–75.

Anonymous (1992) Universal BioVentures introduces astaxanthin-rich *Phaffia* yeast. *Bioprocess. Technol.*, **14**, 1–2.

Arai, S., Mori, T., Miki, W. *et al.* (1987) Pigmentation of juvenile coho salmon with carotenoid oil extracted from antarctic krill. *Aquaculture*, **66**, 255–64.

Barford, J.P. (1987) The technology of aerobic yeast growth, in *Yeast biotechnology*, (eds D.R. Berry, I. Russell and G.G. Stewart), Allen & Unwin, London, pp. 200–30.

Barnett, J.A., Payne, R.W. and Yarrow, D. (1983) *Yeasts: Characteristics and identification.*, Cambridge University Press, Cambridge,

Bernhard, K. (1990) Synthetic astaxanthin. The route of a carotenoid from research to commercialization, In *Carotenoids. Chemistry and Biology (Proceeding of the 8th International Symposium on Carotenoids)*, (eds N.I. Krinsky, M.M. Mathews-Roth and R.F. Taylor), Plenum Press, New York, pp. 337–63.

Bjerkeng, B., Storebakken, T. and Liaaen-Jensen, S. (1990) Response to carotenoids by rainbow trout in the sea: resorption and metabolism of dietary astaxanthin and canthaxanthin. *Aquaculture*, **91**, 153–63.

Bluhm, R., Branch, R., Johnston, P. and Stein, R. (1990) Aplastic anemia associated with canthaxanthin ingested for tanning purposes. *J. Am. Med. Ass.*, **264**, 1141–2.

Borowitzka, L.J. and Borowitzka, M.A. (1990) Commercial production of β-carotene by *Dunaliella salina* in open ponds. *Bull. Marine Sci.*, **47**, 244–52.

Bubrick, P. (1991) Production of astaxanthin from *Haematococcus*. *Bioresource Technol.*, **38**, 237–9.

Buchwald, M. and Jencks, W.P. (1968) Optical properties of astaxanthin solutions and aggregates. *Biochemistry*, **7**, 834–43.

Chen, H.-M. and Meyers, S.P. (1983) Ensilage treatment of crawfish waste for improvement of astaxanthin pigment extraction. *J. Food Sci.*, **48**, 1516–55.

Choubert, G. and Luquet, P. (1983) Utilization of shrimp meal for rainbow trout (*Salmo gairdneri* Rich.) pigmentation. Influence of fat content of the diet. *Aquaculture*, **32**, 19–26.

Choubert, G. and Storebakken, T. (1989) Dose response to astaxanthin and canthaxanthin pigmentation of rainbow trout fed various dietary carotenoid concentrations. *Aquaculture*, **81**, 66–77.

Chun, S.B., Chin, J.E., Bai, S. and An, G.-H. (1992) Strain improvement of *Phaffia rhodozyma* by protoplast fusion. *FEMS Microbiol. Lett.*, **93**, 221–6.

Cygnarowicz, M.L. and Seider, W.D. (1990) Design and control of a process to extract β-carotene with supercritical carbon dioxide. *Biotechnol. Prog.*, **6**, 82–91.

De Mot, R., Demeersman, M. and Verachtert, H. (1984) Comparative study of starch degradation and amylase production by non-ascomycetous yeast species. *Syst. Appl. Microbiol.*, **5**, 421–32.

Evans, C.T., Adams, D. and Wisdom, R.A. (1990) Yeasts and their use in astaxanthin production (Enzymatix, England). *Patent application EP 0 427 405 A1*.

Flenø, B., Christensen, I. and Larsen, R. (1988) Astaxanthin-producing yeast cells, methods for their preparation and their use (Danisco Biotechnology, Denmark). *Patent application PCT/DK88/00068*.

Foss, P., Storebakken, T., Schiedt, K. *et al.* (1984) Carotenoids in diets for salmonids. 1. Pigmentation of rainbow trout with the individual optical isomers of astaxanthin in comparison with canthaxanthin. *Aquaculture*, **41**, 213–26.

Foss, P., Renstrøm, B. and Liaaen-Jensen, S. (1987) Natural occurrence of enantiomeric and *meso* astaxanthin. 7. Crustaceans including zooplankton. *Comp. Biochem. Physiol. B: Comp. Biochem.*, **86B**, 313–14.

Foss, P., Storebakken, T., Austreng, E. and Liaaen-Jensen, S. (1987) Carotenoids in diets for salmonids. V. Pigmentation of rainbow trout and sea trout with astaxanthin, astaxanthin dipalmitate and canthaxanthin. *Aquaculture*, **65**, 293–305.

Gentles, A. and Haard, N.F. (1991) Pigmentation of rainbow-trout with enzyme-treated and spray-dried *Phaffia rhodozyma*. *Prog. Fish. Cult.*, **53**, 1–6.

Golubev, V.I., Bab'eva, I.P. and Novik, S.N. (1977) Yeast succession during birch sap flow. *Ekologiya*, **5**, 23–8.

Goodwin T.W. (1980) *The Biochemistry of the Carotenoids, Vol. I, Plants*, 2nd edn, Chapman & Hall, London,

Goodwin T.W. (1984) *The Biochemistry of the Carotenoids, Vol II, Animals*, 2nd edn, Chapman & Hall, London,

Grung, M., D'Souza, F.M.L., Borowitzka, M. and Liaaen-Jensen, S. (1992) Algal carotenoids. 51. Secondary carotenoids. 2. *Haematococcus pluvialis* aplanospores as a source of (3*S*,3'*S*)-astaxanthin esters. *J. Appl. Phycol.*, **4**, 165–71.

Gueho, E., Kurtzman, C.P. and Peterson, S.W. (1989) Evolutionary affinities of Heterobasidiomycetous yeasts estimated from 18S and 25S ribosomal RNA sequence divergence. *Syst. Appl. Microbiol.*, **12**, 230–6.

Guillou, A., Choubert, G. and de la Noue, J. (1992) Comparative accumulations of labelled carotenoids (^{14}C-astaxanthin, ^{13}H-canthaxanthin and ^{3}H-zeaxanthin) and their metabolic conversions in mature female rainbow trout (*Oncorhynchus mykiss*). *Comp. Biochem. Physiol. B: Comp. Biochem.*, **102**, 61–5.

Gundisch, J. (1991) Once again – are salmon and trout fed with the colorant canthaxanthin marketable. *Arch. Lebensmittel Hygiene*, **42**, 129–30.

Haard, N.F. (1988) Astaxanthin formation by the yeast *Phaffia rhodozyma* on molasses. *Biotechnol. Lett.*, **10**, 609–14.

Herbert, V. (1991) Canthaxanthin toxicity. *Am. J. Clin. Nutr.*, **53**, 573.

Hetherington, P.J., Follows, M., Dunnill, P. and Lilly, M.D. (1971) Release of protein from baker's yeast (*Saccharomyces cerevisiae*) by disruption in an industrial homogenizer. *Trans. Inst. Chem. Engrs.*, **49**, 142–8.

Hundle, B.S., Beyer, P., Kleinig, H. *et al.* (1991) Carotenoids of *Erwinia herbicola* and an *Escherichia coli* HB101 strain carrying the *Erwinia herbicola* carotenoid gene cluster. *Photochem. Photobiol.*, **54**, 89–93.

Ibrahim, A., Shimizu, C. and Kono, M. (1984) Pigmentation of cultured red sea bream, *Chrysophrys major*, using astaxanthin from antartic krill, *Euphausia superba*, and a mysid, *Neomysis* sp. *Aquaculture*, **38**, 45–57.

Inoue, T., Simpson, K.L., Tanaka, Y. and Sameshima, M. (1988) Condensed astaxanthin of pigmented oil from crayfish carapace and its feeding experiment. *Nippon Suisan Gakkaishi*, **54**, 103–6.

Johnson, E.A. and An, G.-H. (1991) Astaxanthin from microbial sources. *Crit. Rev. Biotechnol.*, **11**, 297–326.

Johnson, E.A. and Lewis, M.J. (1979) Astaxanthin formation by the yeast *Phaffia rhodozyma*. *J. Gen. Microbiol.*, **115**, 173–84.

Johnson, E.A., Conklin, D.E. and Lewis, M.J. (1977) The yeast *Phaffia rhodozyma* as a dietary pigment source for salmonids and crustaceans. *J. Fish. Res. Board Can.*, **34**, 2417–21.

Johnson, E.A., Villa, T.G., Lewis, M.J. and Phaff, H.J. (1978) Simple method for the isolation of astaxanthin from the Basidiomycetous yeast *Phaffia rhodozyma*. *Appl. Environ. Microbiol.*, **35**, 1155–9.

Johnson, E.A., Lewis, M.J. and Grau, C.R. (1980) Pigmentation of egg yolks with astaxanthin from the yeast *Phaffia rhodozyma*. *Poult. Sci.*, **59**, 1777–82.

Johnson, E.A., Villa, T.G. and Lewis, M.J. (1980) *Phaffia rhodozyma* as an astaxanthin source in salmonid diets. *Aquaculture*, **20**, 123–34.

Johnson, E.A., Schreiber, D., Hall, W.T. and Geldiaytuncer, B. (1990) Process for *in vivo* production of astaxanthin and *Phaffia rhodozyma* yeast of enhanced astaxanthin content (Igene Biotechnology, USA). *Patent application PCT/UC89/03327*.

Jyonouchi, H., Hill, R.J., Tomita, Y. and Good, R.A. (1991) Studies of immunomodulating actions of carotenoids. I. Effects of β-carotene and astaxanthin on murine lymphocyte functions and cell surface marker expression in *in vitro* culture system. *Nutr. Cancer*, **16**, 93–105.

Kanemitsu, T. and Aoe, H. (1958a) Studies on carotenoids of salmon – I. Identification of the muscle pigments. *Bull. Jpn. Soc. Sci. Fish.*, **24**, 209–15.

Kanemitsu, T. and Aoe, H. (1958*b*) On the studies of carotenoids of salmon – II. Determination of muscle pigment. *Bull. Jpn. Soc. Sci. Fish.*, **24**, 555–8.

Katsuyama, M., Komori, T. and Matsuno, T. (1987) Metabolism of three stereoisomers of astaxanthin in the fish, rainbow trout and tilapia. *Comp. Biochem. Physiol. B: Comp. Biochem.*, **86B**, 1–5.

Khare, A., Moss, G.P., Weedon, B.C. L. and Matthes, A.D. (1973) Identification of astaxanthin in Scottish salmon. *Comp. Biochem. Physiol. B: Comp. Biochem.*, **45B**, 971–3.

Kobayashi, M., Kakizono, T., Nishio, N. and Nagai, S. (1992a) Effects of light intensity, light quality, and illumination cycle on astaxanthin formation in a green alga, *Haematococcus pluvialis*. *J. Ferment. Bioeng.*, **74**, 61–3.

Kobayashi, M., Kakuzobo, T., Yamagucho, K. *et al.* (1992b) Growth and astaxanthin formation of *Haematococcus pluvialis* in heterotrophic and mixotrophic conditions. *J. Ferment. Bioeng.*, **74**, 17–20.

Krinsky, N.I. (1989) Antioxidant functions of carotenoids. *Free Rad. Biol. Med.*, **7**, 617–35.

Kurashige, M., Okimasu, E., Inoue, M. and Utsumi, K. (1990) Inhibition of oxidative injury of biological membranes by astaxanthin. *Physiolog. Chem. Phys. Med. NMR*, **22**, 27–38.

Lewis, M.J. (1990). *Phaffia rhodozyma*: A red yeast with potential as an animal feed in *Biotechnology in the Feed industry: Proceedings of Alltech's Sixth Annual Symposium*, (ed. T.P. Lyons), Alltech Technical Publications, Nicholasville, KY, pp. 391–401.

Lewis, M.J., Ragot, N., Berlant, M.C. and Miranda, M. (1990) Selection of astaxanthin-overproducing mutants of *Phaffia rhodozyma* with β-ionone. *Appl. Environ. Microbiol.*, **56**, 2944–5.

Liaaen-Jensen, S. (1990) Marine carotenoids – selected topics. *New J. Chem.*, **14**, 747–59.

Liaaen-Jensen, S. and Andrewes, A.G. (1985) Analysis of carotenoids and related polyene pigments. *Methods Microbiol.*, **18**, 235–55.

Lim, B.P., Nagao, A., Terao, J. *et al.* (1992) Antioxidant activity of xanthophylls on peroxyl radical-mediated phospholipid peroxidation. *Biochim. Biophys. Acta.*, **1126**, 178–84.

Lin, H.M., Chan, E.C., Chen, C. and Chen, L.F. (1991) Disintegration of yeast cells by pressurized carbon dioxide. *Biotechnol. Prog.*, **7**, 201–4.

Lura, H. and Sægrov, H. (1991a) Documentation of successful spawning of escaped farmed female Atlantic salmon, *Salmo salar*, in Norwegian rivers. *Aquaculture*, **98**, 151–9.

Lura, H. and Sægrov, H. (1991b) A method of separating offspring from farmed and wild Atlantic salmon (*Salmo salar*) based on different ratios of optical isomers of astaxanthin. *Can. J. Fish. Aquat. Sci.*, **48**, 429–33.

Maoka, T., Katsuyama, M., Kaneko, N. and Matsuno, T. (1985) Stereochemical investigation of carotenoids in the Antartic krill *Euphausia superba*. *Bull. Jpn. Soc. Sci. Fish.*, **51**, 1671–3.

Mathews-Roth, M.M. (1991) Canthaxanthin toxicity. Reply. *Am. J. Clin. Nutr.*, **53**, 573–4.

Meyers, S.P. and Bligh, D. (1981) Characterization of astaxanthin pigments from heat-processed crawfish waste. *J. Agric. Food Chem.*, **29**, 505–8.

Miller, M.W., Yoneyama, M. and Soneda, M. (1976) *Phaffia*, a new yeast genus in the Deuteromycotina (Blastomycetes). *Int. J. Syst. Bacteriol.*, **26**, 286–91.

Moore, E.K., Hoare, M. and Dunnill, P. (1990) Disruption of baker's yeast in a high-pressure homogenizer – new evidence on mechanism. *Enz. Microb. Technol.*, **12**, 764–70.

Moore, M.M., Breedveld, M.W. and Autor, A.P. (1989) The role of carotenoids in preventing oxidative damage in the pigmented yeast, *Rhodotorula mucilaginosa*. *Arch. Biochem. Biophys.*, **270**, 419–31.

Mori, T., Okada, S., Yamaguchi, K. *et al.* (1990) Chemistry and utilization of plankton. 21. Pigmentation of coho salmon cultured in sea pens with Antartic krill and littoral mysid. *Bull. Jpn. Soc. Sci. Fish.*, **56**, 935–9.

Nelis, H.J. and De Leenheer, A.P. (1991) Microbial sources of carotenoid pigments used in foods and feeds. *J. Appl. Bacteriol.*, **70**, 181–91.

Nieper, L. (1990) Are salmon and trout fed with the colorant canthaxanthin marketable. *Arch. Lebensmittel Hygiene*, **41**, 127.

Okagbue, R.N. and Lewis, M.J. (1983) Mixed culture of *Bacillus circulans* WL-12 and *Phaffia rhodozyma* on different carbon sources: yeast wall lytic enzyme production and extractability of astaxanthin. *Biotechnol. Lett.*, **5**, 731–6.

Okagbue, R.N. and Lewis, M.J. (1984a) Autolysis of the red yeast *Phaffia rhodozyma*: A potential tool to facilitate extraction of astaxanthin. *Biotechnol. Lett.*, **6**, 247–50.

Okagbue, R.N. and Lewis, M.J. (1984b) Inhibition of the red yeast *Phaffia rhodozyma* by saponin. *Appl. Microbiol. Biotechnol.*, **20**, 278–80.

Okagbue, R.N. and Lewis, M.J. (1984c) Use of alfalfa residual juice as a substrate for propagation of the red yeast *Phaffia rhodozyma*. *Appl. Microbiol. Biotechnol.*, **20**, 33–9.

Okagbue, R.N. and Lewis, M.J. (1985) Influence of mixed culture conditions on yeast-wall hydrolytic activity of *Bacillus circulans* WL-12 and on extractability of astaxanthin from the yeast *Phaffia rhodozyma*. *J. Appl. Bacteriol.*, **59**, 243–55.

Palozza, P. and Krinsky, N.I. (1992) Astaxanthin and canthaxanthin are potent antioxidants in a membrane model. *Arch. Biochem. Biophys.*, **297**, 291–5.

Perry, K.L., Simonitch, T.A., Harrison-Lavoie, K.J. and Liu, S.-T. (1986) Cloning and regulation of *Erwinia herbicola* pigment genes. *J. Bacteriol.*, **168**, 607–12.

Phaff, H.J. (1986) My life with yeasts. *Ann. Rev. Microbiol.*, **40**, 1–28.

Phaff, H.J., Miller, M.W., Yoneyama, M. and Soneda, M. (1972) A comparative study of the yeast florae associated with trees on the Japanese Islands and on the West Coast of North America, in *Proceedings of IV IFS: Fermentation Technology Today*, Osaka, Japan, pp. 759–74.

Polulyakh, O.V., Podoprigora, O.I., Eliseev, S.A. *et al.* (1991) Biosynthesis of torulene and torularhodin in the yeast *Phaffia rhodozyma*. *Prikladnaya Biokhimiya I Mikrobiologiya*, **27**, 541–5.

Prevatt, W.D. (1992) Fusion of *Phaffia rhodozyma* sphaeroplasts – produces astaxanthin for adding to salmonid and crustaceans diet (Phillips Petroleum Company, USA). *Patent application EP 482544A*.

Prevatt, W.D., Dickson, T.D. and Harris, R.L. (1991) Novel strains of *Phaffia rhodozyma* containing high levels of astaxanthin (Phillips Petroleum Company, USA). *Patent application US 467607*.

Robertsen, B., Rørstad, G., Engstad, R. and Raa, J. (1991) Enhancement of non-specific disease resistance in Atlantic salmon, *Salmo salar* L., by a glucan from *Saccharomyces cerevisiae* cell walls. *J. Fish Disease*, **13**, 391–400.

Ros, A.M., Leyon, H. and Wennersten, G. (1985) Crystalline retinopathy in patients taking an oral drug containing canthaxanthin. *Photodermatology*, **2**, 183–5.

Rumsey, G.L., Hughes, S.G., Smith, R.R. *et al.* (1991) Digestibility and energy values of intact, disrupted and extracts from brewer's dried yeast fed to rainbow trout (*Oncorhynchus mykiss*). *Animal Feed Sci. Technol.*, **33**, 185–93.

Saito, A. (1969) Color in raw and cooked atlantic salmon (*Salmo salar*). *J. Fish. Res. Board Can.*, **26**, 2234–6.

Saito, A. and Regier, L.W. (1971) Pigmentation of brook trout (*Salvenius fontinalis*) by feeding dried crustacean waste. *J. Fish. Res. Board Can.* **28**, 509–12.

Schiedt, K., Leuenberger, F.J. and Vecchi, M. (1981) Natural occurrence of enantiomeric and *meso*-astaxanthin. 5. Ex Wild salmon (*Salmo salar* and *Oncorhynchus*). *Helv. Chim. Acta*, **64**, 449–57.

Schiedt, K., Leuenberger, F.J., Vecchi, M. and Glinz, E. (1985) Absorption, retention and metabolic transformation of carotenoids in rainbow trout, salmon and chicken. *Pure Appl. Chem.*, **57**, 685–92.

Schiedt, K., Foss, P., Storebakken, T. and Liaaen-Jensen, S. (1989) Metabolism of carotenoids in salmonids – I. Idoxanthin, a metabolite of astaxanthin in the flesh of Atlantic salmon (*Salmo salar*, L.) under varying external conditions. *Comp. Biochem. Physiol. B: Comp. Biochem.*, **92B**, 277–81.

Schütte, H. and Kula, M.-R. (1990) Pilot-scale and process-scale techniques for cell disruption. *Biotechnol. Appl. Biochem.*, **12**, 599–620.

Sedmak, J.J., Weerasinghe, D.K. and Jolly, S.O. (1990) Extraction and quantitation of astaxanthin from *Phaffia rhodozyma*. *Biotechnol. Tech.*, **4**, 107–12.

Skrede, G., Risvik, E., Huber, M., Enersen, G. and Blumlein, L. (1990) Developing a color card for raw flesh of astaxanthin-fed salmon. *J. Food Sci.*, **55**, 361–3.

Skrede, G., Storebakken, T. and Næs, T. (1990) Color evaluation in raw, baked and smoked flesh of rainbow trout (*Oncorhynchus mykiss*) fed astaxanthin or canthaxanthin. *J. Food Sci.*, **55**, 1574–8.

Sommer, T.R., Potts, W.T. and Morrissy, N.M. (1990) Recent progress in the use of processed microalgae in aquaculture. *Hydrobiologia*, **204/205**, 435–43.

Sommer, T.R., Potts, W.T. and Morrissy, N.M. (1991) Utilization of microalgal astaxanthin by rainbow trout (*Oncorhynchus mykiss*). *Aquaculture*, **94**, 79–88.

Storebakken, T. (1988) Krill as a potential feed source for salmonids. *Aquaculture*, **70**, 193–205.

Storebakken, T. and No, H.K. (1992) Pigmentation of rainbow trout. *Aquaculture*, **100**, 209–29.

Storebakken, T., Foss, P., Austreng, E. and Liaaen-Jensen, S. (1985) Carotenoids in diets for salmonids. II. Epimerization studies with astaxanthin in Atlantic salmon. *Aquaculture*, **44**, 259–69.

Storebakken, T., Foss, P., Schiedt, K. *et al.* (1987) Carotenoids in diets for salmonids. IV. Pigmentation of Atlantic salmon with astaxanthin, astaxanthin dipalmitate and canthaxanthin. *Aquaculture*, **65**, 279–92.

Sørum, U. and Robertsen, B. (1991) Lysis of the astaxanthin-producing yeast *Phaffia rhodozyma* by lytic enzymes from the fungus *Trichoderma harzianum*. Abstract presented at *27th Wintermeeting of the Norwegian Biochemical Society*, Storefjell, Norway.

Terao, J. (1989) Antioxidant activity of β-carotene-related carotenoids in solution. *Lipids*, **24**, 659–61.

Tilseth, S., Hansen, T. and Møller, D. (1991) Historical development of salmon culture. *Aquaculture*, **98**, 1–9.

Torrissen, O.J. (1984) Pigmentation of salmonids. Effect of carotenoids in eggs and start-feeding diet on survival and growth rate. *Aquaculture*, **43**, 185–93.

Torrissen, O.J. (1986) Pigmentation of salmonids. A comparison of astaxanthin and canthaxanthin as pigment sources for rainbow trout. *Aquaculture*, **53**, 271–8.

Torrissen, O., Tidemann, E., Hansen, F. and Raa, J. (1982) Ensiling in acid – a method to stabilize astaxanthin in shrimp processing by-products and improve uptake of this pigment by rainbow trout (*Salmo gairdneri*). *Aquaculture*, **26**, 77–83.

Torrissen, O.J., Hardy, R.W. and Shearer, K.D. (1989) Pigmentation of salmonids – carotenoid deposition and metabolism. *CRC Crit. Rev. Aquat. Sci.*, **1**, 209–25.

Tuveson, R.W., Larson, R.A. and Kagan, J. (1988) Role of cloned carotenoid genes expressed in *Escherichia coli* in protecting against inactivation by near-UV light and specific phototoxic molecules. *J. Bacteriol.*, **170**, 4675–80.

Walker, W.F. (1984) 5S rRNA sequences from Atractiellales and Basidiomycetous yeasts and Fungi Imperfecti. *Syst. Appl. Microbiol.*, 352–9.

Weijman, A.C.M., Rodrigues de Miranda, L. and van der Walt, J.P. (1988) Redefinition of *Candida* Berkhout and the consequent emendation of *Cryptococcus* Kützing and *Rhodotorula* Harrison. *Antonie van Leeuwenhoek: J. Microbiol.*, **54**, 545–53.

Yamada, Y. and Kawasaki, H. (1989) The genus *Phaffia* is phylogenetically separate from the genus *Cryptococcus cryptococcaceae*. *Agric. Biol. Chem.*, **53**, 2845–6.

Yamada, Y., Nagahama, T., Kawasaki, H. and Banno, I. (1990) The phylogenetic relationship of the genera *Phaffia* Miller, Yoneyama et Soneda and *Cryptococcus* Kützing Endend Phaff et Spencer (Cryptococcaceae) based on the partial sequences of 18S and 26S ribosomal ribonucleic acids. *J. Gen. Appl. Microbiol.*, **36**, 403–14.

CHAPTER 17

Fish vaccine: development, production and use of bacterial vaccines, with special reference to salmon

P.M. Aasjord and E. Slinde*

17.1 INTRODUCTION

A limiting factor in cultivation of Atlantic salmon (*Salmo salar* L.) and other fish species has been diseases. Apart from furunculosis and bacterial kidney disease (BKD) in salmon, vibrio infections are by far the most serious problem in aquaculture. If attacked by a pathogen, the fish may develop a disease as illustrated in Figure 17.1. To combat diseases vaccination has been a great success in human and veterinary medicine. In aquaculture the use of vaccines is still in an early developing phase and, as reviewed by Ellis (1988a), it is mainly vaccines against bacterial infections like vibriosis, cold water vibriosis, yersiniosis and furunculosis that have given convincing protection in field trials. Vaccines against other diseases have been less effective, side effects have occurred, and some vaccines have been too expensive both to develop and produce. However, the economical losses due to cold water vibriosis and vibriosis in salmon aquaculture amounted to more than US$ 100 million during the period 1984–88 in Norway (Tilseth, Hansen and Møller, 1991), which justifies a rather high price for fish vaccines, if effective. Research into and development of fish vaccines are in our opinion necessary for a healthy aquaculture industry in the future.

*Recently deceased.

Fisheries Processing: Biotechnological applications.
Edited by A.M. Martin.
Published in 1994 by Chapman & Hall, London. ISBN 0 412 58460 3

In the salmon farming industry, which has grown fast during the last 20 years (Tilseth, Hansen and Møller, 1991), vaccination against bacterial diseases has been an important factor and it is assumed to be just as important for other aquaculture species. In order to develop effective vaccines an understanding of the immune system of the fish is necessary. The present chapter gives a brief description of the immune system in fish and how present vaccines against bacterial diseases are developed, produced and applied. The isolation characterization and reinfection with the pathogenic bacteria, its growth properties, production, inactivation, vaccination procedures, and test and field trials of fish vaccines are described. Since vaccines are mainly used in the salmon farming industry, some salmon diseases where vaccines exist or a disease is already a problem and a vaccine is needed in the near future will be described.

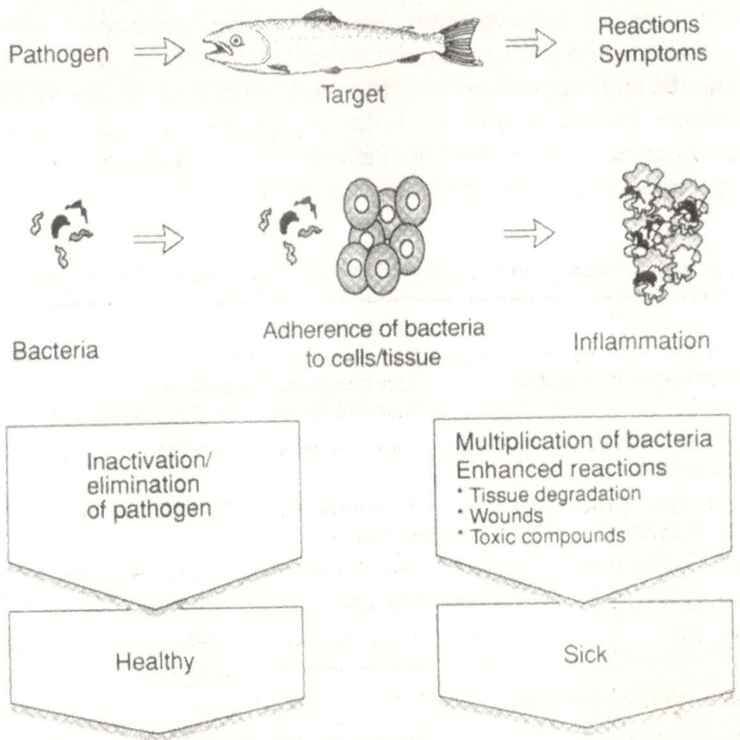

Figure 17.1 Pathogens such as bacteria attack different tissues and organs. Different diseases have their specific characteristics. If the pathogen is inactivated at an early stage the fish is healthy, otherwise a sick fish with wounds and tissue degradation is the result and finally death.

17.2 ANTIBIOTICS VERSUS FISH VACCINES

Numerous fish diseases exist and among them bacterial diseases are a serious threat to the aquacultural industry (Austin and Allan-Austin, 1985). Fish vaccines have become an important tool in the work against infectious diseases. In contrast to antibiotics which are used for treatment of infected fish, vaccines are prophylactic as they prevent the outbreak of diseases. Antibiotics are given to the fish through the feed, while vaccines are either injected into the fish or used as immersion or bath vaccines. To achieve optimum results from vaccination, infected fish or fish treated with antibiotics should not be vaccinated. In Table 17.1 the differences between vaccines and antibiotics are described. In aquaculture there is a need for both vaccines and medical treatment. However, in our opinion the use of effective vaccines is better both from an economical and environmental point of view.

Table 17.1 Differences in the use and effect of vaccines and antibiotics

Vaccines	*Antibiotics*
Treatment prior to infection	Treatment after infection. Fast and correct diagnosis necessary
Prophylactic; minimize losses due to diseases	Loss of fish before treatment
Long-term effect after one or two vaccinations	Short-term effect. Repeated administration necessary
All fishes are treated	Only fish with good appetite are treated
Minimal side effects	Side effects occur
Healthy fish	Accumulation of antibiotics within the fish. Retention before slaughter
Minimal pollution problems	Pollution problems exist
No development of resistance amongst bacteria, mutants may create problems	Several bacteria develop resistance against antibiotics

17.3 FISH IMMUNOLOGY

So far our knowledge of the immune system in fish is limited and big differences between species are likely. They have an immune system for protection against different pathogens like bacteria, viruses and fungi (Ellis, 1989).

The defence mechanism is divided into innate (non-specific) and acquired (specific) protection or immunity against diseases (Figure 17.2). Both mechanical and biochemical barriers, such as unworn skin and

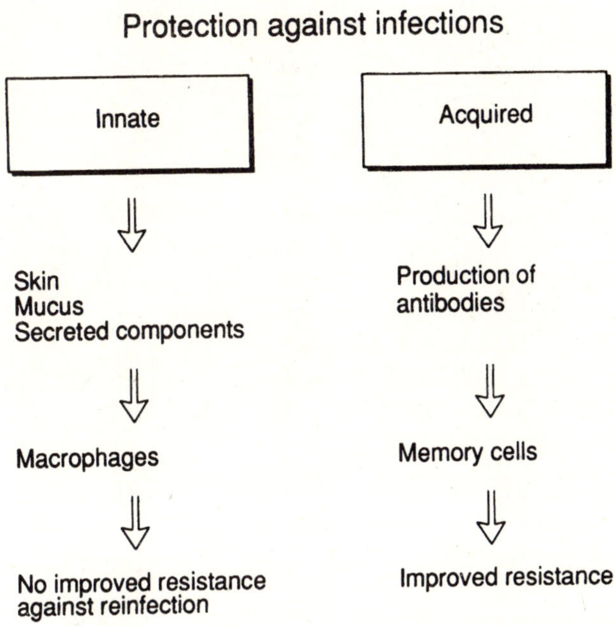

Figure 17.2 A pathogen has to face many obstacles before it is established within the host. Innate protection is the first line of defence. This is both a physical and a chemical barrier. The acquired immune response has the characteristic of remembering pathogens and responding faster to reinfections.

mucus covering skin, gills and intestine, are included in the term innate protection (Ingram, 1980; Fletcher, 1982). Apart from the elimination of infectious agents by macrophages, the production of specific chemical and biochemical compounds (e.g. lysozyme) is of importance for general protection against diseases (Fletcher, 1982).

The principle behind the function of a vaccine is the mechanism of acquired immunity which is activated if the infectious agent breaks through the chemical/biochemical and mechanical barriers. Specificity and memory are two important properties of acquired immunity where specificity means antigenic recognition of the infectious agent or parts of it, and memory is the capacity of the immune system to remember antigens associated with earlier infections.

17.4 HUMORAL AND CELLULAR IMMUNITY

Acquired immunity is divided into humoral (antibody production) and cellular immunity, as illustrated in Figure 17.3 where memory and specificity are key words for both. In humoral immunity B cells (lymphocytes) are stimulated, they divide and develop into antibody producing plasma cells (Corbel, 1975). In contrast to mammals, fish lack bone

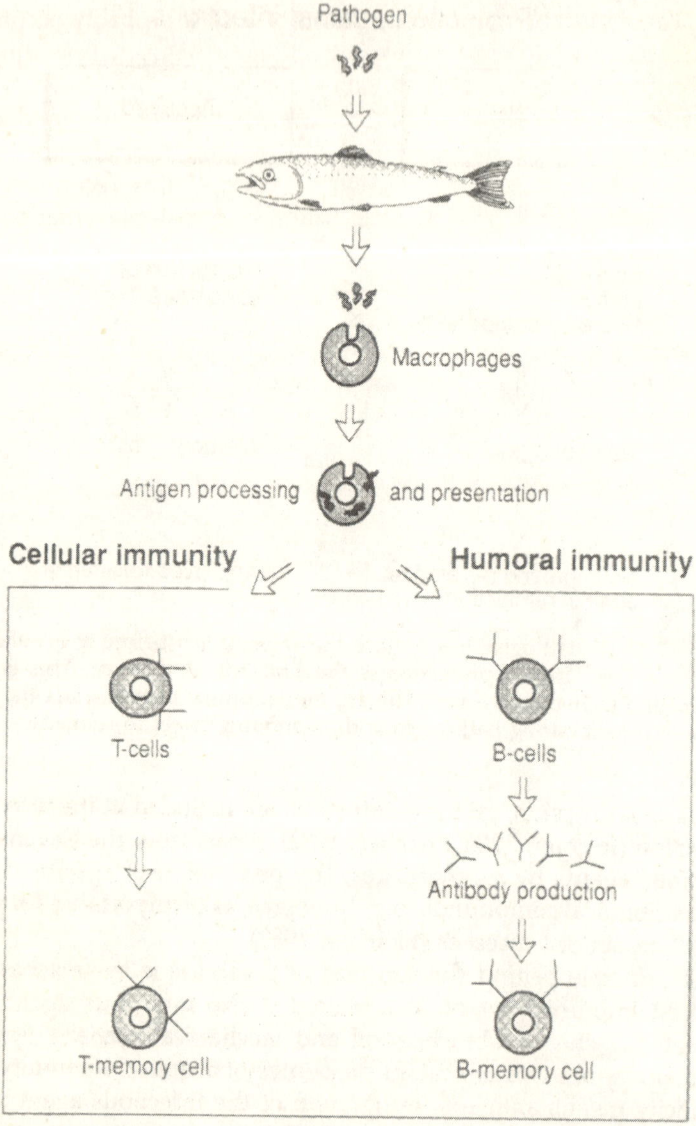

Figure 17.3 Proposed immune reaction in fish. Different cell types participate in the immune response. The macrophages are part of the innate protection, while the B cells (lymphocytes) are a part of the complex acquired immune system. The surfaces of the B cells are covered by immunoglobulins that recognize the pathogen and which give signals to produce more cells. The B cells mature to plasma cells and produce antibodies (immunoglobulins). The antibodies recognize the pathogenic agent which is then destroyed through further reactions. This part of the acquired immunity is termed humoral immunity. If the pathogenic agent is digested by a macrophage, the small pieces now localized on the macrophage surface are recognized by T cells (lymphocytes). In all the processes signal compounds are exchanged between the cells. T cells mature into effector cells (not shown). This is called cellular immunity. Both T and B cells mature into memory cells that form the basis for a fast immune response.

marrow. The maturation and proliferation of B cells probably occurs in the spleen and the anterior part of the kidney instead, resulting in production of antigen specific antibodies.

In the cellular immune response T cells (lymphocytes) are stimulated, cell divisions take place resulting in the development of effector cells. The T cells, which mature in the thymus (Ellis, 1989), recognize the infectious agent, or parts of it, after processing and presentation by macrophages (antigen presenting cells). Further, this is a signal triggering both cell proliferation and extensive production of specific antibodies by other cell populations.

As a part of the acquired immune system, some B and T cell populations act as a 'memory pool', and are responsible for rapid reactions on reinfection by a pathogen against which the fish has been vaccinated.

17.5 UPTAKE OF VACCINES

Fish vaccines are normally administrated by immersion, bath, injection, through the feed or by combinations of these methods. The uptake of vaccine by the immersion and bath methods is not quite understood but it is presumed that the vaccine is taken up mainly through the gills and skin (mucus) through a barrier of cell layers (Ellis, 1988b). Further, some vaccine is swallowed, broken down by acid/enzymes and absorbed by the intestine. Injectable vaccines, on the other hand, once inside the fish, are possessed by immune competent cells entering the injected area. This method has the advantage that each individual receives the same amount of vaccine which is not possible using other procedures. When the immune system has responded to the vaccine, a specific reaction (acquired immunity) upon infection with the pathogenic agent is the result.

It is advisable to vaccinate small fishes because of less consumption of vaccines, early protection and easy handling. However, if the fish is too small, a poor immune response is the result since the pathogen is not adequately processed and recognized (Figure 17.3) due to an immature immune system (Tatner and Horne, 1984; Manning and Mugal, 1985). Our experience is that, in order to achieve an adequate immune response the weight should be above 5 g for salmon (*S. salar* L.), 3 g for trout (*Salmo gairdneri* Richardson), 2 g for cod (*Gadus morhua*) and 1 g for turbot (*Psetta maxima*). However, the response also depends on the water temperature. High temperatures result in a rapid response, but not necessarily a good protection (Lillehaug *et al.*, 1993). In practice one has to follow the recommendation of the vaccine producer.

17.6 FISH DISEASES

The numerous diseases existing in fish aquaculture caused by fungi, algae, protozoa, other endo- and ectoparasites and crustacea, create problems for the fish farmer. Today no commercial vaccines are available against diseases caused by any of these organisms. Therefore, good hygienic standards, reliable diagnostic methods, and efficient chemical and pharmaceutical treatment are necessary to minimize the problems.

However, in the case of bacteria, use of antibiotics which may select for resistant strains (Table 17.1) and laws prohibiting the export of fish products containing even trace amounts of antibiotics has led to the development of efficient bacterial vaccines and elucidation of the immune systems of different fish.

To evaluate the efficiency of a vaccine the term relative percent survival (RPS) (Amend, 1981) is used:

$$RPS = 100\left(1 - \frac{\%\ \text{lost vaccinated}}{\%\ \text{lost unvaccinated}}\right)$$

The RPS gives an indication of the potency of a vaccine in laboratory tests performed under controlled conditions and is not necessarily comparable with field trials or vaccination in fish farms. To evaluate the results from laboratory trials, the infection dosage should ideally be adjusted towards a mortality of the control fish group between 50 and 90%. If the mortality is 100% the RPS cannot be calculated.

17.6.1 Vibriosis

Vibriosis is the most common disease in worldwide aquaculture. It is caused by the bacterium *Vibrio anguillarum* and several serotypes have been demonstrated among fish and fish species (Knappskog *et al.*, 1993). Thus, use of a vaccine against vibriosis based on a strain causing disease in cod would give limited or no protection against vibriosis in another fish species, such as trout. Vibriosis appears regularly in aquaculture, in sea or brackish water, and is a serious threat to economies of the fish farmers.

The presence of *V. anguillarum* in combination with external stress, such as a sudden rise in temperature, contributes strongly to the onset of the disease. Typical signs are hemorrhages (bleeding) at the fin bases followed by open wounds and internal bleeding (Anderson and Conroy, 1970). The disease is treated with antibiotics such as oxolinic acid, flumequine and oxytetracycline. Effective vaccines are available and help to reduce the losses although the RPS has been variable, ranging from 65 to 90% in different fish farms. According to several producers, the vaccines used for immunization of Atlantic salmon today contain a formalin inactivated mixture of the most common serotypes of *V. anguillarum*, O1 and O2. Since vaccines against vibriosis in salmon today are based on these two serotypes the importance of including vibrio strains belonging to other serotypes are not known. The identification of other serotypes in diseased fish (Austin and Allan-Austin, 1985) may explain why the vaccines give a variable degree of protection.

Examination of plasmid-profiles in *V. anguillarum* strains has revealed a plasmid which could be associated to the bacterial virulence (Crosa, Schiewe and Falkow, 1977; Crosa, 1980), but the importance of this finding is yet not known.

Vaccines against vibriosis in turbot (*P. maxima*) and cod (*G. morhua*) have also been developed. Both immersion and bath give good protection in cod if the correct serotypes are included in the vaccines (Knappskog *et al.*, 1993). Turbot has been vaccinated by bath as fry with good results.

17.6.2 Cold water vibriosis

Cold water vibriosis, otherwise known as hemorrhagic syndrome or 'Hitra disease', is a disease in salmon that has caused great economic losses in North Sea fish farms (Tilseth, Hansen and Møller, 1991). The disease, which occurs in late autumn, winter and early spring, is caused by the bacterium *Vibrio salmonicida* (Egidius *et al.*, 1981, 1986; Holm *et al.*, 1985).

Characteristics are general internal bleeding and a yellowish colored liver. Today the disease is controlled by effective vaccines (Holm and Jørgensen, 1987). Field trials show RPS values better than 90% and outbreaks in properly vaccinated fish farms are rare. However, occasional but serious outbreaks of the disease have been recently reported. An explanation could be that the vaccines are less effective due to mutations of the bacteria that have resulted in changes of the protein/antigenic pattern. If so, one has to consider whether new bacterial strains have to be included in future vaccines. It is also possible that the use of antibiotics over the years has caused transfer of plasmid-associated resistance towards antibiotics, resulting in more serious consequences of what started as apparently insignificant outbreaks. In general a vaccine producer should take care to isolate bacteria causing a disease in fish farms, in order to follow any change in serological characteristics. Challenge experiments should be done to verify that the vaccine in production is effective also against new serotypes, if discovered.

17.6.3 Bacterial kidney disease

Bacterial kidney disease (BKD), Dee disease, corynebacterial kidney disease and salmonoid kidney disease are all synonyms for a disease in salmon which is caused by the bacterium *Renibacterium salmoninarum*. The disease is a systemic, usually chronic and frequently fatal infection of salmonoids. It was described for the first time in the 1930s for a condition in Atlantic salmon caught in the rivers Dee and Spey, Scotland (Mackie *et al.*, 1933; Smith, 1964). The pathogen was isolated in 1950 by Earp and since from salmonoid fishes from most countries engaged in fish farming. The disease is a serious problem in Canada and Chile but it is also an increasing problem for European salmon farms.

R. salmoninarum is a small, rod-shaped, non-motile, non-sporulating, Gram-positive bacterium which has its own genus and species (Sanders and Fryer, 1980). Isolates from different parts of the world seem to belong to the same serotype.

Infection is transmitted both horizontally from fish to fish and vertically (most common) to the next generation through the roe. Primarily the kidneys are attacked where white/grey infiltrates of lymphocytes and *R. salmoninarum* can be seen microscopically. Further, roe, spleen, muscle and brain are also infected and several internal organs can be seen covered by a white pseudomembrane (Snieszko and Griffin, 1955) consisting of dead tissue infiltrated with bacteria, leucocytes, macrophages and fibroblasts. In addition exophthalmia, eye lesions, swollen abdomen and ulcers are common (Fryer and Sanders, 1981; Hoffman, Popp and van der Graaff, 1984).

R. salmoninarum is somewhat special since it grows and multiplies inside monocytes and macrophages (Bruno, 1986) resulting in spread of the infection throughout the entire fish. Further, since the bacteria in this way seem to be protected against the immune system, the host is continuously exposed to secreted antigens. In attempts to destroy the pathogen the resulting local immune reaction between antigen and antibody could trigger biochemical reactions whose products in turn cause serious damage to surrounding tissue.

Treatment with antibiotics, erythromycin being the most common, does not terminate the disease, the symptoms being only reduced for a period of time (Austin, 1985). Losses in the range of 2–12% of the stock per month have been reported. This can be reduced by lowering stress through good hygiene and management.

So far no effective vaccine against the disease exists. In a field trial with a formalin-inactivated bacterium we have observed a reduced death rate, but over a limited period of time. This may be explained by the infection pattern of BKD which is different to that of more classical diseases. It is therefore obvious that a future efficient vaccine against

BKD must be based on a more sophisticated technology and a better knowledge of both the disease and the immune system.

17.6.4 Yersiniosis

Enteric red mouth (ERM) disease, Hagerman red mouth disease, salmonoid blood spot and yersiniosis are all synonyms for a disease in salmonoid fishes caused by the Gram-negative bacterium *Yersinia ruckeri* (Ross, Rucker and Ewing, 1966). The bacterium was isolated for the first time in the 1950s from rainbow trout in the US, but has since become a worldwide problem (Austin and Austin, 1987).

At least five serotypes are known (O'Leary, 1977; Bullock, Stuckey and Shotts, 1978; Stevenson and Airdrie, 1984), of which three of them are common: serotype 1, the Hagerman strain; serotype 2, the O'Leary strain; and serotype 3, the Australian strain (Bullock and Anderson, 1984). Because serotypes 1 and 3 are very similar giving cross-protection upon immunization of Atlantic salmon but serotype 2 is quite different, a vaccine is usually composed of serotype 1 or 3 in combination with serotype 2. Serotype 1 is an antigenically homogeneous group while serotype 2 is heterogeneous, primarily due to lipopolysaccharide (LPS) antigens in the cell wall of the bacterium.

As indicated by the name, red mouth disease, red spots are often seen in the mouth, in the eye and in the throat region due to subcutaneous bleeding (Busch, 1973). A yellowish secretion can be seen in the intestine. The fish becomes unconscious and exophthalmia can also be observed.

The fishes can become infected in both fresh and seawater, most seriously at temperatures between 15°C and 18°C. In laboratory experiments, depending on the inoculum size, mortalities start 5–19 days after being exposed to the pathogen (Rucker, 1966; Busch, 1973). By performing latent carrier tests in which stress was induced by injection of corticosteroids or by exposure to a water temperature of 25°C (Hunter, Knittel and Fryer, 1980), stress was shown to be an inducing factor for outbreak of the disease.

Y. ruckeri is often resistant to several antibiotics and thus the treatment of the disease can be difficult (Ceschia *et al.*, 1987). Reinfection a few weeks after termination of treatment is common, also due to bacteria found in the sediments below the sea cage in the fish farm (Rucker, 1966). Oxytetracycline (Rucker, 1966) and oxolinic acid (Rodgers and Austin, 1982) are reported to be effective. Vaccines against yersiniosis are produced both in Europe and the US. The RPS values vary from 60 to 90%, probably because the different vaccines are based on serotypes other than those causing the particular outbreak of the disease. A vaccine producer should always be aware of such problems and test the vaccine's ability to protect against outbreak strains. It is our experience that in some cases vaccines based on local isolates (analog vaccine) are the most efficient and could completely wipe out the disease.

17.6.5 Furunculosis

The disease is caused in salmon by the Gram-negative rod-shaped bacterium *Aeromonas salmonicida* ssp. *salmonicida*. It has been demonstrated in several species of both farmed and wild fish populations (McCarthy, 1975), and can be serologically and antigenically divided into various groups. Rainbow trout seems to be relatively more resistant to furunculosis than salmon. Also an achromogeneous subspecies exists, *A. salmonicida* ssp. *achromogenes* (Schubert, 1974). In addition to stress, outbreaks with high mortality are connected to high seawater temperatures.

After initial general symptoms of inflamed liver and spleen, greyish spots and necrosis develop in the liver followed by blood spots and furuncles with a brown/red content in the muscles of larger fish (McCarthy and Roberts, 1980). The furuncles are the origin of external wounds.

Several vaccines, all of which require an adjuvant (immunostimulating agent) for maximum effect, are available. However, the protection achieved by injection of adjuvanted furunculosis vaccines is lower than that provided by vaccines used against other bacterial infections in fish and RPS values vary from 50 to 85%. The duration of the protection is limited and the fish must be revaccinated within 1 year.

17.6.6 Viral diseases

More than 50 different viruses have been isolated and described from fish, and the most common, causing diseases are listed in Table 17.2. No efficient antiviral drugs exist and slaughter is the only alternative. Efficient viral vaccines would therefore be of great economic value.

The glycoprotein from infectious hematopoietic necrosis virus causing disease in salmon and trout has been cloned in *Escherichia coli* (Kurath *et al.* 1985; Koener *et al.* 1987) and used for immunization. Good protection was achieved using the lysed bacteria (Gilmore *et al.*, 1988). Furthermore, the capsid protein VP2 has been cloned in four different infectious pancreatic necrosis virus strains (Leong, Fryer and Winton, 1988; Wu, 1992) and Manning (1988) has demonstrated protection after immunization in rainbow trout. These two examples illustrate the requirement for large-scale bacterial fermentation for production of cloned virus vaccines, with subsequent purification of the antigen.

Possible types of viral vaccines can be illustrated by those used in human and veterinary medicine (Figure 17.4). In our opinion the future for their use in aquaculture lies in the development of vaccines against surface antigens of the viruses and to produce these using gene techno-

Table 17.2 Viruses causing diseases in fish

Virus	*Family*	*Host*	*Vaccine under development*
Viral hemorrhagic septicemia virus (VHSV)	Rhabdoviridae	salmon, trout	sub-unit attenuated
Infectious hematopoietic necrosis virus	Rhabdoviridae	salmon, trout	sub-unit attenuated
Spring viremia virus	Rhabdoviridae	carp, pike	inactivated attenuated
Infectious pancreatic necrosis (IPNV)	Birnaviridae	salmon, trout, turbot, halibut, eel	inactivated attenuated sub-unit
Herpes virus ichataluri	Herpesviridae	channel catfish	attenuated
Pox herpes virus	Herpesviridae	carp	none
Infectious salmon anemia virus	?	salmon	none

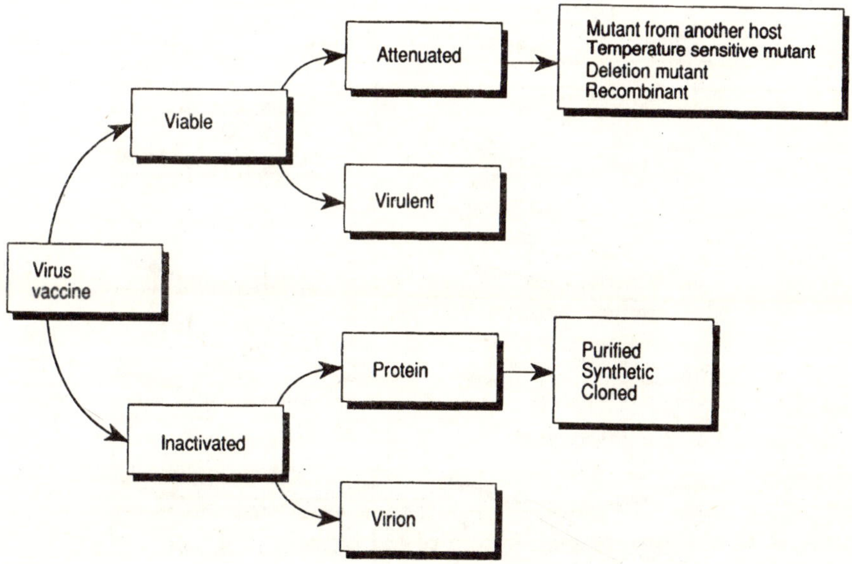

Figure 17.4 Different types of viral vaccines used in human and/or veterinary medicine. In our opinion only inactivated vaccines should be permitted for use in aquaculture.

logy (Leong and Munn, 1991). At present *E. coli* is the most common expression system.

17.7 BACTERIAL VACCINES

A development program for a vaccine consists of two main parts (Figure 17.5). The research phase has the object of obtaining approval of the vaccine from the authorities but it also forms the basis for a production set-up according to standard operating procedures (SOP).

Vaccines are preparations made of living, attenuated or killed infectious agents, parts or products from them, given to humans, animals or fish for protection against infectious diseases. Fish vaccines have so far mainly been based on formalin-killed bacteria. However, as some of the vaccines, i.e. furunculosis vaccines, seem to have limited effect, experiments with additives (adjuvants) and/or subunit vaccines to improve the effect have to be developed. Subunit vaccines may contain important antigens isolated from the pathogen itself or the antigens are produced using gene technology as in the case of viral vaccines.

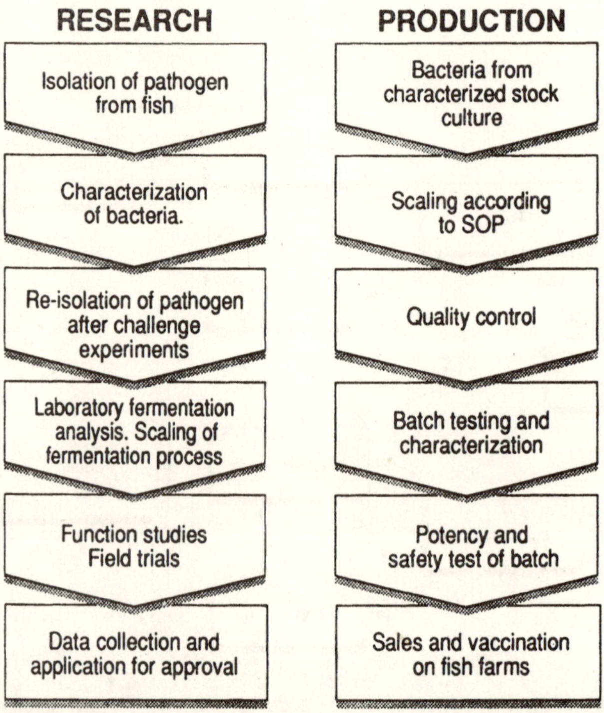

Figure 17.5 Development and production of a bacterial fish vaccine. (SOP, standard operating procedures.)

17.7.1 Adjuvants

Some vaccines give good protection against diseases since they contain strong antigenic substances that activate the immune system. To other vaccines immune stimulating agents (adjuvants) must be added to obtain significant protection against the pathogenic agent. Adjuvants are based on either oils or metal chelates of which mineral oils (Krantz, Reddecliff and Heist, 1964; Paterson and Fryer, 1974; Adams *et al.*, 1988) and aluminum hydroxide/phosphate (Horne *et al.*, 1984; Cossarini-Dunier, 1985), are the most common. In addition a new immune stimulating agent based on β(1,3)-glucan (MacroGard®, Mackzymal, Tromsø, Norway), extracted from *Saccharomyces cerevisiae* cell walls, has been used in fish vaccines for a couple of years (Robertsen *et al.*, 1990; Rørstad *et al.*, 1992).

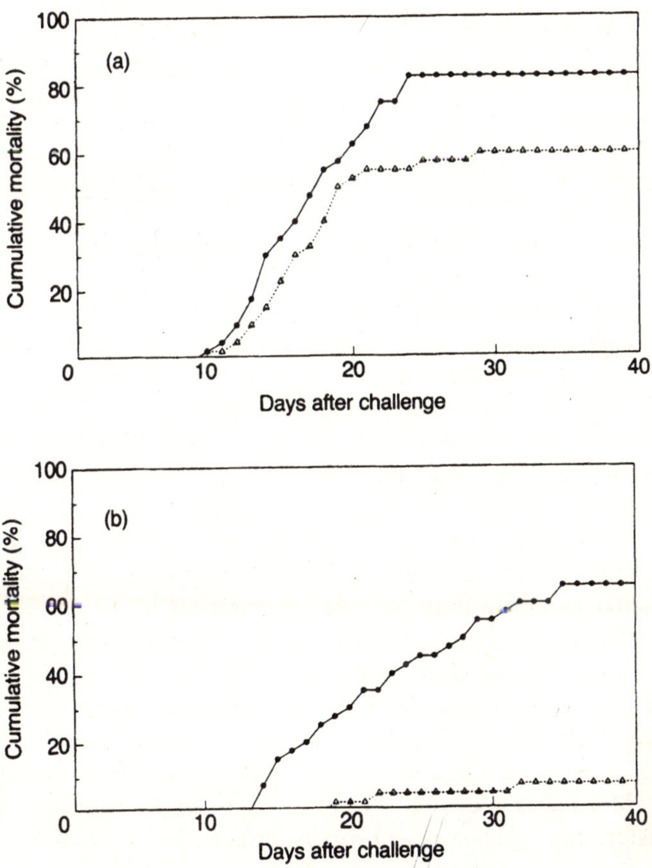

Figure 17.6 The effect of adding β(1,3)-glucan (MacroGard®, Mackzymal, Tromsø, Norway) as adjuvant in a vaccine based on a formalin inactivated bacterin of *A. salmonicida* ssp. *salmonicida*. (a) Vaccine without glucan, compared with (b) vaccine with adjuvant (△) Vaccinated fish; (●) control fish.

Injection may cause quality faults in the slaughtered fish. This problem is enhanced when adjuvants are used, especially in the case of mineral oils where the fish may contain both adherences and pigmentation faults when slaughtered. The results are reduced quality and lower prices. β(1,3)-glucan, on the other hand, seems to be a good adjuvant giving minimal side effects. Figure 17.6 shows the effect of a furunculosis vaccine with and without β(1,3)-glucan. Common for all adjuvant containing vaccines is that they have to be injected to give the desired effect. We have also observed that vaccines containing bacterins for protection against several diseases (i.e. vibriosis, cold water vibriosis and furunculosis) seem to enhance the protection against the individual diseases indicating a form of synergistic effect of the vaccine components.

17.8 ISOLATION AND CHARACTERIZATION OF PATHOGENIC BACTERIA

The principles of isolation and characterization of a fish pathogenic bacteria, which form the basis of SOPs used in production, are shown in Figure 17.7.

Description of the bacterial antigen/culture is of great importance and should be done according to accepted standard procedures. The pathogen is isolated, for example, by inoculating samples from the kidney of diseased fish on blood agar plates containing 1.5% sodium chloride. Bacteria from one single colony regrown in an appropriate broth, checked for purity, dispersed, lyophilized or frozen (−80°C or liquid N_2) in storage tubes are referred to as the master seed or the primary stock culture. The isolated and cultured bacteria, both from the primary and secondary stock (see below), has to be injected into fish and reisolated (Figure 17.7) to show that the purified bacteria causes the same disease as the primary isolate according to Robert Koch (Koch's postulates) who was the first to apply these criteria experimentally. It is preferable to add 10–20% glycerol for cultures for long-term frozen storage. The bacteria must be checked for optimum growth conditions like temperature, pH, media and necessary additives. Bacterial morphology is documented and the organisms characterized by serotyping, biochemical reactions, e.g. using the API system (API system SA, La Balme Les Grottes, Montalieu Vercieu, France), SDS-electrophoresis combined with Western blot of whole-cell protein extracts, as exemplified in Figure 17.8.

The secondary stock culture is generated by culturing an ampoule from the primary stock culture, it is stored in the same manner and used for the vaccine production.

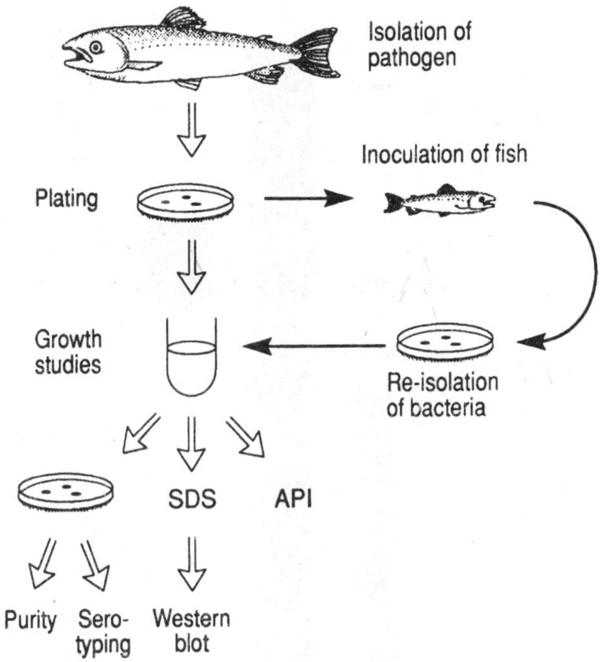

Figure 17.7 Isolation, re-isolation and characterization of a fish bacterial pathogen. SDS, sodium dodecyl sulphate electrophoresis of proteins; API, system for identification of bacteria according to biochemical/fermentation reaction patterns.

17.9 LABORATORY FERMENTATION ANALYSIS AND SCALING UP OF FERMENTATION

The result of the research (Figure 17.5), characterization (Figure 17.7) and scaling of fermentation form the basis for writing out SOPs.

Prior to inoculation of fermentors from the starter culture, bacteria are always checked by microscopy and plating on different types of agar to test viability and purity. The plates are incubated at the optimum growth temperature to give uniform colonies, making detection of foreign contaminants easier. If the results are not satisfactory, the preparation is discarded.

Starter cultures are inoculated and incubated under appropriate conditions of culture and transferred to the next step in the scaling process. In the fermenters the culture conditions (e.g. pH, pO_2, temperature, agitation), are automatically regulated and recorded during the entire fermentation process. To monitor the fermentation scaling process samples are withdrawn for microscopy, cell counts, optical density (OD_{540}) measurements and plating on different agar plates. Figure 17.9 shows examples of growth curves for *V. salmonicida* in a scaling experiment

Figure 17.8 Western blot following SDS-electrophoresis of proteins from *A. salmonicida* ssp. *salmonicida*. Lane A, SDS-gel (12.5%) developed with Coomassie Blue; lane B, Western blot (i.e. the proteins are blotted onto a nitrocellulose sheet) with rabbit anti-*A. salmonicida* serum; lane C, Western blot with salmon anti-*A. salmonicida* serum. Note the absence of rabbit antibodies to the LPS (top lane B) and the absence of salmon antibodies to the LPS cord (lane C bottom) (courtesy Ragnhild Aakre, University of Bergen).

performed in 10, 100 and 1000 l fermenters. Dependent on the micro-organism fermented, the cell density will vary, but under well controlled conditions 10^{12} bacteria/ml are attainable. Batches which have progressed too far into the stationary phase will probably have higher levels of cell debris and degrading enzymes which may have a deleterious effect on the vaccine efficacy. In addition the total and, especially, the viable cell count will be low.

For a statement of product stability, several pilot batches have to be fermented. Finally the vaccines have to be tested for potency (see section 17.11.1) both in controlled laboratory tanks and in field trials.

Total cell count for the inactivated vaccine is performed in a standard counting chamber at different dilutions. The viable counts or colony forming units (c.f.u) are checked by plating on to agar plates from

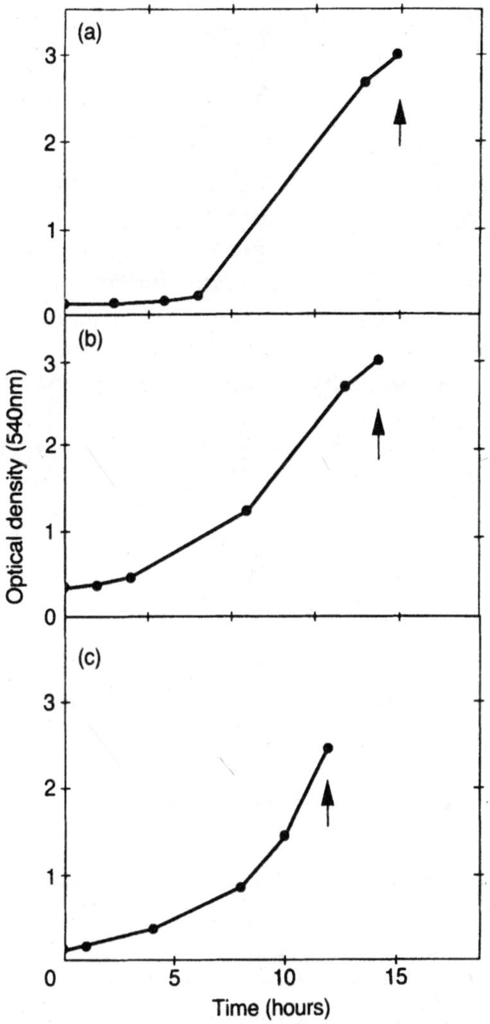

Figure 17.9 Example of growth curves showing scaling up growth of *V. salmonicida* in fermenters: (a) 10 l, (b) 100 l and (c) 1000 l. Inoculation transfer from one fermenter to the next is indicated by arrows (a and b). Harvesting of the final volume of 1000 l is indicated by an arrow (c).

suitable dilutions of the fermentation product, prior to inactivation. Colonies are counted after incubation for 3–5 days.

After 5 days' storage, adjuvant and other additives, all sterility tested, are mixed into the vaccine.

17.10 VACCINE PRODUCTION

The production of a vaccine is based on the SOPs worked out during the research phase. Good manufacturing practice (GMP), good laboratory practice (GLP) and a quality assurance program (QAP) form the basis for the production.

17.10.1 Quality assurance

Generally a properly documented quality program should ensure that all laboratory work is carried out according to GLP and GMP. The quality manager takes care of the quality assurance and quality control, and should be responsible for the production and report directly to the director. The one in charge should not participate in the production.

The quality manager must ensure that the action plan and all SOPs are known and understood by all production participants, that the work is carried out correctly, and that the necessary controls has been performed and signed. In addition, all relevant data must be recorded correctly and filed. Deviation from SOP's should be reported and proper action taken.

SOP should exist but not be limited to:

1. Test and reference substances.
2. Equipment and reagents.
3. Recording of results, reports, storage and retrieval of data.
4. Description of how different tests are performed such as sampling, identification and treatment of microorganisms, or action taken due to production failure.
5. Security and health for employees.

All raw materials used for the fermentation should be available in sealed packaging. Two persons accept and control the raw materials upon arrival. The raw materials should satisfy the quality according to SOP which is also valid for the water, acid, base and antifoam used in the fermenter. All small equipment used during the production from ampoules to stoppered flasks have to be adequately sterilized according to the actual SOP. Proper functioning tests have to be carried out for all equipment. This also includes cleaning of incubators, shakers and laminar flow benches. Instruments for measurement of bacterial growth and process parameters have to be adjusted. Further, electrodes for measurement of pH and oxygen, and the antifoam probe have to be calibrated.

Commercial media dissolved in ultra-filtered or distilled water are commonly used. Prior to use media are either batch sterilized in the fermenter or sterilized by a continuous flow process before entering into the fermenter. The sterilization conditions are 1.1 atm at 121°C for 30 min.

To avoid reactions between sugars and amino acids (Maillard reactions) which may result in toxic reaction products, the sugar is autoclaved separately, at least for large-scale fermentation and the other medium components. Before inoculation the medium is tested for sterility. Samples are withdrawn from the fermenter through the sampler tube and plated on different types of agar plates.

Since foaming is a problem in large batch volumes, antifoam is added during the fermentation process by pulses from the antifoam probe. In small fermenters, on the other hand, antifoam is often added to the medium prior to sterilization.

The flow sheet in Figure 17.10 describes the production of 1000 l vaccines. The example used is for *V. anguillarum* where two serotypes

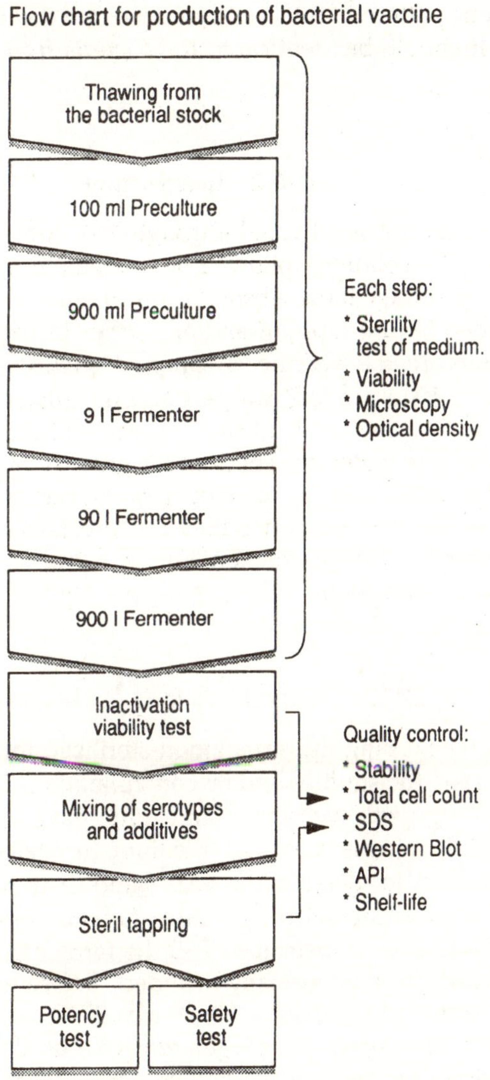

Figure 17.10 Flow chart indicating the scaling up steps and the tests performed during and after fermentation.

are included the final vaccine. The stock culture is thawed from the stored ampoule at −80°C and grown to late exponential growth phase in tryptic soy broth containing 1.5% NaCl. Inoculations are carried out as shown in Figure 17.10. Samples are routinely taken from the fermenters during the fermentation process and analyzed by spectrophotometry (cell density) and light microscopy (morphology and cell number). Viability is determined by incubation of samples (0.1 ml) plated on blood agar plates. Should contamination be detected, the batch is destroyed. If not, prior to inactivation, samples are withdrawn for purity checks using the API system. After fermentation is terminated the electrophoretic pattern of total cell proteins from the inactivated bacteria is checked. It should be identical to the bacteria from the starter inoculum.

17.10.2 Inactivation

The culture may be inactivated through the addition of 0.5% (v/v) formalin (35–37% solution) followed by immediate cooling and transfer to a temporary storage tank. Absolute confirmation that the bacteria are dead is needed before tapping and/or further testing on fish are performed. Therefore the vaccine is sampled, centrifuged to get rid of formaldehyde, resuspended, washed and inoculated into broth. After incubation for 3–4 days, 0.1 ml samples are withdrawn and plated on to agar plates. The plates are checked for growth daily for 1 week.

Because formaldehyde is not sufficient to prevent fungal growth, merthiolate is added to some vaccines as a preservative. However, this is not necessary if the production line is sterile and the tapping is performed in a clean room.

17.10.3 Laboratory tests of vaccines

One week after tapping into irradiation-sterilized infusion bottles, the vaccine is tested for sterility and/or contamination. To do this, media that favor both growth of bacteria and fungi have to be used. Tryptic soy broth (TSB) agar is a typical medium for detection of bacteria. Saboraud medium for detection of yeasts and Czapek Dox medium for detection of environmental fungi (Difco Laboratories, MI), should be included. In addition or instead of TSB the fermentation medium used for vaccine production is normally included. Before use the agar plates are sterility tested by incubation for 48 h at 37°C, contaminated plates discarded and the rest stored refrigerated (4°C) until used.

A test program for the finished tapped and packed product is shown in Figure 17.11. It must be stressed that this is a proposal for how the test could be performed and is not necessarily in accordance with, for

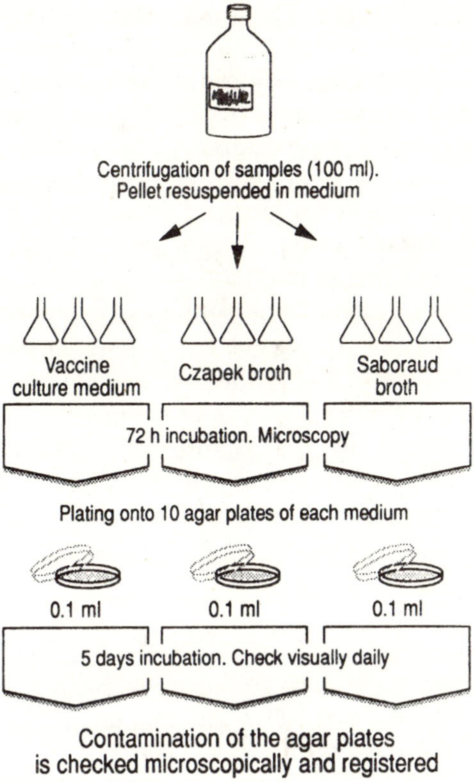

Figure 17.11 Sterility testing of the vaccine product. The medium used for the Ehrlenmeyer bottles and the respective agar plates is identical.

example, the *European Pharmacopeia*. The first and the last bottle in each production sequence are withdrawn from the tapping line and tested.

For the vaccine culture medium the growth temperature is equal to that of the process recipe (SOP). Czapek and Saboraud medium are normally incubated at temperatures between 20 and 30°C. The agar plates are checked visually each day and preparations are made of any colonies appearing. Occasionally some plates may be contaminated due to poor plating technique. All results are written on a quality control form and should be enclosed with the batch documents together with the quality specifications.

The vaccine is stored at 4–8°C, frost-free and never over 15°C for extended periods upon distribution. Unopened vaccine bottles may be kept for several months/years from the production date, given correct storage conditions. Normally the shelf-life is at least 24 months from the production date. The protective effect and toxicity of stored vaccines are tested regularly. The results of vaccination, re-isolation of pathogen, safety tests and other tests and analyses are provided with each vaccine batch.

17.11 TESTS, FIELD TRIALS AND APPROVAL FOR SALE

In addition to documentation of the stability and potency tests during production of several test batches the vaccine has to be tested in field trials, i.e. fish farms, before an application for approval and marketing is possible. The set up for a field trial can be done in several ways, but it is important that the number of fish farms and the number of fish in each test group are high enough for statistical evaluation.

The infection stress associated with the potency test (see below) in the laboratory tanks is relatively much higher than that on fish farms. Therefore, a mortality in the control fish groups of 60–80%, which is desirable for the laboratory test, is not attainable on fish farms. Due to economic losses the fish farmer will order medical treatment of the fishes when they start to die. Our experience is that while the mortality of the vaccinated and control fish in the laboratory tests can be 5 and 80%, respectively, the results from a field trial of the same vaccine could be 0 and 10%.

17.11.1 Potency test

The fish are starved for 1–2 days prior to vaccination or infection and anesthetized in, for example, 0.03% chlorobutanol. Fish are vaccinated once by intraperitoneal injection of 0.1–0.2 ml vaccine. The temperature in the tanks should ideally be around 10°C during the entire vaccination program. At the start of the trial the fish have an average weight of 15–20 g. The fish in the control group are handled in exactly the same manner except that they are fin-clipped and injected with medium containing formalin.

At 6–8 weeks after vaccination the fish are infected. This can be performed by one of three different methods; by injection of the pathogen intraperitoneally into the fish; by bathing the fish in a solution of the pathogen or by cohabitation, i.e. infected fish are placed into tanks together with the vaccinated and the control fishes. Injection and cohabitation (Lillehaug, 1989) are the most common methods because a successful bath infection is dependent on unstressed fish, water salinity (it is best not to put fingerlings or pre-smolts into saline water) and the water temperature (activity of bacteria during infection). Because of the high mortality associated with injection of *A. salmonicida*, the cohabitation method is mainly used when the fish are challenged with furunculosis. In addition *A. salmonicida* can survive in fresh water and is therefore independent of water salinity.

In the following a modification of the cohabitation challenge is briefly described: 50 vaccinated and 50 unvaccinated (control) fish are placed into one tank or into separate tanks, the latter method to avoid continuous cohabitation by sick control fish. Three to ten unvaccinated fish (cohabitants) are anesthetized as described above and infected with *A. salmonicida* by injecting intraperitoneally 2×10^5 bacteria per fish. The cohabitants are released into the tank together with the test fish. The optimum temperature during the furunculosis challenge is 10–11°C. The cohabitants normally start to die after 5–8 days followed by the fish in control and vaccine groups, respectively. Dead fish are removed daily, registered and cumulative mortality and RPS are calculated. The trial is run until there are 5 days without mortality. Reisolation for the identification of *A. salmonicida* from dead fish is done routinely.

A brief description of injection challenge is as follows: 50 vaccinated and 50 unvaccinated control fish are anesthetized, injected with the pathogen (*V. salmonicida*, 5×10^7; *V. anguillarum*, $1–3 \times 10^5$; *Y. ruckeri*, 10^5) intraperitoneally and placed in the same tank. Mortality usually starts in the control group after 5–10 days. The duration of the challenge, the registration of mortality and RPS calculation are done as described earlier. The infection dose must be regarded as an example; however, to attain adequate RPS values, all infection trials should be designed to aim at a mortality in the control group of at least 50–60% but not more than 90%.

17.11.2 Safety test and vaccine side effects

For safety tests an average fish weight of 20 g is adequate. The water temperature should be constant, e.g. around 10°C, during the entire test. A double dose vaccine is injected intraperitoneally into 20 fish. For larger fish the test dosage should be increased. Twenty control fish are injected with vaccine medium. For immersion vaccines the fishes are exposed to a double concentration of vaccine solution for 60 s.

The test results are evaluated after 14 days. With 20 fish per group a maximum mortality of two fish is accepted (due to incorrect injection, etc.). If three or more fish die a new test is performed and if the results are not acceptable, the vaccine batch is discarded. All data are enclosed with the batch documentation.

Vaccinated fishes have to be observed for several months (for salmon preferable in both fresh water and at sea) until slaughter to evaluate side effects, including growth reduction.

17.11.3 Human toxicity by self injection

During injection vaccination there is always the risk of self injection. The symptoms, most often a non-specific toxic effect, are experienced as inflammation, pain and sometimes fever. These symptoms usually occur within 30 min and last between 2 and 48 h. The cause may be either infection or an endotoxin reaction due to, for example, LPS. In addition some adjuvants may cause necrosis. It is advisable to see a doctor.

17.12 VACCINATION METHODS

Our proposal of vaccination rules that may be valid for all fish species are as follow:

1. Vaccination protects against diseases and reduces the use of antibiotics.
2. Vaccines have to be used in accordance with the producers recommendations in order to obtain the protection sought.
3. Better growth of fish and increased profits are the result if the fish are correctly vaccinated.
4. The best protection against diseases is obtained when the fish are vaccinated more than once.
5. Before the fish are transferred from the hatchery to the grow out cage they should always be vaccinated.
6. A vaccination program should be worked out so that vaccination can take place in periods with stable temperatures.
7. Do not vaccinate diseased fish.
8. Avoid stress before, during and after vaccination since stress reduces the efficiency of a vaccine.
9. Fish that have been medicated, stressed by sorting or transported should be rested for three weeks before vaccination.
10. Only trained personnel should be allowed to conduct injection vaccination.
11. If mortality occurs during vaccination, stop immediately and have a veterinarian look at the problem.
12. Correctly conducted vaccination is a quality mark, ensure that a certificate accompanies the fish.

The vaccination methods are described below and are used for vaccination of salmon but may also be valid for other fish species.

Infection trials have shown (Figure 17.12) that injection vaccination gives better protection than immersion vaccination (Baudin-Laurencin and Tangtrongpiros, 1980). Bath vaccination does not appear to generate satisfactory protection. Field trials have shown that different combinations of immersion/bath vaccination have been effective. However, if not recommended by the producer, it must be pointed out that vaccines containing adjuvants are intended for injection procedures only, since adjuvants usually activate immune stimulating cells by this method.

Vaccination in brackish water is unfavorable. The fish are stressed and if the water temperature, in addition, is high (above 12–15°C), the

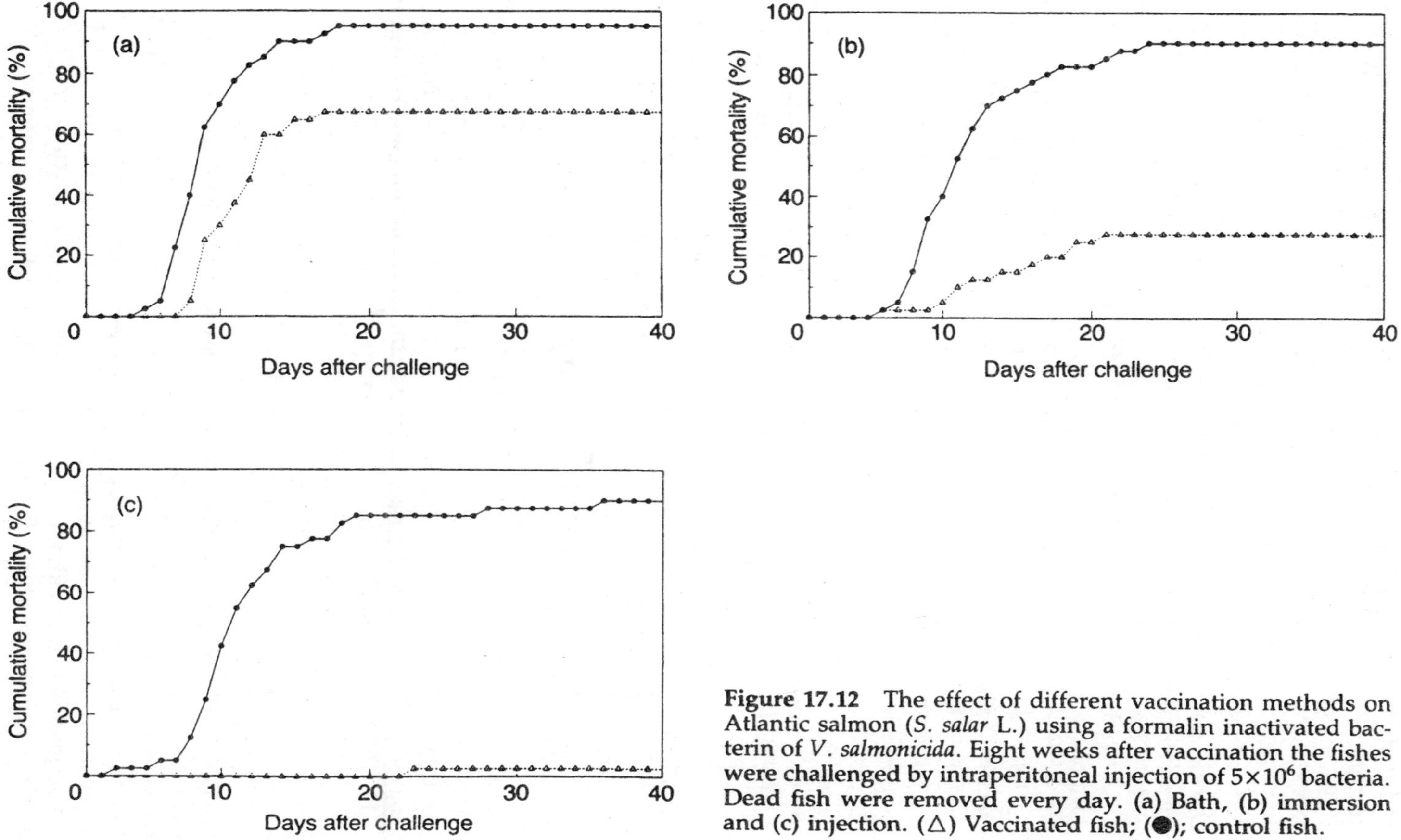

Figure 17.12 The effect of different vaccination methods on Atlantic salmon (*S. salar* L.) using a formalin inactivated bacterin of *V. salmonicida*. Eight weeks after vaccination the fishes were challenged by intraperitoneal injection of 5×10^6 bacteria. Dead fish were removed every day. (a) Bath, (b) immersion and (c) injection. (△) Vaccinated fish; (●); control fish.

danger of vibriosis is imminent. It is assumed, however, that higher water temperature between the first and second vaccinations and following the second vaccination have no significant influence on vaccine efficiency. A minimum temperature of 4°C must be maintained throughout the vaccination program. Below this temperature the vaccine will have reduced efficiency.

Although it has been generally accepted that one should not vaccinate during sorting, several farms have achieved good results with such a vaccination program, but vaccination should be avoided if smoltification has progressed too far.

Generally the fish should be starved for 1–2 days before the vaccination takes place and must not be fed for the first 12–24 h following vaccination. For immersion and bath vaccination it is presumed that feeding immediately after vaccination may result in poor uptake of the vaccine from the intestinal tract and elimination before the immune response has been triggered.

Unused vaccine should never be poured back into the flask in view of possible contamination.

It is inadvisable to attempt to mix vaccines against different diseases since the effectiveness may be reduced. A strong reaction against one of the components may suppress important reactions against other components. In one attempt to make a combination vaccine against vibriosis and yersiniosis, we found that compared to the single vaccines, protection against vibriosis was significantly decreased while that against yersiniosis was not affected at all. However, this may not always be the case and this is an important area of research in the future.

17.12.1 Injection

Injection vaccination is carried out as illustrated in Figure 17.13. The fish should preferably weigh over 10 g and the water temperature should not be below 4°C. The fish are anesthetized and, using an automatic syringe, 0.1–0.2 ml (see the producer's recommendation) vaccine is injected interperitoneally, just anterior to the posterior fin pair. For fish above 500 g, use double the dose of vaccine per fish.

Manual vaccination may be carried out using automatic syringe equipment following the above procedure or one can use a vaccination machine. To avoid penetration of organs or serious wounds in connection with injection, use needles according to fish size, i.e. not larger than necessary. Injection vaccination must be carried out by qualified personnel.

17.12.2 Immersion

The salmon fry must be over 5 g at the time of vaccination. The concentrate is diluted according to the producer's recommendation. A

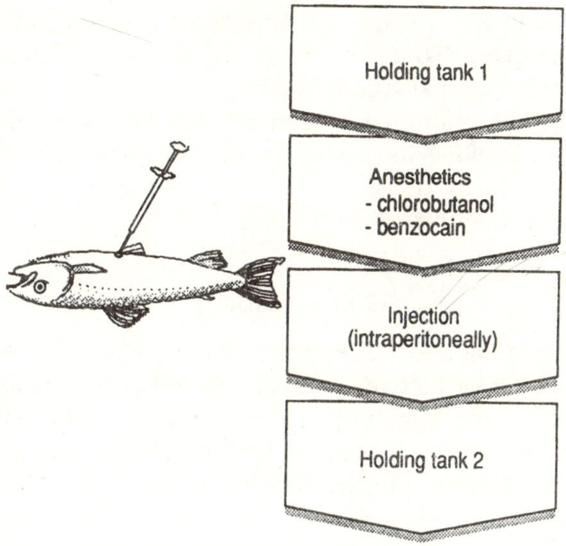

Figure 17.13 Injection vaccination. The dose is dependent on the vaccine, usually 0.1 ml for monovalent and 0.2 ml for multivalent vaccines.

suitable landing net is used to dip 2–3 kg fish in vaccine solution for 30–60 s (Figure 17.14). To reduce a dilution effect of the vaccine, allow water to drip off the net before and after vaccination. Keep the net moving during vaccination, and make sure that the water and vaccine temperature is above 4°C during the entire procedure.

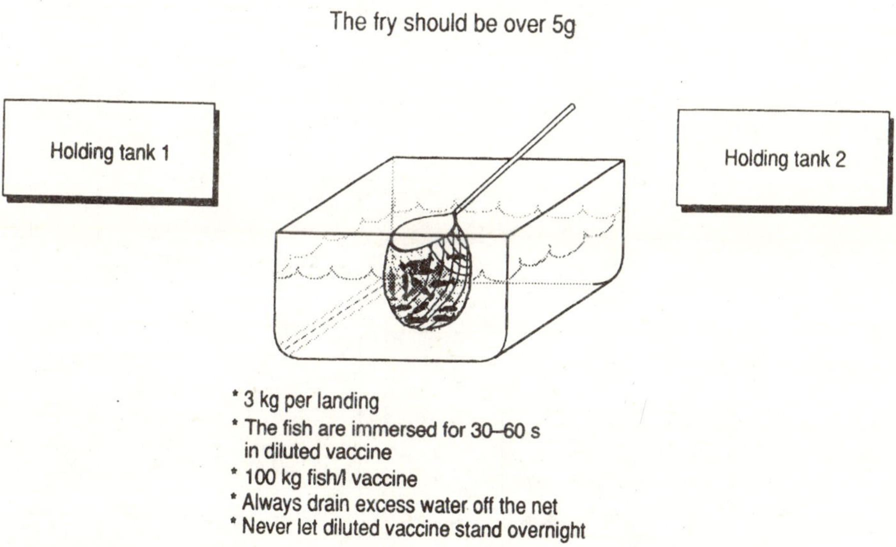

Figure 17.14 Immersion vaccination. The salmon fry should be over 5 g. The dilution of vaccine is generally 1:10 or 2:10, or as defined by the producer.

The vaccine solution is discarded once the recommended weight of fish has been reached (100 kg/l vaccine). To avoid contamination never keep diluted and/or partially used vaccines.

The user must be cautioned against attempts to use more than the recommended ratio of fish to vaccine. During the vaccination procedure the vaccine is gradually diluted, so that exceeding the recommended ratios may lead to insufficient protection.

17.12.3 Bath

The fish are starved as described earlier. The tank water is reduced to a suitable level (100 kg fish/m^3) and vaccine concentrate is added to a final concentration recommended by the producer. The fish should swim in the vaccine solution for 1 h (Figure 17.15). It is of importance to maintain an adequate oxygen supply, so check with an oxygen-measuring device and stir occasionally. The water may also be circulated with a pump. When vaccination is completed, water flow is resumed. Maintain the oxygen supply until the water level has reached normal level.

The user is warned not to drain too much water before bath vaccination. The person in charge must carefully estimate the minimum volume advisable in which the fish can swim.

There is nothing wrong with vaccinating fish of varying sizes at the same time, as long as the total weight bathed does not exceed the

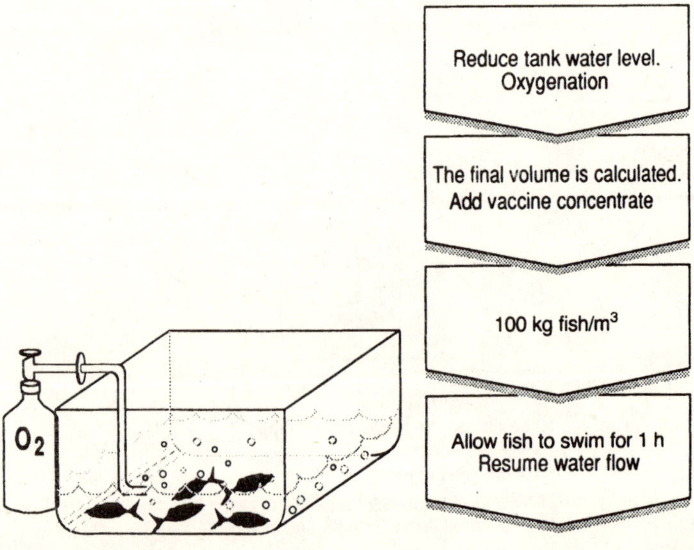

Figure 17.15 Bath vaccination. The dilution is generally 1:1000 or 2:1000, or as defined by the producer.

recommended limit. In addition to the impending danger of oxygen deficiency, the fish become stressed. For this reason the water level must be sufficient. A sign that the fish are not experiencing problems is that they oppose the current generated by stirring or pumping. Danger signs include erratic swimming, often along the surface. Following vaccination, when the water level has returned to normal, the fish will sometimes be agitated and leap. Normally they will quieten down again within a short time.

17.12.4 Oral vaccines

An effective oral vaccine giving RPS values comparable with injection or immersion vaccines would be of high economic importance to the fish farmer. Thus the vaccination could take place whenever he wanted without interrupting the daily routines on the farm.

Several oral vaccines have been tested in field trials, but none of them seems to give satisfactory protection against diseases. For re-vaccination on the other hand, their effects are reported to be adequate. However, to maintain the desired protection level the fish have to be re-vaccinated several times at relatively short intervals. At present the cost of oral vaccination is too high.

17.12.5 Re-vaccination

Re-vaccination (Figure 17.16) enhances the capability of the memory cells (Figure 17.3) to recognize the infective agent and even more memory cells are mobilized. Thus both the level and the duration of the protection are significantly improved. Re-vaccination is therefore assumed to give prolonged protection against the disease.

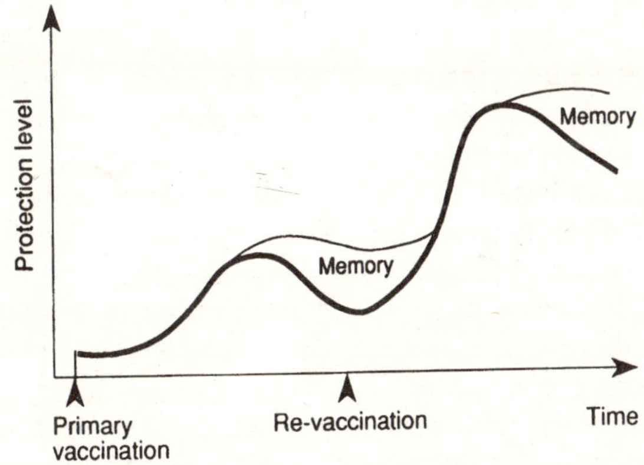

Figure 17.16 The importance of re-vaccination for the level of protection.

If regarded as necessary, re-vaccination may take place after 4–5 weeks. The fish cannot be released into the sea until at least 3 weeks after vaccination. It is recommended that the fish are re-vaccinated after 1 year. As mentioned above, oral vaccines have been tested for this purpose but with limited success.

17.13 FUTURE USE OF VACCINES

One of the main obstacles to the development of efficient fish vaccines is our limited knowledge of the immune systems of different fish species. In the case of oral vaccines our knowledge of the digestive process and thus absorption of vaccine components is scarce.

Today fish vaccines are based mainly on formalin inactivated bacteria. However, the lack of effective vaccines against BKD and several other diseases, and the fact that adjuvants have to be added to some vaccines to give significant protection, all show a need for developing vaccines based on other principles than killed bacteria. Further, there is a need for new vaccine administration methods, which are both cost- and work-effective. The vaccine producers have to develop oral vaccines since these are work-effective, apparently have no side effects and do not stress the fish.

For the development of effective vaccines against diseases like BKD and viruses, and for improvement of today's furunculosis vaccines, research based on recombinant DNA technology will probably give results within the near future. The recombinant technology is based on identification of subunits (protein antigens) from the infectious agent which stimulate the immune system to give protection against the disease. DNA sequences which code for the subunits are incorporated into the genome or a plasmid of a procaryotic or eucaryotic organism, which in turn is used to produce large amounts of the antigens for use in vaccines.

As described earlier the present oral vaccines do not give significant protection against diseases. One of the problems seems to be related to the fact that the vaccine components are a minor part of the feed pellet. Oral vaccines are based on the principle of vaccine uptake through the intestine and stimulation of immune competent cells connected to the anterior part of the intestine. Thus the uptake of vaccine and its immune stimulating effect may be seriously suppressed by the dominant feed uptake. Therefore it is desirable to construct a feed pellet concept where the vaccine components constitute the major part. In addition there are problems connected with a vaccine pellet, since it must not be solubilized or digested until it reaches the anterior part of the fish intestine. In addition it should satisfy all the demands of an ideal fish feed.

The future of aquaculture demands effective vaccines in order to avoid pollution problems due to use of antibiotics. In this context oral vaccines and use of gene technology are probably the solutions.

REFERENCES

Adam, A., Auchinachie, N., Bundy, A. *et al.* (1988) The potency of adjuvanted injected vaccines in Rainbow trout (*Salmo gairdneri* Richardson) and bath vaccines in Atlantic salmon (*Salmo salar* L.) against furunculosis. *Aquaculture*, **69**, 15–26.

Amend, D.F. (1981) Potency testing of fish vaccines. *Dev. Biol. Standard.*, **49**, 447–54.

Anderson, J.W. and Conroy, D.A. (1970) Vibrio diseases in fishes, in *A Symposium on Diseases of Fishes and Shellfishes*, (ed. S.F. Snieszko), American Fisheries Society, Washington, *Special Publication* **5**, 226–72.

Austin, B. (1985) Evaluation of antimicrobial compounds for the control of bacterial kidney disease in rainbow trout, *Salmo gairdneri* Richardson. *J. Fish Dis.*, **8**, 209–20.

Austin, B. and Allen-Austin, D. (1985) A review. Bacterial pathogens of fish. *J. Appl. Bacteriol.*, **58**, 483–506.

Austin, B. and Austin, D.A. (1987) *Bacterial Fish Pathogens: Disease in Farmed and Wild Fish*, Ellis Horwood, New York.

Baudin-Laurencin, F. and Tangtrongpiros, J. (1980) Some results of vaccination against vibriosis in Brittany, in *Fish Diseases*, (ed. W. Ahne), Springer-Verlag, Berlin, pp. 60–8.

Bruno, D.W. (1986) Histopathology of bacterial kidney disease in laboratory infected rainbow trout, *Salmo gairdneri* Richardson, and Atlantic salmon, *Salmo salar* L., with reference to naturally infected fish. *J. Fish Dis.*, **9**, 523–37.

Bullock, G.L., Stuckey, H.M. and Shotts, E.B. (1978) Enteric redmouth bacterium: comparison of isolates from different geographical areas. *J. Fish Dis.*, **1**, 351–4.

Bullock, G.L. and Anderson, D.P. (1984) Immunization against *Yersinia ruckeri*, cause of enteric redmouth disease, in *Symposium on Fish Vaccination; Theoretical Background and Practical Results on Immunization against Infectious Diseases*, (ed. P. De Kinkelin), Office International des Epizooties, Paris, pp. 151–66.

Busch, R.A. (1973) The serological surveillance of salmonoid populations for presumtive evidence of specific disease association. *PhD Dissertation*, University of Idaho.

Ceschia, G., Giorgetti, G., Bertoldini, G. and Fontebasso, S. (1987). The *in vitro* sensitivity of *Yersinia ruckeri* to specific antibiotics. *J. Fish Dis.*, **10**, 65–7.

Corbel, M.J. (1975) The immune response in fish: a review. *J. Fish Biol.*, **7**, 539–63.

Cossarini-Dunier, M. (1985) Effect of different adjuvants on the humoral immune response of rainbow trout. *Dev. Comp. Immunol.*, **9**, 141–6.

Crosa, J.H. (1980) A plasmid associated with virulence in the marine fish pathogen *Vibrio anguillarum* specifies an iron-sequestering system. *Nature*, **283**, 566–8.

Crosa, J.H., Schiewe, M.H. and Falkow, S. (1977) Evidence for plasmid contribution to the virulence of the fish pathogen *Vibrio anguillarum*. *Infect. Immun.*, **27**, 509–13.

Earp, B.J. (1950) Kidney Disease in Young Salmon. *MSc Thesis*, University of Washington.

Egidius, E., Andersen, K., Clausen, E. and Raa, J. (1981). Cold-water vibriosis or 'Hitra disease' in Norwegian salmonoid farming. *J. Fish Dis.*, 4, 353–4.
Egidius, E., Wiik, R., Andersen, K. *et al.* (1986) *Vibrio salmonicida* sp. nov., a new fish pathogen. *Int. J. Syst. Bacteriol.*, **36**, 518–20.
Ellis, A.E. (ed.)(1988a) *Fish Vaccination*, Academic Press, New York.
Ellis, A.E. (1988b) General principles of fish vaccination, in *Fish Vaccination*, (ed. A.E. Ellis), Academic Press, New York, pp. 1–19.
Ellis, A.E. (1989) The immunology of teleosts, in *Fish Pathology*, (ed. R.J. Roberts), Baillière Tindall, London, pp. 135–52.
Fletcher, T.C. (1982) Non-specific defence mechanisms of fish. *Dev. Comp. Immunol. Suppl.*, **2**, 123–32.
Fryer, J.L. and Sanders, J.E. (1981) Bacterial kidney disease of salmonoid fish. *Ann. Rev. Microbiol.*, **35**, 273–98.
Gilmore, R.D., Engelking, H.M., Manning, D.S. and Leong, J.C. (1988) Expression in *E. coli* of an epitope of the glycoprotein of infectious haematopoietic necrosis virus protects against viral challenge. *Bio/Technology*, **6**, 295–300.
Hoffman, R., Popp, W. and van der Graaff, S. (1984) Atypical BKD predominantly causing ocular and skin lesions. *Bull. Eur. Ass. Fish Pathol.*, **4**, 7–9.
Holm, K.O., Strøm, E., Stenvåg, K., Raa, J. and Jørgensen, T. (1985) Characteristics of a *Vibrio* sp. associated with the 'Hitra disease' of Atlantic salmon in Norwegian fish farms. *Fish Pathol.*, **20**, 125–9.
Holm, K.O. and Jørgensen, T. (1987) A successful vaccination of Atlantic salmon, *Salmo salar* L., against 'Hitra disease' or coldwater vibriosis. *J. Fish Dis.*, **10**, 85–90.
Horne, M.T., Roberts, R.J., Tatner, M. and Ward, P. (1984) The effects of the use of potassium alum adjuvant in vaccines against vibriosis in rainbow trout, *Salmo gairdneri* Richardson. *J. Fish Dis.*, **7**, 91–9.
Hunter, V.A., Knittel, M.D. and Fryer, J.L. (1980) Stress-induced transmission of *Yersinia ruckeri* infection from carriers to recipient steelhead trout *Salmo gairdneri* Richardson. *J. Fish Dis.*, **3**, 467–72.
Ingram, G.A. (1980) Substances involved in the natural resistance of fish to infection – a review. *J. Fish Biol.*, **16**, 23–60.
Knappskog, D.H., Rødseth, O.M., Slinde, E, and Endresen, C. (1993) Immunochemical analyses of *Vibrio anguillarum* strains isolated from cod (*Gadus morhua* L.) suffering from vibriosis. *J. Fish Dis.*, **16**, 327–38.
Koener, J.F., Passavant, C.W., Kurah, G. and Leong, J.C. (1987) Nucleotide sequence of cDNA clone carrying the glycoprotein gene of infectious haematopoietic necrosis virus, a fish rhabdovirus. *J. Virol.*, **61**, 1342–9.
Krantz, G.E., Reddecliff, J.M. and Heist, C.E. (1964) Immune response of trout to *Aeromonas salmonicida*. Part 1. Development of agglutinating antibodies and protective immunity. *Progressive Fish Culturist*, **26**, 3–10.
Kurath, G., Ahern, K.G., Pearson, G.D. and Leong, J.C. (1985). Molecular cloning of the six mRNA species of the infectious haematopoietic necrosis virus, a fish rhabdovirus, and gene order determination by R-loop mapping. *J. Virol.*, **53**, 469–76.
Leong, J.C., Fryer, J.L. and Winton, J.R. (1988) Vaccination against infectious haematopoietic necrosis, in *Fish Vaccination*, (ed. A.E. Ellis), Academic Press, New York pp. 193–203.
Leong, J.A. and Munn, C.B. (1991) Potential use of recombinant DNA in the development of fish vaccines. *Bull. Eur. Ass. Fish Pathol.*, **1**, 30–40.
Lillehaug, A. (1989) Oral immunization of rainbow trout, *Salmo gairdneri* Richardson, against vibriosis with vaccines protected against digestive degradation. *J. Fish Dis.*, **12**, 579–84.
Lillehaug, A., Ramstad, A., Bækken, K. and Reitan, L.J. (1993) Protective

immunity in Atlantic salmon (*Salmo salar* L.) vaccinated at different water temperatures. *Fish and Shellfish Immunol.*, **3**, 143–56.

McCarthy, D.H. (1975). Fish furunculosis. *J. Inst. Fish. Management*, **6**, 13–18.

McCarthy, D.H. and Roberts, R.J. (1980). Furunculosis of fish – the present state of our knowledge, in *Advances in Aquatic Microbiology*, (eds M.A. Droop and H.W. Jannasch), Academic Press, London, pp. 293–341.

Mackie, T.J., Arkwright, A.J., Pryce-Tannatt, T.E. *et al.* (1933) *Second Interim Report of the Furunculosis Committee*, HMSO, Edinburgh.

Manning, D.S. (1988) Deletion mapping and expression of the large genomic segment of infectious pancreatic necrosis virus. *Doctoral Thesis*, Oregon State University.

Manning, M.J. and Mugal, M.S. (1985) Factors affecting the immune responses of immature fish, in *Fish and Shellfish Pathology*, (ed. A.E. Ellis), Academic Press, London, pp. 27–40.

O'Leary, P.J. (1977) Enteric redmouth of salmonids: a biochemical and serological comparison of selected isolates. *MS thesis*, Oregon State University.

Paterson, W.D. and Fryer, J.L. (1974) Immune response of juvenile coho salmon (*Oncorhynchus kisutch*) to *Aeromonas salmonicida* cells administered intraperitoneally in Freund's complete adjuvant. *J. Fish. Res. Board Can.*, **31**, 1751–5.

Robertsen, B., Rørstad, G., Engstad, R. and Raa, J. (1990) Enhancement of non-specific disease resistance in Atlantic salmon, *Salmo salar* L., by a glucan from *Saccharomyces cerevisiae* cell walls. *J. Fish Dis.*, **13**, 391–400.

Rodgers, C.J. and Austin, B. (1982) Oxolinic acid for control of enteric redmouth disease in rainbow trout. *Veterinary Record*, **112**, 83.

Ross, A.J., Rucker, R.R. and Ewing, W.H. (1966) Description of a bacterium associated with redmouth disease of rainbow trout (*Salmo gairdneri*). *Can. J. Microbiol.* **12**, 763–70.

Rucker, R.R. (1966) Redmouth disease of rainbow trout (*Salmo gairdneri*). *Bulletin de l'Office International des Epizooties*, **65**, 825–30.

Rørstad, G., Aasjord, P. and Robertsen, B. (1992) Adjuvant effect of a yeast glucan in vaccination against furunculosis in Atlantic salmon (*Salmo salar* L.). *Fish Shellfish Immunol.*, **3**, 179–90.

Sanders, J.E. and Fryer, J.L. (1980) *Renibacterium salmoninarum* gen. nov., the causative agent of bacterial kidney disease in salmonoid fishes. *Int. J. Syst. Bacteriol.*, **30**, 496–502.

Schubert, R.H.W. (1974) Genus II. *Aeromonas* Kluyver and van Niel 1936, 398, in *Bergey's Manual of Determinative Bacteriology*, 8th edn, (eds R.E. Buchanan and N.E. Gibbons), Baltimore, Williams and Wilkins, pp. 345–8.

Smith, I.W. (1964) The occurrence and pathology of Dee disease. *Dept. Agric. Fish. Scot. Freshwater and Salmon Fisheries Research* 34, HMSO, Edinburgh.

Snieszko, S.F. (1957) Genus IV. *Aeromonas* (Kluyver and van Niel, 1936). In *Bergey's Manual of Determinative Bacteriology*, 7th edn, (eds R.S. Breed, E.G.D. Murray and N.R. Smith), Baltimore, Williams and Wilkins, pp. 189–93.

Snieszko, S.F. and Griffin, P.J. (1955) Kidney disease in brook trout and its treatment. *Progressive Fish Culturist*, **17**, 3–13.

Stevenson, R.M.W. and Airdrie, D.E. (1984) Serological variation among *Yersinia ruckeri* strains. *J. Fish Dis.*, **7**, 247–54.

Tatner, M.F. and Horne, M.T. (1984) The effect of early exposure of *Vibrio anguillarum* on the immune response of the fry of the rainbow trout, *Salmo gairdneri* Richardson. *Aquaculture*, **41**, 193–202.

Tilseth, S., Hansen, T. and Møller D. (1991) Historical development of salmon culture. *Aquaculture*, 98, 1–9.

Wu, J.L. (1992) Nucleotide sequences of two cDNA clones of IPNV–Ab type. *Bull. Inst. Zool. Academia Sinica*, **31**(1), 15–22.

Index

Page numbers appearing in *italic* refer to tables and page numbers appearing in **bold** refer to figures.